ALLE ZEIT WACH 1842

W. Ahne E. Kurstak (Eds.)

Viruses of Lower Vertebrates

With 113 Figures

Springer-Verlag Berlin Heidelberg New York
London Paris Tokyo Hong Kong

Prof. Dr. Dr. habil. WINFRIED AHNE
Institut für Zoologie und Hydrobiologie
der Universität München
Kaulbachstrasse 37
D-8000 München 22, FRG

Prof. Dr. EDOUARD KURSTAK
Dept. Microbiology and Immunology
Faculty of Medicine
University of Montreal
Montreal, P.Q. H 3c 3J7
Canada

ISBN-13: 978-3-642-83729-6 **e-ISBN-13: 978-3-642-83727-2**
DOI: 10.1007/978-3-642-83727-2

Library of Congress Cataloging-in-Publication Data. Viruses of lower vertebrates / W. Ahne, E. Kurstak (eds.). p. cm. Papers presented at the First International Symposium on Viruses of Lower Vertebrates held at the Institute of Zoology and Hydrobiology of the University of Munich, August 22–25, 1988. ISBN 0-387-50859-7 (U.S. : alk. paper) 1. Fishes – Microbiology – Congresses. 2. Viruses – Congresses. 3. Amphibians – Microbiology – Congresses. 4. Reptiles – Microbiology – Congresses. 5. Vertebrate Viruses – congresses. 6. Virus Diseases – veterinary – congresses. I. Ahne, W., 1939–. II. Kurstak, Edouard. III. International Symposium on Viruses of Lower Vertebrates (1st : 1988 : Institute of Zoology and Hydrobiology of the University of Munich) [DNLM: 1. Amphibia – immunology – congresses. 2. Fishes – immunology – congresses. 3. Reptiles – immunology – congresses. QW 164 V821 1988] QR321.V57 1989 597'.0234 – dc20 DNLM/DLC

Softcover reprint of the hardcover 1st edition 1989

2131/3145-543210 – Printed on acid-free paper

Preface

Virology of lower vertebrates began in the 20th centrury as a branch of pathology. In 1914 WEISSENBERG opened this field when he postulated that the hypertrophy of lymphocystis cells of several fish could be of viral origin. LUCKE (1934) found similarities between the renal adenocarcinoma of *Rana pipiens* and the herpes simplex infections of man. Prior to 1970 little was known about viral agents and viral diseases of reptiles. Frontline virological studies were concentrated on arbovirus infections in reptiles. The maintenance of the viruses in lower vertebrates proved to be of medical and veterinarian importance. Later, increased studies on viruses of fish, amphibians and reptiles gave greater insight into their occurrence and distribution. In the last 10 years very substantial growth in the knowledge of viruses, viral diseases and immunology of lower vertebrates has been recognized.

As for lower vertebrates under the aspect of virus pathology, up to now attention was paid mainly to viral diseases of economically important fish but much information on molecular virology and oncogenicity was obtained from the viruses occurring in amphibians. The fact that lower vertebrates are involved in multiple host infections brought new views in the field of zoonosis. New information shows that certain viruses like influenza virus or caliciviruses are capable of bridging species lines and even the land-sea interface. There is some evidence for a certain connection between the global developments in aquaculture and the influenza pandemics.

However, information on the virology of lower vertebrates is still sparse and widely scattered. It was therefore enthusiastically greeted that an event was created where scientists, representing more than ten nations, could come together in order to discuss new research findings on viruses of lower vertebrates. The First International Symposium on Viruses of Lower Vertebrates held at the Institute of Zoology and Hydrobiology of the University of Munich, August 22-25, 1988, carried out under the auspices of the International Comparative Virology Organization and the Virology Group of the German Veterinarian Association, was organized by W. Ahne (FRG) and E. Kurstak (Canada). The organizers were assisted by an International Advisory Board consisting of

D. Anderson (USA), H.F. Clark (USA), P. Dobos (Canada), N. Fijan (Yugoslavia), J.L. Fryer (USA), A. Granoff (USA), R.P. Hedrick (USA), B.J. Hill (England), D.C. Kelly (England), T. Kimura (Japan), P. de Kinkelin (France), K. Maramorosch (USA), B.L. Nicholson (USA), T. Sano (Japan), J. Shchelkunov (USSR), K.F. Shortridge (Hong Kong), J.R. Winton (USA), K. Wolf (USA).

During the Symposium the participants expressed their interest and actively discussed problems related to viral nomenclature. With the expansion of viral surveillance and improved detection methods, there has been a significant increase in the number of viruses isolated and characterized from lower vertebrates. It was the consensus of the assembled participants that a working group should be established to prepare general recommendations for the naming of viruses of lower vertebrates. This working group was given the following guidelines:

1. To make its interest known to, and solicit input from, the International Committee on the Taxonomy of Viruses (ICTV).
2. To draft recommendations for nomenclature of viruses from lower vertebrates consistent with the current guidelines of the ICTV or other guidelines issued under the auspices of ICTV.

3. To present the final draft recommendation at the Second International Symposium on Viruses of Lower Vertebrates to be held in 1991 in the USA. If possible, an interim report should be presented at the Sixth International Conference on Comparative and Applied Virology in Banff, Alberta, October 15-21, 1989.

The following committee was formed in order to develop recommendations for the nomenclature of viruses of lower vertebrates:

Chairman:	J.L. Fryer (USA)
Herpesviruses:	R. Goorha (USA), R.P. Hedrich (USA), T. Kimura (Japan)
Iridoviruses:	K. Essani (USA), B. Nicholson (USA)
Birnaviruses:	P. Dobos (Canada), B.J. Hill (England)
Reovirus:	W. Ahne (FRG), J.R. Winton (USA)
Rhabdoviruses:	P. de Kinkelin (France), P.E.V. Jørgensen (Denmark)
Paramyxovirus:	W. Ahne (FRG), W. Neubert (FRG).

The book contains papers presented at the First International Symposium on Viruses of Lower Vertebrates held in Munich, August 22-25, 1988. Unfortunately, not all speakers were able to submit the manuscripts in time.

We thank Dr. D. Czeschlik, Springer Verlag, for making the publication of the Proceedings possible.

Munich, Spring 1989

W. Ahne, E. Kurstak
(Chairmen)

Contents

Section III Viruses of Fishes

List of Contributors

You will find the addresses at the beginning of the respective contribution

Section I Viruses of Amphibia

Viruses of Amphibia: An Historical Perspective

A. Granoff
Department of Virology and Molecular Biology
St. Jude Children's Research Hospital
332 North Lauderdale
Memphis, Tennessee 38101
USA

INTRODUCTION

The first postulation of an amphibian virus infection was over 50 years ago by Lucké (1934) who discovered that renal adenocarcinoma cells of the common leopard frog, *Rana pipiens*, often contained intranuclear acidophilic inclusion bodies that were similar to those found in herpes simplex infections and other viral diseases. Lucké's subsequent transmission experiments (Lucké, 1938), led him to conclude that the frog renal adenocarcinoma (Lucké tumor) was of viral origin. This interpretation was substantiated more than three decades later when a herpesvirus (Fawcett, 1956) was firmly established as the causative agent (Tweedell, 1967; Naegele et al., 1974). Attempts to isolate, cultivate and characterize the herpesvirus associated with the Lucké tumor resulted in the isolation of several other viruses, including members of the families *Iridoviridae (Granoff et al., 1965; 1968; Clark et al., 1968), Herpesviridae (Rafferty, 1965; Gravell et al., 1968)*, and *Adenoviridae* (Clark et al., 1973; Wong and Tweedel, 1974). Other iridoviruses (Wolf et al., 1968; Briggs and Burton, 1973; Desser and Barta, 1984), as well as members of the *Togaviridae* (Spalatin et al., 1964; Burton et al., 1966; Kozuch et al., 1977) and *Caliciviridae* (Smith et al., 1986) families, have also been identified as infecting amphibia.

The most studied amphibian virus is frog virus 3 (FV3), an iridovirus isolated by Granoff et al. (1965) from a frog renal adenocarcinoma of the leopard frog, *R. pipiens*. Similar isolates were made from normal frog kidney and liver (Granoff et al., 1966), but FV3 was chosen for further investigation because of its presence in the tumor. Although FV3 bore no causal relationship to the tumor (Tweedell and Granoff, 1968), it had several interesting and unique characteristics that made it worthy of study in its own right. These features and those of other DNA and RNA viruses are the subject of this paper.

DNA VIRUSES

Herpesviridae. To date, there is only one report (Tweedell and Wong, 1974) of the cultivation of the herpesvirus associated with the Lucké tumor. A second herpesvirus, frog virus 4 (FV4), has been isolated (Rafferty, 1965), and propagated and characterized for several of its properties (Gravell et al., 1968). This virus is distinct from the one associated with the Lucké tumor and Dr. Tweedell describes what is known of these viruses in his presentation.

Iridoviridae. Plaques appearing spontaneously in monolayers of frog kidney cells overlaid with agar were the source of the first virus (FV1-2) isolated from *R. pipiens* (Granoff et al., 1965). Subsequently, additional isolates were made from cultured normal adult frog kidney cells, liver homogenates of normal or tumor-bearing frogs and homogenates of Lucké tumors (FV3, FV9-23) (Granoff et al., 1966). Other iridoviruses (LT1-LT4) have been isolated by Clark et al. (1968) from kidneys of normal and Lucké tumor-bearing *R. pipiens* (LT1-4). Several isolations were made from organs of red newts (*Diemictylus*, formally *Triturus viridescens*), that had been inoculated with viable cell suspensions from tumor-bearing kidneys of *R. pipiens*. Other isolates (T6-20) were recovered from homogenates of normal newt tissues (Clark et al., 1969). Wolf et al. (1968; 1969) have isolated an iridovirus from viscera homogenates of both normal adult *R. catesbeiana* and normal or edematous tadpoles. This agent has been designated tadpole edema virus (TEV). Viruses isolated by Granoff, Clark and Wolf are antigenically related (Came et al., 1968; Clark et al., 1969) and probably represent various strains of the same virus. More definitive data based on molecular characterization of these agents are presented by Essani and Granoff elsewhere in this volume.

Two other iridoviruses have been reported. A frog erythrocytic virus (FEV) has been isolated from *R. catesbeiana* and the mink frog, *R. septentrionalis* (Desser and Barta, 1984). This virus was found in the red cells of frogs, and several of its properties have been investigated by Jasmine Gruia-Gray, who reports on her findings in these proceedings. An amphibian leukocyte virus has been observed by electron microscopy in *R. catesbeiana* leukocytes (except acidophiles) (Briggs and Burton, 1973), but has not been cultured.

Although tests for tumorigenic activity did not indicate a role for FV3 in the origin of the Lucké tumor, the virus was pathogenic in tailbud and hatching stage embryos (Tweedell and Granoff, 1968) and in Fowler's toads (*Bufo woodhousei fowleri*) (Granoff et al., 1969). Tailbud and hatching stage *R. pipiens* embryos were rapidly killed with as few as 900 plaque-forming units (pfu), while larvae were killed by much higher concentrations, about 3 x 10^6 pfu. Animals dying from virus infection were edematous and had dermal hemorrhages. Larvae reaching metamorphosis that had received 10^5 pfu and survived the virus did not develop tumors, nor were there any observable alterations in their development. In Fowler's toads, shown to be susceptible to infection by TEV (Wolf et al., 1968), inoculation with 10^8 pfu of FV3 killed 100% of toads within 9 days and 10^6 pfu killed 50% within 21 days. Virus multiplication occurred in kidney, liver, lung, skeletal muscle and fat body. Toads dead or moribund from FV3 infection were edematous and hemorrhagic. The histopathology was compatible with a generalized virus infection with multiple focal or confluent necrosis and hemorrhage in the tissues examined. Both the gross pathology and histopathology of FV3 infection are similar to, but apparently less severe than, the pathology induced in the same host by TEV (Wolf et al., 1968). Similar pathology has been observed in Fowler's toads inoculated with Clark's isolates LT1 and LT2. However, none of Clark's isolates tested caused obvious disease in efts although the virus multiplied in various organs.

As noted earlier, iridoviruses have several unique properties as exemplified by the best studied member of the group, FV3. Some of these features are summarized below; for others, the interested reader should consult a recent volume dedicated to the *Iridoviridae* family which describes the molecular biology of the replication cycle in detail (Willis et al., 1985; Murti et al., 1985).

FV3 DNA, a linear double-stranded molecule of 1×10^8 daltons, is unique among animal viruses in that it is circularly permuted and terminally redundant. As would be predicted from this fact, the physical map of the genome is circular, although the molecule is linear. Another unique property is that over 20% of cytosine residues of the CpG sequences are methylated by a virus-coded DNA methyltransferase, exceeding the proportion reported for any other animal virus. Heavy methylation of FV3 DNA does not appear to be involved in transcriptional regulation (as discussed by Goorha at this meeting), but it may play a role in virus morphogenesis and assembly and may also be related to restriction-modification systems similar to those encountered among bacteriophages.

The various aspects of transcriptional and translational regulation of FV3 are dealt with elsewhere in this volume and will not be discussed here. However, I will comment briefly on the overall strategy of FV3 DNA replication. The genome of incoming FV3 particles reaches the nucleus where stage 1 replication takes place. Transcription during the early stages of infection is by cellular RNA polymerase II, which likely has been modified by a virion structural protein to transcribe methylated DNA. The parental FV3 genome in the nucleus also serves as the template for stage 1 DNA replication. Progeny DNA synthesized in the nucleus (as well as the parental genomes) may then be used as a template for further transcription or be transported to the cytoplasm, where it can participate in the second stage of DNA replication. During stage 2, large replicative complexes (concatemers) are produced in the cytoplasm and are then cleaved to produce mature viral DNA in assembled virions. Presumably, concatemer processing is intimately associated with DNA packaging and virus assembly to produce circularly permuted and terminally redundant genomes.

A number of FV3 properties resemble those of the bacteriophages, e.g., two stage DNA replication, lack of polyadenylated mRNAs, a restriction-modification system, a terminally redundant and circularly permuted genome, a highly methylated genome and no spliced genes. Thus, phylogenetically FV3 may represent an intermediate between viruses of prokaryotes and eukaryotes. DNA viruses of eukaryotic organisms are usually classified as either cytoplasmic or nuclear, depending on whether the nucleus or the cytoplasm is the site of nucleic acid synthesis and virus assembly. However, the fact that FV3 utilizes both the nucleus and cytoplasm for its replication suggests, that it may also be an intermediate form in the evolution of nuclear and cytoplasmic DNA viruses.

Adenoviridae. There have been two reports of the isolation of an adenovirus from an amphibian. The first, by Clark et al. (1968), describes a virus isolated from cells of a granuloma-bearing kidney of *R. pipiens* that was considered an adenovirus based on its size, structure and nucleic acid type. Replication of this virus, termed frog adenovirus 1 (FAV1), was limited in tissue culture to a turtle heart (TH) cell line at an optimal temperature of 30°C. The virus did not possess group specific, complement-fixing antigens of mammalian adenoviruses and was not neutralized by hyperimmune serum to avian adenoviruses. It did not cause tumors in the appropriate test system for tumor production in *R. pipiens* tadpoles. Thus, FAV-1 appears to be an orphan virus. A second adenovirus was isolated from a Lucké tumor inoculated into a pronephric frog kidney cell line (Wong and Tweedell, 1974). The very restricted host cell susceptibility and similarity of other properties suggested that this virus was likely FAV-1. No further studies on amphibian adenoviruses have been reported.

Papovaviridae. During experiments to isolate the Lucké tumor herpesvirus, two *R. pipiens* embryo cell lines inoculated with homogenates of this tumor showed several alterations in morphology (Granoff et al., 1969). In one case, characteristic stranding and clumping of cells were observed and could be serially transmitted by cell-free extracts from the infected cells. In another experiment, a cell line that was originally epithelioid assumed a more fibroblastic nature and grew in whorls, a property that could also be serially passaged. Electron microscopic examination of thin sections of both *R. pipiens* embryo cell lines failed to demonstrate evidence of viral infection, but negative stains of infected culture medium revealed the presence of icosahedral virus particles approximately 45 nm in diameter, similar to known papovaviruses. Particles of similar size have been observed in thin sections of Lucké tumor cells (Lunger, 1966; Stackpole et al., 1968; Granoff et al., 1969) that also contain herpesviruses and have been considered to be related in some way to herpesvirus biosynthesis. However, they may represent the papova-like virus just described. No additional information is available on the papova-like virus.

RNA VIRUSES

Togaviridae - Evidence of infection with Western equine encephalitis (WEE) virus has been documented by the presence of antibody in the serum of infected *R. pipiens* and by the isolation of virus (Spalatin et al., 1964; Burton et al., 1966). The fact that WEE virus can be isolated from naturally infected *R. pipiens* suggests that this species may serve as a potential reservoir for the virus. *R. pipiens* is susceptible to experimental infection with WEE by subcutaneous, oral, or intramuscular routes as demonstrated by a significant antibody response (Spalatin et al., 1964). Virus could not be recovered from any organs of sacrificed animals. In another study by Burton et al. (1964), virus was isolated from the blood of six *R. pipiens* while a large number of other animals (50/179) had detectable serum neutralizing antibodies to the virus. To my knowledge, there has been no additional information published on WEE in amphibia since these reports.

Sindbis virus. There is a single abstract reporting the isolation of Sindbis virus from *R. ridibunda (Kozuch, 1977)*.

The virus was isolated from the blood and identified by virus neutralization tests with hyperimmune mouse serum. It had properties compatible with known Sindbis viruses, such as causing the death of suckling mice after intracerebral or intraperitoneal inoculation and producing cytopathic effects in 24-hour-old chick embryo cell cultures.

Caliciviridae. Calicivirus was isolated from spleen, kidney, and lymph of two dead Bell's horned toads (*Ceratophrys ornata*) (Smith et al., 1986). The virus could not be neutralized by antisera to other existing calicivirus and therefore is a unique calicivirus serotype. Although similar to picornaviruses in several respects, caliciviruses differ in size, structure and replication strategy. Members of this family include vesicular xanthem of swine, San Miguel sea lion virus, and feline calicivirus. Caliciviruses apparently have a broad host range, including pigs, primates, mink, fish, cattle, swine, dolphins, sea lions, dogs, donkeys, sheep, buffalo, and snakes. Smith et al. (1986) speculated that poikilotherms such as amphibia may serve as reservoirs for caliciviruses and that calicivirus virulence may be modulated by host body temperature. No further information on the amphibian calicivirus is available.

CONCLUSION

Amphibia have proved to be extremely useful animals for solving many basic problems of biology. The viruses to which they are host can contribute important information to our general understanding of viruses as infectious and oncogenic agents. For example, the data generated from studies on FV3 have provided new insights into the versatility and diversity of viral regulation of gene expression and reproduction, as well as into the evolution of viruses. The toxic effects of FV3 on the liver of experimental animals such as mice and rats, as reported by Kirn at this meeting, have provided insights of general importance into liver physiology and pathology. Iridoviruses that infect erythrocytes and leukocytes of amphibia also have the potential for being useful models of viral infection of blood cells and may, like FV3, open many biological research avenues.

Studies of the Lucké tumor herpesvirus have provided evidence that herpesviruses can cause carcinomas in outbred populations strengthening the notion that herpesviruses may cause human cancers. For other viruses, such as togaviruses, amphibia may serve as a natural reservoir for infection of other hosts.

It is not clear what role, if any, the adenovirus or the papova-like virus have in their association with frog tissue. They may truly be passenger viruses, but even in this case, viruses from lower vertebrates offer unusual opportunities for adding relevant information to the repertoire of biological information---from evolution, to regulation of gene expression, to disease.

ACKNOWLEDGEMENTS

The work cited from the author's laboratory was supported by Public Health Service Research Project Grant CA 07055, Cancer Center Support (CORE) Grant CA 21765 from the National Cancer Institute, and the American Lebanese Syrian Associated Charities (ALSAC).

LITERATURE REFERENCES

Briggs RT, Burton PR (1973) Fine structure of an amphibian leukocyte virus. J Submicr Cytol 5:71-78

Burton AN, McLintock J, Rempel JG (1966) Western equine encephalitis virus in Saskatchewan garter snakes and leopard frogs. Science 154:1029-1031

Came PE, Geering G, Old LJ, Boyse EA (1968) A serological study of polyhedral cytoplasmic viruses isolated from amphibia. Virology 36:392-400

Clark HF, Brennan JC, Zeigel RF, Karzon DT (1968) Isolation and characterization of viruses from the kidneys of *Rana pipiens* with renal adenocarcinoma before and after passage in the red eft (*Triturus viridescens*). J Virol 2:629-640

Clark HF, Gray C, Fabian F, Zeigel R, Karzon DT. Comparative studies of amphibian cytoplasmic virus strains isolated from the leopard frog, bullfrog and newt. Mizell M (ed) (1969) Biology of Amphibian Tumors, pp 310-326. Springer-Verlag New York Heidelberg Berlin
Clark HF, Michalski F, Tweedell KS, Yohn D, Zeigel RF (1973) An adenovirus, FAV-1, isolated from the kidney of a frog (*Rana pipiens*). Virology 51:392-400
Desser SS, Barta JR (1984) An intraerythrocytic virus and rickettsia of frogs from Algonquin Park, Ontario. Can J Zool 62:1521-1524
Fawcett DW (1956) Electron microscope observations of intracellular virus-like particles associated with the cells of the Lucké renal adenocarcinoma. J Biophys Biochem Cytol 2:725-742
Granoff A, Came PE, Rafferty KA (1965) The isolation and properties of viruses from *Rana pipiens*: Their possible relationship to the renal adenocarcinoma of the leopard frog. Ann NY Acad Sci 126:237-255
Granoff A, Came PE, Breeze DC (1966) Viruses and renal carcinoma of *Rana pipiens*. I. The isolation and properties of virus from normal and tumor tissue. Virology 29:133-148
Granoff A, Gravell M, Darlington RW. Studies on the viral etiology of the renal adenocarcinoma of *Rana pipiens* (Lucké Tumor). Mizell M (ed) (1969) Biology of Amphibian Tumors, pp 279-295. Springer-Verlag New York Heidelberg Berlin
Gravell M, Granoff A, Darlington RW (1968) Viruses and renal carcinoma of *Rana pipiens*. VII. Propagation of a herpes-type frog virus. Virology 36:467-475
Kozuch O, Labuda M, Nosek J (1977) Isolation of Sindbis virus from the frog *Rana ridibunda*. Acta Virol 22:78
Lucké B (1934) A neoplastic disease of the kidney of the frog, *Rana pipiens*. Amer J Cancer 20:352-379
Lucké B (1938) Carcinoma in the leopard frog: Its probable causation by a virus. J Exp Med 68:457-468
Lunger PD (1966) A new intranuclear inclusion body in the frog renal adenocarcinoma. J Morph 118:581-588
Murti KG, Goorha R, Chen M (1985) Interaction of frog virus 3 with the cytoskeleton. Curr Top Microbiol Immunol 116:107-131
Naegele RF, Granoff A, Darlington RW (1974) The presence of the Lucké herpesvirus genome in induced tadpole tumors and its oncogenicity Koch-Henle postulates fulfilled. Proc Natl Acad Sci USA 71:830
Rafferty KA Jr (1965) The cultivation of inclusion-associated viruses from Lucké tumor frogs. Ann NY Acad Sci 126:3-21
Smith AW, Anderson MP, Skilling DE, Barlough JE, Ensley PK (1986) First isolation of calicivirus from reptiles and amphibians. Am J Vet Res 47:1718-1721
Spalatin J, Connell R, Burton AN, Gollop BJ (1964) Western equine encephalitis in Saskatchewan reptiles and amphibians, 1961-1963. Can J Comp Med Vet Sci 28:131-142

Stackpole CW, Mizell M (1968) Electron microscopic observations on herpes-type virus-related structures in the frog renal.adenocarcinoma. Virology 36:63-72

Tweedell KS (1967) Induced oncogenesis in developing frog kidney cells. Cancer Res 27:2042-2052

Tweedell KS, Granoff A (1968) Viruses and renal carcinoma of *Rana pipiens*. V. Effect of frog virus 3 on developing frog embryos and larvae. J Nat Cancer Inst 40:407-410

Tweedell K, Wong WY (1974) Frog kidney tumors induced by herpesvirus cultured in pronephric cells. J Natl Cancer Inst 52:621

Willis DB, Goorha R, Chinchar VG (1985) Macromolecular synthesis in cells infected by frog virus 3. Curr Top Microbiol Immunol 116:77-106

Wolf K, Bullock GL, Dunbar CE, Quimby MC (1968) Tadpole edema virus: A viscerotropic pathogen for anuran amphibians. J Inf Dis 118:253-262

Wolf K, Bullock GL, Dunbar CE, Quimby MC. Tadpole edema virus: Pathogenesis and growth studies and additional sites of virus infected bullfrog tadpoles. Mizell M (ed) (1969) Biology of Amphibian Tumors, pp 327-336. Springer-Verlag New York Heidelberg Berlin

Wong WY, Tweedell KS (1974) Two viruses from the Lucké tumor isolated in a frog pronephric cell line (37981). Soc Exp Biol Med Proc 145:1201-1206

Herpesviruses: Interaction with Frog Renal Cells

K.S. Tweedell
Department of Biological Sciences
University of Notre Dame
Notre Dame, IN 46556
USA

Introduction

The prevalence of a spontaneous tumor of the kidney in the leopard frog, Rana pipiens, collected from north central and northeastern United States (Lucke 1934; Rafferty 1967; McKinnell 1965) and studies which predicted its viral etiology (Lucke 1938, 1952; reviewed by Rafferty 1964) ultimately led to the identification of the first amphibian herpes virus, the Lucke Tumor Herpes Virus (LTHV) (Fawcett 1956; Lunger 1964a,b; Lunger et al. 1965). Subsequent investigations (Tweedell 1967, 1969; Mizell et al. 1969; McKinnell and Tweedell 1970; Naegele et al. 1974; Tweedell and Wong 1974) firmly established the Lucke tumor herpes virus as the oncogenic agent of the frog renal adenocarcinoma. Another herpesvirus isolated from the pooled urine of frog tumor animals by Rafferty (1965) is commonly called frog virus 4 (FV4) (Rafferty 1967). This virus is non-oncogenic and is distinctly different from the LTHV (Gravell et al. 1968; Gravell 1971).

Both herpesviruses have double stranded linear DNA and are indistinguishable based on size (100 μm), icosehedral symmetry with 162 capsomeres (Lunger 1964,1966; Lunger et al. 1965; Gravell et al. 1968) but differ in molecular weight (Granoff and Naegele 1978; Naegele and Granoff 1980) base composition (Gravell 1971; Wagner et al. 1970) DNA base sequence homology (Gravell 1971) immunological relatedness (Granoff 1972) and physical-chemical resistance (Tweedell 1972; Granoff 1983). Both viruses are compared in Table 1.

The Host Cell. The indigenous (autochthonus) tumor cell for both viruses arises from the transformation of parenchymal cells of the adult mesonephros (Duryee 1956; Fawcett 1956) to form an unencapsulated, locally invasive adenocarcinoma which may metastasize to the lungs, liver, pancreas and other organs (Lucke 1938). The primary tumor proliferates, with frequent mitoses at 20-25° C. This form of the tumor, whether field collected or maintained in the laboratory at 20-25° presents no visible sign of herpesvirus production (the nonpermissive "summer tumor" phase) (Zambernard and McKinnell 1969; Rafferty 1964; Roberts 1963).

Table 1. Herpes viruses of *Rana pipiens*

COMPARISON OF TWO AMPHIBIAN HERPES VIRUSES

HERPES VIRUS:	FROG VIRUS 4	LUCKE TUMOR VIRUS
ISOLATION	POOLED URINE OF TUMOR FROGS	RENAL CARCINOMA IN COLD PHASE
MOLECULAR WT. DNA		
RANGE	73.9 - 79.2	62.5 - 73.4
AVE.	77.2	65.5
BASE COMPOSITION % G + C	54 - 56	45 - 47
BASE SEQUENCE HOMOLOGY (DNA-DNA HYBRIDIZATION)		
FV4	+	-
LTHV	-	N.D.
VIRUS ASSOCIATED CELL ANTIGENS	N.D.	+
CULTURE IN VITRO	FROG EMBRY. CELLS PLAQUE ASSAYED	FROG PRONEPHRIC CELLS SHORT TERM
BIOLOGICAL ACTIVITY	CHRONIC DISEASE IN TADPOLES; LETHAL TO EMBRYOS	ONCOGENIC: RENAL CARCINOMA

Virus Host Tumor Cell Interaction.

Replication of LTHV. In the frog renal carcinoma activation of the LTHV genome and virus replication takes place naturally in frog carcinoma cells that have been exposed to reduced temperatures. The hallmark of LTHV replication in the neoplastic cell is the development of Cowdry Type A intranuclear inclusions first described in the frog tumor by Lucke (1952) who noted their seasonal variation. The tumors always display typical cytological changes: nuclear enlargement, marginated chromatin with a peripheral nucleolus and predominate acidophilic intranuclear inclusions (Rafferty 1964; Tweedell

1967; Mizell 1969; Granoff 1972). This nuclear inclusion phenomenon, consistent under both field and laboratory conditions, is the permissive "winter tumor" phase, since tumor animals collected during hibernation (McKinnell and Ellis 1972a) or after emergence from hibernation or post spawning possess herpesvirus (McKinnell and Zambernard 1968; McKinnell and Ellis 1972b). Electron microscopic examination of these cold induced inclusions reveals a crystalline array of single membraned capsids with or without a nucleoid (Fawcett 1956; Rafferty 1964; Lunger et al. 1965; Zambernard et al. 1966; Mizell et al. 1969; Granoff 1972), double layered capsids with nucleoids (Zambernard and Mizell 1965) and nuclear membranous sacs of nucleated viruses (Lunger et al. 1965; Zambernard et al. 1966; Stackpole 1969; Mizell 1969).

Activation of the viral genome in the tumor cell to invoke viral replication occurs in response to temperatures below 11.5° C. Induction of nuclear inclusion bodies (NIB) and thus LTH virus replication, developed in primary tumors of animals kept below 5° C but they were never found in animals held at 20-25° C (Roberts 1963). Periodic biopsy of tumors held at low temperatures showed that the tumors possessed nuclear inclusions at 4° C after 5 or more months (Rafferty 1965). Transplants of originally virus free tumors to the eyechambers of *Rana pipens* hosts, subsequently placed at 7-8° C also develop nuclear inclusion bodies and typical LTH virus as soon as 9 1/2 weeks (Mizell et al. 1968). Intraocular transplants of non-virus tumors into heterologous hosts established that NIB and LTH virus production at 7.5-9.5° C is independent of the indigenous host. Other intraocular transplants remained inclusion and virus free when maintained from 15 to 26° C (Skinner and Mizell 1972). Simultaneous appearance of LTH virus occurs in the primary tumor and in metastatic tumor cells located in the liver, fat body and bladder after exposure of the tumor animal to 4° C (McKinnell and Cunningham 1982).

Induction of nuclear inclusions and development of virus in tumor cells can occur *in vitro* as well. Explants of inclusion free primary tumor tissue fragments placed in organ culture at 9° C developed intranuclear inclusions after 12 weeks whereas control explants at 23° C remained inclusion free (Morek 1973). Nuclear inclusions and virus induction were verified in similar tissue culture explants from virus free tumors kept at 7.5° after 3 months (Breidenback et al. 1971). Inclusion bodies and virus production, followed by virus recovery, was also obtained from organ culture of virus free induced tumors cultured at 7.5° C (Naegele et al. 1974). Thus, activation of the viral genome and replication of the Lucke tumor herpes virus in the frog renal carcinoma cells is conditioned or determined by low temperatures both _in vivo_ and _in vitro_. The initiation of these events is not dependent upon the host organism but the actual mechanisms await further inquiry. Possible insight into this temperature dependency may be forthcoming. Preliminary investigations with the _myc_ oncogene has revealed distinct patterns in the winter tumor DNA that were absent from the summer tumor and normal kidney (Imseis et al. 1987).

Tumor cells that harbor replicating virus can be identified immunologically. Detection of the Lucke tumor herpes virus antigens in "virus containing" cells of the winter tumor was demonstrated by indirect immunofluorescence with antiserum to virus obtained from winter tumors (Paul et al. 1972; Naegele and Granoff 1977). Nuclear, cytoplasmic and membrane antigens were observed. The degree of indirect immunofluorescence was positively correlated with the presence of intranuclear inclusions and herpes-virus (Naegele and Granoff 1977). Both studies failed to show the presence of viral antigens on fixed, non-virus producing tumor cells.

Viral Latency. Typically, the examination of renal tumors obtained from animals kept at warm temperatures (18 -25° C) or field collected in the summer (Zambernard and McKinnell 1969) the summer phase, are strikingly devoid of intranuclear inclusions (Lucke 1952; Mizell and Zambernard 1965; Zambernard and Mizell 1965; Granoff 1972). Viral particles are not detected by electron microscopy in these tumors (Fawcett 1956; Roberts 1963; Mizell and Zambernard 1965; Zambernard and Mizell 1965; Mizell et al. 1968). Instead, cell nucleoli are enlarged and hyperactive, chromatin is diffuse and the cytoplasm is strongly basophilic (Tweedell 1965). Mitotic figures are usually numerous however, the mitotic coefficient of tumors in frogs collected in December is reported as higher (4.65/1000 cells) than tumor animals captured in July (2.94/1000 cells) (Marlow and Mizell 1972).

Although devoid of virus, the tumor cell extant at warm temperatures still produces LTHV antigens. Both virus free and virus producing tumor cells exhibit membrane antigens in the viable, unfixed state. Antiserum to LTHV coupled with indirect immunofluorescence could distinguish membrane antigens on both virus free and electron microscopic verified tumor virus producers while normal kidney cells were negative (Naegele and Granoff 1977). Transcripts of the viral genome are also present in virus free, warm tumors. Labelled RNA extracted from non virus tumors will hybridize with DNA derived from LTHV isolated from the virus producing cells. No virus specific RNA was detected in normal kidney cells (Collard et al. 1973).

Manifestation of LTHV Oncogenic Activity in Tumor Cells. Other virus associated and virus related changes are manifest in the spontaneous tumors and tumors induced by embryos inoculated with LTHV. Detection of intracellular viral antigens with antiserum to the LTH virus was found only in tumors containing the LTH virus however, both virus free and virus resident tumor cell membranes reacted with the LTHV antiserum when viable tumor cells were examined. Moreover, after adsorption of the antiserum with virus free tumors, virus associated membrane antigens were still detectable by indirect immunofluorescence (Naegele and Granoff 1977) which they suggested may be related to the malignant state (Naegele and Granoff 1980).

LTHV induced oncogenesis does result in new virus related cellular products. Isolated plasma membranes of pronephric tumor cells has yielded two categories of proteins, tumor cell associated and a complex shared with normal cells (Mordan and Tweedell 1978) (Fig 1). When pronephric tumor cell membranes were prelabeled with ^{3}H galactose before isolation of the plasma membranes, six galactose labeled glycoproteins were found unique to the pronephric tumor cell surface, 175, 73, 60, 45 and 39 kD. In addition, an antiserum prepared to pronephric tumor cell plasma membranes differentiated the group of tumor cell specific antigens from a normal cell complex on the surface of viable tumor cells by indirect immunofluorescence. After interaction of the antiserum with tumor cell plasma membranes, antigen-antibody complexes were solubilized from the pronephric tumor cell membranes and recovered on a protein A Sepharose affinity column. Several proteins and glycoproteins were identified as cell surface antigens. Three of these peaks at 39, 60 and 73 kD were labeled immunoprecipitates (Fig 2) unique to the tumor cell membrane (Tweedell and Mordan unpublished data).

Another change in the renal cell resulting from LTHV oncogenesis is in the relative levels of nucleolar and ribosomal RNA nucleotides after transformation. In ribosomal RNA from the virus free Lucke renal carcinoma and in the induced pronephric tumor cell lines (Tweedell 1978) there is a sharp drop in adenylic acid (6.8%) and a moderate decline in guanylic acid when compared to the normal mesonephros. Conversely, there is a sharp increase in cytidylic (7%) and a moderate rise in uridylic acid levels (3.8%) in the tumor cells. Nearly identical quantitative changes in the nucleotides were noted in 40S nucleolar nRNA and in both 18S and 28S rRNA LTHV when induced pronephric tumor cell lines were compared to normal pronephric cells (Tweedell and Malin 1984).

Screening for viral oncogene expression in a LTHV induced frog pronephric tumor cell line (PNKT 48, Tweedell 1978) has suggested the presence of transcripts homologous to several viral oncogene probes whose expression also fluctuate with temperature. Expression of several oncogenes (*abl*, *myc*, *src*, *sis*, *K-ras*, *erb A*) were enhanced at 28° C (invasion permissive temperature) while three oncogenes (*ros*, *H-ras*, *fes*) showed increased expression at 18° C (invasion restrictive) temperature (Sauerbier et al. 1986).

LTHV Tumorigenesis of Normal Kidney Cells.

Viral Oncogenesis in vivo. Numerous attempts to produce renal tumors in the autogenic host by experimental transmission with cell free extracts were conducted by Lucke 1938, 1952; Duryee

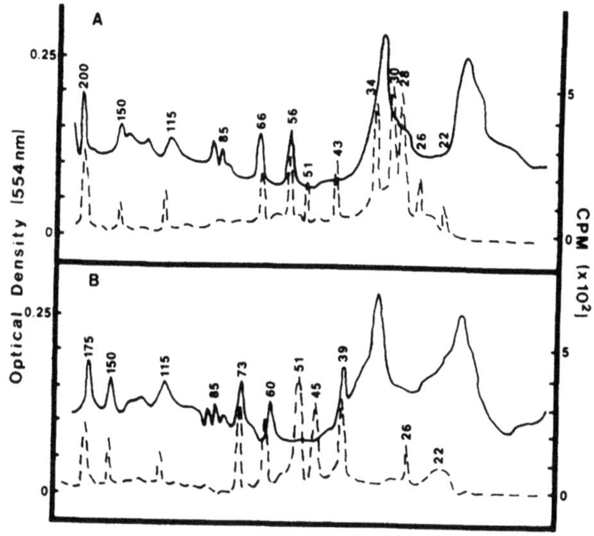

Fig 1 Plasma membrane proteins solubilized after membrane isolation from (A) normal pronephric cells and (B) LTHV induced pronephric tumor cells, compared to chromatographic profiles of similar membrane fractions from cells prelabeled, prior to isolation, with galactose oxidase-tritiated borohydride. The solid line is a densitometric scan of Coomassie blue R stained proteins solubilized from isolated cell membranes and electrophoresed on polyacrylamide gels. The dashed line is counts per minute of cell membrane glycoproteins prelabeled before isolation. The number above each peak corresponds to the molecular weight divided by 10^3.

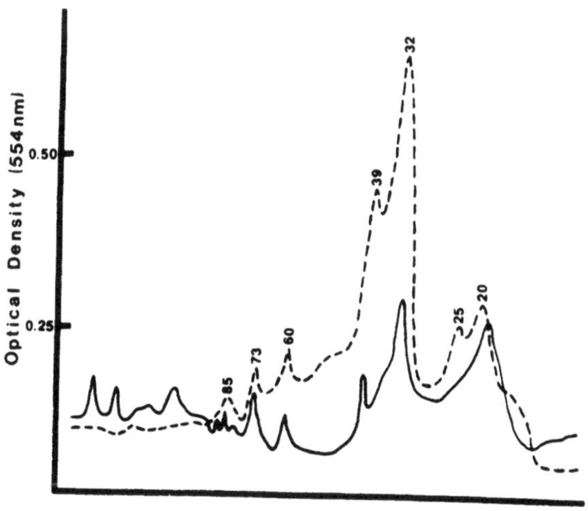

Fig 2 Immunoprecipitates of prelabelled cell surfaces of LTHV induced pronephric tumor cells. Dotted line shows the Protein A sepharose retained prelabeled membrane glycoproteins that were immunoreactive to tumor cell plasma membrane antiserum. The lower solid line is a scan of proteins on PAGE for relative molecular weights. (Mordan and Tweedell, unpublished results)

1956 and Rafferty 1963a) with limited success. More direct evidence for the viral etiology was produced when virus was recovered from various subcellular fractions and filtrates of spontaneous nuclear inclusion tumors and injected into *Rana pipiens* embryos. Injection of embryos from the blastula stage through the early larval stage produced renal adenocarcinomas in

66-92% of the animals as they reached metamorphosis. Similar cell extracts from inclusion free tumors or normal kidneys did not produce tumors (Tweedell 1967). LTH virus derived from tumor nuclear fractions (Tweedell 1972) or after purification by rate zonal centrifugation (Toplin et al. 1971) produced pronephric and mesonephric tumors and identified the infective form of the herpes virus (Mizell 1969; Mizell et al. 1969). The tropic effect of the herpes virus for the differentiating kidney cell alone was evident in that both pronephric and mesonephric kidneys developed tumors with identical pathology. Another site of LHV was discovered in the ascitic fluid from a cold treated, tumor bearing frog. Injection of the LHV from these animals also caused renal tumors in developing frog embryos (Naegele and Granoff 1972). Evidence for viral latency in the induced tumors was seen by the appearance of intranuclear inclusion bodies in the induced tumor animals held at 9° (Tweedell 1972) and by the presence of herpes virus in eyechamber transplants from a pronephric tumor (Mizell 1969). The capstone to establishing the LTHV as the etiological virus was presented in an elegant experiment by Naegele et al. 1974. Induction of renal tumors was obtained by inoculation of embryos with virus from a spontaneous Lucke tumor. Next, tissue explants from the virus free carcinomas were placed in organ culture at 7.5 or 22° C. Each culture group was monitored for both intranuclear inclusions and herpes virus. Only the tumor implants held at 7.5° C developed herpes virus; those at 22° C were virus free. Tissue fractions from each tumor explant group were inoculated again into frog embryos. Only those embryos receiving extracts from the cold treated explants developed tumors in hosts raised at 22° C. These experiments confirmed the dependency of the LTHV on low temperature for the activation and replication in the tumor cell. More importantly, they indicated that tumors induced in embryos by LTHV from cold treated spontaneous tumors would yield an infective agent that had the same oncogenic properties as the original virus.

Tumorigenic Changes in vitro. Little is known about the succession of oncogenic changes as a result of LTHV virus transformation _in vivo_ or _in vitro_. Preneoplastic changes in the in situ kidney cell were described by Duryee (1965) particularly the presence of enlarged hyperactive nucleoli and increased basophilia in the Lucke tumor were linked to RNA proliferation (Tweedell 1965).

Some insight into cell transformation has been derived from in vitro studies on larval kidney tissue fragments cultivated with fragments of non-inclusion tumors separated by a vertical 150μ thick, 0.45u (pore size) filter to prevent cellular exchange (Morek 1973). Other kidney fragments were preincubated in sterile filtrates of LTHV from a previously established oncogenic tumor. The organ cultures were maintained at 9° and 25° C for up to 13 weeks. Nuclear inclusions (type A) were detected in the tumor tissue at 6 weeks and increased steadily with the incubation period. Histopathological changes, based on 14 criteria used in clinical pathological diagnosis, began to appear in the normal kidney after 3-4 weeks at 9°. These included enlarged hyperactive nucleoli, increased cytoplasmic basophilia, increased mitoses, loss of cell polarity, cell migration and columnar growths at 11-12 weeks. Normal kidney controls cultured alone or transfiltered from normal kidney did not exhibit these changes at 9° C whereas stromal cell invasion took place in kidney control at 25°C. The LTHV exposed kidney tissue tubule cells began to incorporate ^{3}H thymidine at 6-7 weeks while the non-virus exposed kidney tissue only showed isotope incorporation into stromal and blood cells.

LTHV Infection and Replication in Permissive Kidney Cells.

Exposure of a cell line of differentiating pronephric kidney cells, derived from embryos of Rana pipiens (Wong and Tweedell 1974) to LTHV derived from tumor cell fractions of an oncogenic tumor resulted in a CPE in 12 days at 25°C. After 3 viral passages, spanning 12 cell subcultures, detaching pronephric cells were examined by electron microscopy. Arrays of intranuclear herpes virus capsids, along with intranuclear and cytoplasmic sacs of nucleated virus, and various stages of virus and virus associated structures, previously described in the primary tumor by Fawcett (1956) were found in the infected pronephric cells (Wong and Tweedell 1975). The oncogenecity of the viral infection was tested by recovering virus from cell fractions of the infected cells and inoculation of frog embryos. Both pronephric and mesonephric tumors were produced in 52% of the survivors (Tweedell and Wong 1974).

Substantiating evidence that the oncogenic virus was replicating in the pronephric cell came when antiserum to purified Lucke tumor herpes virus was applied to both normal and herpes virus infected pronephric cells (Tweedell and Mizell 1979). Only the infected cells exhibited intracellular immunofluorescence after exposure to the LTHV antiserum. While all evidence indicates that herpes virus production in primary tumors, intraocular implants or organ culture of tumor tissue is a low temperature related event, the same parameters are not applicable for virus replication in the permissive kidney cell. The essential difference may well be a low temperature activation of the viral genome in the renal carcinoma cell and an opportunistic replication of the virus in the permissive, transformed non-malignant pronephric cell.

Etiology of LTHV

Perpetuation of the Lucke Tumor Virus. The natural life cycle of the Lucke tumor virus as proposed by Rafferty (1964), that of early viral incorporation into the embryo and ultimate expression in the young adults is still compatible with evidence on the LTHV accumulated to date (Naegele and Granoff 1980). Three optional states of the virus were postulated by Mizell and Zambernard (1965). The virus could exist as an integrated inactive latent form in the normal kidney cell, in a lysogenic state in an actively growing tumor or in a non-integrated vegetative state when the virus replicates in the tumor cell. The complete viral genome needed for replication may indeed utilize oncogenesis for its ultimate perpetuation since genome activation and viral replication only occur naturally in the obligate tumor cell at low temperatures (4-11°C).

The preferred route for the natural transmission of the oncogenic herpes virus has not been established although experimental evidence to date favors vertical transmission. Tumor bearing frogs kept with normal adults show no significant rise in tumor incidence (Lucke 1952). While the incidence of spontaneous tumors in laboratory animals appears to rise over a long period of time (Rafferty and Rafferty 1961; Rafferty 1963a) tumor incidence was not higher in crowded vs. isolated laboratory animals (Rafferty 1963b). Yet, animals pre-stored at 4°C for 4 months and then maintained for several months suggested that as crowding increased the occurrence of tumors increased somewhat (DiBerardino and King, 1965). In all of these early "crowding" experiments none of the groups were kept at the lower virus inducing temperatures which may have influenced the results.

Rafferty (1964) postulated that an early viral infection of eggs or hatched embryos could occur from virus released via the urine. Exposure of fertilized eggs or sperm or newly hatched embryos and larvae to tumor derived oncogenic LTHV placed in the culture medium continuously for over a month failed to produce any tumors in the metamorphosed yearlings (Tweedell 1969). Alternatively, if developing eggs already possess the oncogenic virus, a latent infection of either the sperm or the egg could take place during gametogenesis (Mizell and Zambernard 1965). When post-ovulatory oocytes were exposed _in vivo_ to abdominal injections of LTHV from a Lucke tumor at the time of ovulation, mesonephric renal carcinomas developed in young yearling adults (Tweedell 1969). These typical adenocarcinomas sometimes metastasized to the lung, liver and ovaries, a feature usually seen only in terminal renal tumors of the mature adult. The implication that LTHV transmission could be accomplished by infection of oocytes in a reproductive tumor bearing female was supported by the recovery and subsequent tumor induction by herpes virus from the ascitic fluid of a female having a renal tumor (Naegele and Granoff 1972). These experiments do not rule out other means of natural transmission but they do emphasize the tropic nature of the LTHV for the differentiating kidney cell with its potential for oncogenesis. Secondly, they reaffirm that the LTHV probably exists in a latent form in the normal cell, whether it be in an arrested state or in an integrated form.

The Frog Urine Herpesvirus.

Herpesvirus FV 4 will not cause tumors when embryos or tadpoles are injected with the virus. Instead, it causes a chronic disease when tadpoles are exposed producing later massive edema in adults (Rafferty 1967) or extensive subdermal hemorrhagic lesions when larvae are inoculated and, in addition, edema in the embryos (Tweedell unpublished results). Higher concentrations of the virus will produce surface vesicles, then killing the embryos after 13 days (Granoff 1983).

Virus Isolation and Cytopathic Effect. The herpesvirus FV 4 was originally isolated from the pooled urine of adult frogs with renal tumors when the virus grew slowly in an embryonic kidney cell line from *Rana sylvatica* (Rafferty 1965), two embryonic cell lines of *Rana pipiens* (Gravell et al. 1968) and in a frog kidney cell line (Gravell 1971) but not in a Lucke tumor cell line (Rafferty 1965). The infective virus causes a cytopathic effect in the kidney and embryo cell lines following a rise in viral titer. The nuclei enlarge and develop multiple, small eosinophilic intranuclear inclusions, reported to contain DNA. Both nuclei and the cytoplasm become vacuolated (Rafferty 1965; Gravell et al. 1968, Granoff et al. 1969). When infected cells are examined by electron microscopy, small clumps of crystalline arrays, consisting of single and double ringed virus particles are found in the nucleus. Developmental stages of typical intracellular herpes virus are seen along with extracellular enveloped particles (Granoff and Darlington 1969; Granoff et al. 1969). Cell associated virus yields are first detected at 4-5 days and free virus at 6-7 days. The virus titers ranged from 1 x 10^5 PFU/ml (Gravell et al. 1968). Relatively little is known about the particulars of the life cycle of FV 4 nor the parameters controlling its replication in the renal tumor cell except that the isolation was from urine of frogs with nuclear inclusion tumors (Rafferty 1965).

Discussion.

The LTHV appears to have a natural tropism for the kidney cell and oncogenesis of renal cells occurs at higher temperatures (20°C) *in vivo*. This is based on the observations that exposure of oocytes to LTHV leads to formation of renal carcinomas in the young adults (Tweedell 1969) which means that the viral genome is sheparded through development whether in an episomic or integrated state until its ultimate expression in the kidney cell. The successful production of renal tumors by embryo or larval inoculation (Tweedell 1967; Mizell et al. 1969; Naegele and Granoff 1972; McKinnell and Tweedell 1970) often produced tumors of the embryonic kidney (pronephros) and at later stages, in the newly forming mesonephros (Tweedell and Wong 1974). Normally, the pronephros degenerates early in the larval stage and tumors of the pronephros have never been found in nature. Inoculation of blastula stages (Tweedell 1967) or even earlier embryonic stages (Mizell 1985) causes similar renal tumors and supports the proposed concept of kidney cell tropism. Admittedly, it may be that the specificity may reside in the virus itself as an oncogene (a concept, if not the mechanism) which once proposed that the original Lucke tumor virus was a modified (mutated) tissue differentiated nucleoprotein (Rose 1952; Rose and Rose 1952). Aberrant expression of proto-oncogenes by retroviruses has been cited in many tumors. Although the direct causal relationship between herpesviruses, oncogenes and tumorigenesis appears tenuous (Bishop 1986) an active myc oncogene is present in the EBV related Burkitt's lymphoma. The earlier cited enchancement of several oncogenes in the pronephric tumor cells and in the virus (winter) tumor suggest that one or more oncogenes may be involved in LTHV tropism, replication and maintenance of the oncogenic state.

References

Bishop JM (1985) Viral oncogenes. Cell 42:28-38

Bridenbach GP, Skinner MS, Wallace JH, Mizell M (1971) *In-vitro* induction of a herpes-type virus in "summer-phase" Lucke tumor explants. J Virol 7:679-682

Collard W, Thornton H, Mizell M, Green M (1973) Virus-free adenocarcinoma of the frog (summer phase tumor) transcribes Lucke tumor herpesvirus-specific RNA. Science 181:448-449
DiBerardino MA, King TJ (1965) Renal adenocarcinomas promoted by crowded conditions in laboratory frogs. Cancer Res 25:1910-1912
Duryee WR (1956) Precancer cells in amphibian adenocarcinoma. Ann NY Acad Sci 63:1280-1302
Fawcett DW (1956) Electron microscope observations of intracellular virus-like particles associated with the cells of the Lucke renal adenocarcinoma. J Biophys Biochem Cytol 2:725-741
Granoff A (1969) Viruses of Amphibia. Curr Top Microbiol Immunol 50:107
Granoff A (1972) Herpesvirus and frog renal adenocarcinoma. Fed Proc Am Soc Exp Biol 31:1626-1633
Granoff A (1983) Amphibian Herpesviruses. In: Roizmann B (ed) The Herpesviruses. Plenum Pub Corp pp 367-384
Granoff A, Darlington RW (1969) Viruses and renal carcinoma of Rana pipiens. VIII. Electron microscopic evidence for the presence of herpes virus in the urine of a Lucke tumor-bearing frog. J Virol 38:197-200
Granoff A, Naegele RF (1978) The Lucke tumor: A model for persistent virus infection and oncogenesis. In: Stoveno S, Todaro G, Fox FC (eds) Persistent Viruses. Vol XI, ICN-UCLA Symp Molecular and Cellular Biology, Academic Press, NY pp 15-25
Granoff A, Gravell M, Darlington RW (1969) Studies on the viral etiology of the renal adenocarcinoma of Rana pipiens (Lucke tumor). In: Mizell M (ed) Biology of amphibian tumors. Springer-Verlag, New York pp 279-295
Gravell M (1971) Viruses and renal carcinoma of Rana pipiens. X. Comparison of herpes-type viruses associated with Lucke tumor-bearing frogs. J Virol 43:730-733
Gravell M, Granoff A, Darlington RW (1968) Viruses and renal carcinoma of Rana pipiens. VII. Propagation of a herpes-type frog virus. J Virol 36:467-475
Imseis H, Mizell M, Ranier J, Hoffman L (1987) Oncogene analysis of the Lucke carcinoma and of early development in Rana pipiens. Amer Zool 27:144
Lucke B (1934) A neoplastic disease of the kidney of the frog, Rana pipiens. Am J Cancer 20:352-379
Lucke B (1938) Carcinoma in the leopard frog: its probable causation by a virus. J Exp Med 68:457-468
Lucke B (1952) Kidney carcinoma in the leopard frog: a virus tumor. Ann NY Acad Sci 54:1093-1109
Lunger PD (1964a) Thin-section electron microscopy of the mature Lucke frog kidney tumor virus. J Virol 22:285-286
Lunger PD (1964b) The isolation and morphology of the Lucke frog kidney tumor virus. J Virol 24:138-145
Lunger PD (1966) Amphibia-related viruses. Advan Virus Res 12:1-33
Lunger PD, Darlington RW, Granoff A (1965) Cell-virus relationships in the Lucke renal adenocarcinoma: An ultrastructure study. Ann NY Acad Sci 126:289-314

Marlow PB, Mizell S (1972) Incidence of Lucke renal adenocarcinoma in Rana pipiens as determined by histological examination. J Natl Cancer Inst 48:823-829
McKinnell RG (1965) Incidence and histology of renal tumors of leopard frogs from the north central states. Ann NY Acad Sci 126:85-98
McKinnell RG, Cunningham WP (1982) Herpesviruses in metastactic Lucke renal adenocarcinoma. Differentiation 22:41-46
McKinnell RG, Ellis VL (1972a) Epidemiology of the frog renal tumor and the significance of tumor nuclear transplantation studies to a viral aetiology of the tumor--a review. In: Biggs PM, de-The G, Payne LN (eds) Oncogenesis and herpesviruses. International Agency for Research on Cancer, Lyon pp 183-197
McKinnell RG, Ellis VL (1972b) Herpesviruses in tumors of post-spawning Rana pipiens. Cancer Res 32:1154-1159
McKinnell RG, Tweedell KS (1970) Induction of renal tumors in triploid leopard frogs. J Natl Cancer Inst 44:1161-1166
McKinnell RG, Zambernard J (1968) Virus particles in renal tumors from spring Rana pipiens of known geographic origin. Cancer Res 28:684-633
Mizell M (1969) State of the art: Lucke renal adenocarcinoma In: Mizell M (ed) Biology of Amphibian Tumors. Springer-Verlag, New York pp 1-25
Mizell M (1985) Lucke frog carcinoma herpesvirus: Transmission and expression during early development. In: Klein G (ed) Adv Viral Oncology, vol V. Raven Press, New York pp 129-146
Mizell M, Stackpole CW, Halperen S (1968) Herpes-type virus recovery from "virus-free" frog kidney tumors. Proc Soc Exp Biol Med 127:808-814
Mizell M, Toplin I, Isaacs JJ (1969) Tumor induction in developing frog kidneys by a zonal centrifuge purified fraction of the frog herpes-type virus. Science 165:1134-1137
Mizell M, Zambernard J (1965) Viral particles of the frog adenocarcinoma: Causative agent or passenter virus. II. A promising model system for the demonstration of a "lysogenic" state in a metazoan tumor. Ann NY Acad Sci 126:146-169
Morek DM (1973) An organ culture study of frog renal tumor and its effects on normal frog kidney in vitro. Oncology 28:536-552
Mordan LJ, Tweedell KS (1978) Cell surface proteins from normal and transformed frog kidney cells. Amer Zool 18:629
Naegele RF, Granoff A (1972) Viruses and renal carcinoma of Rana pipiens. XIII. Transmission of the Lucke tumor by herpesvirus-containing ascitic fluid from a tumor-bearing frog. J Natl Cancer Inst 49:299-303
Naegele RF, Granoff A (1977) Viruses and renal carcinoma of Rana pipiens. XV. The presence of virus-associated membrane antigen(s) on Lucke tumor cells. Int J Cancer 19:414-418
Naegele RF, Granoff A (1980) The Lucke tumor and its herpesvirus. In: Rapp F (ed) Oncogenic Herpesviruses. CRC Press, Boca Raton, Florida pp 85-101
Naegele RF, Granoff A, Darlington RW (1974) The presence of the Lucke herpesvirus genome in induced tadpole tumors and its

oncogenicity: Koch-Henle postulates fulfilled. Proc Natl Acad Sci USA 71:830-834

Paul SM, Mizell M, Craige B, Blazek J, Skinner M (1972) Specific immunofluorescence test for detection of herpesvirus antigen in cells of the Lucke renal adenocarcinoma. Proc Soc Exp Biol Med 139:944-948

Rafferty KA, Jr. (1963a) Effect of injected frog kidney tumor extracts on development of tumors under promoting conditions. J Natl Cancer Inst 30:1103-1113

Rafferty KA, Jr. (1963b) Spontaneous kidney tumors in the frog: rate of occurrence in isolated adults. Science 141:720-721

Rafferty KA, Jr. (1964) Kidney tumors of the leopard frog: a review. Cancer Res 24:169-185

Rafferty KA, Jr. (1965) The cultivation of inclusion-associated viruses from Lucke tumor frogs. Ann NY Acad Sci 126:3-21.

Rafferty KA, Jr. (1967) The biology of spontaneous renal carcinoma of the frog. In: King JS (ed) Renal neoplasia. Little, Brown & Co, Boston pp 301-315

Rafferty KA Jr. Rafferty NS (1961) High incidences of transmissible kidney tumors in uninoculated frogs maintained in a laboratory. Science 133:702-703

Roberts ME (1963) Studies on the transmissibility and cytology of the renal carcinoma of _Rana pipiens_. Cancer Res 23:1709-1714

Rose SM (1952) Interaction of tumor agents and normal cellular components in Amphibia. Ann NY Acad Sci 54:1110-1119

Rose SM, Rose F (1952) Tumor agent transformation in Amphibia. Cancer Res 12:1-12

Sauerbier W, McKinnell RG, Okamura CS, Tweedell KS (1986) Differential expression of cellular oncogenes associated with culture of a frog tumor cell line (PNKT-4B) at invasion-permissive and restrictive temperatures. Proc Amer Assoc Cancer Res 27:66

Skinner MS, Mizell M (1972) The effect of different temperatures on herpesvirus induction and replication in Lucke tumor explants. Lab Invest 26:671-681

Stackpole CW (1969) Herpes-type virus of the frog renal adenocarcinoma. I. Virus development in tumor transplants maintained at low temperature. J Virol 4:75-93

Toplin LM, Mizell M, Sottong P, Monroe J (1971) Zonal centrifuge applied to the purification of herpesvirus in the Lucke frog kidney tumor. Appl Microbiol 21:132-139

Tweedell KS (1965) Cytopathology of a Frog Renal Adenocarcinoma _in vitro_ with fluorescence microscopy. Ann NY Acad Sci 126:170-187

Tweedell KS (1967) Induced oncogenesis in developing frog kidney cells. Cancer Res 27:2042-2052

Tweedell KS (1969) Simulated transmission of renal tumors in oocytes and embryos of _Rana pipiens._ In: Mizell M (ed) Biology of Amphibian Tumors. Springer-Verlag, New York pp 254-260

Tweedell KS (1972) Experimental alteration of the oncogenicity of frog tumor cells--viral fractions. Proc Soc Exp Biol Med 140:1246-1251

Tweedell KS (1978) Pronephric tumor cell lines from Herpes virus transformed cells. In: de-The G, Henle W. Rapp F (eds)

Oncogenesis and Herpes Virus. Inter Agen Res Cancer, Lyon pp 609-616
Tweedell KS, Malin JH (1984) Changes in Nucleolar and Ribosomal RNA of the Frog Kidney after Malignant Transformation by the Lucke Tumor Herpes Virus. Exp Cell Biol 52:196-206
Tweedell KS, Mizell M (1979) Detection of Lucke herpesvirus antigens in infected frog pronephric cells. Arch Virol 59:239-249
Tweedell KS, Wong WY (1974) Frog Kidney Tumors Induced by Infected Pronephric Cell Cultures. J Natl Can Instit 52:621-624
Wagner EK, Roizman B, Savage T, Spear PG, Mizell M, Durr E, Sypowicz D (1970) Characterization of the DNA of herpesviruses associated with Lucke adenocarcinoma of the frog and Burkett lymphoma of man. Virol 42:257-261
Wong WY, Tweedell KS (1974) Two Viruses from the Lucke Tumor Isolated in a Frog Pronephric Cell Line. Proc Soc Exp Biol Med 145:1201-1206
Wong WY, Tweedell KS (1975) Viral Carcinogenesis in a Pronephric Cell Line. Amer J Pathology 80:143-152
Zambernard J, McKinnell RG (1969) "Virus-free" renal tumors obtained from pre-hibernating leopard frogs of known geographic origin. Cancer Res 29:653-657
Zambernard J, Mizell M (1965) Viral particles of the frog renal adenocarcinoma: causative agent or passenger virus? I. Fine structure of primary tumors and subsequent intraocular transplants. Ann NY Acad Sci 126:127-145
Zambernard J, Vatter AE, McKinnell RG (1966) The fine structure of nuclear and cytoplasmic inclusions in primary renal tumors of mutant leopard frogs. Cancer Res 26:1688-1700

Frog Virus 3 DNA Replication

R.M. Goorha
Department of Virology and Molecular Biology
St. Jude Children's Research Hospital
332 North Lauderdale
Memphis, Tennessee 38101
USA

INTRODUCTION

Frog virus 3 (FV3) is a large icosahedral virus belonging to the *Iridoviridae* family. The genome is double-stranded, linear DNA with a molecular weight of 100 x 10^6 (Kelly and Avery, 1974; Murti et al., 1982). Among animal viruses, FV3 was the first demonstrated to contain a circularly permuted and terminally redundant genome (Goorha and Murti, 1982; Murti et al., 1982). Another unusual feature of FV3 DNA is its high degree of methylation; over 20% of deoxycytidine residues are methylated at 5-carbon position (Willis and Granoff, 1980). In this paper, I discuss FV3 DNA replication and how the novel features of the FV3 genome, i.e., high degree of methylation and circular permutation, impact on the strategy of DNA replication.

DNA REPLICATION

Cellular sites of synthesis. An unusual feature of FV3 DNA replication is that it occurs both in the nucleus as well as in the cytoplasm. We showed by electron microscopic autoradiography and biochemical techniques that about 30% of the viral DNA was synthesized in the nucleus and majority of viral DNA (~90%) was transported into the cytoplasm as determined by pulse-chase experiments (Goorha et al., 1978). The experiments clearly established that the nucleus is a site for FV3 DNA synthesis and that a precursor-product relationship existed between nuclear and cytoplasmic viral DNA. However, the amount of labeled DNA found in the cytoplasm could not be accounted for solely by transport of labeled DNA from the nucleus. The kinetics of labeling indicated that the cytoplasm was also a site of viral DNA synthesis (Goorha, 1982). Aubertin (1975) also showed that FV3 DNA replicated and accumulated in the nucleus of arginine-starved FV3-infected BHK cells and only after the addition of arginine was viral DNA transported to the cytoplasm for eventual incorporation into infectious virus particles (Martin et al., 1984).

Sizes of replicating viral DNA. Utilization of the nucleus as well as the cytoplasm for viral DNA replication, so far, has been reported only for FV3 although it is possible that other members of *Iridoviridae* family may employ similar strategy of DNA replication. However, viral DNA synthesis in two separate sites raised several interesting questions. For example what are the relationships between the DNA molecules in each of the cellular compartments, and how do these two sites of DNA replication fit into the overall replication scheme of FV3. Examination of the sizes of the replicating viral DNA in nucleus and cytoplasm of the infected cell provided an explanation (Goorha, 1982). In these experiments, infected cells were labeled with $[^3H]$-thymidine at various times after the infection and the sizes of the newly synthesized, replicating viral DNA was determined by neutral sucrose gradient analysis. Data from these experiments established that viral DNA replication occurs in two stages that can be distinguished by the temporal appearance, the site, the size, and the state of DNA methylation of replicating DNA molecules. First stage DNA replication occurs early in infection (0-3 hr). DNA is synthesized exclusively in the nucleus, and the size of the replicating DNA is between genome and twice the genome size. Since in neutral sucrose gradients parental genome remains associated with the progeny DNA, the size of the progeny DNA is genome or less than genome size during the first stage of DNA replication. In contrast, second stage DNA replication occurs only in the cytoplasm, late in infection (after 3 hr) and the replicating DNA is present as large concatemer (more than ten times the genome size). Evidently, the progeny DNA during second stage of DNA replication consist of multiple genome length molecules. Results of pulse-chase experiments showed that the concatemeric DNA served as the precursor for the production of mature viral DNA molecules. The viral DNA in nucleus is not methylated, DNA methylation takes place only after the DNA has been transported to the cytoplasm.

The isolation and characterization of a temperature-sensitive mutant defective in DNA transport (ts 12488) also corroborated the two-site model of FV3 replication (Goorha and Dixit, 1984). At the nonpermissive temperature (30°C), the rate of viral DNA synthesis in ts 12488-infected cells was 10-20% of the rate attained at the permissive temperature (25°C). Biochemical characterization of the DNA synthesized at 30°C showed that (1) newly synthesized DNA was present in the nuclear fraction only, (2) the nuclear viral DNA represented the full complement of the FV3 genome as analyzed by restriction endonuclease HindIII, and (3) the size of the replicative complex (as determined by neutral sucrose gradient analysis) was between genome and twice-genome length. These characteristics established that at the nonpermissive temperature, ts 12488 was arrested in the first stage of DNA replication. Temperature shift-down (30°C → 25°C) of ts 12488 infected cells at 5 hr pi showed that within 30 min of the shift, the replicative complex became very large (more than ten times genome size), and pelleted upon centrifugation in neutral sucrose gradients. Furthermore, newly synthesized DNA was now found in the cytoplasmic fraction also. These results suggest that upon shift-down, ts 12488 entered into the second stage of DNA replication, where progeny DNA was synthesized as a large concatemer.

Mechanism of concatemer production and its role in optimum DNA replication. Two different mechanisms of concatemer production during viral DNA synthesis are known: (1) In some prokaryotic viruses with circular (Eisen et al., 1968; Gilbert and Dressler, 1968) or linear (Skalka, 1977) genomes. DNA replication proceeds first as a circle and then via rolling circles, leading to the formation of concatemers. (2) In the second mechanism, exemplified by phage T4 (Luder and Mosig, 1982; Miller, 1975), DNA replication first occurs through a linear replicating structure to form unit-length progeny molecules. These then recombine through a complex process to form large, branched concatemeric structures. During this stage, DNA replication occurs in conjunction with recombination. The data obtained with the transport-defective mutant (ts 12488) offer indirect clues to the mechanism of concatemer formation in FV3-infected cells. Within 30 min of shift-down to permissive temperature, a substantial fraction of newly synthesized DNA was pelleted and therefore considered concatemeric. Under the centrifugation conditions employed, pelleted DNA had to be at least ten times larger than genome-length molecules. It seems unlikely that concatemers equivalent to ten or more genome lengths could be produced in 30 min via rolling circle replication. Instead, the data favor the interpretation that concatemers are generated by recombination.

We have recently isolated a endonuclease defective mutant that provide strong genetic evidence for a recombination-replication mechanism of FV3 DNA synthesis (Goorha, unpublished results). In mutant infected cells, viral DNA synthesis as well as virus yield was reduced 3-4 fold and the size of concatemeric DNA in the mutant-infected cells was substantially reduced as compared to wild-type FV3 ($FV3^+$)-infected cells. This was a surprising result since one would expect that lack of endonuclease activity would result in increase size of the replicating viral DNA. However, the size of the single-stranded DNA in the replicating DNA complex, determined by alkaline sucrose gradient analysis, was considerably larger in mutant-infected cells in comparison to $FV3^+$-infected cells. Large single-stranded DNA molecules reflect reduced endonuclease activity in the mutant infected cells. It appears, therefore, that lack of endonucleolytic nicking results in longer single-stranded DNA molecules which reduces recombination among replicating molecules, thereby leading to smaller concatemeric DNA. These results strongly support our previous conclusion that concatemeric DNA molecules in FV3-infected cells are produced through recombination. Consistent with this notion, we found that frequency of recombination among the endonuclease-minus mutants derived from previously characterized ts mutants was drastically reduced as compared to recombination frequencies among ts mutants before they were made endonuclease-minus.

Our results are consistent with the interpretation that a high degree of recombination is necessary for optimal second-stage DNA replication, since the rate of viral DNA synthesis in endonuclease$^-$ mutant is reduced by 60-70%. We believe that during the second-stage, recombinational intermediates present in the concatemeric molecule are utilized as primers for viral DNA replication. This mode of replication represents chain elongation and results in a high rate of DNA synthesis. Our recent observations indeed show that concatemeric DNA replication represents chain elongation and not true chain initiation, since the origin of FV3 DNA replication is not utilized during second-stage DNA synthesis (Goorha and Antol, submitted for publication). A high frequency of recombination among temperature-sensitive mutants of FV3 (Naegele and Granoff, 1971) is also consistent with the proposed role of recombination in DNA replication. Endonuclease activity is an essential step in the recombination process. Therefore, endonuclease-minus mutant is highly deficient both in recombination as well as DNA replication. As expected, reconstitution of endonuclease activity in mixedly infected cells (FV3$^+$ and endonuclease$^-$ mutant) restores higher recombination rates, increased size of concatemeric molecules, and optimal DNA synthesis.

It is noteworthy that even in the mutant-infected cells, concatemeric DNA molecules were produced through recombination, since the size of replicating molecules was much larger in neutral sucrose gradient than in alkaline sucrose gradients. Apparently, other endonuclease(s) of viral or cellular origin are utilized, although less efficiently, for recombination. For example, FV3 virions contain single-stranded and double-stranded endonuclease activities (Aubertin et al., 1971; Kamp and McAuslan, 1972) and endonuclease mutants possess virion associated endonuclease activities in normal amounts (unpublished results). These and other endonucleases may serve as a backup system for viral DNA recombination/replication. Additionally, the frequency of recombination may also be controlled by the ligase activity in the infected cells.

Functions of FV3 concatemeric DNA. In viruses that possess linear genomes, the production of concatemers is required for regeneration of the ends of the DNA molecules (Watson, 1971), and FV3 concatemers apparently serve a similar purpose. Concatemeric forms of replicating DNA also serve as precursors for the production of mature viral DNA molecules (reviewed in Becker and Murialdo, 1978). The most thoroughly studied example of such a relationship is phage T4, in which concatemeric DNA is cleaved and then packaged into the virions via a "headful" mechanism. Evidence also suggests that as a consequence of this process, the DNA packaged into the phage heads becomes circularly permuted and terminally redundant (Streisinger et al., 1967). A comparable relationship between concatemer processing and packaging or virus assembly, or both, has not been established in animal viruses. However, FV3 concatemeric DNA is utilized for the production of mature viral DNA (Goorha, 1982). It is, therefore, reasonable to assume that FV3 may also utilize "headful" packaging mechanism to generate circularly permuted and terminally redundant progeny molecules that are found in the infectious virions (Goorha and Murti, 1982). However, experimental evidence for "headful" or any other mechanism of DNA packaging remains to be established.

ROLE OF DNA METHYLATION IN FV3 DNA REPLICATION

As mentioned earlier, FV3 DNA is highly methylated (Willis and Granoff, 1980). To understand the role of DNA methylation in virus replication, we examined the effect of 5-azacytidine, a drug that inhibits DNA methylation (Goorha et al., 1984). 5-azacytidine (10 μM) reduced the production of infectious FV3 by 100-fold or more and inhibited methylation of viral DNA by about 80%. Inhibition of DNA methylation did not affect viral gene expression since there was no detectable inhibition of virus-specific RNA or protein synthesis in 5-azacytidine-treated cells. In contrast, the size of the replicating DNA measured under completely denaturing conditions (alkaline sucrose gradients) was much smaller than that found during infection in the absence of drug. These results suggest that the undermethylated DNA was susceptible to endodeoxyribonuclease(s). Additionally, electron microscopic examination of FV3-infected, 5-azacytidine-treated cells revealed that preformed capsids remained empty or were only partially filled with viral DNA. Based on these data, it is suggested that methylation of DNA protects it from endonucleolytic cleavage and that the integrity of genomic DNA is required for its proper packaging into virions.

CONCLUSIONS

From our previous and current data, we hypothesize that FV3 DNA replication is initiated at one or more preferred origin (Goorha and Antol, submitted for publication) probably on a linear DNA molecule (Goorha, 1982). Soon afterwards, DNA replication enters a second stage (Goorha and Dixit, 1984) during which extensive recombination events create many more replication forks leading to a rapid increase in the rate of DNA synthesis. Since FV3 DNA is circularly permuted and terminally redundant (Goorha and Murti, 1982), it greatly facilitates recombination and most likely promotes the formation of a complex, branched network of DNA. Inhibition of recombination greatly reduces DNA synthesis, suggesting that recombination and replication at this stage are interdependent and as mentioned earlier, the large replicating complex serves as the precursor for the production of mature viral DNA (Goorha, 1982).

ACKNOWLEDGEMENTS

This research was supported by Research Grant GM 23638 and by Cancer Center Support (CORE) grant CA 21765 from the National Cancer Institute, and by American Lebanese Syrian Associated Charities.

REFERENCES

Aubertin AM, Palese P, Tan KB, Vilagines R, McAuslan BR (1971) Proteins of a polyhedral cytoplasmic deoxyvirus. III. Structure of FV3 and location of virus-associated adenosine triphosphate phosphohydrolase. J Virol 8:643-648

Becker A, Murialdo H (1978) Head morphogenesis of complex double-stranded DNA bacteriophages. Microbiol 42:529-576

Eisen H, Pereira, De Silva LH, Jacob F (1968) The regulation and mechanism of DNA synthesis in bacteriophage lambda. Cold Spring Harbor Symp Quant Biol 33:753-564

Gilbert W, Dressler D (1968) DNA replication: The rolling circle model. Cold Spring Harbor Symp Quant Biol 33:473-484

Goorha R (1981) Frog virus 3 requires RNA polymerase II for its replication. J Virol 37:496-499

Goorha R (1982) Frog virus 3 DNA replication occurs in two stages. J Virol 43:519-528

Goorha R, Murti KG (1982) The genome of frog virus 3, an animal DNA virus, is circularly permuted and terminally redundant. Proc Natl Acad Sci USA 79:248-252

Goorha R, Dixit P (1984) A temperature-sensitive (ts) mutant of frog virus 3 (FV3) is defective in second-stage DNA replication. Virology 136:186-195

Goorha R, Murti G, Granoff A, Tirey R (1978) Macromolecular synthesis in cells infected by frog virus 3. VIII. The nucleus is a site of frog virus 3 DNA and RNA synthesis. Virology 84:32-50

Goorha R, Willis DB, Granoff A, Naegele RF (1981) Characterization of a temperature-sensitive mutant of frog virus 3 defective in DNA replication. Virology 112:40-48

Goorha R, Granoff A, Willis D, Murti KG (1984) The role of DNA methylation in virus replication: Inhibition of frog virus 3 replication by 5-azacytidine. Virology 138:94-102

Kang HS, McAuslan BR (1972) Virus-associated nucleases: Location and properties of deoxyribonucleases and ribonucleases in purified frog virus 3. J Virol 10:202-210

Kelly DC, Avery RJ (1974) Frog virus 3 deoxyribonucleic acid. J Gen Virol 24:339-348

Luder A, Mosig G (1982) Alternate mechanism for initiation of DNA replication forks in bacteriophage T4-priming by RNA polymerase and by recombination. Proc Natl Acad Sci USA 79:1101-1105

Martin JP, Aubertin AM, Londre L, Kirn A (1984) Fate of frog virus 3 DNA replicated in the nucleus of arginine-deprived CHO cells. J Gen Virol 65:721-732

Miller RC (1975) Replication and molecular recombination of T4 phage. Ann Rev Microbiol 29:355-376

Murti KG, Goorha R, Granoff A (1982) Structure of frog virus 3: Genomic size and arrangement of nucleotide sequences as determined by electron microscope. Virology 116:275-283

Naegele RF, Granoff A (1971) Viruses and renal carcinoma of *Rana pipiens*. XI. Isolation of FV3 temperature-sensitive mutants: Complementation and genetic recombination. Virology 44:286-295

Skalka AM (1977) DNA replication-bacteriophage lambda. Curr Top Microbiol Immunol 78:201-237

Streisinger G, Emrich J, Stahl MM (1967) Chromosome structure in phage T4. III. Terminal redundancy and length determination. Proc Natl Acad Sci USA 57:292-295

Watson JD (1971) Origin of concatemeric T7 DNA. Nature 239:197-201

Willis D, Granoff A (1980) Frog virus 3 DNA is heavily methylated at CpG sequences. Virology 107:250-257

Transcriptional Regulation of Frog Virus 3

D.B. Willis, J.P. Thompson[1], and W. Beckman[2]
Department of Virology and Molecular Biology
St. Jude Children's Research Hospital
Memphis, TN, 38101
USA

INTRODUCTION

Transcriptional regulation of eukaryotic promoters is now thought to be mediated by the interaction of specific _trans_-acting factors with _cis_-regulatory DNA sequences and RNA polymerase II (Kingston et al., 1985). Large DNA viruses such as the iridovirus frog virus 3 (FV3) provide an excellent model system for the study of such regulatory processes, for they carry out a complex transcriptional program in which different classes of mRNA are synthesized in sequential order as infection progresses, and in which viral proteins play a role as _trans_-acting regulatory factors (Willis and Granoff, 1985). Investigation into the mechanism of transcriptional regulation in FV3 offers an additional dimension into transcriptional control mechanisms, because unlike that of the more commonly studied eukaryotic DNA viruses (poxvirus, adenovirus, and herpesvirus), the genomic DNA of FV3 is highly methylated by a virus-encoded DNA methyltransferase (Willis et al., 1984a). Thus, not only does FV3 display a temporal regulation of transcription that depends on specific promoter sequences and _trans_-acting viral proteins, but it has also evolved a mechanism that allows the host RNA polymerase II to transcribe highly methylated DNA (Thompson et al., 1984b). In light of the current controversy over whether DNA methylation blocks transcription or whether DNA becomes methylated because transcription ceases (Bird, 1986), the elucidation of the mecha-

[1]Department of Medicine
University of Tennessee
Memphis, TN, USA

[2]Department of Life and Health Sciences
University of Delaware
Newark, DE
USA

nism that FV3 employs to transcribe methylated DNA has become of considerable importance.

TEMPORAL REGULATION OF FV3 TRANSCRIPTION

Because of the complete switch-off of host cell mRNA synthesis (Goorha and Granoff, 1974), and the lack of variable-length poly (A) tails on the FV3 mRNA (Willis and Granoff, 1976), it was possible to resolve very sharp bands of viral RNA on denaturing gels (Willis et al., 1977). Characteristic patterns that corresponded to each viral RNA class were readily apparent, and allowed us to subdivide FV3 mRNA into three classes--immediate-early (IE), delayed-early (DE), and late. The IE class consisted of those RNA species synthesized in the absence of protein synthesis, and represented approximately 1/3 of the single-stranded genome equivalent (Willis and Granoff, 1978). Depending on the cell line, IE synthesis took place in the first 1-2 hr post-infection (hpi). An additional third of the genome was transcribed in the presence of the amino acid analog fluorophenylalanine (FPA) or by a temperature-sensitive (ts) mutant, ts 9467, that is defective in late transcription (Willis et al., 1979a). DE mRNA was normally synthesized 2-4 hpi, prior to DNA replication, and by 5-6 hpi, a complete single-strand equivalent of the genome was transcribed.

The ability to distinguish three temporal classes of mRNA with inhibitors of protein synthesis suggested that at least two distinct viral proteins were necessary for a complete cycle of mRNA synthesis (Goorha et al., 1979). The first protein regulated the transition from the IE stage to the DE stage, and a second was required for the transition from DE to late. In addition, the lack of infectivity of naked FV3 genomic DNA and its nongenetic reactivation by virion proteins (Willis et al., 1979b) clearly indicated that a third, virion-associated protein was required to initiate IE transcription. After Goorha (1981) demonstrated that FV3 used the host RNA polymerase II for IE transcription, it appeared that the role of the FV3 virion protein was that of a _trans_-acting transcription factor.

Transcriptional mapping has confirmed that the various classes of FV3 mRNAs, as well as mRNAs within classes, are all transcribed from separate promoters; the gene for the major late capsid protein, for example, lies snugly nestled between two IE genes (Mesnard et al., 1988). None of the FV3 genes that have been sequenced show any evidence of splicing (Willis et al., 1984b; Beckman et al., 1988).

The proposition that unlinked genes can be coordinately induced by _cis_-responsive sequences with a common _trans_-regulatory molecule has an attractive simplicity. This hypothesis requires that some form of homologous sequence exists in order for certain classes of genes to be turned on in concert. Recombinant DNA technology has made it possible to determine the sequences of selected representative genes and their promoters, and to observe the interaction of these sequences with _trans_-acting proteins present in a cellular extract.

Immediate-early FV3 mRNA synthesis. To examine the nucleotide sequences involved in transcriptional regulation, we have recently begun the cloning and sequencing of specific FV3 genes, using restriction fragments from a nonmethylated DNA mutant lacking DNA methyltransferase activity (Willis et al, 1984a). The use of this mutant was necessary because the highly methylated wild-type FV3 DNA could not survive in the normal _E. coli_ hosts used for replicating plasmids (R. Goorha, personal communication). The first FV3 gene to be sequenced was that of infected cell RNA (ICR) 169, a major IE gene located on XbaI fragment K, with its 5' end approximately 100 base pairs from the XbaI site; the gene contains an open reading frame of 471 nucleotides, and codes for a protein of 17,640 daltons (Willis et al., 1984b). The start and stop sites of transcription were determined by S1 nuclease analysis, and the transcriptional start site mapped 18 nucleotides upstream from the translational start site. Although there was no canonical TATA box (Benoist et al, 1980), there was a 7 base pair AT-rich region (TATTTTA) at -29 from the transcriptional start site, a position that could be expected to serve a similar function.

To assay for the sequences important in promoter function, we constructed a plasmid containing 78 bp of ICR 169 promoter sequence (from the 5' XbaI site to the DraI site at the start site of transcription) 5' to the coding sequences for the bacterial enzyme chloramphenicol acetyltransferase (CAT). We used this chimeric plasmid, mp18-169pCAT, to transfect fathead minnow (FHM) cells, and assayed for CAT activity by the method of Gorman et al. (1982). No CAT activity was observed unless the cells were subsequently infected with FV3 (Willis and Granoff, 1985); the kinetics of CAT synthesis were identical to those of the authentic IE gene product, appearing at 2 hr and peaking at 4 hr. The amount of CAT activity present in transfected cells exposed to UV-inactivated FV3, or the amount of CAT-specific mRNA synthesized in the presence of inhibitors of protein synthesis was equal to that of cells infected with wild-type virus, showing that virion-associated, rather than virus-induced, proteins were sufficient to initiate transcription of this gene.

We have now carried out a detailed site-directed mutagenesis analysis of the sequences in the ICR 169 promoter (Willis, 1987). There are two areas of importance in the regulation of this gene. First the TATTTTA sequence, which positions the RNA polymerase at the correct start site; when this 7 bp sequence is deleted, the promoter can still be induced (albeit at a reduced efficiency) by infection with FV3, but the start site has moved to a cryptic site 65 bases downstream of the normal initiation site. The critical sequences for FV3 *trans*-activation are the 14 bp immediately 5' to the TATTTTA region. Elimination of this region completely abolishes the ability of the promoter to be turned on by FV3 infection. A sequence with over 75% homology to the FV3 *cis*-responsive region in ICR 169 has also been found in two other IE FV3 genes (A.M. Aubertin, personal communication), lending credence to the belief that more than one IE gene responds to the same virion *trans*-acting protein.

However, another FV3 IE gene, ICR 489 (coding for a 46 kD protein), has also been recently sequenced in our laboratory (Beckman et al., 1988). This gene was selected for study because its mRNA is vastly overproduced in the absence of protein synthesis, implying a negative regulatory control under the customary conditions of infection (Willis and Granoff, 1978). The AT-rich sequence at -41 bp from the start site of transcription (AAATAAAA), was different from that of the ICR 169 promoter. Although the ICR 489 promoter lacked the 14 bp pair cis-responsive sequence shown to be crucial for ICR 169 induction by FV3, it did contain a standard eukaryotic gene CCAAT box (Raymondjean et al., 1988), and a GC-rich sequence identical to the eukaryotic SP1 binding site (Dynan and Tijian, 1983). Preliminary site-directed mutagenesis indicates that removal of the CCAAT and GGCGGG sequences eliminates the response of an ICR 489 promoter-CAT plasmid construct to induction by FV3. In addition, the amount of CAT mRNA synthesized in the absence of protein synthesis is much greater than that synthesized when viral-induced proteins are present, mimicking the situation seen with ICR 489 in vivo (Willis and Granoff, 1978).

Role of virion protein(s) in induction of FV3 IE genes. The most obvious role for the virion trans-acting protein(s) is to serve as a transcription factor that binds directly or indirectly to the DNA and thereby directs the host cell polymerase to the promoter. Using the gel retardation assay of Carthew et al., (1985), we have been unable to demonstrate binding of either the ICR 169 or ICR 489 promoter to proteins extracted from purified virions. However, a similar assay employing nuclear extracts from uninfected or FV3-infected HeLa cells and a double-stranded radioactive oligonucleotide of the crucial 14 bp sequence from the ICR 169 promoter displayed a sequence-specific binding of this sequence to a protein in both extracts (Willis, 1988). Interestingly, this sequence bears a strong homology to a GA-rich sequence in the promoter of the mouse and human major histocompatibility gene, MHC-I (Baldwin and Sharp, 1988). In the gel retardation assay a double-stranded 14 bp MHC I promoter sequence bound in a sequence-specific manner to a protein the same size as the one that bound the 14 bp ICR 169 promoter.

At the present time, our working hypothesis is that FV3 IE promoters contain one or more sequences resembling sequences that bind to cellular transcription factors. However, binding of these cellular factors is of itself insufficient to initiate transcription. A virion-associated protein is absolutely required before any viral transcription takes place. What might the role of the virion protein be? One possibility is that of a protein-protein interaction with cellular transcription factors, either to enhance the cellular factor's activity, like the herpesvirus tegument protein (Kristie and Roizman, 1987), or to remove a negative influence, as has recently been shown for the adenovirus Ela protein and the product of the retinoblastoma gene (Whyte et al., 1988). Another possibility is that the FV3 virion protein modifies the cellular transcription factors without binding; the highly active protein kinase associated with the virus particle could play such a role (Silberstein and August, 1976). Further experimentation is necessary to determine which, if either, of these two hypotheses is correct.

<u>Delayed-early and late transcription.</u> Although a virus-induced (as contrasted to a virion-associated) protein is clearly necessary for DE transcription (Willis et al., 1979a), we have only begun to map and identify DE genes, and none of the promoters have as yet been sequenced. This group of FV3 genes is of considerable importance, because some of the most interesting functional genes--those associated with DNA synthesis and late transcription--are DE genes.

Sequencing of the gene for the major late capsid protein is in progress in the laboratory of A.M. Aubertin; she has graciously sent us the sequence of the promoter before publication. It, too, has a TATTTTA box at -35 from the transcriptional start site, but it lacks the 14 bp IE <u>cis</u>-responsive sequence. We have placed 480 nucleotides from this promoter 5' to the CAT coding region and observed the following: when the promoter clone contains 300 nucleotides 3' to the established start site of transcription (Tham et al., 1986), the plasmid is induced with the kinetics of the typical late FV3 gene--i.e, it appears at 4 hpi and peaks at 8 hpi (unpublished observations). No CAT mRNA is induced by FV3 when transfected cells are infected in the presence of either cycloheximide (IE mRNA), FPA (DE mRNA), or cytosine arabinoside, an inhibitor of viral DNA synthesis (Goorha and Granoff, 1974). On the other hand, when the 300 bp stretch of 3' sequences are deleted from the ICR 534 promoter-CAT construct, a small amount of CAT is induced by FV3 under these conditions, implying that sequences 3' to the transcriptional start site are required to maintain the promotional integrity of this gene as a late promoter.

ROLE OF DNA METHYLATION IN THE REGULATION OF TRANSCRIPTION

Site-specific DNA methylation has a strong silencing effect on genes transcribed by eukaryotic RNA polymerase II (Razin and Riggs, 1980), yet FV3 uses this enzyme to transcribe FV3 mRNA from its highly methylated genome (Goorha, 1981). Since virtually every cytosine in the dinucleotide sequence dCpdG is methylated in the FV3 genome (Willis and Granoff, 1980), either the promoters do not contain such sequences in critical sites, or the virus has evolved a mechanism for overriding the normal inhibitory effect of cytosine methylation on RNA polymerase II activity. The evidence to be described below supports a combination of these two mechanisms, in which the IE promoters are not methylated at critical sites, and an FV3 early gene product allows transcription of genes normally inhibited by methylation.

To test for the existence of an FV3 <u>trans</u>-acting protein that would permit transcription from methylated templates, we made use of a plasmid containing the adenovirus 12 E1A promoter attached to the CAT gene, pAd12E1A-CAT. This plasmid had previously been shown to be incapable of promoting transcription when it was methylated in vitro with the bacterial HpaII and HhaI methylases before being transfected into susceptible cells (Kruczek and Doerfler, 1983). We found that if cells were transfected with the methylated pAd12E1A-CAT and then infected with FV3, CAT activity equivalent to that seen in cells transfected with the unmethylated plasmid could be induced (Thompson et al., 1986). Since the introduced plasmid remained methylated after FV3 infection, and the transcriptional start sites were the same as those in cells transfected with the unmethylated pAd12E1A-CAT, we concluded that an FV3 gene product somehow allowed RNA polymerase II to transcribe promoters that it would normally avoid.

Further experiments showed that no CAT enzyme or CAT specific mRNA was made in FV3-infected cells that had been transfected with the methylated plasmid in the absence of protein synthesis, or in transfected cells treated with UV-inactivated virus (Thompson et al., 1986). Therefore, the FV3 protein responsible for overriding the inhibitory effect of methylation on transcription was not the same as the virion protein required to initiate IE viral mRNA synthesis, but was instead a viral-induced gene, probably an FV3 IE gene product.

Since the synthesis of IE mRNA takes place in the absence of protein synthesis or after infection with UV-inactivated virus--conditions under which the input genomic DNA remains methylated (Willis et al., 1984a)--the methylated genome must serve as a template for IE FV3 mRNA synthesis. Therefore, the IE genes cannot require the protein that overrides the inhibitory effect of methylation--either because the promoters are not methylated, or because the dCpdG sequences are located at sites that are not critical to transcription. The promoter of ICR 169 contains three dCpdG doublets (Willis et al., 1984b), the two proximal of which have been shown by genomic sequencing to be methylated (Thompson et al., 1988). By changing one base surrounding each of the three dCpdG sequences, we were able to convert them to either HpaII or HhaI restriction sites. Similar to what we saw with the wild-type promoter, the mutated mp18-169P-CAT plasmid was transcribed only when the transfected cells were subsequently infected with live or UV-inactivated virus, showing that the mutations did not alter the ability of the virion-associated transcriptional activator to promote transcription from the ICR 169 promoter (Thompson et al., 1988). More important, methylation of the 169 promoter with bacterial methylases before transfection did not affect the ability of these genes to be activated by FV3. Therefore, the dCpdG sequences are not in critical positions in the ICR 169 promoter, and the response of the promoter to the virion-associated protein, although sequence-specific, is independent of its methylation status.

We are thus left with the problem of what role the *trans*-acting protein that overrides the inhibitory effect of methylation plays in the regulation of FV3 transcription, and what its mechanism of action might be. We know that separate viral-induced proteins are needed to turn on DE transcription and late transcription (Willis et al., 1977). A direct effect on specific promoters is dubious, since the nonmethylated FV3 mutant undergoes a normal temporal transcriptional program (Goorha and Antol, 1988). However, the growth cycle of the mutant is substantially prolonged, and the final yield is often less than 10% of that obtained with wild type virus. The role of this *trans*-acting protein may be simply to facilitate transcription from methylated promoters either by direct interaction with methylated DNA or by altering the host RNA polymerase II so that it is no longer inhibited by methylation of the template.

CONCLUSIONS

There are numerous examples of cellular transcription factors that bind to regions in viral promoters--adenovirus E2 (Kovesdi et al., 1986), adenovirus major late (Carthew et al., 1985), herpesvirus alpha genes (Kristie and Roizman, 1987) and SV40 early promoters (Dynan and Tijian, 1983). On the other hand, these same three viruses also have regulatory sequences which bind viral gene products (Kingston et al., 1985). From the information that has accumulated so far, it is clear that, although all transcriptional activators accomplish the same end result, they do so by a wide variety of means. Therefore, a study of one gene or one virus cannot provide the complete story on transcriptional regulation. FV3, which possesses both a sequence-specific regulatory cascade and an ability to override the inhibitory effect of DNA methylation on transcription, offers an exceptional opportunity to examine eukaryotic transcriptional regulatory mechanisms from a novel point of view.

ACKNOWLEDGMENTS

This work was supported by research grant CA07055 and Cancer Center Support (CORE) grant CA 21765 from the National Cancer Institute and by the American Lebanese Syrian Associated Charities (ALSAC).

LITERATURE REFERENCES

Baldwin AS, Sharp PA (1986) Two transcription factors NF-κB and H2TF1, interact with a single regulatory sequence in the class I major histocompatibility complex promoter. Proc Natl Acad Sci 85:723-727

Beckman W, Tham TN, Aubertin AM, Willis DB (1988) Structure and regulation of the immediate-early frog virus 3 gene that encodes ICR 489. J Virol 62:1271-1277

Benoist C, O'Hare K, Breathnach R, Chambon P (1980) The ovalbumin gene--sequence of putative control regions. Nuc Acids Res 8:127-142

Bird, A (1986) CpG rich islands and the function of DNA methylation. Nature 321:209-213

Carthew RW, Chodosh LA, Sharp PA (1985) An RNA polymerase transcription factor binds to an upstream element in the adenovirus major late promoter. Cell 43:439-448

Dynan W, Tijian, R (1983) The promoter-specific transcription factor SP1 binds to upstream sequences in the SV40 early promoter. Cell 35:79-87

Goorha R (1981) Frog virus 3 requires RNA polymerase II for its replication. J Virol 37:496-499

Goorha R, Antol K (1988) Origin of frog virus 3 DNA replication. Virology (in press)

Goorha R, Granoff A (1974) Macromolecular synthesis in cells infected by frog virus 3. I. Virus specific protein synthesis and its regulation. Virology 60:237-250

Goorha R, Willis DB, Granoff A (1979) Macromolecular synthesis in cells infected by frog virus 3. XII. Viral regulatory proteins in transcriptional and post-transcriptional controls. J Virol 32:442-448

Gorman CM, Moffat LF, Howard BH (1982) Recombinant genomes which express chloramphenicol acetyltransferase in mammalian cells. Mol Cell Biol 2:1044-1051

Kingston RE, Baldwin AS, Sharp PA (1985) Transcription control by oncogenes. Cell 41:3-5

Kovesdi I, Reichel R, Nevins JR (1986) Identification of a cellular transcription factor involved in E1A _trans_-activation. Cell 45:219-228

Kristie TM, Roizman B (1987) Host cell proteins bind to the _cis_-acting site required for virion-mediated induction of herpes simplex 1 alpha genes. Proc Natl Acad Sci USA 84:71-75

Kruczek I, Doerfler W (1983) Expression of the chloramphenicol acetyltransferase gene in mammalian cells under the control of adenovirus type 12 promoters: effect of promoter methylation on gene expression. Proc Natl Acad Sci USA 80:7586-7590

Mesnard JM, Tham TN, Tondre L, Aubertin AM, Kirn A (1988) Organization of RNA transcripts from a 78 kb region of the frog virus 3 genome. Virology 165:122-133

Raymondjean M, Cereghini S, Yaniv M (1988) Several distinct "CCAAT" box binding proteins coexist in eukaryotic cells. Proc Natl Acad Sci USA 85:757-761

Razin A, Riggs AD (1980) DNA methylation and gene function. Science 210:604-610

Silberstein H, August JT (1976) Characterization of a virion protein kinase as a virus-specified enzyme. J Biol Chem 251:3185-3190

Tham TN, Mesnard JM, Tondre L, Aubertin AM (1986) Mapping of the gene coding for the major late structural polypeptide on the frog virus 3 genome. J Gen Virol 67:301-308

Thompson JP, Granoff A, Willis DB (1986) Trans-activation of a methylated adenovirus promoter by a frog virus 3 protein. Proc Natl Acad Sci USA 83:7688-7692

Thompson JP, Granoff A, Willis DB (1988) Methylation of the promoter for an immediate-early frog virus 3 gene does not inhibit transcription. J Virol (in press)

Whyte P, Buchkovick KJ, Horowitz JM, Friend SH, Raybuch M, Weinberg RA, Harlow, E (1988) Association between an oncogene and an anti-oncogene: the adenovirus E1A proteins bind to the retinoblastoma gene product. Nature 334:124-129

Willis DB (1987) DNA sequences required for _trans_-activation of an immediate-early frog virus 3 gene. Virology 161:1-7

Willis DB (1988) Regulatory sequences of frog virus 3 immediate-early gene ICR 169 contain septamer region of host MHC binding site. Abstracts of 7th Poxvirus-Iridovirus Workshop, Heidelberg

Willis D, Granoff A (1976) Macromolecular synthesis in cells infected by frog virus 3. V. The absence of polyadenylic acid in the majority of virus-specific RNA species. Virology 73:543-547

Willis D, Granoff A (1978) Macromolecular synthesis in cells infected by frog virus 3. IX. Two temporal classes of early viral RNA. Virology 86:443-453

Willis D, Granoff A (1980) Frog virus 3 DNA is heavily methylated at CpG sequences. Virology 107:250-257

Willis D, Granoff A (1985) trans-activation of an immediate-early frog virus 3 promoter by a virion protein. J Virol 56:495-501

Willis DB, Goorha R, Miles MF, Granoff A (1977) Macro-molecular synthesis in cells infected by frog virus 3. VII. Transcriptional and post-transcriptional regulation of virus gene expression. J Virol 24:326-342

Willis DB, Goorha R, Granoff A (1979a) Macromolecular synthesis in cells infected by frog virus 3. XI. A ts mutant of frog virus 3 that is defective in late transcription. Virology 98:328-335

Willis DB, Goorha R, Granoff A (1979b) Nongenetic reactivation of frog virus 3 DNA. Virology 98:476-479

Willis DB, Goorha R, Granoff A (1984a) DNA methyltransferase induced by frog virus 3. J Virol 49:68-91

Willis DB, Foglesong D, Granoff A (1984a) Nucleotide sequence of an immediate-early frog virus 3 gene. J Virol 53:905:912

Translation Regulation of Frog Virus 3

A.M. Aubertin, L. Tondre, and T.N. Tham

Laboratoire de Virologie
Unité INSERM U 74
3 rue Koeberlé
67000 Strasbourg
France

INTRODUCTION

Frog virus 3 (FV 3), the amphibian virus the most extensively analyzed belongs to the family Iridoviridae. The study of the multiplication cycle of this virus, isolated by Granoff et al (1966), has revealed several original features such as i) the replication occurs in the nucleus and the cytoplasm of the host cell, ii) the RNA polymerase II is required for immediate early transcription, iii) a structural protein is necessary as transacting factor (for review, Willis et al, 1985) iv) the virus or structural proteins are highly toxic for cells or animals (Kirn, 1970 ; Aubertin et al, 1973 ; 1976 ;Drillien et al, 1977). As expected for a virus with a large genome, its expression is regulated at several levels and in this paper, attention will be focused on the control of translation and the mechanisms will be discussed.

INHIBITION OF HOST PROTEIN SYNTHESIS

Studies of FV 3 effects on cellular protein synthesis have revealed the rapid decrease which follows FV 3 infection.

The first experiments carried out either with inactivated virus particles or under non permissive or restrictive conditions for virus gene expression have shown that the inhibition takes place independently of virus multiplication.

The overall results can be summarized as follows : i) the inhibition is dependent on the multiplicity of infection ii) cellular mRNAs are not degraded but polyribosomes disaggregate iii) the rate of chain elongation remains unaffected (Elharrar and Kirn, 1977 ; Guir et al, 1970 ; Raghow and Granoff, 1979 ; Zylber-Katz and Weisman, 1975).

In fact, proteins solubilized from the virions are sufficient to impair the cell translational machinery. The inhibition of protein synthesis persists in extracts of cells preincubated with the viral proteins and by reconstitution experiments, it was shown that, in these extracts, the ribosomal fraction (the ribosomes plus the factors associated with them) is defective. Moreover the inhibition can be reproduced by addition of viral proteins to an **in vitro** translation system, indicating that some virus structural proteins may block the translational machinery by a direct interaction with one of its components (Cordier et al, 1981).
There is no evidence that this inhibition is a necessary step in the course of infection to diminish the competition between viral and host messages and to facilitate the translation of the former.

SEQUENTIAL SYNTHESIS OF VIRAL PROTEINS

Soon after infection, polypeptides, displaying different electrophoretic mobilities or different rates of synthesis are produced ; those species were assumed to be virus encoded. The general scheme accepted for the viral genome to be fully expressed is that it takes place in three phases. The classification of the viral polypeptides was based either on the time of maximum synthesis (Goorha and Granoff, 1974 ; Willis et al, 1977) or on the time of appearance when a clear sequential induction was noticed (Elliott and Kelly, 1980 ; Cordier et al, 1986).
An illustration of this expression in cascade observed in CHO infected cells is given in Fig. 1.
First immediate early proteins are synthesized right at the beginning of infection (pp42K, 16K), then delayed early proteins after 1 h (pp 90K, 31K) and by 3-4 h late species (pp 48K, 70K, 56K). The cascade may even be more complex as by using restrictive growth conditions, either supraoptimal temperature or different aminoacids analogs, the number of delayed early proteins synthesized was not identical indicating that more than a single event is probably necessary for inducing this class of molecules (Cordier et al, 1986). Comparison of the synthesis at 29°

and 37° shows that once delayed early genes are expressed, the shut off of host mRNA translation is complete while at 37°, the synthesis of cellular proteins is not totally abolished.

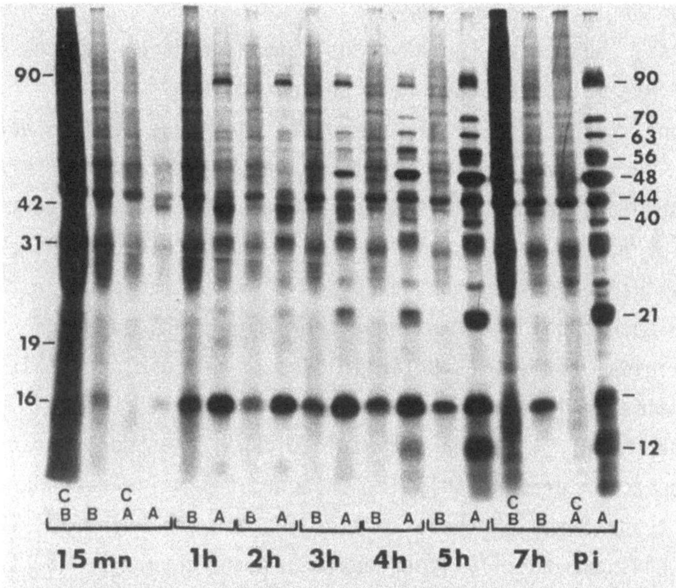

Figure 1 : Fluorogram of FV 3 infected CHO cell polypeptides synthesized at optimal (29°, samples A) and supraoptimal (37°, samples B) temperatures. Proteins were labeled for 30 min with ^{35}S methionine at the time indicated. Identical fractions of cell lysate were analyzed by electrophoresis on a SDS-polyacrylamide (12.5 %) gel.C, uninfected cells. Molecular weights are indicated in kilodaltons.

The primary step in the regulation of viral polypeptide synthesis is at the level of transcription. The transcription of the immediate early messages takes place in the absence of protein synthesis (Willis and Granoff, 1978 ; Lopez et al, 1986). Upon removal of the inhibitor of protein synthesis, the appearance of delayed early polypeptides is dependent on the presence of functional immediate early proteins and transcription of delayed early genes (Elliott and Kelly, 1980). To produce the late proteins, functional early polypeptides and additional transcription are necessary. Furthermore, DNA replication has certainly an influence on late viral protein synthesis but it is not clear whe-

ther it has only an effect on the quantity (Goorha et al, 1981 ; Martin et al, 1981) or is necessary for late protein synthesis (Elliott and Kelly, 1980 ; Chinchar and Granoff, 1984). There is no definite explanation for these differences but they more likely reflect a variability in the requirement for the transcription of late genes not for the translation of late message per se.
Nevertheless, evidence for a translational control of the different classes of RNA molecules have been presented.
Although the transcription of immediate early and delayed early genes continues concomitantly with that of late genes, their translation products are not any more detected, suggesting that a post transcriptional event may be responsible for this shut off (Goorha et al, 1979).
Indirect evidence for a participation of viral proteins in these controls have been obtained by using either amino acid analogs to impair their biological activities or inhibitors of protein synthesis added at different time of infection.
On the other hand, it appears that late mRNAs are poorly translated **in vitro** ; a factor present only in infected cells is necessary for an efficient translation of these species (Raghow and Granoff, 1983) whether some RNAs are totally dependent on viral factors to be translated is not excluded as late transcripts can be detected by nothern blot analysis while when added to a cell free translation system, no translation products can be obtained (Tondre et al, 1988).

FUNCTIONAL DOMAINS IN mRNAs

Our recent studies on the organization of the viral genome and transcriptional mapping have elicited two facts which may be related to the regulation of the translation of FV 3 messages (Mesnard et al, 1988). First, by northern blot analysis, the transcripts of two immediate early genes (p46 located in Sal IF and p42B located in Hind IIIF) were found to have slightly different sizes at the beginning and at late time of infection. The determination of the extremities of the molecules by S1 mapping, has shown that only the 5' ends were modified. While these immediate early transcripts were efficiently translated in a

reticulocyte lysate, no translation products of the late transcripts could be revealed. This suggests that, in these cases, the modified leader regions impair the message activity of the transcripts.
Second, transcripts complementary to the 5' end of the immediate early message encoded by gene p42 (located in Sal IF) are synthesized at late time of infection. Although these late transcripts may serve as messages (sequence data show that there is an ORF in this region), they are produced in larger amount than those transcribed from the immediate early strand. So, one would expect if a hybrid is formed between these two molecules, that the expression of the immediate early gene can be greatly diminished if not abolished.
Thus, these observations point to a control of immediate early gene expression which could take place, for the genes analyzed, via a modification of the transcription pattern.

Several genes have now been sequenced and the 5' end of the transcripts localized by nuclease mapping. Does the examination of the sequence data reveal differences which could be linked to the translational control of the messages ? According to the scanning hypothesis, the 40S ribosomal subunit proceeds from the 5' terminal cap structure and initiates translation a the first AUG codon encountered ; if the flanking sequences around the AUG codon are not ideal (consensus sequence ACCAUGG), the ribosome will migrate downstream to the next AUG and will initiate there (Kozak, 1981). Do we find variations in the leader region which may influence this initiation step ? Differences in the internal methylation of the three classes of FV 3 messages were reported but all have the cap structure (Raghow and Granoff, 1979). For the four immediate early mRNAs, the open reading frame starts at the first AUG codon. The flanking sequences established for FV 3 messages are given in Fig 2. An A (the most usual case) occurs three residues before the AUG codon and G or A follows it (except for P18). In most cases the sequence is optimal with respect to the nucleotides in position - 3 and + 4, but not for the two late mRNAs.

	GENE	A/GCCATGG	LEADER LENGTH
IMMEDIATE EARLY	P 46	ATCATGG	146
	P 42	AACATGG	32
	P 42 B	ACAATGA	42
	P 18	ACAATGC	19
LATE	P 70	GACATGT	16
	P 48	GAAATGT	10
		CTTATGA	64
		GCAATGT	84
		AAGATGT	173

Figure 2 : Comparison of the sequences around the codon which starts the ORF or the ATG codons located upstream, to the consensus sequence. The position of the ATG codons versus the 5' end is given in the right column (published sequences : P18, Willis et al 1984 ; P42 = ICR 489, Beckman et al, 1988).

In the case of the P48 late message which encodes the major capsid protein (48Kda or 55Kda depending on publications), an ORF begins with the fourth AUG (calculated protein, 44Kda) but a small ORF starts with the first AUG (coding capacity, 16 Kda) and overlaps the long ORF. Can ribosomal frameshifting be excluded ? Signals recognized at the frameshifting sites have been reported for retroviruses (Jacks et al, 1987) and coronaviruses (Brierley et al, 1987). A combination of these elements, a sequence TTTA preceding a moderately stable (ΔG : - 14.8 Kcal/mole) stem loop are located upstream the stop codon TGA of the first ORF. Whether this signal is used to produce ribosome slipping from the small ORF into the - 1 frame (the long ORF) has not to been investigated. However frameshifting would be an inefficient mechanism to generate the most abundant protein at late time of infection, except if a factor favors the event. The other mechanisms, internal initiation at the 4th AUG or scanning through the leader to the 4th AUG codon cannot be rejected, both may require a viral protein for a proper initiation.

Secondary structure in the messages, particularly in the leader region, may have various effect. Although, theoretically, a structure may serve to determine which AUG will initiate translation, the most documented effect for such structures is a decrease in the translational efficiency. One or several inverted repeats are found in the 5' region of late messages. However the hairpins never exced - 25 Kcal/mole and the 40S ribosomal subunits may be able to melt such a secondary structure. Nevertheless, **in vivo**, the ionic environment (concentration of monovalent ions is increased at late time of infection, unpublished data) or the interaction with a factor could stabilize the structure.
An other domain of the mRNAs which has to be considered is the 3' terminus. The untranslated region is never long (40-100 nucleotides). The most characteristic feature is the hyphenated dyad symmetry which precedes the 3'end, recalling the structure present at the end of prokaryotic messages. By analogy, we have suggested that this structure may be a signal for transcription termination. However, FV 3 messages lack poly A at 3'end (Willis and Granoff, 1976), and the inverted repeat may allow an intramolecular pairing which could protect the RNA from an exonucleolytic attack. The stability of the hairpin varies for the different mRNA species (ΔG = - 10.6 to - 28.6 Kcal/mole), but the less stable structures are not found for a given class of molecules. One may wonder whether the differences are a reflect of the half life of the messages or not.
The available data allow some speculation on the eventual involvement of certain sequences in the control of gene expression and particularly of translation. The number of genes sequenced is limited (6) and additional informations are awaited in order to extend the comparison and to design experiments probing the mechanism of translational control.

ACKNOWLEDGMENTS

We are extremely grateful to A. Kirn for his constant interest and encouragement throughout this work. We cordially thank O. Cordier for her contribution in the study of protein synthesis.

REFERENCES

Aubertin AM, Hirth C, Travo C, Nonnenmacher H, Kirn A (1973) Preparation and properties of an inhibitory extract from Frog Virus 3 particles. J Virol 11:694-701.

Aubertin AM, Anton M, Bingen A, Elharrar M, Kirn A (1977) Solubilized viral proteins produce fatal hepatitis in mice. Nature 265:456-457.

Beckman W, Tham TN, Aubertin AM, Willis DB (1988) Structure and regulation of the immediate-early Frog virus 3 gene that encodes ICR 489. J Virol 62:1271-1277.

Brierley I, Boursnell MEG, Binns MM, Bilimoria B, Block VC, Brown TDK, Inglis SC (1987) An efficient ribosomal frame-shifting signal in the polymerase-encoding region of the Coronavirus IBV. The Embo J 6:3779-3785.

Chinchar VG, Granoff A (1984) Isolation and characterization of a Frog Virus variant resistant to phosphonoacetate : genetic evidence for a virus-specific DNA polymerase. Virology 138:357-361.

Cordier O, Aubertin AM, Lopez C, Tondre L (1981) Inhibition de la traduction par le FV 3 : Action des protéines virales de structure solubilisées sur la synthèse protéique in vivo et in vitro. Ann Virol Inst Pasteur 132E:25-39.

Cordier O, Tondre L, Aubertin AM, Kirn A (1986) Restriction of Frog Virus 3 polypeptide synthesis to immediate early and delayed early species by supraoptimal temperatures. Virology 152:355-364.

Drillien R, Spehner D, Kirn A (1977) Cell killing by FV 3 : evidence for cell killing by single viral particles or single viral subunits. Biochem Biophys Res Commun 79:105-111.

Elharrar M, Kirn A (1977) Effect of Frog Virus 3 infection on protein synthesis activity of mouse liver ribosomes. FEMS Lett 1:13-16.

Elliott RM, Kelly DC (1980) Frog Virus 3 replication : Induction and intracellular distribution of polypeptides in infected cells. J Virol 33:28-51.

Goorha R, Granoff A (1974) Macromolecular synthesis in cells infected by Frog Virus 3. I. Virus-specific protein synthesis and its regulation. Virology 60:237-250.

Goorha R, Willis DB, Granoff A (1979) Macromolecular synthesis in cells infected by Frog Virus 3. XII - Viral regulating proteins in transcriptional and post-transcriptional controls. J Virol 32:442-448.

Goorha R, Willis DB, Granoff A, Naegele RF (1981) Characterization of a temperature-sensitive mutant of Frog Virus 3 defective in DNA replication. Virology 112:40-48.

Granoff A,, Came PE, Breeze DC (1966) Viruses and renal carcinoma of Rana pipiens. I. The isolation and properties of virus from normal and tumor tissue. Virology 29:133-148.

Guir J, Braunwald J, Kirn A (1970) Inhibition de la synthèse du DNA et des RNA cellulaires dans les cellules KB infectées avec le virus 3 de la grenouille (FV 3). C R Acad Sci (Série D) Paris 270:2605-2608.

Kirn A (1971) Pouvoir léthal pour la souris du virus 3 de la grenouille (FV 3). C R Acad Sci (Série D) Paris 272:2504-2506.

Jacks T, Townsley K, Varmus HE, Mators J (1987) Two efficient ribosomal frame shifting events are required for synthesis of mouse mammary tumor virus gap-related polyproteins. P N A S 84:4298-4302.

Kozak M (1981) Possible role of flanking nucleotides in recognition of the AUG initiator codon by eukaryotic ribosomes. N A R 9:5233-5252.

Lopez C, Aubertin AM, Tondre L, Kirn A (1986) Thermosensitivity of Frog virus 3 genome. Virology 152:365-374.

Martin JP, Aubertin AM, Lecerf F, Kirn A (1981) Intracellular distribution and phosphorylation of virus-induced polypeptides in Frog Virus 3-infected cells. Virology 110:349-365.

Mesnard JM, Tham TN, Tondre L, Aubertin AM, Kirn A (1988) Organisation of RNA trasncripts from a 7,8 kilobase region of the Frog Virus 3 genome. Virology 165:122-133.

Raghow R, Granoff A (1979) Macromolecular synthesis in cells infected by Frog Virus 3. X. Inhibition of cellular protein synthesis by heat-inactivated virus. Virology 98:319-327.

Raghow R, Granoff A (1980) Macromolecular synthesis in cells infected by Frog Virus 3. XIV. Characterization of the methylated nucleotide sequences in viral messenger RNAs. Virology 107:283-294.

Raghow R, Granoff A (1983) Cell-free translation of Frog Virus 3 messenger RNAs : Initiation factors from infected cells discriminate between early and late viral mRNAs. J Biol Chem 258:511-578.

Tondre L, Tham TN, Mutin PH, Aubertin AM (1988) Molecular cloning and physical and translation mapping of Frog Virus 3 genome. Virology 162:108-117.

Willis D, Granoff A (1976) Macromolecular synthesis in cells infected by Frog Virus 3. V. The absence of polyadenylic acid in the majority of virus-specific RNA species. Virology 73:543-547.

Willis D, Granoff A (1978) Macromolecular synthesis in cells infected by Frog Virus 3. IX. Two temporal classes of early viral RNA. Virology 86:443-453.

Willis DB, Goorha R, Miles M, Granoff A (1977) Macromolecular synthesis in cells infected by Frog Virus 3. VII. Transcriptional and post-transcriptional regulation of virus gene expression. J Virol 24:326-342.

Willis DB, Goorha R, Chinchar VG (1985) Macomolecular synthesis in cells infected by frog virus 3. Curr Top Microbiol Immunol 116:77-106.

Willis DB, Foglesong D, Granoff A (1984) Nucleotide sequence of an immediate-early Frog Virus 3 gene. J Virol 53:905-912.

Zylber-Katz E, Weisman P (1975) Effects on host cell polyribosomes following infection with Frog Virus 3 at non-permissive temperature. Arch Virol 47:181-185.

Murine Hepatitis Induced by Frog Virus 3 (FV 3)

A. Kirn, J.L. Gendrault, A.M. Steffan, J.P. Gut, and A. Bingen
Laboratoire de Virologie
Unité INSERM U 74
3 rue Koeberlé
67000 Strasbourg
France

INTRODUCTION

In the course of a virus infection of the liver, hepatocellular damage may result either directly from the replication of virus particles in the parenchymal cells, as is the case in mouse hepatitis, (Bang and Warwick, 1959) or indirectly from the effect of immunological factors in response to an antigenic virus-induced transformation of the hepatocyte membrane. The toxic degenerative hepatitis produced by Frog Virus 3 in mice or rats (Kirn, 1971 ; Kirn et al., 1972) constitutes a new model of hepatocellular injury in which the damage to the parenchymal cells is secondary to sinusoidal cell necrosis produced by FV 3 particles or FV 3 proteins.

CHARACTERISTICS OF FV 3 HEPATITIS

Intravenous inoculation of a suspension of crude or purified virus into rats or mice kills the animals within 19 to 30 hours after a final phase of hypothermia, respiratory disorders and convulsions (Gut et al., 1981). The effect is dose-dependent and specific, since animals immunized against FV 3 resist a lethal challenge of virus (Kirn et al., 1973/74), it is not influenced by interferon and may be produced by solubilized virus proteins (Aubertin et al., 1977). The effect is a toxic one since FV 3 does not multiply at temperatures above 30°C (Granoff et al., 1966).

Pathological examinations 30 hours after the inoculation of 1 LD100 into rats reveal acute degenerative lesions in the liver, the other organs showing only slight congestion. Studies of the patterns of distribution of radioactive-labelled virus in the organism show that two third of the radioactivity are found in the liver but that the spleen and the lungs also contain considerable amounts of FV 3 (Gut et al., 1981). The divergence between the distribution of the virus in the organs and the localization of the lesions suggests that non-viral factors may be involved in liver damage.

ULTRASTRUCTURAL AND BIOLOGICAL CHANGES IN THE LIVER

Ultrastructural examinations of liver specimens in infected rats or mice reveal that the first lesions appear in the Kupffer (KC) and the endothelial cells (EC). A few minutes after inoculation, numerous virus particles are found inside vacuoles in the cytoplasm of KC (Fig. 1). One hour after administration, obvious cell damage occurs whereas after 2 hours the KC are destroyed and breaks appear in the endothelial lining (Fig. 2).

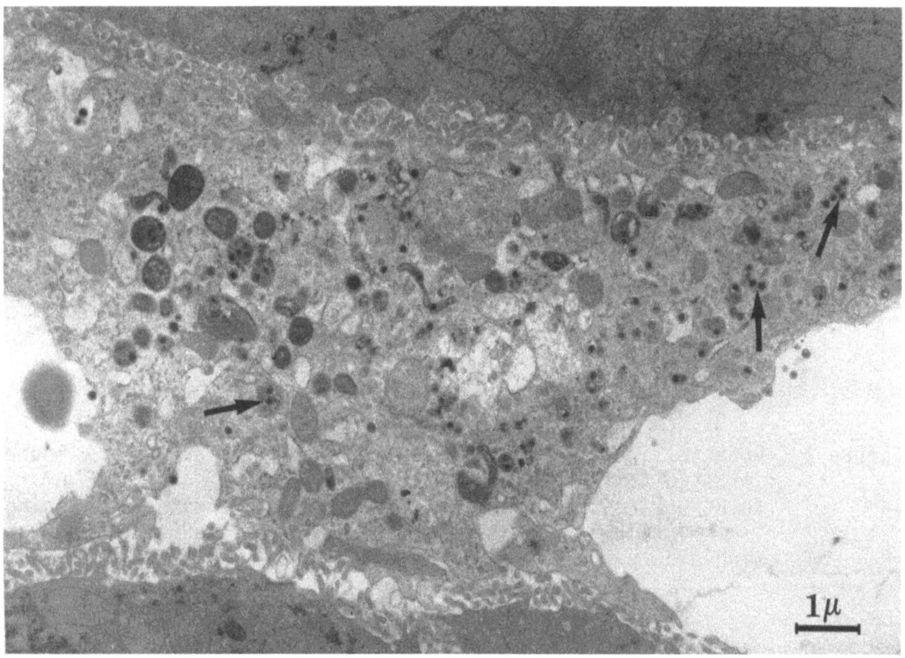

Figure 1 : Mouse Kupffer cell 15 min after the infection with FV 3. Numerous virus particles may be observed within the cellular cytoplasm (⟵).

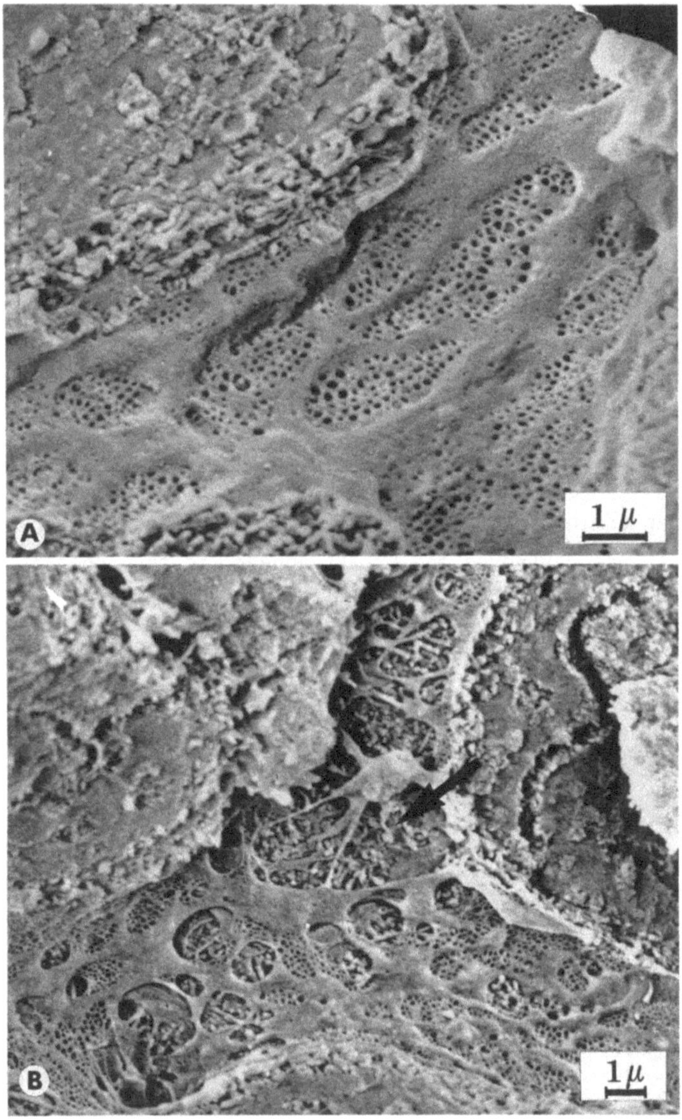

Figure 2 : Hepatic sinusoid from a control mouse (**A**) and from a mouse infected for 2 h with 1 LD100 of FV 3 (**B**).
In B one may observe the breaks in the endothelial lining (⟵) allowing the hepatocyte microvilli to bulge into the sinusoid.

At this time, parenchymal cell damage affects only the nuclei which show : a modification in shape, chromatin condensation, a change in the amount of perichromatin and interchromatin granules, the appearance of nuclear bodies composed of small filaments and an increase in the fibrillar material of the nucleolus. There is a direct correlation between the number of damaged nuclei and the amount of virus administered (Gut et al., 1981). Obvious cytoplasmic lesions occur later and only become generalized 14 hours after the infection. Virus particles are never observed in the hepatocytes although autoradiographic studies reveal viral proteins, but not DNA, inside the parenchymal cells (Bingen et al., 1982). In the livers of infected animals, the typical biochemical perturbations produced by FV 3 in cell cultures may be found. There is, namely, an inhibition in orotic acid incorporation, in the activity of RNA polymerases, in the DNA polymerase activity, and in protein synthesis (Elharrar et al., 1975 ; Elharrar and Kirn, 1974 & 1977).

In the course of FV 3 hepatitis perturbations in lipid metabolism also occur (Elharrar-Pessah et al., 1982). One may thus find a fall in plasma lipid concentration as well as a 50 % drop in non-esterified fatty acids, an alteration in plasma lipoprotein profiles, a decreased palmitic and leucin incorporation into lipid and protein moieties of the three plasma lipoprotein fractions and an increase in palmitic acid incorporation into liver phospholipids.

Finally, FV 3 hepatitis is characterized by the release of cytolytic enzymesin the serum : first, cathepsin D, a lysosome hydrolase mainly contained in the KC, which reaches a peak 3 hours after infection of the animals ; second, transaminases which attain a maximum just before death (Kirn et al., 1983).

PATHOGENESIS OF THE HEPATITIS

Two factors at least seem to be involved in the pathogenesis of FV 3 hepatitis : the destruction of the sinusoidal cells and the effect of toxic substances and mediators on the parenchymal cells.

A - DESTRUCTION OF THE SINUSOIDAL CELLS

Under normal conditions the sinusoidal cells constitute a continuous lining interrupted only by the fenestrae of the endothelial cells. The destruction of the KC and EC allows the hepatocytes and blood in the sinusoid to come into direct contact. This enables material in the blood, which, under physiological conditions, does not attain the parenchymal cells, to reach these cells and to be taken up. Accordingly, in FV 3 infected mice, colloidal carbon inoculated intravenously may be found in the cytoplasm of the hepatocytes (Kirn et al., 1978). Not only do the hepatocytes become able to take up colloidal carbon but they also phagocytose large particles such as latex (Fig. 3), viruses or erythrocytes.

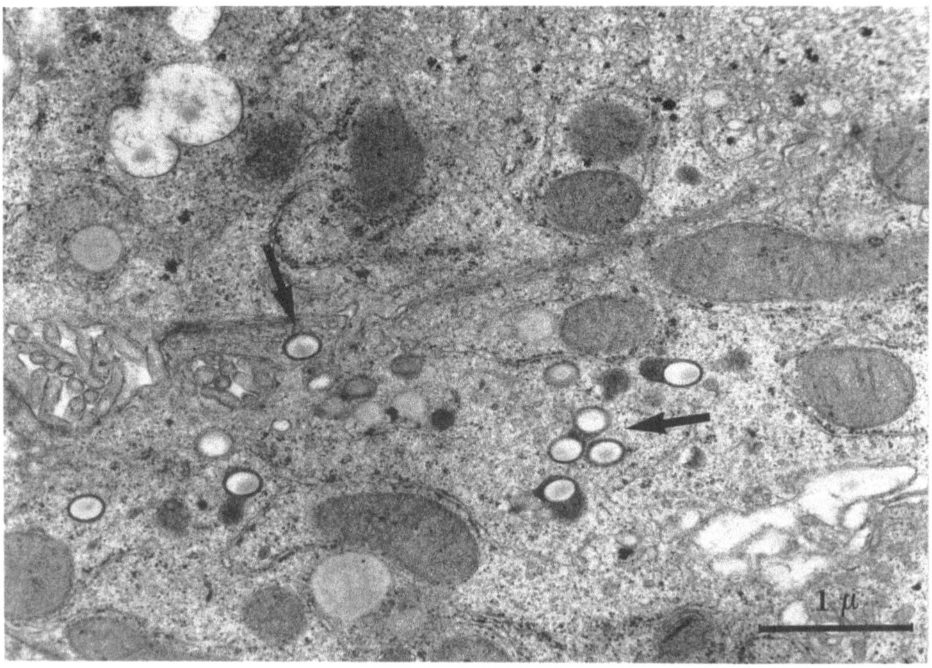

Figure 3 : Section of a mouse liver infected with a sublethal dose of FV 3 4 hours before the intravenous administration of a suspension of latex grains (312 um). Numerous latex grains have been taken up by the parenchymal cells (←). BC bile canaliculi.

Accordingly, in animals infected with sublethal doses of FV 3 and super-infected 4 hours later with non hepatotropic viruses like vaccinia or herpes, a fatal hepatitis occurs (Steffan and Kirn, 1979) ; the destruction of the KC and the breakages in the endothelial lining allow the virus particles to reach the hepatocytes, to be taken up and to multiply within these cells. However, if the delay between the inoculation of FV 3 and the non-hepatotropic virus is long enough to permit the regeneration of the sinusoidal cells, the hepatitis does not develop.

B - NECROSIS OF THE PARENCHYMAL CELLS

Sinusoidal cell damage may be related to the uptake of large amounts of FV 3 particles by the KC, and of smaller amounts by the EC, thus leading to the destruction of these cells. However, since FV 3 does not reach the hepatocytes, one may wonder how parenchymal cell degeneration occurs. Inhibition of macromolecular cell metabolism cannot be the cause of the hepatocellular necrosis since i) there is no direct relationship between metabolic insufficiency and hepatocellular degeneration, (Kirn et al., 1978) ii) the cellular lesions are localized in the liver although FV 3 is found in large quantities in the spleen and lungs. This evokes the effect of a second series of etio-pathogenic factors. Until now two of them, endotoxins and leukotrienes, have been shown to play a role.

a) Endotoxins

Endotoxins, which are released by gram-negative bacteria of the digestive tract, are continuously absorbed, constitute a physiological component of the portal blood and are cleared and detoxified by the sinusoidal cells (Nolan, 1981). In FV 3-infected animals, the damaged sinusoidal cells might no longer take up the endotoxins which could then exert their hepatotoxic potentialities as demonstrated by three arguments.

i) An increase in the pool of endotoxin by the inoculation of exogenous endotoxin after FV 3 aggravates the issue of the hepatitis. Accordingly, mice inoculated with a sublethal dose of virus are 20 times more sensitive to endotoxins than the control animals (Gut et al., 1980).
ii) The decrease or suppression of endogenous endotoxin strengthens the resistance of the animals. Germ-free mice are twice as resistant to FV 3 (Gut et al., 1984) and colectomy protects rats against a lethal dose of endotoxin. However, inoculation of 0.01 LD100 of endotoxin into colectomized rats suppresses their resistance to the virus (Gut et al., 1982).
iii) The inhibition in certain physiopathological effects of endotoxin by treatment of mice with polymyxin B or indomethacin also attenuates the hepatitis. Moreover, the sensitivity of mice to FV 3 is decreased in animals rendered endotoxin-tolerant or genetically so (Gut et al., 1984).

Furthermore, the inhibition in macromolecular synthesis produced by viral proteins, which creates irreversible organic lesions, may sensitise the hepatocytes to endotoxin toxicity.

b) Effects of leukotrienes

Two lines of evidence based on inhibitor studies and leukotriene measurement (Hagmann et al., 1987) bear out the role played by the pro-inflammatory leukotrienes in mediating acute liver injury.

Direct evidence of leukotrienes as mediators in FV 3 hepatitis by measuring rat bile leukotriene **in vivo** substantiates the suggested close relationship between the action of the virus and the increase in leukotriene production. The amount of cysteinyl leukotrienes produced and secreted into the bile subsequent to FV 3 inoculation is at least as high as that observed after severe tissue damage.

Blockade of leukotriene biosynthesis by the 5-lipoxygenase inhibitor AA 861 leads to a reduction in hepatocellular injury after FV 3 infection. This indicates that an early inhibition in leukotriene formation and activity is able to prevent the sequence of events leading to evident hepatocyte damage following FV 3 administration. On the other hand, the protection given against FV 3-induced hepatitis by an inhibitor of both cyclooxygenase and lipoxygenase supplies further evidence of the primary role played by arachidonate metabolites and confirms the observations that many leukotriene effects are further mediated by cyclooxygenase products. We have also found that, when stimulated, the KC itself can produce and metabolize pro-inflammatory leukotrienes. Finally, it should be stressed that the bacterial endotoxins themselves actively stimulate leukotriene production (Hagmann et al., 1984).

CONCLUSIONS

FV 3 hepatitis brings to light new data concerning the pathogenesis of liver diseases.

1) The pathogenicity of a virus for the liver is not restricted to viruses able to multiply within the parenchymal cells. FV 3 induces sinusoidal cell damage without the virions replicating.

2) Destruction of the sinusoidal lining has dramatic consequences for the parenchymal cells which become able to take up infectious particles or toxic substances directly from the blood.

3) When detoxification is altered, endotoxin may exert an hepatotoxic effect.

4) Leukotrienes, which are produced by activated KC, also seem to play a role in the pathogenesis of the hepatitis.

REFERENCES

Aubertin AM, Anton M, Bingen A, et al. (1977) Solubilised viral proteins produce fatal hepatitis in mice. Nature 265:456-457.

Bang FB, Warwick A (1959) Macrophages and mouse hepatitis. Virology 9:715-717.

Bingen A, Bouteille M, Kirn A (1982) Localization by autoradiography of viral proteins in the parenchymal cells of the liver during Frog Virus 3 induced hepatitis of mice. J Submicrosc Cytol 14:55-62.

Elharrar M, Bingen A, Gendrault JL, et al. (1975) Pathogenesis of Frog Virus 3 induced hepatitis in mice. In: Keppler D, ed. Pathogenesis and Mechanisms of liver cell necrosis. Lancaster: Medical and Technical Publishers, 75-86.
Elharrar M, Kirn A (1974) Inhibition of DNA synthesis by isolated liver nuclei from Frog Virus 3 infected mice. Biochem Biophys Res Commun 57:801-807.
Elharrar M, Kirn A (1977) Effect of Frog Virus 3 infection on protein synthesis activity of mouse liver ribosomes. FEMS Letters 1:13-16.
Elharrar-Pessah M, Petit D, Kirn A, et al. (1982) Alterations of plasma lipoproteins and hepatic lipid biosynthesis in FV 3-induced hepatitis in mice. Exp Mol Pathol 36:118-128.
Granoff A, Came PE, Breeze DC (1966) Viruses and renal carcinoma of Rana pipiens. I. The isolation and properties of virus from normal and tumor tissue. Virology 29:133-148.
Gut JP, Anton M, Bingen A, et al. (1981) Frog Virus 3 induces a fatal hepatitis in rats. Lab Invest 45:218-228.
Gut JP, Schmitt S, Bingen A, et al. (1982) Protective effect of colectomy in Frog Virus 3 hepatitis of rats. Possible role of endotoxin. J Infect Dis 146:594-605.
Gut JP, Schmitt S, Bingen A, et al. (1984) Probable role of endogenous endotoxins in hepatocytolysis during murine hepatitis caused by Frog Virus 3. J Infect Dis 149:621-629.
Gut JP, Steffan AM, Anton M, et al. (1980) Kupffer cell functions and Frog Virus 3 hepatitis in mice and rats. In: Liehr H, Grün M, eds. The Reticuloendothelial System and the Pathogenesis of Liver Disease. Amsterdam: Elsevier/North Holland Biomedical Press, 211-219.
Hagmann W, Denzlinger C, Keppler D (1984) Role of peptide leukotrienes and their hepatobiliary elimination in endotoxin action. Circ Shock 14:223-235.
Hagmann W, Steffan AM, Kirn A, et al. (1987) Leukotrienes as mediators in Frog Virus 3-induced hepatitis in rats. Hepatology 7:732-736.
Kirn A (1971) Pouvoir léthal pour la souris du virus 3 de la grenouille (FV 3). CR Acad Sci (D) (Paris) 272:2504-2506.
Kirn A, Gut JP, Bingen A, et al. (1972) Acute hepatitis produced by Frog Virus 3 in mice. Arch Ges Virusforsch 36:394-397.
Kirn A, Gut JP, Bingen A, et al. (1983) Murine hepatitis induced by Frog Virus 3 : a model for studying the effect of sinusoidal cell damage on the liver. Hepatology 3:105-111.
Kirn A, Gut JP, Steffan AM, et al. (1973/74) Immunization of mice against the toxic hepatitis produced by FV 3 : inhibition of virus penetration into the liver. Intervirology 2:244-252.
Kirn A, Steffan AM, Anton M, et al. (1978) Phagocytic properties displayed by mouse hepatocytes after virus-induced damage of the sinusoidal lining. Biomedicine 29:25-28.
Nolan JP (1981) Endotoxin, reticuloendothelial function, and liver injury. Hepatology, 1:458-465.
Steffan AM, Kirn A (1979) Multiplication of vaccinia virus in the livers of mice after Frog Virus 3-induced damage to sinusoidal cells. J Reticuloendothel Soc 26:531-538.

Ultrastructural, Biochemical, and Biological Properties of an Erythrocytic Virus of Frogs from Ontario, Canada

J. Gruia-Gray, M. Petric*, and S.S. Desser
Department of Zoology
University of Toronto
25 Harbord St.
Toronto, Ontario M5S 1A1
Canada

INTRODUCTION

Intraerythrocytic inclusions were first recorded from amphibians and reptiles by Labbé (1894) and subsequently have been described as different protozoans or even artifacts (Laveran, 1903; Billet, 1904; França, 1911). Electron microscopy revealed that these inclusions contained viruses which measured up to 380 nm in diameter. Some of the largest were reported in fish and frogs (Bernard et. al., 1968; Johnston, 1975; Reno et. al., 1978; Sousa and Weigl, 1976; Desser and Barta, 1984).

Aside from their ultrastructural features, little is known about the biochemical and biophysical properties of the intra-erythrocytic viruses. In addition, the interrelationship between these viruses and the erythrocyte has been examined only to a small extent.

LIGHT AND ELECTRON MICROSCOPICAL OBSERVATIONS

"Uninfected" erythrocytes stained with Giemsa's stain and were found to be elliptical, contained a central, ovoid nucleus and a homogeneous cytoplasm (Fig. 1). In contrast, "infected" cells were more rounded. The nucleus was pushed to one side and the nuclear chromatin was condensed. The cytoplasm contained a red inclusion which was often surrounded by a cloud of smaller in-

---

*Department of Virology
Hospital for Sick Children
Toronto, Ontario
Canada

clusions. The cytoplasm also contained an albuminoid vacuole (Fig. 2). Some frogs had up to 90% of the erythrocytes distorted in this manner.

To determine the nature of these inclusions, infected red blood cells were prepared for electron microscopy. The red inclusion corresponded to a granular mass, the viroplasm. Budding from the viroplasm were virus particles (Fig. 3). Aggregates of the virus particles likely corresponded to the cloud of inclusions observed at the light microscope level. The icosahedral viral core measured 300 - 370 nm from vertex to vertex (Fig.4).

The virus particles were also associated with stacks of lamellar membranes. Extensions of these membranes appeared to surround the virus, forming a viral envelope (Fig.5). The membranes appeared to coincide with the albuminoid vacuole observed at the light microscope level. The enveloped virus measured up to 450 nm in diameter, making it one of the largest icosahedral viruses known. The virus was named Frog Erythrocytic Virus (FEV) and was tentatively classified in the family Iridoviridae.

Infected erythrocytes often contained a variety of inclusions and a membrane-bound, elongate body. The latter occurred in the cytoplasm and often extended the length of the cell (Fig. 6). The inclusion was more dense than the erythrocyte cytoplasm and contained striations parallel to its long axis (Fig. 7). The elongate bodies associated with FEV may be composed of residual core protein, as was postulated by Sousa and Weigl (1976). It is also possible that these bodies might be a condensed form of hemoglobin (Fawcett, 1969).

A typical grouping of inclusions commonly seen in the cytoplasm of infected erythrocytes (Fig. 8) are similar to ones observed in Piscine Erythrocytic Necrosis virus (PEN) infected erythrocytes of Atlantic herring (Reno et. al., 1978). The origin and nature of the cytoplasmic inclusions associated with FEV and PEN infections are unknown, but it is likely that these inclusions are cytopathological alterations of infected erythrocytes.

To date, all the intraerythrocytic viruses and iridoviruse observed are icosahedral, but they do not all possess an envelope. Among the viruses that are enveloped, such as African Swine Fever

Virus (ASFV) and Frog Virus 3 (FV 3), the envelope is acquired either by budding through the host cell membrane or *de novo* in the cell cytoplasm (Willis and Granoff, 1974; Carrascosa *et. al.*, 1984). The viral envelope may also be acquired from lamellar membranes as was observed with FEV and the intraerythrocytic viruses in *Rana pipiens* (from Bernard *et. al.*, 1968) and erythrocytes infected with *Toddia* spp. (from Sousa and Weigl, 1976). As viral (FEV) inclusions were found primarily in the cytoplasm of mature erythrocytes, which generally lack membranes, it is likely that the lamellar membranes associated with the virus particles are formed in response to infection of the host cell by the virus.

Although the ultrastructural characteristics of FEV are similar to those of the family Iridoviridae, these features alone cannot be used to determine whether a virus belongs to a family. The biochemical and biophysical properties must also be considered in order to classify the virus.

BIOCHEMICAL AND BIOPHYSICAL PROPERTIES

When FEV infected erythrocytes were stained with acridine orange, the inclusions appeared apple-green, similar to the internal control, the nucleus. This reaction was consistent with double stranded nucleic acid. When similarly infected erythrocytes were stained with Feulgen stain, the inclusion and the nucleus appeared pink, suggestive of DNA.

In keeping with the characteristics of the family Iridoviridae, the viroplasm and FEV contained double stranded DNA. Although DNA synthesis for most DNA viruses occurs in the nucleus of the host cell, it has recently been shown that DNA replication in FV 3 and ASFV occurs initially in the nucleus and continues in the cytoplasm (Goorha *et. al.*, 1978; Tabarés *et. al.*, 1980; Devauchelle *et. al.*, 1985). Devauchelle *et. al.* (1985) suggested that as the icosahedral core is formed, it "sequesters" a portion of the viroplasm, eventually resulting in a complete nucleocapsid. Evidence for this type of developmental sequence for FEV can be seen in Fig. 3. However, there is no evidence to suggest how FEV enters and exits the erythrocytes, nor whether the nucleus plays a role in the replication of FEV.

The viral DNA had a buoyant density of 1.69 +/- 0.005 g/ml in a cesium chloride density gradient. This buoyant density corresponded to a molar fraction of (G + C ratio) 25 - 36% and a melting temperature of 82 - 86 C. The buoyant density and G + C content of the double stranded DNA of FEV are similar to those of the iridescent viruses (Bellet and Inman, 1967; Kelly and Avery, 1974) and ASFV (Enjuanes et. al., 1976).

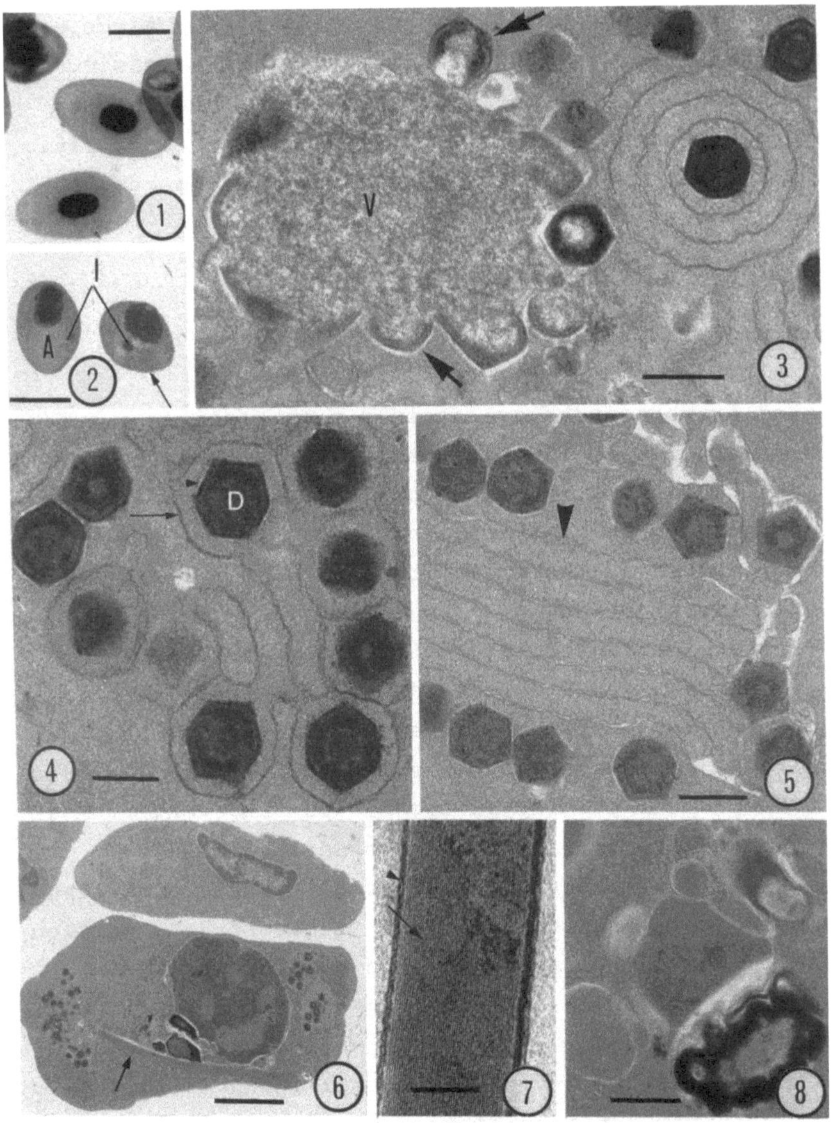

Fig. 1-8

The virus contained at least 16 virus specific proteins, as determined by a 10% polyacrylamide gel (Fig.9). The proteins ranged in relative molecular mass from 19.5 to 91 kilodaltons (kDa). The polypeptides with a relative molecular mass of 31 and 43 kDa appeared to comprise a large percentage of the total protein. These proteins fall within the same range as those of FV 3 (Tan and McAuslan, 1971; Goorha and Granoff, 1974) and ASFV (Tabarés et. al., 1980).

Since FEV cannot be propagated in any of the amphibian, fish, insect and mammalian cell lines attempted, only the structural sensitivity of the virus could be determined. FEV, unlike the iridoviruses, was structurally sensitive to the components of cesium chloride and potassium tartrate density gradients. The physiochemical properties of FEV vary slightly from those of the members of the family Iridoviridae. Similar to other enveloped iridoviruses, FEV is sensitive to lipid solvents such as ether and freon. This resulted in the degradation of the viral envelope

Figure 1 Photo micrograph of uninfected frog erythrocytes with typically central nucleus and homogeneous cytoplasm. Bar = 10 um

Figure 2 Photo micrograph of infected frog erythrocytes containing circular cytoplasmic inclusions (I) often surrounded by smaller ones (arrow) and an albuminoid vacuole (A). Bar = 10 um

Figure 3 A viroplasm (V) and associated virus particles (arrow) seen at the electron microscope level. Note concentric lamellar membranes surrounding a single virion. Bar = 400 nm

Figure 4 FEV is composed of a lipid envelope (arrow), an icosahedral core (arrow head) and a central dense region containing DNA (D). Extensions of lamellar membranes surround the virus core, resulting in an enveloped virus. Bar = 250 nm

Figure 5 Lamellar membranes (arrow head) are often associated with virus particles. Bar = 300 nm

Figure 6 A membrane bound, elongate body (arrow) often spans the length of an FEV infected erythrocyte. Bar = 3 um

Figure 7 The membrane bounded (arrow head), elongate body contains striations (arrow) which run parallel to its length. Bar = 200 nm

Figure 8 The cytoplasm of infected erythrocytes often contain a variety of inclusions. Bar = 1 um

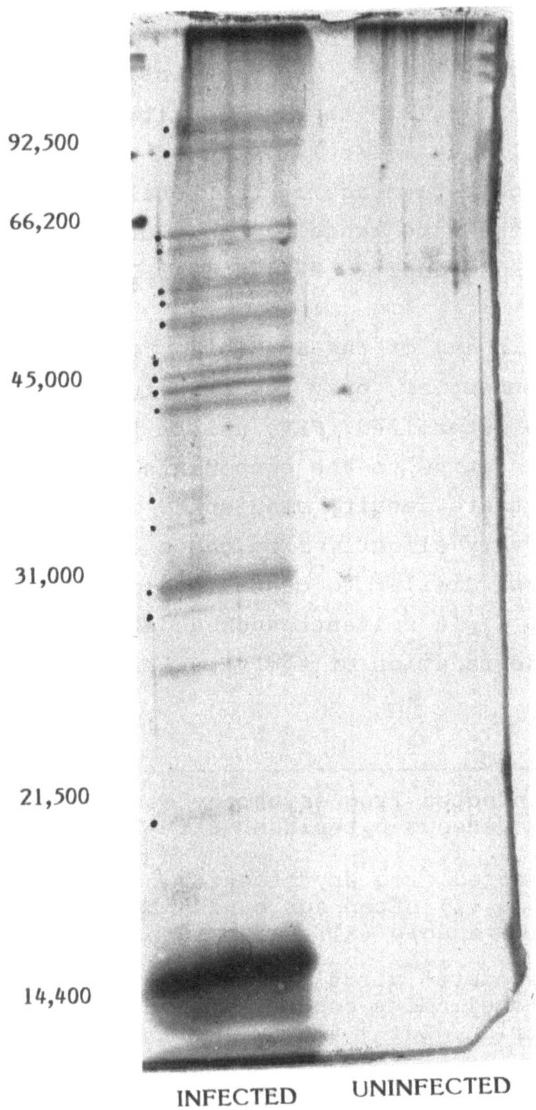

Figure 9 At least 16 virus-specific proteins (dots) are observed in a polyacrylamide gel of infected (viral) and uninfected erythrocyte lysate.

leaving only naked viral cores. FEV appeared structurally stable at pH 7-9, whereas the iridoviruses were sta-ble at pH 3-10. Both FEV and the iridoviruses were stable at -20 to 4C, but sensitive to at least 15 min. at 56C (Matthew, 1979).

Based on the similarities in morphology, biochemical and biophysical properties between FEV and the iridoviruses, FEV is tentatively classified as a member of the Iridoviridae. However, to include FEV in this classification, the size of the members of this family, which ranges from 125 to 300 nm (Matthews, 1979), would have to be extended to 450 nm.

PATHOLOGY

Aside from the presence of intraerythrocytic inclusions, the virus has numerous other effects on the erythrocytes. FEV-infected erythrocytes were more spherical than the flattened elliptical uninfected red blood cells. In addition, the infected red blood cells were significantly smaller in surface dimensions (mean = 13.7 X 19.5 um) than uninfected erythrocytes (mean = 15.3 X 23.1 um).

When stained with Giemsa's stain, the cytoplasm of infected erythrocytes appeared darker than that of uninfected erythrocytes. This dense or "hyperchromic" appearance is probably due to the increased volume of the erythrocyte. When the mean corpuscular volume of the erythrocyte (ratio of packed cell volume to red blood cell count) was determined there was an increase in the volume (mean = 950 u^3) of erythrocytes in frogs with the highest intensity of infection. With the lower intensities of infection, the volume of the erythrocyte was within the normal range (mean = 800 u^3). This suggested that heavily infected frogs possessed prehemolytic spherocytes (Williams et. al., 1972). It is possible that the increase in volume is necessary to accommodate FEV, its associated lamellar membranes, elongated striated bodies and clusters of inclusions.

Since a change in the size and volume of infected erythrocytes occurred, how did this affect the number of erythrocytes? Frogs with the highest intensity of infection exhibited a significantly reduced red blood cell count (mean = .25 X $10^6/mm^3$) with respect to uninfected frogs (mean = .35 X $10^6/mm^3$). The red blood cell count of lightly infected frogs was closer to normal. Therefore, heavily infected animals appeared to be suffering from a spherocytic anemia.

Clearly FEV has an effect on the structure of the erythrocytes, but do these distortions affect the function of the erythrocytes in terms of their oxygen carrying capacity? The mean corpuscular hemoglobin concentration (the ratio of hemoglobin to packed cell volume) was similar to that of uninfected erythrocytes, regardless of the intensity of infection.

Despite the apparent anemia, there were no detectable differences in behaviour or morbidity and mortality between infected and uninfected bullfrogs maintained in the laboratory.

EPIZOOTIOLOGY

FEV occurred primarily in bullfrogs (Rana catesbeiana) of Algonquin Park, while green (Rana clamitans) and mink frogs (Rana septentrionalis) were rarely infected. FEV appeared to infect young bullfrogs, but was never observed in tadpoles. Sexually mature frogs had a much lower prevalence of FEV infection than sexually immature animals. It appeared that sexually mature frogs had acquired an immunity to FEV, since they were refractory to experimental infections.

From the incidence data two peaks of infection were observed. It appeared that the animals acquired the infection in late July or early August. The frogs would overwinter with the infection which would subsequently be seen as the second peak of infection in May of the following year.

The mode of transmission of FEV is currently unknown. Three possible vectors are being investigated, the amphibian feeding leech Batracobdella picta, and two amphibian specific dipterans, a mosquito (Culex territans) and a ceratopogonid (Lasiohelea fairfaxensis).

The least likely vector is the leech which shares a close association with tadpoles; yet tadpoles are never seen to be infected. When tadpoles were inoculated with FEV-infected blood however, they become infected. In addition, no development of FEV was observed in the leech.

The likelihood of either of the two dipterans acting as the vector of FEV is currently being investigated.

SUMMARY

Frog erythrocytic virus, one of the largest icosahedral viruses, is enveloped, measures up to 450 nm in diameter and contains double stranded DNA. The virus occurs in the cytoplasm of erythrocytes of bullfrogs from Algonquin Park, Ontario. The larger acidophilic inclusion stained with Giemsa's stain corresponds to a viroplasm from which FEV buds and forms aggregates of virus particles as seen in the electron microscope. The virus appears to acquire its envelope from lamellar membranes which surround the virus particles.

Frog erythrocytic virus is structurally sensitive to cesium chloride, potassium tartrate and glycerol. It is also sensitive to acidic pHs and a temperature of 56 C for 15 minutes.

The virus contains at least 16 proteins which range in relative molecular mass from 19.5 to 91 kilodaltons. The viral DNA has a buoyant density of 1.690 +/- 0.005 g/ml, a guanine plus cytosine ratio of 25 to 36%, and a melting temperature of 82 to 86 C. The morphological, biochemical and biophysical data on FEV indicate that FEV should be included in the family Iridoviridae.

In heavily infected frogs there is a reduction in the size and number of erythrocytes, and an increase in mean corpuscular volume - indicative of prehemolytic spherocytic anemia. Surprisingly the hemoglobin concentration of infected erythrocytes does not appear to be altered. Frogs with up to 90% of their erythrocytes infected, exhibited neither perceptible behavioural differences nor greater morbidity or mortality than uninfected animals under laboratory conditions.

Epizootiological data suggest that FEV may be transmitted by *Culex territans* and/or *Lasiohelea fairfaxensis*.

LITERATURE REFERENCES

Bellet JD and RB Inman (1967) Some properties of deoxyribonucleic acid preparations from *Chilo*, *Sericesthis* and *Tipula* iridescent viruses. J Mol Biol 25:425-432

Bernard GW, Cooper EL and ML Mandell (1968) Lamellar membrane encircled viruses in the erythrocytes of *Rana pipiens*. J Ultrastruct Res 26:8-16

Billet M (1904) Sur les corpuscules paranucleaires des hematies de la tarente d'Algerie. C R Soc Biol 57:160-161
Carrascosa JL, Carazo, JM, Carrascosa, AL, Garcia, N, Santisteban, A and E Vinuela (1984) General morphology and capsid fine structure of African Swine Fever virus particles. Virol 132:160-172
Desser SS and JR Barta (1984) An intraerythrocytic virus and ricketssia of frogs from Algonquin Park, Ontario. Can J Zool 62:1521-1524
Devauchelle G, Stoltz, DB and F Darcy-Tripier (1985) Comparative ultrastructure of Iridoviridae. CTMI 116:1-18
Enjuanes L, Carrascosa, AL and E Vinuela (1976) Isolation and properties of the DNA of African Swine Fever (ASF) Virus. J Gen Virol 32:469-492
Fawcett DW (1969) An atlas of fine structure - The cell, its organelles and inclusions London: W.B. Saunders Co. pp 332-335
França C (1911) Notes sur les hematozoaires de la Guinee Portugaise. Inst Bact Camara-Pestana Arq 3:229-238
Goorha R and A Granoff (1974) Macromolecular synthesis in cells infected by Frog Virus 3. I. Virus specific protein synthesis and regulation. virol 60:237-250
Goorha R, Murti G, Granoff A and R Tirey (1978) Macromolecular synthesis in cells infected by Frog Virus 3. VIII. The nucleus is a site of Frog Virus DNA and RNA synthesis. Virol 84:32-50
Johnston MRL (1975) Distribution of Pirhemocyton Chatton and Blanc and other, possibly related, infections of poikilotherms. J Protozool 22:529-535
Kelly DC and RJ Avery (1974) The DNA content of four small iridescent viruses: genome size, redundancy, and homology determined by renaturation kinetics. Virol 57:425-435
Labbé A (1894) Recherches zoologiques et biologiques sur les parasites endoglobulaires du sang des vertebres. Arch Zool Exp Gen 3e serie 2:55-258
Laveran MA (1903) Pseudo-hematozoaires endoglobulaires. C R Soc Biol 55:504-507
Matthews REF (1979) Classification and nomenclature of viruses. third report of the International Committee on Taxonomy of viruses. Intervirol 12:129-296
Reno PW, Philippon-Fried M, Nicholson BL and SW Sherburne (1978) Ultrastructural studies of Piscine Erythrocytic Necrosis in Atlantic Herring (Clupea harengus harengus). J Fish B Can 35:148-154
Sousa MA and DR Weigl (1976) The viral nature of Toddia Franca, 1912. Mem Inst Oswaldo Cruz 74:213-230
Tabarés E, Marcotegui MA, Fernandez M and C Sanchez-Botija (1980) Proteins specified by African Swine Fever Virus. I. Analysis of viral structural proteins and antigenic properties. Arch Virol 66:107-117
Tan KN and BR McAuslan (1971) Proteins of polyhedral cytoplasmic deoxyviruses. I. The structural polypeptides of Frog Virus 3. Virol 45:200-207
Williams J, Beutler E, Erslev AJ and RW Rundles (1972) Hematology. New York: McGraw-Hill Book Co. pp. 11-90.
Willis D and A Granoff (1974) Lipid composition of Frog Virus 3. Virol 61:256-269

Properties of Amphibian and Piscine Iridoviruses: A Comparison

K. Essani and A. Granoff
Department of Virology and Molecular Biology
St. Jude Children's Research Hospital
332 North Lauderdale
Memphis, Tennessee 38101
USA

INTRODUCTION

Iridoviruses have been identified in a variety of vertebrate and invertebrate hosts. Amphibian iridoviruses, discussed by Granoff elsewhere in these proceedings, have been isolated from normal or diseased adult leopard frogs (*Rana pipiens*), normal larval or adult newts (*Triturus viridescens*), and diseased bullfrogs (*Rana catesbeiana*). Isolation of piscine iridoviruses from normal adult goldfish (*Carassium auratus*), some freshwater fish (e.g., *Micropogon undulatus* and *Cynosciau arenius*) with papilloma-like lesions, and redfin perch (*Perca fluviatilis*) with hepatocellular and hematopoietic necrosis has been reported by Berry *et al*. (1983), Christmas and Howse (1970) and Langdon *et al*. (1986), respectively. These viruses have been placed in the family *Iridoviridae* based on icosahedral morphology, cytoplasmic site of replication and a large DNA genome. Among iridoviruses, frog virus 3 (FV3), isolated from the renal adenocarcinoma of the leopard frog *Rana pipiens* (Granoff *et al*., 1966) is the only member of the family *Iridoviridae* that has been the subject of detailed molecular analyses (reviewed in Willis *et al*., 1985).

Using immunological criteria (Lehane *et al*., 1968; Clark *et al*., 1969; Kaminski *et al*., 1969) suggested that iridoviruses isolated by Granoff, Clark and Wolf were antigenically related and different from the fish lymphocystis disease virus (FLCDV). In support of this conclusion, genomic DNA from FV3 and FLCDV has no homology (Darai *et al*., 1983).

The purpose of this study was to determine the phylogenetic relationship between selected amphibian and piscine iridoviruses. Such information may be useful in taxonomy and in placing these viruses in ecological niches. The viruses chosen for study were three amphibian iridoviruses and a piscine virus. These were FV3 (Granoff *et al.*, 1966), LT-1 (Clark *et al.*, 1968), tadpole edema virus (TEV) (Wolf *et al.*, 1968) and goldfish virus (GFV) (Berry *et al.*, 1983).

PHYSICAL PROPERTIES

Since iridoviruses from different hosts such as insects and fish may vary greatly in size, as well as among amphibian iridoviruses, as described by Gruia-Gray in these proceedings, we examined purified virus for possible differences. None was found; the three amphibian viruses and the piscine one had identical sizes (~130 x 145 nm) and, as expected, each had icosahedral symmetry.

VIRAL DNAs

To determine genetic relatedness of amphibian and piscine iridoviruses, two techniques were used to analyze their genomic DNA - restriction endonuclease analysis and liquid hybridization. Restriction endonuclease analyses with three different endonucleases (KpnI, HindIII, XbaI) gave similar patterns for FV3, LT-1, and TEV. Although similar, it was possible to differentiate TEV and LT-1 DNA with KpnI for a 5.8 kbp fragment present in FV3 DNA was absent in TEV and LT-1 DNAs. The restriction enzyme XbaI could further distinguish differences between DNAs of the three amphibian iridoviruses where a 16 kbp fragment present in TEV DNA was absent in both FV3 and LT-1 DNAs. Additionally, a 9.8 kbp fragment present in both FV3 and LT-1 DNA was absent in TEV DNA. LT-1 DNA could further be differentiated from FV3 and TEV DNA by the presence of a 7.1 kbp fragment that was not detected in FV3 and TEV DNA. Instead, an 8.3 kbp fragment was detected in FV3 and TEV DNA.

Of the three enzymes used, only KpnI could restrict GFV DNA, and the pattern was distinctly different from any seen with the three amphibian viruses. An interesting observation that deserves comment deals with GFV DNA restricted endonuclease analysis where we found the DNA refractory to digestion with HindIII and XbaI. An examination of the recognition sequence of restriction endonucleases HindIII and XbaI revealed that both of these sequences contain CpT and that the presence of a 5-methylcytosine residue inhibits both HindIII and XbaI endonucleases. This suggests that perhaps GFV DNA has 5-methylcytosine in every CpT sequence, while the cytosine residues in sequence CpT are not methylated in the genomic DNAs of FV3, LT-1 and TEV. Additional piscine iridoviruses must be analyzed to determine whether this methylation pattern is shared among them.

Since the restriction enzyme analyses detected some differences between the amphibian viruses and major ones with the piscine virus, we tested further their relatedness by nucleic acid hybridization. We hybridized [^{32}P]-labeled FV3 DNA to each of the unlabeled viral DNAs. FV3, LT-1 and TEV DNAs were found to be closely related with homologies greater than 90%. In contrast, only 24% of the total radioactive FV3 DNA hybridized to GFV DNA, suggesting a relatively distant relationship between the amphibian viruses and the piscine virus, GFV.

These results were corroborated using [^{3}H]-thymidine-labeled GFV DNA as a probe and allowing it to hybridize to FV3, LT-1 and TEV DNA. Results were again similar with only 23-24% of the GFV DNA hybridizing to amphibian virus DNA, indicating that GFV and the three amphibian iridoviruses share 25% homology.

Taken together, these results indicate that FV3, LT-1 and TEV are closely related and may be considered different strains of the same prototype amphibian iridovirus. On the other hand, GFV appears to be only distantly related to the amphibian iridoviruses.

A characteristic feature of the genomic DNAs of FV3 and the piscine virus, FLCDV, is the high degree of methylation of cytosine residues. To test the genomic DNA of LT-1, TEV and GFV for methylation, we digested purified viral DNA with restriction endonucleases, MspI and HpaII. Both of these enzymes recognize the same sequence "CCGG" and are therefore isoschizomers. However, HpaII is inhibited if the internal cytosine residue in the sequence CCGG is methylated, whereas MspI is not. Analysis of genomic DNA from each of the four viruses indicated DNA of each was methylated. The restriction endonuclease HpaII failed to cleave any of the DNAs while each was readily digested with MspI, suggesting that every internal cytosine residue in the sequence CCGG is methylated. Thus, methylation of genomic DNA appears to be a common property of both amphibian and piscine iridoviruses.

Since each of the viruses tested had a methylated genome, it was likely that methylation was accomplished by a virus-induced DNA methyltransferase as found in the cytoplasm of FV3-infected cells (Willis and Granoff, 1980). This was confirmed when we tested cytoplasmic extracts of infected FHM cells (FV3, LT-1 and TEV) and GFV-infected CAR cells; each extract had substantial DNA-methylase activity.

BIOLOGICAL PROPERTIES

Growth characteristics. We tested FV3, LT-1, TEV, and GFV for growth and cytopathology in piscine [fathead minnow (FHM), brown bullhead catfish (BB), goldfish (CAR)] and mammalian [baby hamster kidney (BHK), human fetal lung (WI-38) cell lines]. FV3, LT-1 and TEV replicated equally well (10^3-10^4 pfu increase, 18 hrs, 30°C)--in FHM, BHK and WI-38 cells, producing a cytopathic effect (CPE) characterized by initial rounding of cells with progression to degeneration and detachment. The amphibian viruses also formed well-defined plaques on FHM monolayers under either agar or methyl cellulose. None of the amphibian viruses multiplied in CAR or BB cells.

In contrast, GFV failed to replicate in FHM, BHK and WI-38 cells, but did replicate in CAR cells where it produced the diffused CPE with cell rounding and cytoplasmic vacuolization 18 hr after infection at 25°C. Only after 6-7 days incubation did cells detach. Although GFV produced plaques, they were not well-defined. As with the amphibian iridoviruses, BB cells were completely refractory to infection with GFV.

Non-genetic reactivation. Evidence supporting the relatedness among the amphibian iridoviruses was also obtained by testing each for non-genetic reactivation (Gravell and Cromeans, 1971). A temperature-sensitive mutant of FV3 (ts 9467) that does not grow at the non-permissive temperature of 30°C was used as the source of reactivating protein. FHM cells were transfected with purified genomic DNAs from FV3 (which is not infectious by itself), LT-1 and TEV, followed by infection with ts 9467 (30°C, 18 hr). The virus yields from these cultures were then assayed and in each instance large numbers of plaques were produced at 30°C where none was detected in controls with the mutant or viral DNA alone. Non-genetic reactivation could not be demonstrated under similar experimental condition with GFV in CAR or FHM cells, providing additional evidence for differences between these viruses.

CONCLUSIONS

Our results indicate that the three amphibian iridoviruses, FV3, LT-1 and TEV, are closely related. However, restriction endonuclease XbaI was able to differentiate between their genomic DNAs, suggesting that they likely represent different strains of the same prototype "Ranavirus". These viruses could, therefore, be referred to as Rana virus type I, II and III. This nomenclature will accommodate further amphibian isolates and simplify the task of classification. By contrast, GFV is only distantly related to the amphibian viruses since it shares only 25% DNA homology with them. This is in sharp contrast to the piscine iridovirus FLCDV DNA that has no homology to FV3 DNA (Darai *et al.*,

1983). Taken together, these data suggest that piscine iridoviruses may be divided into two major groups, one represented by FLCDV and the other by GFV.

ACKNOWLEDGEMENTS

This research was supported by Research Grant CA 07055 and by Cancer Center Support (CORE) grant CA 21765 from the National Cancer Institute, and by American Lebanese Syrian Associated Charities. Janis Gablik kindly provided GFV.

REFERENCES

Berry ES, Shea TB, Gablik J (1983) Two iridovirus isolates from *Carassium auratus*. J Fish Dis 6:501-510

Christmas JY, Howse HD (1970) The occurrence of lymphocystis in *Micropagon undulatus* and *Cynoscian arenius* from Mississippi estuaries. Gulf Res Rep 3:131-154

Clark HF, Brennan J, Zeigel RF, Karzon DT (1968) Isolation and characterization of viruses from kidneys of *Rana pipiens* with renal adenocarcinoma before and after passage in the red eft *Triturus viridescens*. J Virol 2:629-640

Clark HF, Gray G, Fabian F, Zeigel R, Karzon DT. Comparative studies of amphibian cytoplasmic virus strains isolated from the leopard frog, bullfrog and newts (1969) Mizell M (ed) Biology of Amphibian Tumors: Recent results in cancer research, pp 310-326. Springer-Verlag New York

Darai G, Anders K, Koch H-G, Delius H, Gelderblom H, Samalecos, Flügel RM (1983) Analysis of the genome of fish lymphocystis disease virus isolated directly from epidermal tumours of pleuronectes. Virology 126:466-479

Devauchelle G, Stoltz DB, Darcy-Tripier F (1985) Comparative ultrastructure of *Iridoviridae*. Curr Top Microbiol Immunol 116:1-21

Granoff A, Came PE, Breeze DC (1966) Viruses and renal carcinoma of *Rana pipiens*. I. The isolation and properties of virus from normal and tumor tissue. Virology 29:133-148

Gravell M, Cromeans T (1971) The mechanism involved in nongenetic reactivation of frog polyhedralcytoplasmic deoxyribovirus: Evidence for RNA polymerase in the virion. Virology 46:39-49

Kaminski S, Clark HF, Karzon DT (1969) Comparative immune response to amphibian cytoplasmic viruses assayed by the complement fixation and gel immunodiffusion techniques. J Immunol 103:260-267

Langdon JS, Humphrey JD, Williams LM, Hyatt AD, Westbury HA (1986) Fish virus isolation from Australian fish: An iridovirus-like pathogen from redfin perch, *Perca fluviatilis L.* J Fish Dis 9:263-268

Lehane DE Jr, Clark HF, Karzon DT (1968) Antigenic relationship among frog virus demonstrated by the plaque reduction and neutralization kinetics tests. Virology 34:590-595

Willis DB, Goorha R, Chinchar VG. (1985) Macromolecular synthesis in cells infected by frog virus 3. Curr Top Microbiol Immunol 116:77-106

Willis DB, Granoff A (1980) Frog virus 3 DNA is heavily methylated at CpG sequences. Virology 107:250-257

Wolf K, Bullock GL, Dunbar CE, Quimby MC (1968) Tadpole edema virus: A viscerotropic pathogen for anuran amphibians. J Inf Dis 118:253-262

Section II Viruses of Reptilia

Viruses of Reptiles

K.F. Shortridge
Department of Microbiology
University of Hong Kong
Pathology Building
Queen Mary Hospital Compound
Hong Kong

INTRODUCTION

Virus diseases of reptiles have received relatively little attention. The limited number of reviews this decade reflects this situation (Ahne, 1977, Clark and Lunger, 1980; Hoff et al., 1984: Jacobson, 1986). This Symposium has come about because of increased interest in virus diseases of lower vertebrates generally, a trend which is bound to increase particularly for reptiles.

While not comprehensive, this report takes account of current information available on viruses isolated from reptiles or associated with disease in them.

NON-ARTHROPOD BORNE VIRUSES

Much of the information on the occurrence of viruses other than vector transmitted ones in reptiles comes from the examination of these animals in collections. Clinical signs range from readily observable skin lesions (e.g. Rebell et al., 1975) to various forms of debilitation (e.g. Ahne, 1987; Jacobson et al., 1980 and 1985) including poor quality venom (Simpson et al., 1979).

As can be seen from Table 1, much of this information has emerged in the last decade and is based more on the pathology of diseased tissues rather than virus isolation. Classical histopathology is usually supported by thin-section electron microscopy. The clinical picture of the disease and the recognition of virus-like particles in these tissues is strong evi-

Table 1. A list of Viruses (other than Arboviruses) and Reptilian Host Literature

Order	Virus[a]	Evidence[b]	Reference
ORDER SQUAMATA			
Colubridae			
Lampropeltis getulus californiae	Retro	p	Jacobson et al., 1980
Elaphe longissima	Herpes, Parvo	p	Heldstab & Bestetti, 1984
E. quatuorlineata	Adeno	p	Heldstab & Bestetti, 1984
E. guttata	Retro	p	Lunger et al., 1974
E. guttata	Parvo	i	Ahne (pers. comm.)
E. obsoleta	Retro	p	Zschiesche et al., 1988
E. o. quadrivittata			
Gonosoma oxycephala	Paramyxo	i	Ahne et al. 1987b
Elapidae			
Bungarus fasciatus	Herpes	p	Padgett & Levine, 1966; Monroe et al., 1968
Naja naja	Herpes	p	Padgett & Levine, 1966; Monroe et al., 1968
N.n. kaouthia	Herpes	p	Simpson et al., 1979
Crotalidae			
Bothrops atrox	Paramyxo	p, i	Foelsch & Leloup, 1976; Clark et al., 1979
B. schlegeli	Calici	i	Smith et al., 1986
Crotalus lepidus	Paramyxo, Calici	p, i	Jacobson et al., 1980; Smith et al., 1986;
C. unicolor	Calici	i	Smith et al., 1986
Viperidae			
Bitis gabonica	Adeno	p	Heldstab & Bestetti, 1984
Azemiops feae	Reo	p	Jacobson, 1986
Vipera russelli	Retro	p	Ziegler & Clark, 1969; Lunger et al., 1974
Bitis nasicornis	Paramyxo	i	Ahne & Neubert (this symp.)
V. xanthina xanthina	Paramyxo	i	Potgieter et al., 1987
V. spp	Paramyxo	i	Jacobson et al., 1981
Boidae			
Boa constrictor	Herpes, Adeno	p	Hauser et al., 1983; Jacobson et al., 1985
Python regius	Reo	i	Ahne et al., 1987
Python regius	Paramyxo	i	Ahne (pers. comm.)
Morelia argus	Paramyxo	i	Ahne & Neubert (this symp.)
Lacertidae			
Lacerta viridis	Herpes, Papova	p	Raynaud & Adrian, 1976; Cooper et al., 1982
Teiidae			
Ameiva a. ameiva	Rhabdo	i	Causey et al., 1966; Woodall, 1967; Monath et al., 1979
Kentropyx calcaratus	Paramyxo	i	Ahne (pers. comm.)
Kentropyx calcaratus	Rhabdo	i	Causey et al., 1966 Monath et al., 1979
Gekkonidae			
Gehyra variegata	Irido	p	Stehbens & Johnston, 1966

Table 1. A List of Viruses (other than Arboviruses) and Reptilian Host Literature (continued)

Order	Virus[a]	Evidence[b]	Reference
Iguanidae			
Iguana iguana	Herpes	i	Clark & Karzon, 1972
Chamaeleonidae			
Chamaeleo fischeri	Pox	p	Jacobson, 1988
Other			
Spp (Session II)	Myxo/Paramyxo ?	i	Ahne, 1988
ORDER TESTUDINATA (CHELONIA)			
Emydidae			
Chrysemys picta	Herpes	p	Cox et al., 1980
Clemmys marmorata	Herpes	p	Frye et al., 1977
C. sp.	Herpes	p	Rebell et al., 1975
Graptemys pseudogeographica	Herpes	p	Jacobson et al., 1982
G. barbouri	Herpes	p	Jacobson et al., 1982
Chelidae			
Platemys platycephala	Papova	p	Jacobson et al., 1982
Testudinidae			
Gopherus agasszii	Herpes	p	Harper et al., 1982
Testudo hermanni	Irido	p	Heldstab & Bestetti, 1982; Müller et al., 1988
T. graeca	Herpes	p	Cooper et al., 1988
Geochelone chilensis	Herpes	p	Jacobson et al., 1985
G. carbonaria	Herpes	p	Jacobson et al., 1985
Cheloniidae			
Chelonia mydas	Herpes	p, i	Haines & Kleese, 1977; Jacobson, 1986; Rebell et al., 1975
ORDER CROCODYLIA			
Crocodylidae			
Crocodylus niloticus	Adeno, Pox	p	Jacobson et al., 1984; Foggin, 1984
Alligatoridae			
Caiman crocodilus	Pox	p	Jacobson et al., 1979

[a]Viruses are given as family: DNA viruses Herpes = herpesvirus; Adeno = adenovirus; Pox= poxvirus; Irido = iridovirus; Papova = papovavirus; Parvo =parvovirus: RNA viruses: - Myxo = myxovirus (influenza); Paramyxo = paramyxovirus; Reo = reovirus; Retro = retrovirus; Rhabdo = rhabdovirus; Calici = Calicivirus.

[b]p=evidence based on pathology including histology and electron microscopy; i=virus isolation.

NB. Certain of the nomenclature and classification of reptiles is in a state of flux and under review. Some names have been amended to reflect this and to bring old terminology up to date.

dence of a causal association. A number of reptile families is affected.

A wide range of viruses has been recognized:

DNA viruses: herpes-, adeno-, pox-, irido-, papova- and parvoviruses.

RNA viruses: myxo-, paramyxo-, reo-, retro-, rhabdo- and caliciviruses.

About one-third of the isolates are herpes viruses which is perhaps not too surprising because of their association with skin lesions. Other manifestations attributable to herpes viruses include conjunctivitis, pneumonia and general debilitation and malaise (Hauser et al., 1983; Jacobson et al., 1982, 1985 and 1986).

The recent isolation of caliciviruses from intestinal and rectal material of symptomatically and asymptomatically infected rattlesnakes (and also from amphibians) extends the range of viruses associated with reptiles (Smith et al., 1986). Presumably, it is only a matter of time before rotaviruses are detected.

While paramyxoviruses occur widely in nature in both higher and lower vertebrates, but mainly in avian species as described elsewhere in this Symposium, their isolation from reptiles may be of importance for evolutionary considerations. Perhaps more interesting is the report by Ahne (1988), covered in this Symposium, of the isolation of agents sharing some features of ortho- and paramyxoviruses. Recent thinking has been that influenza is primarily a virus of aquatic avian species with a restricted mammalian host range; thus, this finding requires further clarification.

ARTHROPOD BORNE VIRUSES

The information on arbovirus infections of reptiles is in stark contrast to that on other viruses. Free-living animals, rather than those in collections, have provided this. This is perhaps not too surprising because transmission between hosts is mediated by arthropod vectors, mainly the mosquito, and therefore infection can occur over a wide area especially as reptiles do not generally live in close communities. Table 2 gives a guide to the range of mosquitoes known to feed on reptiles. These are mainly Aedes and Culex species. Clearly, there are host preferences which could affect virus occurrence and distribution. Although ticks can be a reservoir for certain arboviruses there is little evidence to show that there is a significant association with reptiles in virus transmission (Shortridge and Oya, 1984).

Frontline virological and second line serological evidence of arbovirus infection of reptiles is supported by experimental

Table 2. A List of Mosquito Species and Reptilian Hosts[a]

Mosquito Species	Host Species	Reference
Aedes aegypti	Lizards	Yuill, 1969
A. albopictus	Snakes	Miyagi, 1972
A. atlanticus	Turtles	Crans & Rockel, 1968
A. canadensis	Turtles, snakes	Crans & Rockel, 1968; DeFoliart, 1967, Hayes, 1961; Nolan et al., 1965
A. sollicitans	Turtles, snakes	Murphy et al., 1967
A. sticticus	Turtles, snakes	Wright & DeFoliart, 1970
A. togoi	Snakes	Miyagi, 1972
A. triseriatus	Turtles	Nolan et al., 1965
A. trivittatus	Turtles	Wright & DeFoliart, 1970
A. vexans	Turtles, snakes	Miyagi, 1972
Anopheles quadrimaculatus	Turtles, snakes	Murphy et al., 1967
Armigeres subalbatus	Turtles, snakes	Miyagi, 1972
Culex infantulus	Turtles, lizards snakes	Miyagi, 1972
C. pipiens	Turtles, snakes	Hayes, 1961; Miyagi, 1972; Wright & DeFoliart, 1970
C. resturans	Turtles, snakes	Hayes, 1961
C. salinanius	Turtles, snakes	Hayes, 1961
C. tarsalis	Snakes	Gebhardt et al., 1966; Thomas & Eklund, 1962
C. territans	Turtles, snakes	Murphy et al., 1967
C. tritaeniorhynchus	Turtles, snakes	Miyagi, 1972
Culesita melanura	Snakes	Hayes, 1961
C. minnesotae	Turtles	Hayes, 1961
C. morsita	Snakes	Hayes, 1961
Mansonia perturbans	Snakes	Hayes, 1961; Murphy et al., 1967; Wright & DeFoliart, 1970
Tripteroides bambusa	Turles, snakes	Miyagi, 1972

[a]After Hoff and Trainer, 1973; Shortridge and Oya, 1984.

infection. These viruses are almost exclusively of the Togaviridae and Flaviviridae families (Table 3). Whether or not this reflects the actual range of arboviruses that occurs in free-ranging reptiles remains to be seen. In fact, there have been very few studies on such animals in recent years most of the effort having been made in the 60's and 70's to explain the natural history of arboviruses of medical and veterinarian importance, particularly overwintering in temperate and cold climates (Shortridge and Oya, 1984). The possible role of reptiles in the maintenance of these viruses in nature is considered elsewhere in this Symposium.

COMMENT

The viruses of reptiles have been a somewhat neglected area. Given the special position of reptiles in the evolutionary scheme, much is to be learnt about the evolution of viruses themselves through increased virus surveillance of reptiles. The following observations offer an insight into how doors might be opened up -

i. the nucleocapsid of Fer-de-Lance virus, a paramyxovirus, is smaller than that of known paramyxoviruses (Clark et al., 1979)

ii. pigs experimentally infected with a calivivirus isolated from a rattlesnake were able to transmit virus to contact pigs (Smith et al., 1986)

iii. studies on the antigenic relationship of reptilians and avian paramyxovirus may provide useful evolutionary information especially as avian and mammalian paramyxovirus do not show any significant relationship (Alexander, personal communication).

iv. rubella virus is a non-arthropod human togavirus (Tesh and Rosen, 1975); it would be fascinating if an antigenically related virus were to be isolated from a reptile.

While the time is ripe for fruitful studies on reptiles there are many problems. Not the least of these is old axiom of "the need to collect the right specimen at the right time and

Table 3. A List of Arboviruses and Reptilian Host Literature[a]

Order	Virus[b]	Evidence[c]	Reference
ORDER SQUAMATA			
Colubridae			
Coluber constrictor	EEE, WEE	ex, i, s	Dalrymple et al., 1972; Gebhardt & Stanton, 1967; Gebhardt et al., 1964; Hayes et al., 1964; Karstad, 1961
Diadophis punctatus	EEE	ex	Karstad, 1961
Virginia valeriae	EEE	ex	Karstad, 1961
Heterodon platyrhinos	EEE	ex	Karstad, 1961
Lampropeltis getulus	EEE, WEE	s	Dalrymple et al., 1972
Nerodia erythrogaster	VSV	s,	Hoff & Trainer, 1973
N. sipedon	EEE, POW	ex, s	Bast et al., 1973; Hayes et al., 1964; Karstad, 1961
Opheodrys aestivus	EEE	ex	Karstad, 1961
Pituophis melanoleucus	SLE, WEE	i, s	Gebhardt & Stanton, 1967; Gebhardt et al., 1964; Spalatin et al., 1964
Ptyas korros	JE	s	Shortridge et al., 1977
Rhabdophis tigrina	JE	ex	Lee, 1968
Storeria occipito-maculata	EEE, POW, SLE	s	Whitney et al., 1968
Thamnophis elegans	WEE	ex, i, s	De St. Jeor, 1969; Gebhardt & Stanton, 1967
T. e. vagrans	WEE	i	Gebhardt & Hill, 1960
T. radix haydeni	WEE	ex, i, s	Burton et al., 1966; McLintock et al., 1967; Spalatin et al., 1964
T. sirtalis	EEE, WEE	ex, i	Gebhardt & Hill, 1960; Hayes et al., 1964
T. s. parietalis	WEE	i, s	Burton et al., 1966; McLintock et al., 1967
T. s. sirtalis	EEE, WEE, SLE, POW	s	Whitney et al., 1968
Thamnophis sp.	WEE	ex, i	Gebhardt et al., 1973; Prior & Agnew, 1971; Thomas et al., 1959, 1980
Elaphe guttata	EEE	ex, s	Karstad, 1961
E. obsoleta	WEE	s	Hoff & Trainer, 1973
E. rufodorsata	JE	ex, i	Lee et al., 1972
E. schrenckii	JE	ex	Lee, 1968
Elapidae			
Bungarus fasciatus	JE	s	Shortridge et al., 1977
Naja naja	JE	s	Shortridge et al., 1974
Crotalidae			
Agkistrodon piscivorus	EEE	s	Karstad, 1961
Bothrops alternata	WEE	ex	Rosenbusch, 1939
Crotalus atrox	WEE, VEE	s	Hoff & Trainer, 1973; Smart et al., 1975

Table 3. A List of Arboviruses and Reptilian Host Literature[a] (continued)

Order	Virus[b]	Evidence[c]	Reference
Lacertidae			
Lacerta agilis	TBE, UUK	ex, s	Sekeyova et al., 1970; Sixl et al., 1971
L. viridis	TBE	ex	Rehacek et al., 1961; Sekeyova et al., 1970
Tachydromus tachydromoides	JE	ex	Doi et al., 1968; Oya et al., 1983
Scincidae			
Cryptoblepharus virgatus	ALM	i, s	Graf, 1967
Eumeces laticeps	EEE	ex	Karstad, 1961
E. latiscutatus	JE	ex	Oya et al., 1983
E. barbouri	JE	ex	Oya et al., 1983
E. marginatus oshimensis	JE	ex	Oya et al., 1983
Teiidae			
Ameiva sp.	EEE	ex, s	Craighead et al., 1962
Cnemidophorus sp.	EEE	s	Craighead et al., 1962
Agamidae			
Calotes versicolor	GPB	s	Chastel, 1966
Anguidae			
Ophisaurus attenuatus	EEE	ex	Karstad, 1961
Gekkonidae			
Gekko gecko	GPB	s	Chastel, 1966
Gehyra australis	CHA	i	Doherty et al., 1973
Hemidactylus frenatus	JE	ex	Yuill, 1969
Phyllurus platurus	JE	ex	Yuill, 1969
Iguanidae			
Anolis carolinensis	EEE	ex	Karstad, 1961
Basiliscus sp.	EEE	s	Craighead et al., 1962
Tropidurus torquatus hispidus	MAY	i	Woodall, 1967
ORDER TESTUDINATA (CHELONIA)			
Emydidae			
Chrysemys picta	EEE, WEE	ex, s	Dalrymple et al., 1972; Hayes et al., 1964
C. p. picta	BUN, EEE, WEE, SLE, POW	s	Whitney et al., 1968
C. scripta elegans	WEE	s	Hoff & Trainer, 1973
Clemmys caspica	GPA, GPB	s	Nir et al., 1972
C. guttata	EEE, WEE	ex, s	Dalrymple et al., 1972; Hayes et al., 1964; Smith & Anderson, 1980

Table 3. A List of Arboviruses and Reptilian Host Literature[a] (continued)

Order	Virus[b]	Evidence[c]	Reference
C. insculpta	EEE	ex	Hayes et al., 1964
Malaclemys t. terrapin	WEE	i	Goldfield & Sussmann,1964
Terrapene carolina	EEE, WEE	ex, i, s	Dalrymple et al., 1972; Goldfield & Sussmann,1964 Hayes et al., 1964; Karstad, 1961
Testudinidae			
Gopherus berlandieri	WEE	ex, i	Bowen, 1977; Sudia et al., 1975
G. polyphemus	EEE	ex	Karstad, 1961
Chelydridae			
Chelydra serpentina	BUN, EEE, WEE, SLE, POW	ex, i, s	Dalrymple et al., 1972; Goldfield & Sussmann,1964 Hayes et al., 1964; Whitney et al., 1968
Trionychidae			
Trionyx ferox	EEE	ex	Karstad, 1961
T. sinensis	JE	s	Shortridge et al., 1975
T. spinifer	VSV	s	Cook et al., 1965
T. s. emoryi	BUN, WEE	i, s	Hoff & Trainer, 1973
Kinosternidae			
Kinosternon subrubrum	EEE, WEE	ex, s	Dalrymple et al., 1972; Karstad, 1961
Sternotherus carinatus	EEE	ex	Karstad, 1961
ORDER CROCODYLIA			
Alligatoridae			
Alligator mississippiensis	EEE	ex, s	Karstad, 1961

[a]After Hoff and Trainer, 1973; Shortridge and Oya, 1984.

[b]EEE = eastern equine encephalitis; WEE = western equine encephalitis; VEE = Venezuelan equine encephalitis; JE = Japanese encephalitis; SLE = St. Louis encephalitis; VSV = vesicular stomatitis; POW = Powassan; BUN = Bunyamwera group; TBE = tick-borne encephalitis; UUK = Uukuniemi; ALM = Almpiwar; MAY = Mayaro; GPB = group B arbovirus; CHA = Charleville; GPA = group A arbovirus

[c]ex = experimental infection; i = isolation from natural infection; s = serological evidence

to have the right isolation system" for the confident interpretation of attempted isolation. Since isolations are often made from apparently asymptomatically infected animals, it will be sometime before the pathogenic status of many of them is clear. More recently, however, Smith et al., (1986) were able to show that it was possible to confirm the pathogenicity of a reptile calicivirus by infection of a closely related species.

There is reasonable evidence to indicate that the ability to detect arboviruses in reptiles is largely dependent upon body temperature (see Shortridge and Oya, 1984; this Symposium). Whether the same situation occurs with other viruses remains to be seen. While the isolation of non-arthropod viruses from wild caught, apparently healthy reptiles is uncommon (Clark and Karzon, 1972; Causey et al., 1966), recent evidence has shown that viruses are more likely to be detected when reptiles are housed in close or crowded conditions (Jacobson et al., 1981) or when new ones are introduced into an apparently healthy collection (Jacobson et al., 1985). This raises further questions about latency, asymptomatic infection and facultative pathogenicity of viruses of reptiles. It would appear that in addition to the role of body temperature, exemplified by arboviruses, stressors may be a major factor for the expression and detection of viruses in reptiles.

Certainly, captive reptiles will provide many data on the occurrence and distribution of viruses in these animals. Relating these to free-living animals and their effect as causative agents of disease and on the ecology of reptiles will be less than easy. As a step towards this, virological studies on the densely populated reptiles of the Galapagos Islands may prove useful.

ACKNOWLEDGEMENT

Thanks are due to Mr. J.E. Cooper and Mr. D. Ball for helpful discussions.

REFERENCES

Ahne, W. (1977) Bei Reptilien vorkommende Viren. In: Krankheiten der Reptilien, H.H. Reichenbach-Klinke (ed.) Gustav Fischer-Verlag, Stuttgart.

Ahne, W., Thomsen, I. and Winton, J. (1987a) Isolation of a reovirus from the snake Python regius. Arch. Virol., 94: 135-139.

Ahne, W., Neubert, W.J. and Thomsen, I. (1987b) Reptilian Viruses: Isolation of myxovirus-like particles from the snake Elaphe oxycephala . J. Vet. Med. B. 34, 607-612.

Ahne, W. (1988) Personal communication; this Symposium.

Bast, T.F., Whitney, E. and Benach J.L. (1973) Considerations on the ecology of several arboviruses in eastern Long Island. Amer. J. Trop. Med. Hyg., 22: 109-115.

Bowen, G.S. (1977) Prolonged western equine encephalitis viremia in the Texas tortoise (Gopherus berlandieri). Amer.J.Trop.Med. Hyg., 26: 171-175.

Burton, A.N., McLintock, J. and Rempel, J.C. (1966) Western equine encephalitis virus in Saskatchewan garter snakes and leopard frogs. Science, 154: 1029-1031.

Causey, O.R., Shope, R.E. and Bensabeth, G. (1966) Marco, Timbo and Chaco, newly recognized arboviruses from lizards of Brazil. Amer. J. Trop. Med. Hyg., 15: 239-243.

Chastel, C. (1966) Infections a arbovirus au Cambodge. Enquete serologique chez les reptiles. Bull. World Health Organ., 34: 701-707.

Clark, H.F. and Karzon, D.T. (1972) Iguana virus, a herpes-like virus isolated from cultured cells of a lizard, Iguana iguana. Infec. Immun., 25: 559-569.

Clark, H.F. and Lunger, P.D. (1980) Viruses of reptiles. In Cooper, J.E., Jackson, O.F. (eds). Diseases of the Reptilia. Vol. 1, Academic Press NY.

Clark, H.F., Lief, F.S., Lunger, P.D., Waters, D., Leloup, P., Foelsch, W.D., and Wyler, R.W. (1979). Fer-de-lance virus: a probable paramyxovirus isolated from a reptile. J.Gen. Virol. 44: 405-418.

Cook, R.S., Trainer, D.O., Glazener, W.C. and Nassif, B.D. (1965) A serological study of infectious diseases of wild populations in South Texas. Trans. North Amer. Wildlife Natur. Res. Conf., 30: 142-155.

Cooper, J.E., Gschmeissner, S. and Holt, P.E. (1982) Viral particles in a papilloma from a green lizard (Lacerta viridis). Lab. Animals, 16: 12-13.

Cooper, J.E., Gchmeissner, S. and Bone, R.D. (1988) Herpes-like virus particles in necrotic stomatitis of tortoises. Vet. Rec., in press.

Craighead, J.E., Shelekov, A. and Peralta, P.H. (1962) The lizard: A possible host for eastern equine encephalitis virus in Panama. Amer. J. Hyg., 76: 82-87.

Crans, W.J. and Rockel, E.G. (1968) The mosquitoes attracted to turtles. Mosq. News, 28: 332-337.

Cox, W.R., Rapley, W.A. and Barker, I.K. (1980) Herpesvirus-like infection in a painted turtle. J. Wildl. Dis. 16:445-449.

Dalrymple, J.M., Young, O.P., Eldridge, B.F. and Russell, P.K. (1972) Ecology of arboviruses in a Maryland freshwater swamp. Amer. J. Epidemiol., 96: 129-140.

DeFoliart, G.R. (1967) Aedes canadensis (Theobald) feeding on Blanding's turtle. J. Med. Entomol., 4: 31.

Doi, R., Oya, A. and Telford, S.R. (1968) A preliminary report on infection of the lizard, Takydromus tachydromides, with Japanese encephalitis virus. Jap. J. Med. Sci. Biol., 21: 205-207.

Foelsch, D.W. and Leloup, P. (1976) Fatale endemische Infektion in einem Serpentarium. Tierärztl. Praxis, 4: 527-536.

Foggin, C.M. (1984) Personal communication: see Jacobson (1986).

Frye, F.L., Oshiro, L.S., Dutra, F.R. and Carney, J.D. (1977) Herpesvirus-like infection in two Pacific pond turtles. J. Amer. Vet. Med. Assoc., 171: 882-884.

Gebhardt, L.P. and Hill, D.W. (1960) Overwintering of western equine encephalitis virus. Proc. Soc. Exp. Biol. Med., 104: 695-698.

Gebhardt, L.P. and Stanton, G.J. (1967) The role of poikilothermic hosts as virus reservoirs. Jap. J. Med. Sci. Biol.,20: 30-34.

Gebhardt, L.P., Stanton, G.J. and Year, S. (1966) Transmission of WEE virus to snakes by infected Culex tarsalis mosquitos. Proc. Soc. Exp. Biol. Med., 123: 233-235.

Gebhardt, L.P., De St. Jeor, S.C., Stanton, G.J. and Stringfellow, D.A. (1973) Ecology of western encephalitis virus. Proc. Soc. Exp. Biol. Med., 142: 731-733.

Gebhardt, L.P., Stanton, G.J., Hill, D.W. and Collett, G.C. (1964) Natural overwintering hosts of the virus of western equine encephalitis. New England J. Med., 217: 172-177.

Graf, P.A. (1967) In: "International catalogue of arboviruses", 2nd ed., T.O. Berge, ed., DHEW Publ. No. (CDC) 75-8301, United States Government Printing Office, Washington.

Haines, H. and Kleese, W.C. (1977) Effect of water temperature on a herpesvirus infection of sea turtles. Infec. Immun., 15; 756-759.

Hauser, B., Mettler, F. and Rubel, A. (1983) Herpes-like infection in two young boas. J. Comp. Path., 93: 515-519.

Hayes, R.O. (1961) Host preferences of Culiseta melanura and allied mosquitos, Mosp. News. 21: 179-187.

Hayes, R.O., Daniels, J.B., Manfield, H.K. and Wheeler, R.E. (1964) Field and laboratory studies on eastern encephalitis in warm- and cold-blooded vertebrates. Amer. J. Trop. Med. Hyg., 13: 595-606.

Heldstab, A. and Bestetti, G. (1982) Spontaneous viral hepatitis in a spur-tailed Mediterranean land tortoise (Testudo hermanni) J. Zoo Anim. Med., 13: 113-120.

Heldstab, A. and Bestetti, G. (1984) Virus associated gastrointestinal diseases in snakes. J. Zoo Anim. Med. 15: 118-128.

Hoff, G.L. and Trainer, D.O. (1973) Arboviruses in reptiles: Isolation of a bunyamwera group virus from a naturally infected turtle. J. Herpetol., 7: 55-62.

Hoff, G.L., Frye, F.L. and Jacobson, E.R. (1984) Diseases of Amphibians and Reptiles. Plenum Press, New York and London, pp. 784.

Jacobson, E.R. (1986) Viruses and viral associated diseases of reptiles. In: Maintenance and Reproduction of Reptiles in Captivity. Vol. II. (V.L. Bels and A.P. Van den Sande, eds) Acta Zoologica et Pathologia Antverpiensia, No. 79, pp.73-90.

Jacobson, E.R. (1988) Personal communication.
Jacobson, E.R., Gardiner, C.H. and Foggin, C.M. (1984) Adenovirus-like infection in two Nile crocodiles. J. Amer. Vet. Med. Assoc., 185: 1421-1422.
Jacobson, E.R., Clubb, S.L., Gaskin, J.M. and Gardiner, C. (1985) Herpesvirus-like infection in Argentine tortoises. J. Amer. Vet. Med. Assoc., 187: 1227-1229.
Jacobson, E.R., Gaskin, J.M. and Gardiner, E.H. (1985) Adenovirus-like infection in a Boa constrictor. J. Amer. Vet. Med. Assoc., 187: 1226-1227.
Jacobson, E.R., Gaskin, J.M. and Clubb, S.L. (1982) Papilloma-like virus infection in Bolivian side-neck turtles. J. Amer. Vet. Med. Assoc., 181: 1325-1328.
Jacobson, E.R., Gaskin, J.M., Page, D. and Calderwood, M.B. (1981) Paramyxo-like virus associated illness in a zoologic collection of snakes. J. Amer. Vet. Med. Assoc. 179:1227-1230.
Jacobson, E.R., Gaskin, J.M., Simpson, C.F. and Terrell, T.G. (1980) Paramyxo-like virus infection in a rock rattlesnake. J. Amer. Vet. Med. Assoc., 177: 769-799.
Jacobson, E.R., Gaskin, J.M., Wahlquist, H. (1982) Herpesvirus-like infection in map turtles. J. Amer. Vet. Med. Assoc., 181: 1322-1324.
Jacobson, E.R., Popp, J., Shields, R.P. and Gaskin, J.M. (1979) Pox-like virus associated with skin lesions in captive caimans. J. Amer. Vet. Med. Assoc., 175: 937-940.
Jacobson, E.R., Seely, J.C. and Novilla, M.N. (1980) Lymphosarcoma associated with virus-like intranuclear inclusions in a California king snake (Colubridae: Lampropeltis). J. Natl. Canc. Inst., 65: 577-583.
Karstad, L. (1961) Reptiles as possible reservoir hosts for eastern encephalitis virus. Proc. North Amer. Wildlife Natur. Res. Conf., 26: 186-202.
Lee, H.W. (1968) Multiplication and antibody formation of Japanese encephalitis virus in snakes: Proliferation of the virus. Seoul J. Med., 9: 157-161.
Lee, H.W., Min, B.W. and Lim, Y.W. (1972) Isolation and serologic studies of Japanese encephalitis virus from snakes in Korea. J. Korean Med. Ass., 15: 69-74.
Lunger, P.D., Hardy, W.D. and Clark, H.F. (1974) C-type particles in a reptilian tumor. J. Natl. Cancer Inst., 52: 1231-1335.
McLintock, J., Burton, A.N. and Rempel, J.G. (1967) Inter-epidemic hosts of western encephalitis virus in Saskatchewan. Proc. Ann. Meet. New Jersey Extermination Ass., 54: 97-104.
Miyagi, I. (1972) Feeding habits of some Japanese mosquitos on cold-blooded animals in laboratory. Trop. Med., 14: 203-217.
Monath, T.P., Cropp, C, Frazier, C, Murphy, F. and Whitfield, S. (1979) Viruses isolated from reptiles: Identification of three new members of the family Rhabdoviridae. Arch. Virol., 60:1-12.
Monroe, J.H., Shibley, G.P. and Schidlovsky, T. (1968) Action of snake venom on Rauscher virus. J. Natl. Cancer Inst., 40: 135-145.

Müller, M. Zanger, N. and Denzler, T. (1988) Iridovirus-epidemie bei der Griechischen Landschildkröte (Testudo Lermanni Lermanni. In Erkrankungen der Zootiere. Proc. Int. Symp. Zoo and Wild Animals. (H. Iffen and H.-D. Schröder, eds). Academie-Verlag, Berlin, DDR. pp. 721-724.
Murphy, F.J., Burbutis, P.P. and Bray, D.F. (1967) Bionomics of Culex salinarius Coquillet. II Host acceptance and feeding by the adult females.
Nir, Y.R., Avivi, A., Lasovski, Y., Margalit, J. and Goldwasser, R. (1972) Arbovirus activity in Israel. Israeli J. Med. Sci., 8: 1695-1701.
Nolan, M.P., Moussa, M.A. and Haynes, D.E. (1965) Aedes mosquitoes feeding on turtles in nature. Mosq. News, 25: 218-219.
Oya, A., Doi, R., Shirasaka, A, Yabe, S. and Sasa, M. (1983) Studies on Japanese encephalitis virus infection of reptiles. I. Experimental infection of snakes and lizards. Japan J. Exp. Med., 53:117-123.
Padgett, F. and Levine, A.S. (1966) Fine structure of the Rauscher leukemia virus as revealed by incubation in a snake venom. Virology, 30: 623-630.
Potgieter, L.N.D., Sigler, R.E. and Russell, R.G. (1987) Pneumonia in Ottoman vipers (Vipera xanthena xanthena) associated with a parainfluenza 2-like virus. J. Wildl. Dis. 23: 355-360.
Prior, M.G. and Agnew, R.M. (1971) Antibody against western equine encephalitis virus occurring in the serum of garter snakes (Colubridae: Thamnophis) in Saskatchewan. Can. J. Comp. Med., 35: 40-43.
Raynaud, M.M.A. and Adrian, M. (1976) Lesions cutanees a structure papillomateuse associees a des virus chez les lezard vert (Lacerta viridis Laur) C.R. Acad. Sci., Paris, 283: 845-847.
Rebell, H., Rywlin, A. and Haines, H. (1975) A herpesvirus-type agent associated with skin lesions of green sea turtles in aquaculture.. Amer. J. Vet. Res., 36: 1221-1224.
Rehacek, J., Nosek, J. and Gresikova, M. (1961) Study of the relation of the green lizard (Lacerta viridis Lavr.) to natural foci of tick-borne encephalitis. J. Hyg. Epidemiol. Microbiol. Immunol., 5: 366-371.
Rosenbusch, F. (1939) Equine encephalomyelitis in the Argentine and its experimental aspects. Proc. Pacific Sci. Congress, 6: 209-214.
Sekeyova, M., Gresikova, M. and Leska, J. (1970) Formation of antibody to tick-borne encephalitis virus in Lacerta viridis and L. agilis lizards. Acta Virol., 14: 87-90.
Shortridge, K.F., Oya, A., Kobayashi, M. and Duggan, R. (1977) Japanese encephalitis antibody in cold-blooded animals. Trans. Royal Soc. Trop. Med. Hyg., 71: 261-262.
Shortridge, K.F., Ng, M.H., Oya, A., Kobayashi, M.,Munro, R., Wong, F. and Lance, V. (1974) Arbovirus infections in reptiles: Immunological evidence for a high incidence of Japanese encephalitis virus in the cobra, Naja naja. Trans. Royal Soc. Trop. Med. Hyg., 68: 454-460.

Shortridge, K.F., Ng, M.H., Oya, A. and Yip, D.Y. (1975) Arbovirus infections in reptiles: Studies on the presence of Japanese encephalitis antibody in the plasma of the turtle, Trionyx sinensis. Southern Asian J. Trop. Med. Public Health, 6: 161-169.
Simpson, C.F., Jacobson, E.R. and Gaskin, J.M. (1979) Herpesvirus-like infection of the venom gland of Siamese cobras. J. Amer. Vet. Med. Assoc., 175: 941-943.
Sixl, W., Sekeyova, M. and Riedl, H. (1971) Discovery of antibodies to arboviruses in Lacerta agilis. Arch. Hyg. Bakteriol., 154: 609.
Smart, D.L., Trainer, D.O, and Yuill, T.M. (1975) Serologic evidence of Venezuelan equine encephalitis in some wild and domestic populations of southern Texas. J. Wildl. Dis., 11: 195-200.
Smith, A. and Anderson, C.R. (1980) Susceptibility of two turtle species to eastern equine encephalitis virus. J. Wildl. Dis., 16: 615-617.
Smith, A.W., Anderson, M.P., Skilling, D.E., Barlough, J.E. and Ensley, P.K. (1986) First isolation of calicivirus from reptiles and amphibians. Amer. J. Vet. Res., 47: 1718-1721.
Spalatin, J., Connell, R, Burton, A.N. and Gollop, B.J. (1964) Western equine encephalitis in Saskatchewan reptiles and amphibians, 1961-1963. Can. J. Comp. Med. Vet. Sci., 28: 131-142.
Stehbens, W.E. and Johnston, M.R.L. (1966) The viral nature of Pirhemocyton tarentolae. J. Ultrastruc. Res., 15: 543-554.
Sudia, W.D., Newhouse, V.F., Beadle, L.D., Miller, D.L., Johnson Jr., G.J., Young, R., Calisher, C.H. and Maness, K. (1975) Epidemic Venezuelan equine encephalitis in North America in 1971: Vector studies. Amer. J. Epidemiol., 101: 17-35.
Tesh, R.B. and Rosen, L. (1975) Failure of rubella virus to replicate in mosquitos. Intervirology, 5: 216-219.
Thomas, L.A. and Eklund, C.M. (1962) Overwintering of western equine encephalomyelitis virus in garter snakes experimentally infected by Culex tarsalis. Proc. Soc. Exp. Biol. Med., 109: 421-424.
Thomas, L.A., Eklund, C.M. and Rush, W.A. (1959) Susceptibility of garter snakes (Thamnophis spp.) to western equine encephalomyelitis virus. Proc. Soc. Exp. Biol. Med., 99: 698-701.
Thomas, L.A., Patzer, E.R., Cory, J.C. and Coe, J.E. (1980) Antibody development in garter snakes (Thamnophis spp.) experimentally infected with western equine encephalitis virus. Amer. J. Trop. Med. Hyg., 29: 112-117.
Whitney, E., Jamnback, H., Means, R.G. and Matthew, T.H. (1968) Arthropod-borne-virus survey in St. Lawrence County, New York: Arbovirus reactivity in serum from amphibians, reptiles, birds and mammals. Amer. J. Trop. Med. Hy., 17: 645-650.
Woodall, J.P. (1967) In: "International catalogue of arboviruses including certain other viruses of vertebrates," 2nd ed., T.O. Berge, ed., DHEW Publ. No. (CDC) 75-8301, United States Government Printing Office, Washington.

Wright, R.E. and DeFoliart, G. R., (1970) Associations of Wisconsin mosquitoes and woodland vertebrate hosts. Ann. Entomol. Soc., Amer., 63: 777-786.
Yuill, T.M. (1969) Mosquitoes for drawing blood from small reptiles. Trans. Royal Soc. Trop. Med. Hyg., 63: 407-408.
Zeigel, R.F. and Clark, H.F. (1969) Electron microscopic observations on a <C>-type virus in cell cultures derived from a tumor-bearing viper. J. Natl. Cancer Inst., 43: 1097-1102.
Zschiesche, W., Konstantinov, A., Ippen, R. and Mladenov, Z. (1988) Lymphoid leukemia with presence of type C virus particles in a four-lined chicken snake: Elaphe obsoleta quadrivittato. In Erkrankungen der Zootiere. Proc. Int. Symp. Zoo and Wild Animals (H. Ippen and H.- Schröder, eds.) Academie-Verlag, Berlin, DDR. pp. 275-277.

Antigenetic Relationship Between Three Members of Paramyxoviridae Isolated from Different Snakes

W. Ahne and W. J. Neubert*
Institute of Zoology and Hydrobiology,
University of Munich
D-8000 München

Introduction

Viruses of poikilothermic vertebrates have not been studied extensively up to now. Apart from the fish-pathogenic viruses (Ahne, 1985), there are some reports on viruses occurring in Amphibia (Granoff, 1969), but the literature concerning Reptilia-associated viruses are very few in number (Ahne, 1977; Lunger and Clark, 1978; Clark and Lunger, 1980; Hoff et al., 1984; Jabobson, 1986). Viruses associated with Reptilia are usually described as "arboviruses" because such viruses multiply in hematophagous arthropods and the poikilothermic vertebrates serve only as reservoirs. However, a disease outbreak in snakes showing respiratory signs and nervous symptoms was reported in Switzerland (Foelsch and Leloup, 1976) and the causal agent has been identified as a paramyxovirus isolated from lung tissue of a dead (Bothrops atrox) (Clark et al., 1979). Previously we presented data of the isolation of a myxovirus like particles from the snake Gonosoma oxycephala (Ahne et al., 1987). Here we report on the comparison of three pleomorphic RNA viruses isolated from different snakes (Gonosoma oyxcephala, Morelia argus, Bitis nasicornis).

Material and Methods

Isolation of virus

Internal organs (liver, spleen and kidney) of three different snakes (Gonosoma oxycephala (30/85); Morelia argus (30/86); Bitis nasicornis (115/87) were homogenized, passed through a 300 nm pore size membrane

---

*Max-Planck-Institute for Biochemistry,
Department of Virus Research
D-8033 Martinsried

filter and dilutions were inoculated into monolayer cultures of IgH2 cells (Clark et al., 1970). Infected cells incubated at 28° C were checked daily for cytopathic effects (CPE).

Hamagglutination test (HA)

The hamagglutination test was performed according to standard procedures in microtitre plates using 0.1 ml of virus dilutions and 0.1 ml of 1% human type O erythrocytes at 20° C (Neubert and Hofschneider, 1983).

Treatment of the isolate with chloroform and with 5-iodo-2-deoxyuridine (IUdR)

Infectious tissue culture fluid was treated with 50% chloroform for 12 hr at 4° C. After centrifugation (300xg, 4° C) the aquaous phase of the treated material as well as control virus were inoculated into IgH2 cells in tenfold serial dilutions. In order to determine the type of nucleic acid of the isolate, IgH2 cells were treated with tissue culture medium containing 50 ug/ml JUdR. Triplicate cell cultures treated with JUdR and untreated cells were infected with tenfold serial dilutions of the isolates. After 12 hr of incubation at 28° C the inocula were discarded and fresh medium without JUdR was added. The frog virus 3 served as a DNA control virus.

Electron microscopy

For electron microscopic studies infectious tissue culture medium was centrifuged at 85,000xg for 3 hr. The pellet of virus was purified by centrifugation through a sucrose gradient (10-50%) at 58,000xg for 1 hr. The preparations were examined by a Zeiss EM 109 transmission electron microscope after negative staining with phosphotungstic acid.

Antisera

Antisera directed against Sendai virus and the virus isolate 30/85 were prepared by subcutaneous immunization of guinea pigs with purified, SDS-disrupted virus (Neubert and Hofschneider, 1983). Antisera directed against mumps virus and measles virus were purchased from Behring-Werke, F.R.G.

SDS-PAGE

Sodium dodecyl sulphate polyacrylamide gel electrophoresis (SDS-PAGE) was performed in a discontinuous slab gel by the method of Laemmli (Laemmli, 1970). Electrophoresis was done at 8 mA (constant currant) for 16 h (13% gel). The protein bands were stained with Coomassie brilliant blue (Sigmund et al., 1988).

Western-blotting and Immunodetection

Viral proteins were separated by SDS-PAGE in 13% gels and transferred eletrophoretically to nitrocellulose sheets (Millipore) as described by Towbin et al. (1979). Immunodetection was performed as essential described by Sigmund et al. (1988).

Results

Isolation of virus

IgH2 cells infected at 28° C showed CPE consisting of syncytia and lysis 24-36 hr after infection. Destruction of the cell monolayer was complete 5 days later. Infectious tissue culture medium showed an infectivity titre in the range of $10^{4.5}$ up to $10^{7.8}$ $TCID_{50}$ per ml.

Hamagglutination

All three isolates possessed a hamagglutination activity and had a HA titre of 1:128.

Treatment of the isolate with chloroform and with JUdR

Treatment with chloroform completely destroyed the infectivity of the isolates, indicating the presence of a lipid- containing envelope.
JUdR did not have any effect on the multiplication of the isolates. In contrast, the DNA control virus did not replicate under the same conditions. The results demonstrated RNA as the genome of the isolates.

Characterization of proteins

Virus particles isolated from the cell supernatant by high speed centrifugation in a sucrose gradient (25-60%) were disrupted by SDS treatment. The protein patterns of the three viral isolates were compared by SDS polyacrylamid electrophoresis (Fig.1A).

Antigenic characterization of the virus isolates

The antigenic relationship within the three virus isolates and to other myxoviruses (Sendai virus, mumps virus, measles virus, respiratory syncitial virus) was tested by immunoblotting. The serum directed against the virus isolate 30/85 reacted specifically with two protein bands of each virus isolate, but not with proteins from Sendai virus (Fig.1B). None of the antisera directed against other paramyxoviruses reacted with proteins of the isolate 30/85 (Ahne et al., 1987) and with the proteins of the two other isolates (unpublished results).

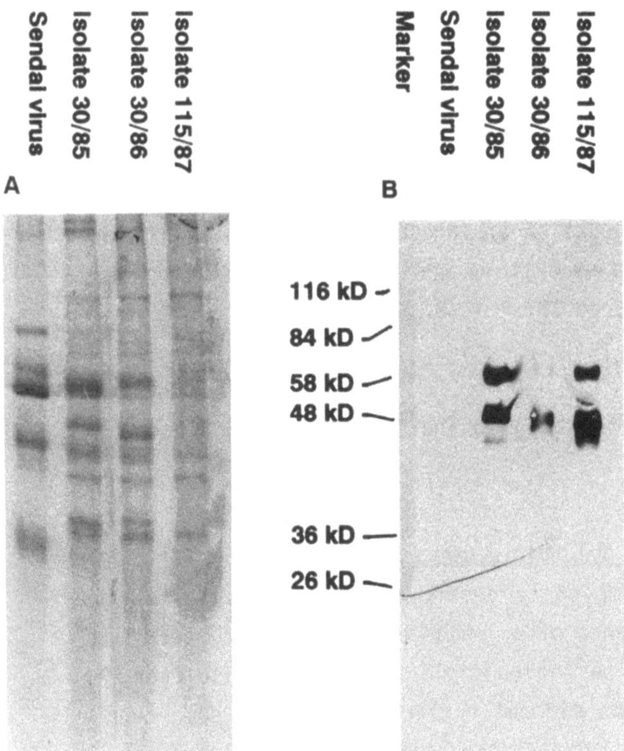

Fig.1: Characterization of the proteins of the three virus isolates. Purified virus was disrupted by SDS and the proteins separated on a 13% polyacrylamide gel. A: The protein bands were stained by Coomassie brilliant blue; Sendai virus (strain D52) as marker. B: Immunoblotting using a serum directed against isolate 30/85. Marker positions from a prestained SDS molecular weight marker (Sigma).

Electron microscopy

Negatively stained preparations of the isolates revealed pleomorphic spheroidal particles measuring about 140-300 nm in diameter. Virions exhibited an outer membrane envelope with projections or spikes. Typical strands of the nucleocapsid are seen within the virion (Fig.2-4).

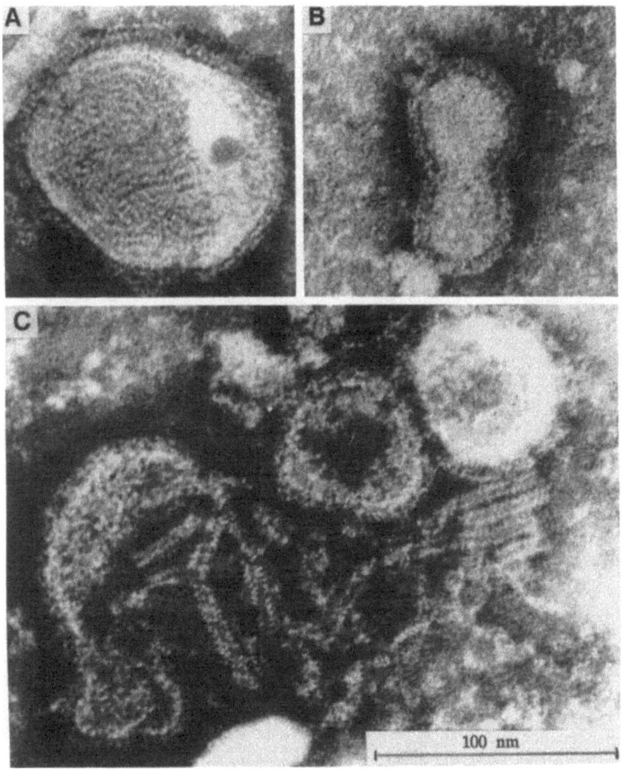

Fig.2: Negative staining of isolate 30/86 from Morelia argus.
Viral nucleocapsid within the pleomorphic particle (A); virus particle exhibiting an unique membrane envelope with spikes (B); non segmented nucleocapsid stained in the virion (C).

Discussion

The virus isolates from Gonosoma oxycephala (30/85), from Morelia argus (30/86) and from Bitis nasicornis (115/87) share jointly some characteristics: i) the temperature needed for virus replication below 37° C, ii) the morphology of the virion and the strands of nucleocapsids, iii) the sensitivity to lipid solvents, iv) the ability to induce syncytia formation and v) the ability for hamagglutination.

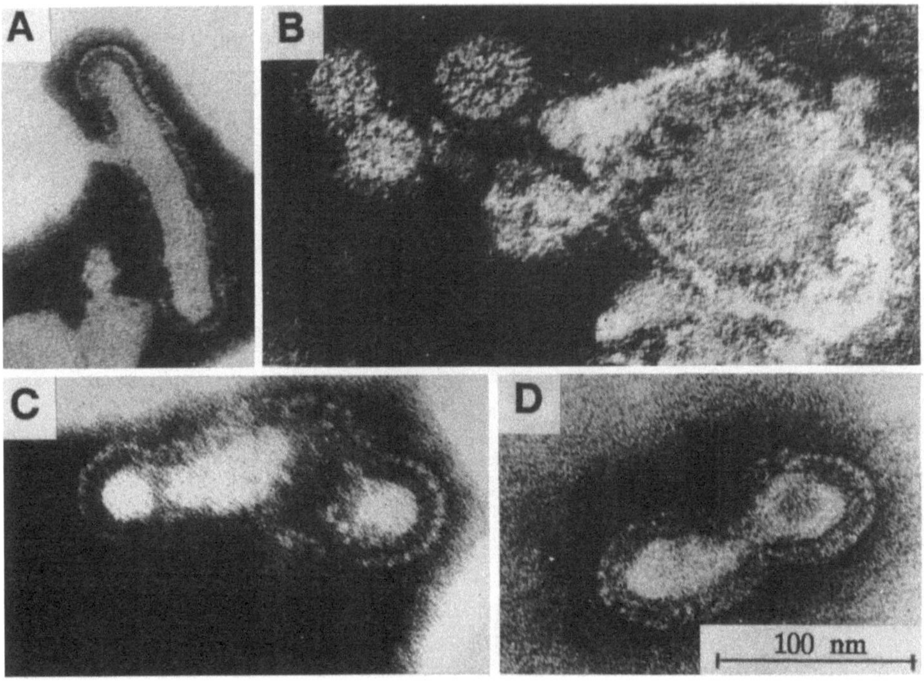

Fig.3: Negative staining of isolate 30/85 from Gonosoma oxycephala. Speroidal morphology of virus particles (A); nucleocapsid stained in the virion (B); virus particles exhibiting an unique membrane envelope with spikes (C,D).

In the SDS-PAGE the proteins patterns of the virus isolates 30/85 and 30/86 are almost identical; isolate 115/87 shows a similar pattern with respect to the major bands. A similar migration pattern, but no identity, is reported for the proteins of respiratory syncitial virus (RSV), a member of the pneumovirus group (McIntosh and Channock, 1985). Thus, there is a first evidence for the isolated viruses to be probably members of the pneumovirus group.

In contrast, by immunoblotting no homology or crossreactivity was found between proteins of each of the virus isolates and an antiserum directed against human respiratory syncitial virus. In addition, the isolate 30/85 has been reported to be not antigenetically related to the mammalian ortho- and paramyxoviruses (Ahne et al., 1987); this is in aggrement with data published for the Fer de Lance virus (Clark et al., 1979).

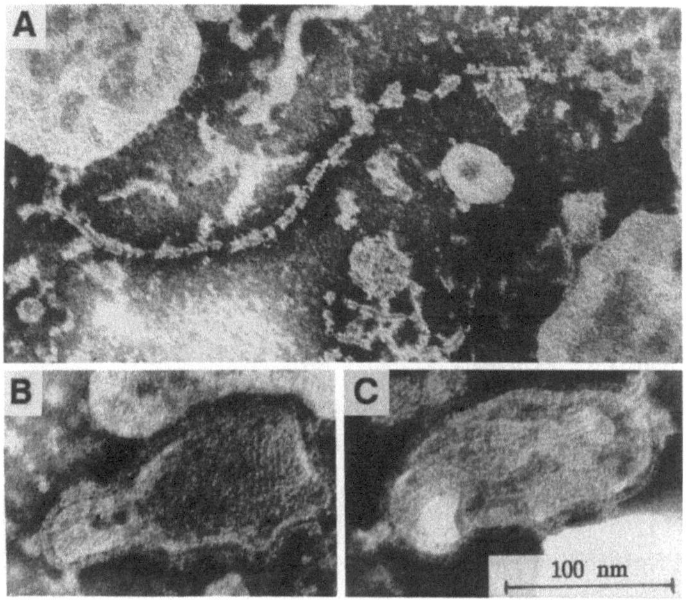

Fig.4: Negative staining of isolate 115/87 from Bitis nasicornis. Negatively stained nucleocapsid (A); speroidal morphology of a virus particle (B); virus paticles and nucleocapsids (C).

However, there is a remarkable antigenetic relationship between the three isolates. Although the viruses were isolated from different species of snakes at different times and they came from different places in West-Germany they show a antigenetic relationship in at least two bands of proteins. Compared to the protein pattern published for respiratory syncitial virus one can speculate that related proteins in the three virus isolates are the F and NP proteins.

Central nervous diseases associated with paramyxoviruses have been reported in reptiles (Foelsch & Leloup, 1979; Jacobson et al., 1980) as well as in mammals (Shimokata et al., 1976) and humans (Baczko et al., 1984). Therefore it is of great epidemiological importance to clarify the relationships between paramyxoviruses originating from the different classes of vertebrates.

Acknowledgements

We wish thank Prof.W.Sachsse (Mainz, F.R.G.) for providing organs of Bitis nasicornis and to Ch.Schommer for excellent technical assistance.

References

Ahne W (1977) Bei Reptilien vorkommende Viren. In: Krankheiten der Reptilien, ed. Reichenbach-Klinke HH, Gustav Fischer Verlag Stuttgart

Ahne W (1985) Virusinfektionen bei Fischen: Ätiologie, Diagnose und Bekämpfung. Zbl Vet Med B 32:237-264

Ahne W Neubert WJ Thomsen I (1987) Reptilian viruses: Isolation of Myxovirus-like particles from snake Elaphe oxycephala. J Vet Med B 34: 607-612

Baczko K Carter MJ Billeter M ter Meulen V (1984) Measles virus gene expression in subacute sclerosing panencephlitis. Virus Res 1: 585-595

Clark HF and Lunger PD (1980) Viruses of Reptiles. In: Diseases of the Reptilion, eds. Cooper JE Jackson OF, Academic Press New York

Clark HF Cohen MM Karzon DT (1970) Characterization of reptilian cell lines established at incubation temperatures of 23° C to 36° C. Proc Soc Exp Biol Med 133:1039-1047

Clark HF Lief FS Lunger PD Waters D. Leloup P Foelsch DW Wyler RW (1979) Fer de Lance virus (FDLV): a probable paramyxovirus isolated from a reptile. J gen Virol 44:405-418

Foelsch DW Leloup P (1976) Fatale endemische Infektion in einem Serpentarium. Tierärztl Prax 4:527-536

Granoff A (1969) Viruses of Amphibia. Curr Top Microbiol Immunol 50:107-137

Hoff GL Frye FL Jacobson ER (1984) Diseases of Amphibions and Reptiles. pp 784, Plenum Press, New York/London

Jacobson ER Gaskin JM Simpson CF Terrell TG (1980) Paramyxo-like virus infection in a Rock Rattlesnake. J Am Vet Med Ass 177:796-799

Jacobson ER (1986) Viruses and viral associated diseases of reptiles. In: Maintenance and reproduction of reptiles in captivity. Vol II eds. Bels VL Van de Sande AP, Acta Zoologica et Pathologica Antverpiensia, 79:73-90

Laemmli UK (1970) Cleavage of structural proteins during the assembly of the head of bacteriophage T4. Nature 227:307-310

Lunger PD Clark HF (1978) Reptilia-related viruses. Adv Virus Res 23: 159-204

Neubert WJ Hofschneider PH (1983) Transient rescue of Sendai-6/94 virus from the persistently infected cell line Cl-E-8 by cocultivation. Virology 125:445-453

McIntosh KM Channock RM (1985) Respiratory syncitial virus. In: Virology, eds. Fields BN et al., pp 1285-1304 Raven Press New York

Shimokata K Nishiyama Y Ito Y Kimura Y Nagata J Ilda M Sobue J (1976) Pathogenesis of Sendai virus infection in the central nervous system of mice. Infect Immun 13:1497-1502

Sigmund M Einberger H Neubert WJ (1988) Simple method for rapid and highly sensitive detection of antiviral-antibodies in serum and cerebrospinal fluid of small laboratory animals. J Virol Methods 22:231-238

Towbin H Staehelin T Gordon J (1979) Electrophoretic transfer of proteins from polyacrylamide gels to nitrocellulose sheets. Proc Natl Acad Sci USA 76:4350-4354

Paramyxovirus Infection in Snakes

M. Müller, N. Zangger, and H.P. Jakob
Institut of Animal Pathology
University of Bern
Länggassstraße 122
3012 Bern
Switzerland

Introduction

Parvo-, Adeno-, Herpes- and Picornavirus (gastroenteritis and encephalitis; Berger et al.,1984; Heldstab et al.,1984), Leukovirus (myxofibroma; Lunger et al.,1978) and Paramyxovirus (pneumonia with encephalitis; Foelsch et al.,1976; Jacobson et al.,1980,1981) have been reported to be pathogenic for snakes. In this report, we describe spontaneous generalized Paramyxovirus infection in juvenil and adult snakes of the families Boidae, Colubridae and Crotalidae.

Materials and Methods

11 Boidae (7 Boa constrictor, 2 Python reticulatus, 2 Corallus caninus), 1 Colubridae (Elaphe guttata) and 2 Crotalidae (Crotalus sp.) were submitted for necropsy because of suggested virus associated infection.
Tissue samples from 5 juvenil and 9 adult animals, 6 males and 5 females, which died spontaneously or were euthanazed (3), were fixed in 4% formalin, embedded in paraffin, sectioned at 4 nm and routinely stained with hematoxylin-eosin. Selected specimens were also stained with von Gieson and periodic acid-Schiff (PAS). Specimens from lung, pancreas and brain were post-fixed in 1% osmium tetroxide and embedded in Spurr. Semithin sections were stained with toluidine blue. Ultrathin sections were contrasted with uranyl acetate and lead citrate and examined by means of a Philips EM 300 electron microscop.
Routine bacteriological cultures were made on lung, liver, spleen, kidney and intestine.

Results

History

In 6 snakes (3 Boa, 2 Python, 1 Elaphe) the symptoms of bronchopneumonia were observed: forced respiration, open mouth and shaking of the head were common. Frequently, there was an expulsion of homogenous and dull fluid from the glottis. In bacterial complication, the fluid was flocky. 2 Boas and 1 Python had CNS-symptoms characterized by head tremor, loss of equilibrium and muscle tone, righting reflex, superficial skin reflex and pupillary reflex. Motion of tongue was retarted. There was a typical spiraling of the entire body. In one Boa we observed a thin skin, which broke spontaneously, and segmental dysecdysis. 2 Corallus and 2 Crotalus died without clinical symptoms.

Pathologic lesions

Histologic examination of respiratory tract demonstrated obstruction of air spaces with moderate amount of exudate, cellular debris and, in case of bacterial complication, with large number of heterophils. Hyperplasia of the mucosa, ballooning of epithelial cells, small cytoplasmic inclusions and mononuclear infiltration demonstrated the massive rhinitis. The epithelial layer in bronchi proliferated to a stratified epithelium. There was metaplasia to cuboidal or columnar profile of all pneumocytes and hyperplasia to thick columnar forms. Squamous ballooning, desquamation and necrosis of pneumocytes were common. Occasional small eosinophilic intracytoplasmic inclusions were seen within bronchus, lining pneumocytes, lymphocytes and myofibres. The myofibres with inclusion bodies degenerated with ballooning and loss of striation. Diffuse thickening of the respiratory interstitium manifested oedema, mononuclear cells and minimal fibrosis.

As well multifocal as disseminated gliosis, status spongiosus and minimal perivascular cuffing of blood vessels were seen in whole brain. The cytoplasmic inclusions of glia cells and neurons were accumulated in hindbrain. Hemorrhage per rhexin was rare. In the brainstem and the peripheral nerves were moderate ballooning of axon sheats and demyelination. In

ependym and plexus cells with large inclusions the nucleus drifted to the apical side. Non purulent and necrotizing choriomeningitis often was massive, likewise with inclusions.
The different sized eosinophilic inclusions in epithelial cells had regularly been associated with ballooning degeneration, desquamation, focal necrosis and moderate exocytosis of lymphocytes and plasmacells (respiratory and gastrointestinal tract, ductus choledochus, tubules of kidney, Bowman's capsule, testicule, ductus seminiferus, oviduct, exo- and endocrine pancreas, chromaffin and medullary cells of adrenal gland, lacrimal and salivary gland).
Although miliary necrosis in liver was rare, large cytoplasmic inclusions in hepatocytes were common.
Pulpitis lead to loss of teeth and to gingivitis. Diapedesis of erythrocytes, necrosis with macrophages, loaded by hemosiderin and inclusion bodies, and lymphocytes were prominent in the follicle of pulpa. The large number of inclusions in pulpa cells was impressive.
In one Boa, focal dermis areas were spongy in case of inter- and intracellular oedema, acanthosis and paracerathosis. Pycnosis of the nucleus with the typical cytoplasmic inclusions were found in basal layers.
Further cytoplasmic inclusions were regularly present in endothelial cells of all arteries and arterioles, endocardium, lymphocytes and reticulum cells in case of viremia.
Ultrastructural examination of pneumocytes, exo- and endocrine pancreas and neurons revealed a large number of extracellular mature virus particles of 140 to 320 nm in diameter. The filamentous and spheroidal particles were enclosed in a cell membrane-derived envelope, studded with slender spikes. Presumptive virions were characteristically budding from cell membrane. Membrane of virion was 22-23 nm in width.
Cytoplasmic inclusion bodies demonstrated accumulation of nucleocapsid strands. Unarranged strands measured 15-16 nm in diameter. Portions of inclusions consisted of highly-ordered membranes in parallel array. Membranes were 7-8 nm in width, separeted by a 11 or 17 nm space.

Bacteriologic examination revealed Salmonella-Arizona (1) and Pseudomonas aeruginosa (3) in lung tissue. Salmonella-Arizona (1) and Salmonella-Arizona combined with Proteus rettgeri (1) were found in septicemia.

Discussion

Histopathological findings, including rhinitis, proliferative bronchopneumonia and degenerative foci in several glands and epithelia, often associated with intracytoplasmic inclusions, supported the diagnosis of viral infection. Ultrastructural appearance confirmed the presence of paramyxo-like particles. Characteristics of the virus coincided with those described of the Fer-de-Lance virus (Clark et al.,1979; Lunger et al.,1979).

Paramyxovirus (PMV) infection is associated with pneumonia in the families Viperidae and Crotalidae (Jacobson et al., 1981,1982) and as pneumonia combined with encephalitis in a rock rattle-snake (Jacobson et al.,1980). Jacobson, 1981, showed in a reticulated Python whithout having clinical disease, a very high titer to the PMV and suggested, that nonviperid snakes may be a host for PMV. Our findings included rhinitis, bronchopneumonia, meningochorioencephalitis and viremia in the families Boidae, Colubridae and Crotalidae.

Proliferative bronchopneumonia corresponded to the findings in literature (Jacobson et al.,1980,1981,1982).

Non purulent meningitis and chorioencephalitis in reptiles have rarely been described in literature (Fankhauser et al., 1985). An Adenovirus was responsible for an encephalitis in a Boa (Heldstab et al.,1984). Zwart, 1984, has recorded a figure of a Boa constrictor with CNS symptoms of a probably PMV infection, but no further comment is given. CNS symptoms and encephalitis caused by a PMV is reported in a rock rattle-snake (Jacobson et al.,1980). Frequently, in PMV infection we got in the family of Boidae encephalitis, meningoencephalitis and degeneration of peripheral nerves combined with bronchopneumonia (43%).

Stomatitis ulcerosa is a well known disease in snakes and tortoises. A Herpesvirus associated glossitis has been documented in a Mediterranean land Tortoise (Heldstab et al.,1984). In PMV infection the epithelium of tongue was intact, but demyelination and ballooning of axon sheats of nervus linguae was characteristic. No report exists on specific disease of teeth in reptiles. Pulpitis, gingivitis and loss of teeth developed to bacterial complicated stomatitis ulcerosa.

Inclusion bodies and ballooning degeneration of epithelium in gastrointestinal tract accorded with the suggestion, that virus associated gastroenteritis is common in snake (Heldstab et al.,1984).

Clinical signs of a Boa constrictor included thin skin and segmental dysecdysis. By handling, the skin has broken spontaneously. For the first time we got virus associated skin lesions in snake.

Adult human suffering of mumps disease often have orchitis. So the first detection in reptiles of virus associated degeneration in male and female genital tract is interesting. Besides, two juvenil Corallus died of a PMV associated bronchopneumonia. After birth there was no more contact with adult snakes. We support a transovarial transmission of the PMV as known in Newcastle Disease (NCD) virus in birds (Brandly et al.,1965). These facts point to the possibility of a vertical virus transmission in snakes.

In snakes with signs of viremia a great number of inclusion bodies were found in lymphocytes, reticulum- and endothelial cells. These cell types are known in NCD viremia as virus breeding station (Brandly et al.,1965).

Corresponding the pathologic lesions the PMV will be spread out by the secrets of salivary and lacrimal gland, of the respiratory and genital tract and by the feces.

These 14 cases suggest that PMV infection in snakes must occur rather frequently, although secondary necrotizing bacterial infections or laking of inclusions often hinder detection of viral changes.

Acknowledgment

We thank Dr.M.Weiss and Mr.J.Di Lullo for technical assistance.

Literature

Berger R (1984) Virusisolierung aus dem Gastrointestinaltrakt von Schlangen. Lizentiatarbeit Bern

Brandly CA, Ranson RP (1965) Newcastle Disease. In: Disease of Poultry, eds. Biester HE, Schwarte LH, 5th ed, pp 633-674 University Press Ames Iowa

Clark HF, Lief FS, Lunger PD, Waters D, Leloup P, Foelsch DW, Wyler RW (1979) Fer de Lance virus (FDLV): a Probable Paramyxovirus Isolated from a Reptile. J Gen Virol 44: 405-418

Fankhauser R, Heldstab A (1985) Nervensysteme. In: Handbuch der Zootierkrankheiten, eds. Ippen R, Schröder HD, Elze K, Vol 1, pp 241-249 Akademie Verlag Berlin

Foelsch DW, Leloup P (1976) Fatale endemische Infektion in einem Serpentarium. Tierärztl Praxis 4:527-536

Heldstab A, Bestetti G (1984) Virus associated gastrointestinal diseases in Snakes. J ZooAn Med 15: 118-128

Heldstab A, Bestetti G (1984) Herpesviridae causing glossitis and meningoencephalitis in land tortoise (Testudo hermanni). Sec internat Coll on Pathol of Reptiles and Amphibians Nothingham

Jacobson E, Gaskin JM, Simpson CF, Terrell TG (1980) Paramyxo-like Virus Infection in a Rock Rattlesnake. JAVMA 177:796-799

Jacobson E, Gaskin JM, Page D, Iverson WO, Johnson JW (1981) Illness Associated with Paramyxo-like Virus Infection in a Zoologic Collection of Snakes. JAVMA 179:1227-1230

Jacobson E, Gaskin JM (1982) Paramyxo-like Virus associated respiratory Disease of Viperid Snakes. First Coll on Pathol of Reptiles and Amphibians Angers

Lunger PD, Clark HF (1978) Reptilia-related viruses. Advances in virus research 23:163-188

Lunger PD, Clark HF (1979) Morphogenesis of Fer-de-Lance virus (FDLV) cultured at optimal (30°c) cell growth temperature. J Comp Pathol 89:265-279

Lunger PD, Clark HF (1979) Morphogenesis of Fer-de-Lance virus (FDLV) cultured at sub-(23°c) and supra-(36°c) optimal cell growth temperatures. J Comp Pathol 89:281-291

Zwart P (1984) Schlangen. In: Krankheiten der Heimtiere, eds. Gabrisch K, Zwart P, pp 287-311 Schlütersche Hannover

Do in Reptiles Togaviruses and Flaviviruses Persist?

K.F. Shorthridge
Department of Microbiology
University of Hong Kong
Pathology Building
Queen Mary Hospital Compound
Hong Kong

INTRODUCTION

At the beginning of this Session, it was noted that captive animals in collections provided evidence that reptiles are associated with a wide range of viruses. In contrast, isolations from wildcaught reptiles have been mainly arboviruses of the Togaviridae and Flaviviridae families, i.e. the former Group A and Group B arboviruses, respectively. These isolations derive mainly from attempts in the 60's and 70's to understand the ecology of certain arboviruses of medical and veterinary importance.

Central to the question of ecology is the need to explain how these arboviruses remain endemic in temperate climates where virus transmission is interrupted by cold winters when there is no mosquito activity. It seems unlikely that higher vertebrates could maintain continuity of virus as infection leads to an immune response resulting in its elimination. Experimental infection suggests that togaviruses and flaviviruses may persist in reptiles and has provided an insight into how they might be maintained in nature from one year to the next, i.e. overwinter, in temperate climates. Aspects of this, initially reviewed elsewhere (Shortridge and Oya, 1984), are considered here.

EFFECT OF TEMPERATURE ON VIRUS RECOVERY

Early work showed that the western equine encephalitis (WEE) virus could be isolated from experimentally infected garter snakes (Thamnopsis spp) emerging from an induced torpor (Thomas et al., 1958; Thomas and Eklund, 1960; Gebhardt and Hill, 1960) and from wild snakes caught in the early spring (Gebhardt et al., 1964; Burton and McLintock, 1966).

Gebhardt et al., (1973) subsequently consolidated these findings experimentally (Figure 1). They showed that, at a temperature of 5° C, for there to be virus sufficient for a culicine mosquito to obtain an infectious blood meal at the end of torpor, the snake had to enter torpor between 18 hours and 11 days following infection. In a somewhat similar experiment on the Texas tortoise (Gopherus Berlandieri) using WEE virus, Bowen (1977) observed that there could be prolonged viraemia at a milder temperature of 10° C; elevation of the temperature to 30° C briefly increased the magnitude of the viraemia but led to its subsequent termination (Figure 2). Infection thresholds for different mosquito species affeared to vary considerably. More recently, Oya et al., (1983) and Doi et al., (1983) made similar findings for Japanese encephalitis (JE) virus and the lizards Tachydromus tachydromoides and Eumeces latiscutatis. Thus, to maintain the arbovirus in nature, it would appear that a reptile would need to be bitten by an infected mosquito in late atumn for successful overwintering and to be bitten again at the end of torpor or relative inactivity in the early spring by newly hatched mosquitoes for onward transmission and possible amplification in other reptiles and vertebrates. In other words, virus maintenance is probably defendent on persistance in reptiles.

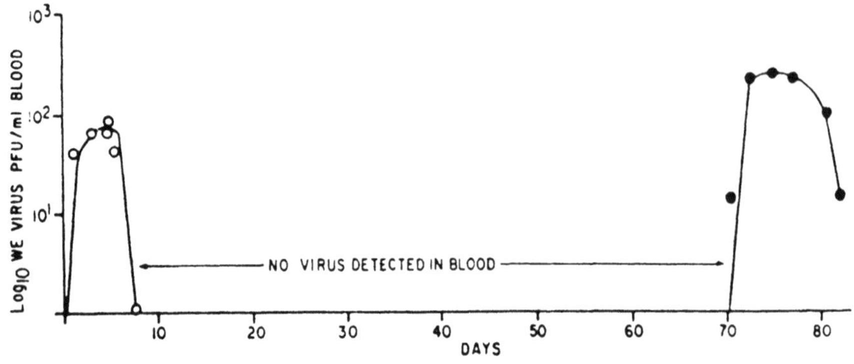

Figure 1

Viremia response of representative snakes, Thamnohis sp., in torpor for 60 days at 5° C beginning 18 hours after western equine encephalitis virus infection and then removed to room temperature (23° C) (from Gebhardt et al., 1973).

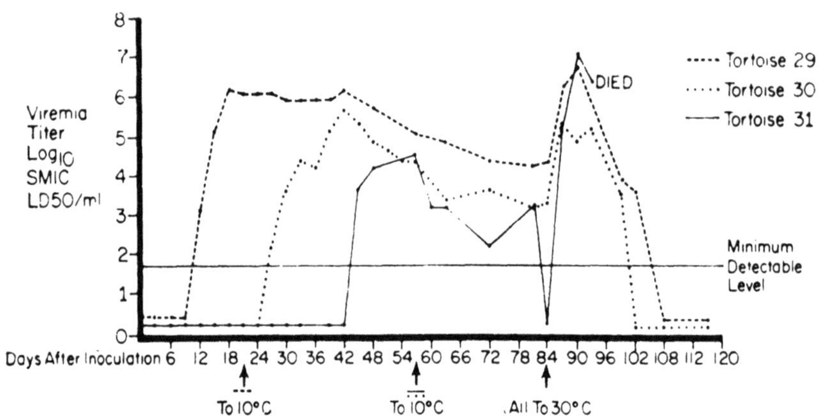

Figure 2

Viremia response of three Texas tortoises, Gopherus berlandieri, held at 20° C until the appearance of viremia and then at 10° C and 30° C following inoculation with western equine encephalitis virus (from Bowen, 1977).

HIDDEN VIRUS ?

The fact that infection can be induced experimentally with relative ease contrasts with relative paucity of isolations from wild-caught reptiles in both warmer (Doherty et al., 1973; Causey et al., 1966; Hoff and Trainer, 1973) and colder climates (Doi et al., 1968; Lee et al., 1972; Mifune et al., 1969). The question arises whether these largely negative findings and presumably many other negative ones not recorded, are really valid. The use of more sophisticated in vitro and in vivo systems for the detection and isolation of the wide range of viruses now recorded for reptiles (see earlier report in this session) suggests not. Two further lines of evidence, albeit indirect, suggest that virus may be present in a reptile for longer than the overwintering period.

1. Using a snake globulin precipitation test (SGP) Thomas et al., (1980) were able to show that an experimentally infected snake was able to maintain a high titre antibody to WEE virus over a working period of 4-5 years. These snakes were resistant to WEE viraemia on challenge, furthermore, the virus-antibody complex in the mouse was found to be non-protective suggesting that there may be residual virus capable of periodic antigenic stimulation.

2. In a study on the occurrence of antibody to JE virus in snakes and turtles of southern China, (Shortridge et al., 1974 and 1975) showed that not only is there a seasonal difference in antibody levels but that there is variation in the class of immunoglobulin detected (Table 1). This provides further support for the idea of a cyclical occurrence of viraemia in nature, at least in the temperate climates, as has been suggested by Karstad back in 1961.

Table 1 Seasonal Characteristics of Japanese Encephalitis Virus Hemagglutination Inhibition (HI) Antibody in the Sera or Plasma of Representative Reptiles[a]

Species	Season[b]	Mean[c] HI titers	Immunoglobulin detected[d]	
			IgG	Macro-globulin
Naja naja	Spring/Summer	712 +/- 651	+	-
	Autumn/Winter	127 +/- 181	+	+
Bungarus fasicatus	Spring/Summer	145 +/- 111	+	+
	Autumn/Winter	137 +/- 70	+	+
Ptyas korros	Spring/Summer	31 +/- 38	+	-
	Autumn/Winter	36 +/- 16	+	+
Trionys sinensis	Spring/Summer	35 +/- 39	+	-
	Autumn/Winter	29 +/- 50	+	-

[a] Abridged from Shortridge et al., (1974 and 1975).

[b] Spring/Summer = April-September; Autumn/Winter = October-March.

[c] Based on a positive titer of 10 or greater.

[d] Based on pools of serum collected mid-season and submitted to sucrose density gradient centrifugation.

The possible site of localization of virus in reptiles is unclear. In what was perhaps one of the earliest experimental infection studies on this field, Rosenbusch (1939) was able to isolate WEE virus from the brain but not from the blood of the snake, Bothrops alternata. A similar observation was made by Gebhardt and Hill (1960) who isolated the virus from the brain of one of four garter snakes following torpor at 4° C.

In the case of free-ranging animals, Charleville virus was isolated from the heart, liver and lung of G. australis (Doherty et al., 1973) while Marco, Timbo and Chaco viruses were isolated from the liver and heart of A.a.ameiva (Causey et al., 1966). On the other hand, Hoff and Trainer (1973) were unable to isolate virus from the blood, heart, lung and liver of 47 snakes and two lizards collected in Texas in the summer, although they were able to show the presence of WEE virus antibody in these animals.

Using fluorescent antibody technique, Doi and Oya (1978) were unable to locate a focus of virus multiplication in experimentally infected lizards. Thus, there is no clear-curt evidence whether or not there arbovirus may be hidden in reptiles.

COMPLEMENTARY MAINTENANCE FACTORS ?

Various hypotheses have been advanced to explain the survial of arboviruses through adverse periods and their reappearance under favourable conditions (see Reeves, 1974). Since then, virus maintenance in nature through persistence in reptiles (and probably amphibians) as outlined above has achieved some credibility. At least three other factors may be pertinent to virus maintenance:

1. Transovarial transmission of JE virus has been demonstrated in mosquitoes but is rare (Rosen et al., 1978). In other words, viruses might be maintained in nature in the absence of overwintering in reptiles. Indeed, it is possible that a high infection rate may be detrimental to virus survival by exhausting the supply of susceptible hosts particularly if they carry virus for long periods of time.

2. Congenital WEE virus infection has been demonstrated in asymptomatically infected Thamnopsis snakes some time after being caught in the wild (Gebhardt et al., 1964). This suggests that it may be possible for viruses to be transmitted in silent form from one generation to the next.

3. The demonstration by Oya et al., (1983) that lizards (Tachydromus tachydromoides and Eumeces latiscutatus) could be infected with JE virus by peroral feeding on only one infected mosquito suggests that reptiles could be infected by arboviruses in the absence of vector engorgement. The importance of this finding is that it overcomes objections to reptiles being arbovirus reservoirs because of restricted and limited feeding preferences of primary vectors.

COMMENTS

As might be inferred from the above, the roles of a haematophagous insect vector in facilitating virus persistance in reptiles would offer to be crucial. An alternative or perhaps complementary mechanism is now suggested, i.e. through catching infected mosquitoes in the mouth. This in turn raises interesting questions about the pathogenesis of arboviruses in reptiles and indeed other viruses associated with them. Whether these viruses might enter a state of persistence as appears to be the case for certain toga- and flaviviruses is open to study. Given the fact that many of them would not have been detected had it not been for the effects of various stressors in crowded conditions in collections is a good indication that they might.

The factors that give rises to persistance are little, if at all, understood. However, experimental evidence did suggest that in the colder climates at least, the time of infection prior to a reptiles enterring a torpid state seems to be important. There is certainly no clear-cut information whether virus becomes localized or hidden particularly over the winter period.

Many questions remain unanswered and until they are answered, this special association between virus and host persistance, will remain open to question. It is unfortunate that very little work has been done in this field since the experimental studies of the 60's and 70's to explain the successful overwintering of certain toga- and flavivirus of medical and veterinary importance. It is appropiates that this type of work be resumed in new sets of carefully designed experiments to take advantages of better virus detection and isolation system now available. The use of manipulative procedures such as immunosuppression should prove useful in this respect. They may also lead to the recognition of hitherto, unrecognized agents.

Studies on persistance should also consider the situation in tropical areas. As pointed out by Reeves (1974), prolonged dry or rainy seasons or other environmental factors may disrupt continous transmission of virus by their effect on vector populations. Thus, evidence for virus persistance as a means of maintaining arboviruses in nature needs to cover tropical, subtropical and temperate climates.

All in all, renewed studies in this field are needed. Much new information will no doubt emerge and may provide a useful insight into the evolution of viruses.

REFERENCES

Burton A N , McLintock J and Rempel J G (1966) Western equine encephalitis virus in Saskatchewan garter snakes and leopard frogs. Science, 154: 1029-1031

Causey O R , Shope R E and Bensabeth G (1966) Marco, Timbo and Chaco, newly recognized arboviruses from lizards of Brazil. Amer. J. Trop. Med. Hyg., 15:239-243

Doherty R L, Carley J G, Standfast H A, Dyce A L, Kay B H and Snowdon W A (1973) Isolation of arboviruses from mosquitoes biting midges, sandflies and vertebrates collected in Queensland 1969 and 1970. Trans. Royal Soc. Trop. Med. Hyg., 67:536-543

Doi R and Oya A (1978) Unpublished data, National Institute of Health, Tokyo

Doi R, Oya A and Telford S R (1968) A preliminary report on infection of the lizard, Takvdromus tachvdromides, with Japanese encephalitis virus, Jap. J. Med. Sci. Biol., 21:205-207

Doi R, Oya A, Shirasaka A, Yabe S and Sasa M (1983) Studies on Japanese encephalitis virus infection of reptiles II. Role of lizards on hibernation of Japanese encephalitis virus. Jap. J. Exp. Med., 53:125-134

Gebhardt L P and Hill D W (1960) Overwintering of western equine encephalitis virus, Proc. Soc. Exp. Biol. Med., 104:695-698

Gebhardt L P, Stanton G J, Hill D W and Collett G C (1964) Natural overwintering hosts of the virus of western equine encephalitis, New England J. Med., 217:172-177

Gebhardt L P, De St Jeor S C, Stanton G J and Stringfellow D A (1973) Ecology of western encephalitis virus, Proc. Soc. Exp. Biol. Med., 142:731-733.

Hoff G L and Trainer D O (1973) Arboviruses in reptiles: Isolation of a bunyamwera group virus from a naturally infected turtle, J. Herpetol., 7:55-62

Karstad L (1961) Reptiles as possible reservoir hosts for eastern encephalitis virus, Proc. North Amer. Wildlife Natur. Res. Conf., 26:186-202

Mifune K, Shichijo A, Neda Y, Shuenaga O and Miyagi I (1969) Low susceptibility of common snakes in Japan to Japanese encephalitis virus. Trop. Med., 11:27-32

Oya A, Doi R, Shiasaka A, Yabe S and Sasa M (1983) Studies on Japanese encephalitis virus infection in reptiles. I. Experimental infection of snakes and lizards. Jap. J. Exp. Med., 53:117-123

Reeves W C (1974) Overwintering of arbovirus. Prog. Med. Virol., 17:193-220.

Rosen L, Tesh R B, Lien J C and Cross J (1978) Transovarial transmission of Japanese encephalitis virus by mosquitoes, Science, 199:909-911.

Shortridge K F and Oya A (1984) Arbovirus in diseases of Amphibians and Reptiles. (G.L. Hoff, F.L. Frye and E. Jacobson eds) Plenum Press. New York and London, pp. 107-148

Shortridge K F, Ng M H, Oya A, Kobayashi M, Munro R, Wong F and Lance V, (1974) Arbovirus infections in reptiles: Immunological evidence for a high incidence of Japanese encephalitis virus in the cobra, Naja naja, Trans. Royal Soc. Trop. Med. Hyg., 68:454-460

Shortridge K F, Ng M H, Oya A and Yip D Y (1975) Arbovirus infections in reptiles: Studies on the presence of Japanese encephalitis antibody in the plasma of the turtle, Trionyx sinensis. Southeast Asian J. Trop. Med. Public Health, 6:161-169

Thomas L A and Eklund C M (1962) Overwintering of western equine encephalomyelitis virus in garter snakes experimentally infected by Culex tarsalis, Proc. Soc. Exp. Biol. Med., 109:421-424

Thomas L A and Eklund C M and Rush W A (1959) Susceptibility of garter snakes (Thamnophis spp.) to western equine encephalomyelitis virus, Proc. Soc. Exp. Biol. Med., 99:698-701

Thomas L A, Patzer E R, Cory J C and Coe J E (1980) Antibody development in garter snakes (Thamnophis spp.) experimentally infected with western equine encephalitis virus. Amer. J. Trop. Med. Hyg., 29:112-117

Differential Virus Susceptibility of Reptilian Organ Cultures to Human Viruses

R.R. Bhonde
National Facility for Animal Tissue & Cell Culture
c/o Department of Zoology
University of Poona
Ganeshkhind
Pune 411 007
India

INTRODUCTION

Organ cultures of ciliated respiratory epithelium from a variety of sources have provided a useful system for propogation of viruses (Hoorn and Tyrrell, 1969). Organ cultures of trachea obtained from different mammalian hosts have been shown to support replication of respiratory viruses such as Influenza types A, B and C (Hoorn and Tyrrell, 1965) and nonrespiratory viruses such as Vaccinia, Polio-1, Herpes simplex types 1 and 2, and West Nile (Arbovirus group B) (Bhonde, Wagh and Gupta, 1983). Similarly tracheal cultures from other vertebrates such as birds have been studied for their ability to support Influenza A2 (Hong Kong) virus replication (Blaskovic, *et al*, 1972). Our earlier report on the virus susceptibility of tracheal cultures from reptiles to mammals revealed that reptilian tracheal cultures did not support replication of human viruses such as Herpes simplex types 1 and 2, Vaccinia, Chandipura (Rhabdoviridae), West Nile and Japanese Encephalitis (Arbovirus group B) (Bhonde and Wagh, 1982). Hence present studies were undertaken to ascertain how far tracheal cultures from different species of lizards would support viruses.

MATERIAL AND METHODS

Animals:. Two different species of lizards, viz., *Hemidactylus flaviviridis* (commonly known as house lizard or wall lizard) and <u>Calotes versicolor</u> (garden lizard) were obtained from the Poona serpentorium, Katraj, Pune, India.

<u>Organ cultures</u>: Donor animals were anaesthetized with chloroform. Tracheas were dissected out in strictly aseptic conditions and were cut into circular rings, which were kept immersed for 20 minutes in prewarmed (37 °C) phosphate buffered saline (PBS) (pH 7.2), containing 200 U penicilline/ml and 200 ug/ml streptomycin. Following Cherry, TailorRobinson's technique (1970), tracheal rings were placed in 10 ml screw capped vials, with one ring in each tube containing one milliliter of nutrient medium. The nutrient medium consisted of Eagle's minimum essential medium with Earle's salts (MEM) supplemented with 2% (v/v) heatinactivated goat serum. The culture tubes were incubated at 28 °C. The culture fluid was renewed every alternate day and cultures were observed daily for ciliary activity.

<u>Virus inoculation</u>: Tracheal cultures exhibiting vigorous ciliary activity were selected for virus inoculation. Togaviruses as listed below were used in the present experiments.

a) Sindbis strain AR 339 (Arbovirus group A),
b) Chikungunya strain 634029 (Arbovirus group A),
c) West Nile strain E 101 (Arbovirus group B),
d) Japanese Encephalitis strain P 20778 (Arbovirus group B).

All these viruses were obtained from the virus stock registry of National Institute of Virology, Pune, India. Each culture tube was inoculated by dropping 0.2 ml of 2 $\log_{10}$ $TCID_{50}$/ml virus suspension prepared in nutrient medium without serum. Virus was allowed to adsorb on to the tracheal ring explants up to 3 hours at 37 °C. Virus suspension was removed and the explants were washed three times with PBS (pH 7.2) to remove unadsorbed virus. The cultures were fed with fresh nutrient medium and incubated at 28 °C. Mockinfected (medium without virus) controls and uninoculated control cultures were run simultaneously in each set of experiments. Experimental observations were made on pools of medium obtained from a group of 3 to 4 tubes for each virus. Virus and mockinfected control cultures were observed daily for changes in ciliary activity, if any.

Virus assays: Culture fluids from infected and mock-infected control cultures were collected every alternate day for 10 days and were stored at 70°C till the end of the experiment. All the fluid samples collected at different postinoculation days were assayed simultaneously in Vero (African green monkey kidney) cells, a susceptibile cell line. Serial tenfold dilutions of the fluid samples were made in Eagle's MEM. For each dilution, 4 tubes were used with 0.2 ml of the inoculum per tube. These tubes were incubated at 37°C and observed daily for the appearance of cytopathic effect up to 7 days. The titre of the virus was determined in terms of $TCID_{50}$/ml by the method of Reed and Muench (1938).

RESULTS

It was possible to recognise virus multiplication in tracheal cultures by the following criteria: (1) changes in ciliary activity, such as production of ciliostasis (stationary cilia) and/or ballooning effect; (2) production of histopathological lesions or denudation of ciliated epithelium; (3) rise in the virus titre of the medium at a certain time higher than that of the inoculum and (4) detection of virus in culture fluid for long periods despite repeated medium change..

The data obtained are summarized in Tables I and II. Referring to Table I, it is seen that tracheaL Cultures of *Hemidactylus flaviviridis* supported multiplication of Sindbis and Chikungunya viruses, however, there was no indication of multiplication of West Nile and Japanese encephalitis viruses. It was observed that ciliary activity

TABLE I

REPLICATION OF TOGAVIRUSES IN TRACHEAL CULTURES OF *Hemidactylus flaviviridis*

Viruses	Virus titre* at different days after infection				
	2	4	6	8	10
Sindbis	2.5	3.5	4.5	4.5	4.5
Chikungunya	2.0	2.5	4.0	5.0	5.0
West Nile	1.5	0.0	0.0	0.0	0.0
Japanese encephalitis	2.0	0.0	0.0	0.0	0.0
Uninoculated control	0.0	0.0	0.0	0.0	0.0

* Virus titre expressed in $\log_{10}$ $TCID_{50}$/ml.
Initial inoculum, 2 $\log_{10}$ $TCID_{50}$/ml.

of virus infected tracheal cultures remained unaffected; and it was comparable to those of uninoculated control cultures. Sindbis and Chikungunya viruses replicated in these cultures exhibiting a steady rise in virus titre from 2nd to 6th postinoculation days. Thereafter, virus titre remained unaltered up to the 10th postinoculation day (studied so far). Presence of West Nile and Japanese Encephalitis viruses was detected in these cultures on 2nd postinoculation day only, thereafter there were no signs of virus replication; probably exhibiting residual virus.

Referring to Table II, it is seen that *Calotes versicolor* tracheal cultures supported replication of Sindbis and Chikungunya viruses, but these cultures did not show any evidence of multiplication of West Nile and Japanese Encephalitis viruses. No change was observed in the ciliary activity of virus infected tracheal cultures. Although Sindbis and Chikungunya viruses replicated in *Calotes versicolor* tracheal cultures, they could be recovered from the culture

TABLE II

REPLICATION OF TOGAVIRUSES IN TRACHEAL CULTURES OF *Calotes versicolor*

Viruses	Virus titre* at different days after infection				
	2	4	6	8	10
Sindbis	3.0	3.0	2.0	0.0	0.0
Chikungunya	3.5	3.0	3.0	2.0	0.0
West Nile	0.0	0.0	0.0	0.0	0.0
Japanese encephalitis	0.0	0.0	0.0	0.0	0.0
Uninoculated control	0.0	0.0	0.0	0.0	0.0

* Virus titre expressed in $\log_{10}$ $TCID_{50}$/ml. Initial inoculum, 2 $\log_{10}$ $TCID_{50}$/ml.

fluid only up to 6 and 8 postinoculation days, respectively. No virus could be detected in the culture fluid sample of 10th postinoculation day.

DISCUSSION

The results obtained in the present studies indicate that tracheal organ cultures of *Hemidactylus flaviviridis* and *Calotes versicolor* support replication of selected Togaviridae, Arbovirus group A viruses, Sindbis and Chikungunya, and do not support replication of selected Arbovirus group B viruses, West Nile and Japanese Encephalitis, suggesting a common denominator. However, the virus susceptibility pattern of *Hemidactylus flaviviridis* and *Calotes versicolor* tracheal cultures differs significantly indicating their differential sensitivity. *Hemidactylus flaviviridis* tracheal cultures appeared to be highly susceptible to Sindbis and Chikungunya viruses as compared to those of

Calotes versicolor, as the former supported continued and undiminished virus replication persisting up to 10 days or more (data not shown in the table); whereas, the latter supported replication of these viruses for a short period. Thus the two species of lizards exhibited differential susceptibility to Sindbis and Chikungunya viruses indicating inherent difference attributed to species and at the same time revealing a common feature of not supporting multiplication of West Nile and Japanese encephalitis viruses suggesting specificity of a group as lizards (Lacertilia).

Numerous reptilian cell culture systems have been used for avian and mammalian virological investigations, Cell lines from Gecko heart and lung have been tested for replication of Vaccinia virus, Herpes simplex virus, Avian tracheitis virus, Newcastle disease virus, vesicular stomatitis virus and Sindbis virus (Clark et al, 1970 and Michalski et al, 1974). Most of these viruses have been shown to replicate in these cell lines producing cytopathic effect. Lunger and Clark (1975) employed Tokay Gecko lung, turtle heart and viper spleen cell cultures for replication of vesicular stomatitis virus and showed that this virus replicates in these reptilian cell cultures efficiently. These reports support the present studies on susceptibility of Hemidactylus flaviviridis to Arbovirus group A viruses.

It may be inferred from the reports cited and the present studies that at least certain reptiles may normally represent an integral part of the cycle of vertebrate-mosquito transmission of specific togaviruses. There is an evidence that a single Western Equine Encephalitis virusinfected mosquito is capable of inducing viremia in garter snake, Thamnophis sirtalis (Gebhardt et al, 1966). Other hosts which may be involved in normal mosquito-borne virus transmission cycle include Western Equine Encephalitis virusinfected Texas tortoises, Chrysemys picta (Bowen, 1977). With regard to Western Equine Encephalitis virus, it has been proposed that garter snakes represent the primary and birds the secondary host (Gebhardt and Hill, 1960). Another example of primary host includes Japanese encephalitis virusinfected Chinese cobra, Naja naja (Shortridge et al, 1974). The incidence of Japanese encephalitis virus infection in cobra Naja naja and the distribution of Naja naja in those geographical areas that display a high incidence of human Japanese Encephalitis virus infection, it has been suggested that the cobra, Naja naja may act as a natural host of Japanese encephalitis virus. Such situation does not exist with regard to Japanese encephalitis in Hemidactylus flaviviridis and Calotes versicolor as indicated in present studies. However, the role of Hemidactylus flaviviridis in acting as a reservoir for Togaviridae Arbovirus group A viruses cannot be ignored as evidenced by the present data. Thus, it may be inferred that different reptilian species exhibit differential susceptibility to togaviruses.

REFERENCES

Bhonde RR, Wagh UV (1982) Differential virus susceptibility exhibited by tracheal organ cultures from reptiles to mammals. Ind J Exptl Biol 20:814-816

Bhonde RR, Wagh UV, Gupta NP (1983) Replication of non-respiratory viruses in tracheal organ cultures. Br J Exptl Path 64:1-5

Blascovic P, Rhodes AJ, Labzoffsky NA (1972) Infection of chick embryo tracheal organ cultures with Influenza A2 (Hong Kong) virus. Arch Ges Virusforsch 37:104-113

Bowen GS (1977) Prolonged Western Equine Encephalitis viremia in the Texas Tortoise. Am J Trop Med Hyg 26:171-175

Cherry JD, Taylor-Robinson D (1970) Large quantity production of chicken embryo tracheal organ cultures and use in virus and mycoplasma studies. Applied Microbiology 19: 658-662

Clark HF, Cohen MM, Karzon DT (1970) Characterization of reptilian cell lines established at incubation temperatures of 23 to 36°C. Proc Soc Exp Biol Med 133:1039-1047

Gebhardt LP, Hill DW (1960) Overwintering of western equine encephalitis virus. Proc Soc Exp Biol Med 104:695-698

Gebhardt LP, Stanton GJ, de St Jeor S (1966) Transmission of WEE virus to snakes by infected Culex tarsalis mosquitoes. Proc Soc Exp Biol Med 123:233-235

Hoorn B, Tyrell DAJ (1965) On the growth of certain 'newer' respiratory viruses in organ culture. Bri J Exp Pathol 46: 109-118

Hoorn B, Tyrrell DAJ (1969) Organ cultures in virology. Prog Med Virol 11:408-450

Lunger PD, Clark HF (1975) Host effect on vesicular stomatitis virus morphogenesis and "T" particle formation in reptilian, avian and mammalian cell lines. In Vitro 11:239-246

Michalaski F, Cohen MM, Clark HF (1974) Adult and exmbryonic Gecko cells in vitro : Growth characteristics, Infection by Rabies, Sindbis and Polyoma viruses, and transformation by SV-40. Proc Soc Exp Biol Med 146:337-348

Reed LJ, Muench H (1938) A simple method of estimating fifty per-cent end points. Am J Hyg 27:493-497

Shortridge KF, Ng MH, Oya A, Kobayashi M, Munro R, Wong F, Lance V (1974) Arbovirus infections in reptiles: Immunological evidence for a high incidence of Japanese Encephalitis virus in the cobra Naja naja. Trans R Soc Trop Med Hyg 68:454-460

Section III Viruses of Fishes

Lymphocystis Disease of Fishes

K. Anders
Institut für Meereskunde
Düsternbrooker Weg 20
2300 Kiel 1
West-Germany

Lymphocystis disease is a well-known viral infection of teleost fish species. It is characterized by conspicuous disease signs (Fig. 1A): small nodules which are located mostly on the body and fins. Occasionally, lymphocystis nodules occur in or on the surface of the body cavity and the internal organs as well as on the gills. Each nodule represents one single hypertrophied cell which may reach a size up to 2 mm in diameter and a volume of about a million-fold that of the normal cell, for example in European flounder (Platichthys flesus). There are infected fish with just one lymphocystis cell but in most cases several hypertrophied cells occur singly or grouped together in raspberry-like clusters with a tumorous appearance. Therefore, these signs are considered as "pseudo"tumors.

Lymphocystis disease was first noted over 100 years ago in European flounder from British coastal waters. Until today the condition has been found in 141 teleost fish species representing 45 families and 11 orders (Tab. 1). Perciformes and Pleuronectiformes dominate the host range representing 75% and 10% of the total number of species, respectively. It is a striking fact that so far no reports are available from wild and cultured salmonid and cyprinid species although many of them were subjects of thorough investigations on fish health in general.

Lymphocystis disease seems to be ubiquitous in the aquatic environment. It has been found in fish from the Arctic, temperate, and tropical regions (Fig. 2). The disease has been reported from such different habitats as the freshwater Great Lakes of North America and the Red Sea with salinities of about 40‰. Most reports concerning the marine distribution of the disease are from the North Atlantic continental shelf areas of Europe and North America which are identical with the areas where most field surveys on fish diseases have been carried out.

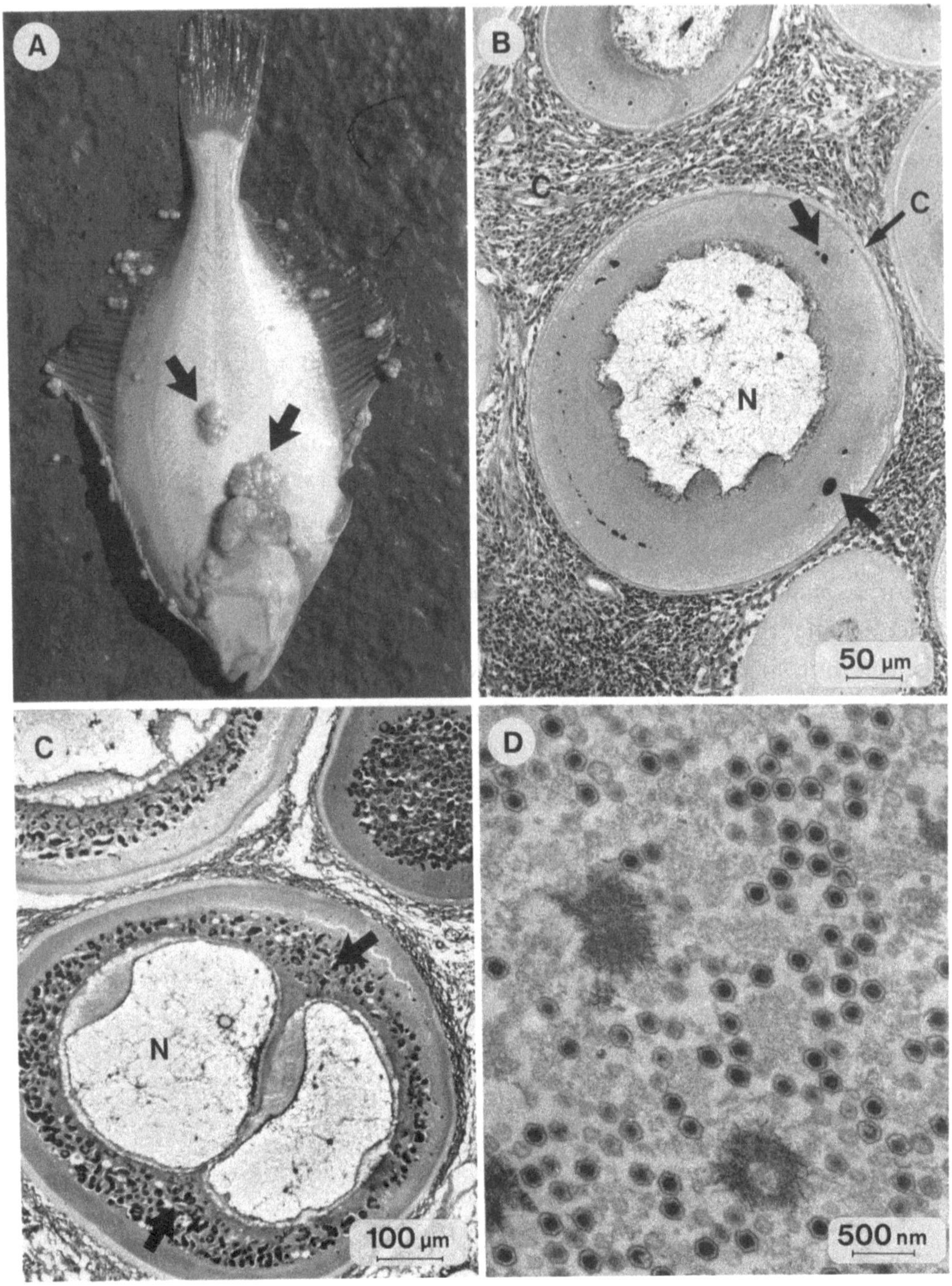

Fig. 1: Lymphocystis disease of European flounder. A: Large pseudotumors on the body surface; B: young lymphocystis cell with a clearly visible hyaline capsule (C) and small cytoplasmic inclusion bodies (arrows), H&E stain; C: older lymphocystis cell with well-differentiated cytoplasmic inclusion bodies (arrows), H&E stain; D: cytoplasmic iridovirus particles.

Tab. 1: Checklist of lymphocystis host fish species.

ORDER FAMILY Species	Diagn. techn.	Occur- rence		Original reference
CLUPEIFORMES				
CLUPEIDAE				
Clupea harengus / herring	E	W	M,B	Aneer and Ljungberg, 1976
SALMONIFORMES				
ESOCIDAE				
Esox lucius / pike	M	W	F	Amin, 1979
SILURIFORMES				
ICTALURIDAE				
Ictalurus punctatus / channel catfish	M	F?	F	Harshbarger, 1977 (RTLA No. 1565)
LOPHIIFORMES				
OGCOCEPHALIDAE				
Ogcocephalus nasutus / shortnose batfish	M	?	B?	Harshbarger, 1973
ATHERINIFORMES				
CYPRINODONTIDAE				
Fundulus heteroclitus / mummichog	L	?	B?	Weissenberg, 1939
BERYCIFORMES				
HOLOCENTRIDAE				
Holocentrus rubrum / red squirrelfish	E	P	M	Spitzer et al., 1982
ZEIFORMES				
ZEIDAE				
Zeus faber / John Dory	M?	W?	M	Durham and Anderson, 1981
SCORPAENIFORMES				
SCORPAENIDAE				
Sebastodes constellatus / starry rockfish	M	P	M	McCosker et al., 1976
Sebastodes maliger / quillback rockfish	E	P	M	McCosker et al., 1976
Sebastodes nebulosus / yellowstripe rockfish	M	P	M	McCosker et al., 1976
Sebastodes schlegeli / Kurosoi	E	W	M	Tanaka et al., 1984
TRIGLIDAE				
Eutrigla gurnardus / grey gurnard	E	W	M	Möller, 1981
HEXAGRAMMIDAE				
Ophiodon elongatus / lingcod	M	W	M	Walker, 1947

Tab. 1 (continued)

PERCIFORMES				
CENTROPOMIDAE				
Centropomus undecimalis / snook	E	W	B	Howse, 1972
Lates calcarifer / sea bass, giant perch	M	F	B	Leong, pers. comm., 1987
PERCICHTHYIDAE				
Morone americanus / white perch	M	P	B	Nigrelli and Ruggieri, 1965
Morone saxatilis / striped bass	L	W	M	Nigrelli, 1952
Doederleinia berycoides	M	?	M	Harshbarger, 1973
SERRANIDAE				
Hypoplectrus nigricans / black hamlet	L	W	M	Williams et al., 1984
Lateolabrax japonicus / Japan sea bass	L	F?	M	Miyazaki and Egusa, 1972
Serranus atricauda / striped sea perch	M	P	M	Porter, 1953
GRAMMIDAE				
Gramma loreto / fairy basslet	M	P	M	Kingsford, 1975
CENTRARCHIDAE				
Lepomis cyanellus / green sunfish	M	P,E	F	Wolf, 1962
Lepomis gibbosus / pumpkinseed	L	W	F	Weissenberg, 1945
Lepomis gulosus / Warmouth	M	?	F	Harshbarger, 1972
Lepomis humilis / orange-spotted sunfish	M	P	F	Nigrelli, 1943
Lepomis macrochirus / bluegill	E	W	F	Weissenberg, 1945
Lepomis megalotis / longear sunfish	L	W	F	Weissenberg, 1945
Lepomis pallidus	M	P	F	Nigrelli and Ruggieri, 1965
Micropterus dolomeiui / smallmouth bass	M	W	F	Crites and Snyder, 1986
Micropterus pseudaplites / Kentucky black bass	L	W	F	Weissenberg, 1945
Micropterus punctulatus / spotted bass	M	W	F	Crites and Snyder, 1986
Micropterus salmoides / largemouth bass	E	W	F	Wolf, 1962
Pomoxis annularis / white crappie	L	W	F	Weissenberg, 1945
Pomoxis nigromaculatus / black crappie	L	W	F	Weissenberg, 1945
APOGONIDAE				
Apogon quadrisquamatus / sawcheek cardinalfish	L	W	M	Williams et al., 1984
PERCIDAE				
Gymnocephalus cernuus / ruffe	L	W	B,F	Weissenberg, 1914
Perca flavescens / yellow perch	L	W	F	Walker, 1947
Stizostedion canadense / sauger	L	W	F	Bangham and Hunter, 1939
Stizostedion vitreum glaucum / blue pike	M	W	F	Bangham and Hunter, 1939
Stizostedion vitreum vitreum / walleye	E	W	F	Mavor and Feinberg, 1918
CARANGIDAE				
Seriola quinqueradiata / yellowtail	L	F	M	Matsusato, 1975
LUTJANIDAE				
Lutjanus griseus / gray snapper	M	W	M	Sigel et al., 1966

Tab. 1 (continued)

SPARIDAE				
Archosargus probatocephalus / sheepshead	M	W	M	Smith and Taylor, 1972
Diplodus annularis / white bream (?)	L	W	M	Joseph, 1917
Pagrus major / red sea bream, Madai	?	F	M	Yasunaga, 1986
Sparus auratus / gilthead seabream	E	F	M	Paperna et al., 1982
SCIAENIDAE				
Bairdiella chrysoura / silver perch	E	W	B	Lawler et al., 1974
Cynoscion arenarius / sand seatrout	L	W	B	Christmas and Howse, 1970
Cynoscion nothus / silver seatrout	L	W	B	Smith, 1973
Cynoscion regalis / weakfish	L	W	B	Smith, 1970
Larimus fasciatus / banded drum	M	E	B?	Lawler et al., 1974
Micropogonias undulatus / Atlantic croaker	E	W	B	Christmas and Howse, 1970
Pogonias cromis / black drum	M	E	B?	Cook, 1972
Stellifer lanceolatus / star drum	L	W	B	Smith, 1970
MULLIDAE				
Mullus surmuletus / red mullet	L	W	M	Alexandrowicz, 1951
EPHIPPIDAE				
Platax orbicularis / orbiculate batfish	E	P	M	Lawler et al., 1978
SCATOPHAGIDAE				
Scatophagus argus / spotted scat	E	P	B	Nigrelli and Ruggieri, 1965
Scatophagus rubrifrons	L	P	F	Giavenni, 1981
CHAETODONTIDAE				
Chaetodon auriga / threadfin butterflyfish	M	P	M	McCosker et al., 1976
Chaetodon capistratus / foureye butterflyfish	E	P	M	Lawler et al., 1978
Chaetodon trifasciatus / purple butterflyfish	M	P	M	McCosker et al., 1976
Chaetodon miliaris	M	P	M	McCosker et al., 1976
Chaetodon ocellatus / spotfin butterflyfish	L	W	M	Williams et al., 1984
Chaetodon striatus / banded butterflyfish	M	P	M	Nigrelli, 1943
Chelmon rostratus / copperbanded angelfish	L	P	M	McCosker et al., 1976
Forcipiger longirostris / forceps fish	M	P	M	Nigrelli and Ruggieri, 1965
Holacanthus bermudensis / blue angelfish	L	P	M	Weissenberg, 1937
Holacanthus ciliaris / queen angelfish	E	P	M	Nigrelli and Ruggieri, 1965
Pomacanthus annularis / blue-ringed angelfish	M	P	M	McCosker et al., 1976
Pomacanthus aureus	M	W	M	Lopez et al., 1969
Pomacanthus arcuatus / black angelfish	L	P	M	Nigrelli, 1943
Pomacanthus paru / French angelfish	E	W	M	Weissenberg, 1951
Pomacanthus semicirculatus / Koran angelfish	E	P	M	Lawler et al., 1978
OPLEGNATHIDAE				
Oplegnathus fasciatus / Japanese parrot fish	?	F	M	Yasunaga, 1986
CICHLIDAE				
Aequidens portalegrensis / black acara	M	P	F	Nigrelli, 1952
Aequidens pulcher / blue acara	M	P	F	Nigrelli and Ruggieri, 1965
Aequidens rivulatus / green terror	M	P	F	Huizinga, 1977
Apistogramma ramirezi / Ramirez'dwarf cichlid	M	P?	F?	Harshbarger, 1972

Tab. 1 (continued)

Cichlasoma synspilum	E	W	F	Weissenberg, 1951
Haplochromis elegans	L	W	F	Paperna, 1973
Hemichromis bimaculeatus / jewelfish	M	E	F	Nigrelli and Ruggieri, 1965
Platinum gourami	M	P	F?	Leibovitz, 1980
Pterophyllum scalare / scalare	M	?	F	Harshbarger, 1972
Sarotherodon amphimelas	L	W	F	Paperna, 1973
Sarotherodon esculentus	L	W	F	Paperna, 1973
Sarotherodon melanotheron /black-chinned mouthbreeder	M	E	F	Nigrelli and Ruggieri, 1965
Sarotherodon variabilis	L	W	F	Paperna, 1973
Symphysodon discus / discus, Pompadour fish	M	P	F	Porter, 1952
Tilapia ovalis / oval tilapia	M	E	F	Nigrelli and Ruggieri, 1965
Tilapia sparrmanii / Sparrman's tilapia	M	E	F	Nigrelli and Ruggieri, 1965
POMACENTRIDAE				
Amphiprion ocellaris / clownfish	M	P	M	Benisch, 1937
Dascyllus aruanus / white-tailed damselfish	L	P	M?	Lawler et al., 1977
Dascyllus melanurus / black-tailed damselfish	L	P	M?	Lawler et al., 1977
Pomacentrus partitus / bicolor damselfish	L	W	M	Williams et al., 1984
Premnas biaculeatus / spiny clownfish	L	P	M	Benisch, 1937
LABRIDAE				
Bodianus mesothorax / blackbelt wrasse	M	P	M	McCosker et al., 1976
Ctenolabrus rupestris / rock cook	M	P	M	Nieschulz, 1921
Lachnolaimus maximus / hogfish	L	P	M	Weissenberg, 1937
BLENNIIDAE				
Hypsoblennius gentilis / bay blenny	L	W	M	Weissenberg, 1951
Hypsoblennius gilberti / rockpool blenny	M	W	M	McCosker, 1969
Hypsoblennius jenkinsi	L	W	M	Weissenberg, 1955
ANARHICHADIDAE				
Anarhichas lupus / striped wolffish	L	W	M	own observation
CLINIDAE				
Acanthemblemaria crockeri	L	W	M	McCosker, 1969
ELEOTRIDAE				
Dormitator maculatus / fat sleeper	M	P	B	Nigrelli, 1943
GOBIIDAE				
Gillichthys seta / mudsucker	L	W	M	McCosker and Nigrelli, 1971
Gobiosoma paradoxum	M	W	B	McCosker and Nigrelli, 1971
ACANTHURIDAE				
Zanclus cornutus / Moorish idol	E	P	M	Lawler et al., 1978
Zebrasoma veliferum / sailfin tang	E	P	M	Lawler et al., 1978
SIGANIDAE				
Lo vulpinus / fox-faced rabbitfish	M	P	M	Lawler et al., 1977
Siganus guttatus / golden spinefoot	M	F	M	Jones and Hine, 1983

Tab. 1 (continued)

SCOMBRIDAE				
Scomberomorus maculatus / spotted Spanish mackerel	?	W?	M	Overstreet, 1988
ANABANTIDAE				
Anabas testudineus / climbing perch	M	P?	F	Thakur and Nasar, 1977
BELONTIIDAE				
Colisa fasciata / giant gourami	M	P	F	Rasin, 1928
Colisa labiosa / giant gourami (?)	M	P	F	Nieschulz, 1921
Macropodus opercularis / paradise fish	L	P	F	Weissenberg, 1931
Trichogaster microlepis / moonlight gourami	E	P	F	Spitzer et al., 1982
Trichogaster pectoralis / snakeskin gourami	E	P	F	Durham and Anderson, 1981
Trichogaster trichopterus sumatranus / blue gourami	E	P	F	Durham and Anderson, 1981
Trichopsis pumilus	L	P	F	Geus, 1963
PLEURONECTIFORMES				
BOTHIDAE				
Paralichthys dentatus / summer flounder	M	F	B	Stickney and White, 1974
Paralichthys lethostigma / southern flounder	M	F	B	Stickney and White, 1974
Paralichthys olivaceus / Hirame	E	W,F	M	Tanaka et al., 1984
PLEURONECTIDAE				
Hippoglossoides platessoides / long rough dab	L	W	M	Templeman, 1965
Limanda aspera / yellowfin sole	E	W	M	Alpers et al., 1977
Limanda herzensteini / brown sole	?	F?	M	Yasunaga, 1986
Limanda limanda / dab	E	W	M	Weissenberg, 1931
Limanda punctatissima	M	W	M	Peters and Kranz, 1980
Liopsetta glacialis / arctic flounder	M?	W	M	Stich et al., 1977
Parophrys vetulus / English sole	E	W	M	McCain et al., 1982
Platichthys flesus / European flounder	E	W	M,B,F	Lowe, 1874
Platichthys stellatus / starry flounder	M	W	B	Peters et al., 1980
Pleuronectes platessa / European plaice	E	W	M	McIntosh, 1885
Pseudopleuronectes americanus / winter flounder	L	W	M,B?	Murchelano and Bridges, 1976
SOLEIDAE				
Solea solea / common sole	L	W,F	M	Johnstone, 1907
TETRAODONTIFORMES				
OSTRACIONTIDAE				
Acanthostracion quadricornis / West Indian cowfish	M	P	M	Nigrelli, 1943
Lactophrys cornutus / East Indian cowfish	M	P	M	Nigrelli, 1943
Lactoria cornuta / long-horned cowfish	M	P	M	Nigrelli, 1943
BALISTIDAE				
Alutera schoepfi / orange filefish	L	W	M	Weissenberg, 1938
DIODONTIDAE				
Diodon holocanthus / spiny puffer	L	W	M	McCosker and Nigrelli, 1971
Diodon hystrix / porcupinefish	M	E	M	McCosker and Nigrelli, 1971

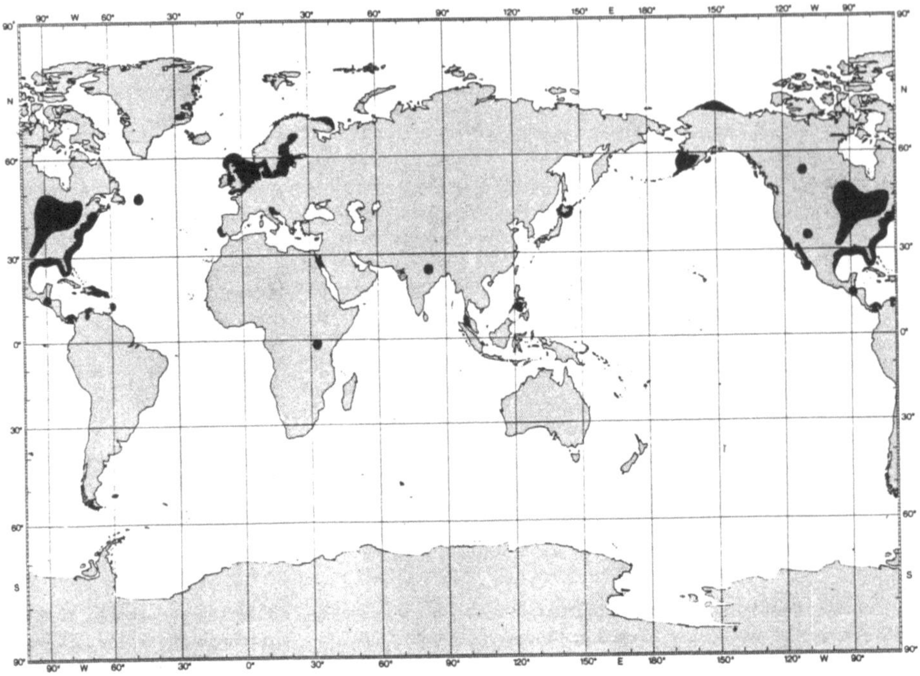

Fig. 2: Geographical distribution of lymphocystis disease: areas from which infected fish have been reported (marked in black).

Lymphocystis disease is observed frequently in wild-living fish populations. Obviously, highest prevalences occur in various flatfish species of estuarine and marine shelf regions of the North Atlantic and the Bering Sea. In the lower Elbe River lymphocystis prevalences of European flounder showed distinct regional differences within a few kilometers with highest values in the central estuary (Tab. 2, Möller, 1984). It is supposed that highly fluctuating salinities in this central part of the river are at least partly responsible for the reduction of the zoobenthos biomass which serves as food especially for demersal fish species (Möller, 1988, Fiedler, in prep.). Consequently, flounder in this area feed less and show significantly reduced body condition factors. This may lead to a lowered resistence to lymphocystis virus.

The disease occurs in cultured fish as well. Outbreaks of infection are known from tropical ornamental fish held in petfish aquaria and as yet from 10 marine fish species which were kept under farming conditions (Tab. 3).

Tab. 2: Lymphocystis prevalences in European flounder. Lower Elbe River, November to December 1984. Data are length-adjusted and related to 18 cm long specimens (recalculated after Anders, 1988). Description of regions: I = outer estuary near Cuxhaven, VII = near Hamburg harbour.

region	I	II	III	IV	V	VI	VII
average prevalence	2.4	3.9	4.1	1.7	1.2	0	0
n investigated	207	473	274	159	282	103	426

Tab. 3: Lymphocystis disease in cultured fish: reported cases.

Fish species	Location		Reference
Lates calcarifer	Estuary in West-Malaysia	B	Leong, pers. comm., 1987
Seriola quinqueradiata	Japanese coastal waters	M	Egusa, 1983, Matsusato, 1975
Pagrus major	Japanese coastal waters	M	Yasunaga, 1986
Sparus auratus	Red Sea (Israel)	M	Paperna et al., 1982
Oplegnathus fasciatus	Japanese coastal waters	M	Yasunaga, 1986
Siganus guttatus	Philippine coastal waters	M	Jones and Hine, 1983
Paralichthys dentatus	Estuary in Georgia, U.S.A.	B	Stickney and White, 1974
Paralichthys lethostigma	Estuary in Georgia, U.S.A.	B	Stickney and White, 1974
Paralichthys olivaceus	Japanese coastal waters	M,B	Tanaka et al., 1984
Solea solea	Scottish coastal waters	M	Anderson and Conroy, 1968

In certain wild-living fish populations the prevalence of infection was found to increase with increasing length (age) of the host fish and to decrease in very old specimens (Fig. 3). In European flounder the smallest specimen afflicted with lymphocystis disease measured 9 cm in total body length. There is evidence that highest prevalences and intensities of infection occur during or shortly before the spawning season in several fish species, for example in North Sea dab (Limanda limanda) (Wolthaus, 1984) and Elbe flounder (Möller and Anders, 1986).

The debilitating effect of lymphocystis disease on its host is generally low. Reported mortalities among spontaneously infected pet fishes are likely due to secondary infections but this assumption remains to be proven. Almost

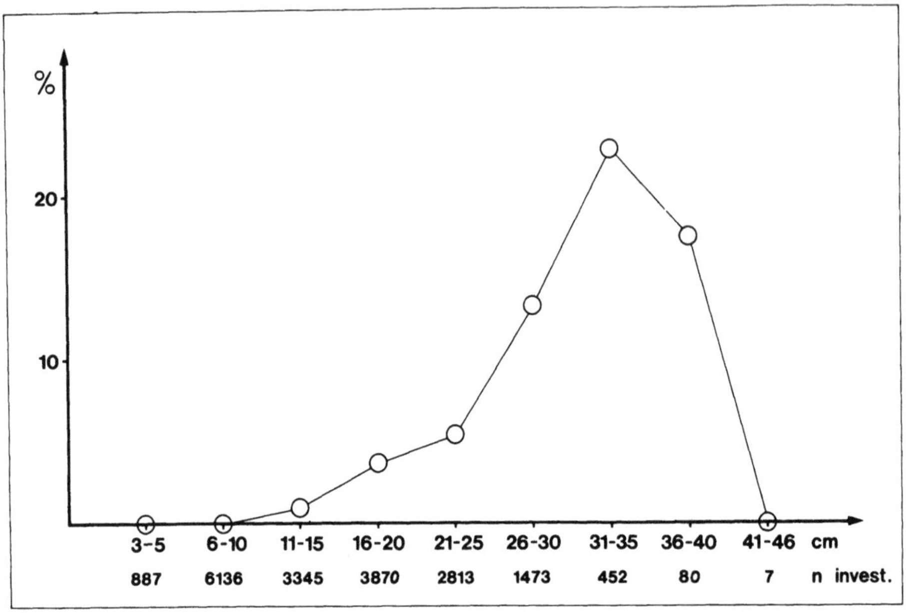

Fig. 3: Relationship between lymphocystis prevalence and total fish length in European flounder from the lower Elbe River (November 1984 to August 1986, summary of all stations).

no mortalities have been documented from disease outbreaks among farmed marine fish. Lesions normally regress totally. In severe cases only when pseudotumors especially those located in the mouth region grow that large that food intake is handicapped the affected fish may starve to death. Own experimental studies with European flounder have shown that artificial transmission routes like injection and wound treatment with purified virus material led to emaciation, mortality and less pronounced disease signs than in specimens infected via natural routes like the gills (Anders, 1984).

The mode of lymphocystis virus transmission is relatively well understood. It is known that "horizontal" transmission happens via respiratory organs, digestive tract or natural and artificial skin wounds between specimens of same or different fish species. Transmission via fish ectoparasites seems likely but so far no direct proofs could be given. Nothing is known about whether lymphocystis virus can be transmitted "vertically" from one generation of fish to the next.

Several infection experiments have shown a more or less pronounced host specifity. In several cases transmissions between specimens of the same fish species and between different fish species within a genus or a family were successful. The only two reports of successful transmissions between fish species of different orders are doubtful since only very few specimens could be infected.

The virus is probably spread within the organism via the blood. Nothing is known definitely about the chronology of organ infection with lymphocystis virus. However, external tissues like the skin, fins and gill lamellae are more frequently afflicted than internal organs such as spleen, heart, kidney, liver, and ovaries.

The histological structures and cytological changes of developing lymphocystis cells are unique. Each single lymphocystis nodule represents one enormously enlarged connective tissue cell which has an unmistakable histological structure. Characteristic are a thickened hyaline capsule (Fig. 1B) which is PAS-positive and therefore easy to recognize. This capsule itself is surrounded by a layer of collagenous tissue. Other striking cell components are the intracytoplasmic inclusion bodies which totally fill out the cytoplasm of a mature cell (Fig. 1C) and stain strongly Feulgen-positive. These are the sites of virus assembly. Electron microscopic autoradiography experiments have confirmed that (Spitzer et al., 1982). By transmission electron microscopy large amounts of cytoplasmic virus particles mainly with hexagonal profiles can be seen lying inbetween and around the inclusion material (Fig. 1D) and sometimes arranged in crystalloid arrays. Furthermore the nucleus of an infected cell highly enlarges as well and develops a convolutive surface structure. It may contain one to several nucleoli. Virus particles are probably released into the environment by rupture of mature lymphocystis cells whilst most probably being still a part of the fish. Neither lytic processes nor budding of virus particles from cells have been observed. Inclusion bodies and nuclei eventually disappear from degenerated cells and the lymphocystis cell is replaced by connective tissue. The host tissue response generally is minimal and restricted to a congestion of blood vessels and fibroblast proliferation (Russell, 1974). Almost nothing is known about the immunologic responses of lymphocystis infected fish.

The cytological changes of developing lymphocystis cells are in accordance with the propagation of virus particles (Dunbar and Wolf, 1966, Midlige and Malsberger, 1968, Wolf and Carlson, 1965) and with the localization of

DNA (inclusion bodies), RNA and proteins (Pritchard and Malsberger, 1968). The following characteristics of cell development were gained from experimentally at 25°C infected bluegill sunfish (Lepomis macrochirus) (Fig. 4). The ecliptic stage 3 to 5 days post infection (p.i.) was characterized by decrease of virus titer and increase of size as well as basophilic reaction of the infected cells. Intracytoplasmic inclusion bodies could first be demonstrated on day 5 p.i. They were especially localized in perinuclear regions. 6 days p.i. the virus titer reached its original value. The highest propagation rates of lymphocystis disease virus (LDV) from 6 to 12

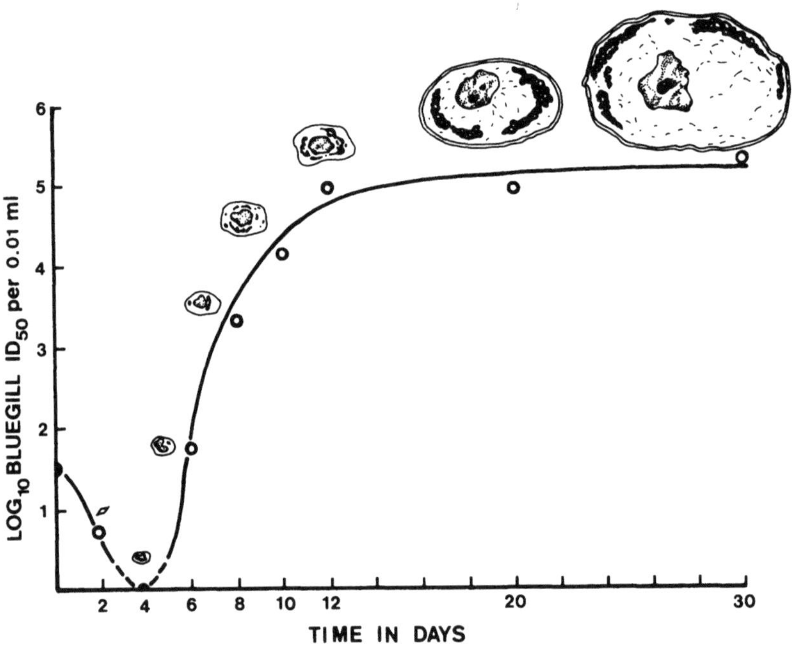

Fig. 4: Lymphocystis cell development in relation to multiplication of the virus in experimentally infected bluegill sunfish (redrawn after Wolf and Carlson, 1965).

days p.i. were accompanied by enlargement, differentiation, and localization of inclusion bodies at the cell periphery. The virus titer reached its maximum value about 2 weeks p.i. No further increase could be observed in the following two weeks although cell enlargement continued to "maturity". The inclusion bodies did not show any changes at this time. About one month p.i. pronounced degenerative changes of lymphocystis cells were noted. During the whole developmental cycle most of the RNA was concentrated in perinuclear regions while protein concentrations were associated with the development of the inclusion bodies and the formation of the hyaline capsule.

According to its properties lymphocystis disease virus (LDV) belongs to the Iridoviridae where it represents the only member of the "lymphocystis disease virus group" (Matthews, 1982).

It was first isolated from bluegill sunfish in 1966 by Wolf and co-workers. Since then River's postulates could be fulfilled for other fish species as well and there is no doubt that the iridovirus is the causative agent.

A typical LDV virion seen by electron microscopy is characterized by an electron-dense concentric core, a more or less prominent electron-translucent intermediate region, and a complex inner membrane, all of which are included in an icosahedral capsid. Globular subunits and external filaments were observed at the surface of the particles (Fig. 5). Virions are clearly visible as tiny granules even under the light microscope at high magnifications.

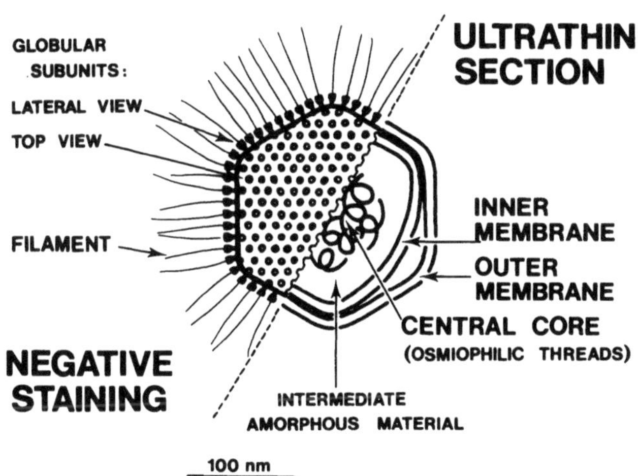

Fig. 5: Possible ultrastructure of lymphocystis virus (redrawn after Berthiaume et al., 1984).

Sizes of virus particles which are reported in the literature vary considerably. Many authors claim that these differences might be species-specific.

However, it seems evident that these differences are at least partly due to different methods of measurements and calibration as well as to variations in tissue processing. To emphasize the latter problem, Zwillenberg and Wolf (1968) showed a marked difference in size of lymphocystis virus particles from bluegill sunfish when embedded and sectioned (250 nm) in comparison to non-embedded negatively-stained tissue (300 nm). Further support to the species-specifity of particle sizes was given by Spitzer et al. (1982) who found significant differences between particle sizes from different fish species which were processed and measured under the same conditions.

Molecular biological studies of the viral DNA genome of three flatfish species from the North Sea identified two different types of LDV: LDV-1 is usually associated with European flounder and European plaice (Pleuronectes platessa), whereas LDV-2 is found in dab (Darai et al., 1983). These differences might reflect the phylogenetic relationship between the three host species (Anders and Darai, 1985) and indicate that there might be distinct differences in genome structure of lymphocystis viruses or virus strains from other fish species as well.

Further studies will have to show whether differences in genome structure are responsible for the ability of the virus or the virus strains to "conquer" the world with its different kinds of aquatic environments.

The biochemistry and molecular biology of LDV from bluegill sunfish and European flounder has been studied thoroughly in recent years by several French, Canadian, and German working groups (Darai et al., 1983, 1985, Flügel et al., 1982, Robin and Berthiaume, 1981, Robin et al., 1983, 1984, 1986, Wagner et al., 1985). A review on the properties of LDV has been prepared by Flügel (1985).

Virtually all characterization work has been done on virus in fish tissue or after direct extraction and purification of virus material. The reason is that there are striking difficulties of growing LDV in tissue culture. As an example, infective bluegill LDV was obtained at maximum yields only in bluegill cell cultures at low passage numbers with all the virus remaining cell-associated (Walker and Hill, 1980). Fortunately, mature lymphocystis cells produce such large amounts of virus particles so that sufficient quantities of virus are gained simply by direct extraction. Recently, successful cloning of the genome of LDV from European flounder led to further characterization of the DNA on the molecular level (Darai et al., 1985).

So far, no serious attempts have been made to develop control methods for lymphocystis disease. Presently available recommendations are restricted to single advices such as keeping fish in good nutritional condition, separating diseased fish from the stock and thoroughly controlling newly introduced fish species. However, the inclusion of new, highly-priced percid species in aquaculture systems, as progressively done in some tropical countries, may lead at least to problems in marketing lymphocystis diseased fish. Therefore, in future there might be call for more sophisticated and successful treatments. The same demand may arise from tropical pet fish culture which is thought to increase in importance due to future restrictions to exportation of wild catches.

Preliminary experimental vaccination studies have shown that susceptible fish can be readily protected but there is little likelyhood of commercial vaccine development in the absence of simple cultivation methods for the virus.

References

Alexandrowicz JS (1951) Lymphocystis tumours in the red mullet (Mullus surmuletus L.). J Mar Biol Assoc UK 30:315-332
Alpers CE, McCain BB, Myers MS, Wellings SR (1977) Lymphocystis disease in yellowfin sole (Limanda aspera) in the Bering Sea. J Fish Res Bd Can 34:611-616
Amin OM (1979) Lymphocystis disease in Wisconsin fishes. J Fish Dis 2:207-217
Anders K (1984) Die Lymphocystis-Krankheit der Fische. Möller Kiel 102pp
Anders K (1988) Biologie von Tumor- und tumorähnlichen Krankheiten der Elbfische. Möller Kiel 173pp
Anders K, Darai G (1985) Genome analysis of fish lymphocystis disease virus. In: Ellis AE (ed) Fish and shellfish pathology. Academic Press London:301-306
Anderson JIW, Conroy DA (1968) The significance of disease in preliminary attempts to raise flatfish and salmonids in sea water. Bull Off int Epizoot 69:1129-1137
Aneer G, Ljungberg O (1976) Lymphocystis disease in Baltic herring (Clupea harengus var. membras L.). J Fish Biol 8:345-350
Bangham RV, Hunter GW (1939) Studies on fish parasites of Lake Erie. Distribution studies. Zoologica 24:385-448
Benisch J (1937) Über das Auftreten der Lymphocystis-Krankheit bei einigen Korallenfischarten. Wschr Aquarien Terrarienkunde 34:380-382
Berthiaume L, Alain R, Robin J (1984) Morphology and ultrastructure of lymphocystis disease virus, a fish iridovirus, grown in tissue culture. Virology 135:10-19
Christmas JY, Howse HD (1970) The occurrence of lymphocystis in Micropogon undulatus and Cynoscion arenarius from Mississippi estuaries. Gulf Res Rep 3:131-154
Cook DW (1972) Experimental infection studies with lymphocystis virus from Atlantic croaker. Proc 3rd Ann Workshop World Maricult Soc. St. Petersburg Florida:329-335
Crites JL, Snyder FL (1986) Lymphocystis and other virus-related skin diseases in walleye. Ohio Sea Grant Progr, Fact Sheet 9. Ohio State Univ Columbus (USA)
Darai G, Anders K, Koch HG, Delius H, Gelderblom H, Samalecos C, Flügel RM (1983) Analysis of the genome of fish lymphocystis disease virus isolated directly from epidermal tumours of Pleuronectes. Virology 126:466-479
Darai G, Delius H, Clarke J, Apfel H, Schnitzler P, Flügel RM (1985) Molecular cloning and physical mapping of the genome of fish lymphocystis disease virus. Virology 146:292-301
Dunbar CE, Wolf K (1966) The cytological course of experimental lymphocystis in the bluegill. J Infect Dis 116:466-472
Durham PJK, Anderson CD (1981) Lymphocystis disease in imported tropical fish. N Z vet J 29:88-91
Egusa S (1983) Disease problems in Japanese yellowtail, Seriola quinqueradiata, culture: a review. Rapp P v Réun Cons int Explor Mer 182: 10-18
Fiedler M (in prep) Die Bedeutung von Zoobenthos und Zooplankton der Unterelbe als Fischnahrung. PhD thesis University of Kiel (West-Germany)
Flügel RM (1985) Lymphocystis disease virus. In: Willis DB (ed) Iridoviridae. Curr Top Microbiol Immunol 116:133-150

Flügel RM, Darai G, Gelderblom H (1982) Viral proteins and adenosine triphosphate phosphohydrolase activity of fish lymphocystis disease virus. Virology 122:48-55
Geus A (1963) Über das Auftreten der Lymphocystis-Krankheit bei Trichopsis pumilus (Arnold). Aqu Terr Zeitschrift 16:315-316
Giavenni R (1981) Considerazioni sulle pin diffuse forme morbose riscontrabili a carico dei pesci ornamentali. 1°-Pesci tropicali di acqua dolce. Riv Ital Piscicol Ittiopatol 16:130-140
Harshbarger JC (1972) Work of the registry of tumours in lower animals with emphasis on fish neoplasms. Symp zool Soc Lond 30:285-303
Harshbarger JC (1973) Activities report. Registry of Tumors in Lower Animals. Smithsonian Institution Washington DC, 141pp
Harshbarger JC (1977) Activities report. Registry of Tumors in Lower Animals: 1976 supplement. Smithsonian Institution Washington DC 42pp
Howse HD (1972) Snook (Centropomus: Centropomidae): new host for lymphocystis, including observations on the ultrastructure of the virus. Am Midl Nat 88:476-479
Huizinga HW (1977) Your fishes health: lymphocystis disease in the green terror. Trop Fish Hobbyist 25:47-60
Johnstone J (1907) Internal parasites and disease conditions of fishes. Proc Trans Liverpool Biol Soc 21:270-303
Jones JB, Hine PM (1983) Ergasilus rotundicorpus n. sp. (Copepoda: Ergasilidae) from Siganus guttatus (Bloch) in the Philippines. Syst Parasit 5:241-244
Joseph H (1917) Über Lymphocystis, einen fraglichen protozoischen Parasiten. Verh K K Zool Bot Ges Wien 67:64
Kingsford E (1975) Treatment of exotic marine fish diseases. Palmetto Publish Comp St. Petersburg Florida, 90pp
Lawler AR, Howse DR, Cook DW (1974) Silver perch, Bairdiella chrysura: new host for lymphocystis. Copeia 1:266-269
Lawler AR, Ogle JT, Donnes C (1977) Dascyllus spp.: new hosts for lymphocystis and a list of recent hosts. J Wildl Dis 13:307-312
Lawler AR, Ogle JT, Donnes C (1978) New hosts for lymphocystis. Gulf Res Rep 6:183-184
Leibovitz L (1980) Lymphocystis disease. J Am Vet Assoc 176:202
Leong TS (pers comm 1987) Universiti Sains Malaysia Penang
Lopez DM, Sigel MM, Beasley AR, Dietrich LS (1969) Biochemical and morphologic studies of lymphocystis disease. Natl Canc Inst Monogr 31:223-236
Lowe J (1874) Fauna and flora of Norfolk. Part IV. Trans Norfolk Norwich Natural Soc, Fishes Vol 1:21-56
Matsusato T (1975) On the lymphocystis disease in cultured yellowtail. Fish Pathol 10:90-93
Matthews REF (1982) Classification and nomenclature of viruses. Fourth Report of the International Committee on Taxonomy of Viruses. Intervirology 17:4-199
Mavor JW, Feinberg SM (1918) Lymphocystis vitrel. A new sporozoan from the pike-perch, Stizostedion vitreum, Mitchill. Trans Wisc Acad Sci, Arts Lett 19:559-561
McCain BB, Myers MS, Varanasi U, Brown DW, Rhodes LD, Gronlund WD, Elliott DG, Palsson WA, Hodgins HO, Malins DC (1982) Pathology of two species of flatfish from urban estuaries in Puget Sound. Interagency Energy/Environment, R&D Program Report. Seattle USA
McCosker JE (1969) A behavioral correlate for the passage of lymphocystis disease in three blennioid fishes. Copeia 3:636-637

McCosker JE, Lagios MD, Tucker T (1976) Ultrastructure of lymphocystis virus in the quillback rockfish (Sebastes maliger) with records of infection in other aquariumheld fishes. Trans Am Fish Soc 105:333-337
McCosker JE, Nigrelli RF (1971) New records of lymphocystis disease in four eastern Pacific fish species. J Fish Res Bd Can 28:1809-1810
McIntosh WC (1885) Multiple tumours in plaice and common flounders. 3rd Ann Rept Fish Bd Scotl 1884:66-67
Midlige FH, Malsberger RG (1968) In vitro morphology and maturation of lymphocystis virus. J Virol 2:830-835
Miyazaki T, Egusa S (1972) Lymphocystis disease in Japan sea bass (Lateolabrax japonicus (Cuvier and Valenciennes)) (In Japanese) Fish Pathol 6:83-89
Möller H (1981) Fish diseases in German and Danish coastal waters in summer 1980. Meeresforsch 29:1-16
Möller H (1984) Dynamics of fish diseases in the lower Elbe River. Helgol Meeresunters 37:389-413
Möller H (1988) Fischbestände und Fischkrankheiten in der Unterelbe 1984-1986. Möller Kiel 344pp
Möller H, Anders K (1986) Diseases and parasites of marine fishes. Möller Kiel 365pp
Murchelano RA, Bridges DW (1976) Lymphocystis disease in the winter flounder, Pseudopleuronectes americanus. J Wildl Dis 12:101-103
Nieschulz O (1921) Die Lymphocystis-Erkrankung der Fische. Fischerbote 13:105-107
Nigrelli RF (1943) Causes of disease and death of fishes in captivity. Zoologica 28:203-216
Nigrelli RF (1952) Virus and tumors in fishes. Ann NY Acad Sci 54:1076 -1092
Nigrelli RF, Ruggieri GD (1965) Studies on virus diseases of fishes. Spontaneous and experimentally induced cellular hypertrophy (lymphocystis disease) in fishes of the New York Aquarium, with a report of new cases and an annotated bibliography (1874-1965). Zoologica 50:83-95
Overstreet RM (1988) Aquatic pollution problems, southeastern US coasts: histopathological indicators. Aquat Toxicol 11:213-239
Paperna I (1973) Lymphocystis in fish from East African lakes. J Wildl Dis 9:331-335
Paperna I, Sabnai I, Colorni A (1982) An outbreak of lymphocystis in Sparus aurata L. in the Gulf of Aqaba, Red Sea. J Fish Dis 5:433-437
Peters N, Kranz H (1980) On the relationship between lymphocystis disease and X-cell papillomatosis of flatfish. Int Counc Explor Sea, Spec Meet Dis Comm Import Mar Fish Shellfish, Copenhagen 1980, 28
Peters N, Stich HF, Kranz H (1980) The relationship between lymphocystis disease and X-cell papillomatosis in flatfish. In: Dawe CJ et al (eds) Phyletic approaches to cancer. Proc 11th Int Symp Princess Takamatsu Cancer Res Fund Tokyo
Porter A (1952) Report of the honorary parasitologist for 1951. Proc zool Soc Lond 122:535-536
Porter A (1953) Report of the honorary parasitologist for 1952. Proc zool Soc Lond 123:253-257
Pritchard HN, Malsberger RG (1968) A cytochemical study of lymphocystis tumor cells in vivo. J expl Zool 169:371-380
Rasin K (1928) Prispevek k pathogenesi Lymphocystis johnstonei Woodcock. II. Biol Spisy Vysoké Skoly Zvérolekarské Brno (Publ Biol Ecole Vet Brno) 7:1-14 (Biol Abstr No 29631, 1931)
Robin J, Berthiaume L (1981) Purification of lymphocystis disease virus (LDV) grown in tissue culture. Evidences for the presence of two types of viral particles. Rev Can Biol 40:323-329

Robin J, Berthiaume L, Laperrière A (1984) Etude des polypeptides de structure du virus de la maladie lymphokystique des poissons (lymphocystis disease virus) multiplié en culture cellulaire. Ann Virol (Inst Pasteur) 135 E:67-80
Robin J, Laperrière A, Berthiaume L (1986) Identification of the glycoproteins of lymphocystis disease virus (LDV) of fish. Archs Virol 87:297-305
Robin J, Larivière-Durand C, Bernard J (1983) The chemical composition of lymphocystis disease virus of fish. Revue Can Biol exp 42:173-176
Russell PH (1974) Lymphocystis in wild plaice (Pleuronectes platessa L.) and flounder (Platichthys flesus L.) in British coastal waters. A histopathological and serological study. J Fish Biol 6:771-778
Sigel MM, Beasley AR, Launer MC (1966) Cytological and cytochemical (AO) studies on lymphocystis in tissue culture. In Vitro 2:153
Smith AC, Taylor RL (1972) Preliminary results of marine pathology study off the Florida coast. Mar Technol Soc J 6:51-52
Smith FG (1970) A preliminary report on the incidence of lymphocystis disease in the fish of the Sapelo Island, Georgia, area. J Wildl Dis 6:469-471
Smith FG (1973) Lymphocystis disease in Cynoscion nothus, Cynoscion regalis and Stellifer lanceolatus from Georgia estuaries. Proc 26th Ann Conf Southeastern Assoc Game Fish, Comm :598-608
Spitzer RH, Koch EA, Reid RB, Downing SW (1982) Metabolic-morphologic characteristics of the integument of teleost fish with mature lymphocystis nodules. Cell Tissue Res 222:339-357
Stich HF, Acton AB, Dunn BP, Oishi K, Yamazaki F,Harada T, Peters G, Peters N (1977) Geographic variations in tumor prevalence among marine fish populations. Int J Cancer 20:780-791
Stickney RR, White DB (1974) Lymphocystis in tank-cultured flounder. Aquaculture 4:307-308
Tanaka M, Yoshimizu M, Kusakari M, Kimura T (1984) Lymphocystis disease in Kurosoi Sebastes schlegeli and Hirame Paralichthys olivaceus in Hokkaido, Japan. Bull Jap Soc Sci Fish 50:37-42
Templeman W (1965) Lymphocystis disease in American plaice of the Eastern Grand Bank. J Fish Res Bd Can 22:1345-1356
Thakur NK, Nasar SAK (1977) On the occurrence of lymphocystis in Anabas testudineus (Bloch). Curr Sci 46:150-151
Wagner H, Simon D, Werner E, Gelderblom H, Darai G, Flügel RM (1985) Methylation pattern of fish lymphocystis disease virus DNA. J Virol 53:1005-1007
Walker DP, Hill BJ (1980) Studies on the culture, assay of infectivity and some in vitro properties of lymphocystis virus. J gen Virol 51:385-395
Walker R (1947) Lymphocystis disease and neoplasia in fish. Anat Rec 99:559-560
Weissenberg R (1914) Über infektiöse Zellhypertrophie bei Fischen (Lymphocystiserkrankung). Sitzungsber Kgl Preuss Akad Wiss Berlin, Sitz Physik-Math Classe 1914:792-804
Weissenberg R (1931) Lymphocystiskrankheit der Fische. In: v Prowazek S, Nöller W (eds) Handbuch der pathogenen Protozoen 3. Leipzig: 1344-1380
Weissenberg R (1937) Intracellular parasitism in fish producing a gigantic growth of the infected cells. Anat Rec 70:68
Weissenberg R (1938) Studies on virus diseases of fish. I. Lymphocystis disease of the orange filefish (Aleutera schoepfii). Am J Hygiene 28: 455-462
Weissenberg R (1939) Studies on virus diseases of fish. II. Lymphocystis disease of Fundulus heteroclitus. Biol Bull 76:251-255

Weissenberg R (1945) Studies on virus diseases of fish. IV. Lymphocystis disease in Centrarchidae. Zoologica 30:169-184
Weissenberg R (1951) Four additions to the list of host fish in which lymphocystis tumours have been observed as the result of spontaneous viral infection. Anat Rec 111:581
Weissenberg R (1955) The third spontaneous case of lymphocystis virus disease of fish from the Pacific coast of North America. Anat Rec 122: 434-435
Williams EH, Bunkley-Williams L, Grizzle JM (1984) Lymphocystis from West Indian marine fishes. J Wildl Dis 20:51-52
Wolf K (1962) Experimental propagation of lymphocystis disease of fishes. Virology 18:249-256
Wolf K, Carlson CP (1965) Multiplication of the lymphocystis virus in the bluegill (Lepomis macrochirus). Ann NY Acad Sci 126:414-419
Wolf K, Gravell M, Malsberger RG (1966) Lymphocystis virus: isolation and propagation in centrarchid fish cell lines. Science 151:1004-1005
Wolthaus BG (1984) Seasonal changes in frequency of diseases in dab, Limanda limanda, from the southern North Sea. Helgol Meeresunters 37:375-387
Yasunaga N (1986) Lymphocystis. Review lecture held at a short course on viral diseases of maricultured fish in March 1986. Tokyo Nihon Suisan Hogo Kyokai, Jap Fish Health Academy (In Japanese; not to be cited)
Zwillenberg LO, Wolf K (1968) Ultrastructure of lymphocystis virus. J Virol 2:393-399

Herpesviruses of Fishes

R. P. Hedrick and T. Sano*
Aquaculture and Fisheries Program
Department of Medicine
School of Veterinary Medicine
University of California
Davis California 95616
USA

Introduction

Herpesviruses are the cause of diseases among many vertebrates including fishes. The earliest description of a fish virus disease was recorded as early as 1563 as "carp pox". The presence of typical herpesvirus particles associated with carp pox however, were not observed until 400 years later by Schubert (1966). The first herpesvirus to be isolated from fish was the agent causing channel catfish virus disease (CCVD) by Fijan et al. 1970. Several years passed before herpesviruses were first isolated from salmonid fishes (Wolf and Taylor 1975). A herpesvirus was isolated from percid fishes by Kelly et al. (1980) and from cyprinid fishes by Sano et al. (1985) and more recently from anguillid fishes (Sano et. al. 1988). Herpesviruses have been observed but not isolated from an additional six families of fishes. Considering the number and diversity of fish species (at least 20,000) the total number of herpesviruses described to date is bound to increase as virological examinations using newly developed cell lines of fishes of less economic emphasis are conducted.

The herpesviruses of fish mimic their counterparts in higher vertebrates in the diversity of diseases they can induce in the host. These range from mild subclinical infections to severe hemorrhagic syndromes accompanied by high mortalities. The oncogenic potential of certain salmonid and cyprinid herpesviruses have also been demonstrated (Kimura et al. 1981b; Sano et al. 1983; Sano et al. 1985) further emphasizing the spectrum of changes associated with this group of viruses. Unfortunately, much is yet to be understood about the basic

---

*Laboratory of Aquatic Pathology
Tokyo University of Fisheries
Tokyo
Japan

biological and molecular relationships of the viruses to these diverse disease processes.

The purpose of this review is not to cover in detail previously known information about the herpesviruses but instead to describe some of the more recent studies on the known agents and discoveries of new members in the last three years. This review is perhaps timely due to recent increases in members of this important group of fish pathogens and new publications that give some perspectives on the biological and taxonomic relationships of several of these agents. Those herpesviruses that are known from fishes are listed in Tables 1 and 2.

Host Range

Recent studies on the host range of the ictalurid herpesvirus (channel catfish virus) were recently reported. In the first report, the virus was found to replicate poorly and disappeared within 3 d following injection or waterborne exposure of European catfish, Silurus glanis (Plumb et al. 1985; Chumnongsitathum et al. 1988). The resistance of two clarid catfish, the Asian catfish (Clarias batrachus) and the African catfish (C. gariepenis) to waterborne exposures to the ictalurid herpesvirus was demonstrated as virus was recovered only from fish after the first day of exposure (Boon et al. 1988).

Two additional herpesviruses have been isolated from salmonid fishes bringing the total number of isolates to six. Hedrick et al. (1987a) reported isolations of herpesviruses from steelhead trout (Salmo gairdneri) in California, United States and Sano (1988) has recovered a virus from coho salmon (Oncorhynchus kisutch) held in seawater net pens in Japan. Although the two newer isolates were from new geographic regions or hosts, they are closely related to previously described viruses from North America by Wolf and Taylor (1975) and Japanese salmonids by Kimura et al. (1981a) respectively.

Herpesviruses were isolated by Sano et al. 1988 from Japanese (Anguilla japonica) and European (A. anguilla) eels. These agents have yet to be fully characterized but appear to be identical and replicate selectively in cell lines of anguillid origin.

A herpesvirus was isolated directly from affected epidermis of Asagi carp (Cyprinus carpio) with carp pox or papilloma by Sano et al. (1985a). Following injection of healthy 1.5 year old Asagi carp with the virus some fish died and epidermal papilloma characteristic of carp

Table 1. Herpesviruses of fishes that have been isolated in cell culture

NAME	HOST	REFERENCE
SALMONID HERPESVIRUSES		
Type I		
Herpesvirus salmonis	Rainbow Trout (*Salmo gairdneri*)	Wolf, and Taylor, 1975
Steelhead Herpesvirus	Steelhead trout (*Salmo gairdneri*)	Hedrick et al. 1987a
Type II		
NeVTA	Kokanee salmon (*Oncorhynchus nerka*)	Sano 1976
OMV	Masou salmon (*O. masou*)	Kimura et al. 1981a
YTV	Yamame salmon (*O. masou*)	Sano et al. 1983
CSTV	Coho salmon (*O. kisutch*)	Sano (1988)
ICTALURID HERPESVIRUS		
Channel Catfish Virus (CCV)	Channel Catfish (*Ictalurus punctatus*)	Fijan et al. 1970
CYPRINID HERPESVIRUS		
Herpesvirus cyprini	Fancy carp (*Cyprinus carpio*)	Sano et al. 1985a
PERCID HERPESVIRUS		
Herpesvirus vitreum	Walleye (*Stizostedion vitreum*)	Kelly et al. 1980
ANQUILLID HERPESVIRUS		
Eel herpesvirus	European eel (*Anguilla anguilla*)	Sano et al. 1988
	Japanese eel (*A. japonica*)	

Table 2. Herpesviruses of fishes observed by electron microscopy

NAME	HOST	REFERENCE
Carp pox (*Herpesvirus cyprini*)?	Common Carp (*Cyprinus carpio*)	Schubert 1966
Golden Ide Virus (*Herpesvirus cyprini*)?	Golden Ide (*Leuciscus idus*)	McAllister et al. 1985
Pacific Cod Herpesvirus	Pacific Cod (*Gadus macrocephalus*)	McArn et al. 1978
Pike Epidermal Proliferative Herpesvirus	Northern Pike (*Esox lucius*)	Yamamoto et al. 1983
Herpesvirus scophthalmi	Turbot (*Scophthalmus maximus*)	Buchanan and Madeley, 1978.
Sheatfish Herpesvirus	Sheatfish (*Silurus glanis*)	Békési et al. 1981
Smooth Dogfish Herpesvirus	Smooth Dogfish (*Mustelus canis*)	Leibovitz and Leboutiz, 1985
Smelt Papillomatous Virus	Smelt (*Osmerus eperlanus*)	Anders and Moller, 1985
Viral Epizootic Epitheliotropic Virus	Lake Trout (*Salvelinus namaycush*)	Bradley et al. 1988
Flounder Herpesvirus	Japanese Flounder (*Paralichthys olivaceus*)	Iida et al. 1988
Eel herpesvirus	European eel (*A. anguilla*)	Békési et al. 1986

pox occurred among survivors (Sano et al, 1985b). McAllister et al. (1985) observed a condition identical to carp pox in golden ide (*Leuciscus idus*) imported to the United States from Germany. Attempts to isolate the virus were unsuccessful. We have also observed classical cases of carp pox among koi carp (*C. carpio*) in California and have also been unable to isolate the causative agent (Hedrick et al. unpublished data). The condition is seasonal with the highest incidence recorded in the late spring and early summer. There is no mortality directly attributable to the condition and the skin growths eventually are sloughed.

Nine additional skin diseases associated with herpesviruses from different species of fish have been described for which no virus could be isolated (Table 2). Five of these diseases (and the accompanying observation of virions) were reported in the last three years and these are discussed below.

Similar epidermal changes to those describe for carp papilloma have been observed among spawning smelt (Osmerus eperlanus) by Anders and Möller (1985). Herpesvirus particles were observed within cells of affected tissues and the condition again appeared to be seasonal and perhaps related to hormonal changes associated with spawning. Herpesvirus-like particles have also been observed among European eels with similar signs to those in smelt as reported by Békési et al. (1986).

The first viral disease to be observed among elasmobranchs was described by Leibovitz and Lebouitz (1985). A dermatitis of the smooth dogfish shark (Mustelus canis) held in an aquarium as well as wild animals was associated with the presence of herpesviruses found in affected cells. Spontaneous regression of skin lesions was also reported.

The most recently described epidermal infections associated with a herpesvirus were described by Bradley et al. (1988) and Iida et al. (1988) from lake trout (Salvelinus namaycush) and Japanese flounder (Paralichthys olivaceus) respectively. In contrast to the other herpesviruses associated with epidermal lesions, both diseases were characterized by high mortality among affected juveniles or larvae. In the lake trout infections, termed viral epizootic epitheliotropic disease, many of the infected fish succumb to concurrent infections with fungal agents. Although no viruses were isolated from diseased fish in cell cultures, abundant nucleocapsids can be recovered directly from infected tissues and these have been used to transmit the infection to previously healthy lake trout. Further studies on the relationship of the virus to mortality are underway. Transmission using filtrates from affected flounder tissues have also been achieved with the flounder herpesvirus.

Biological and Immunological Studies

Recent studies on the herpesviruses from fish have examined the biology and immune response of the salmonid and ictalurid herpesviruses. Yet unresolved however, is an understanding as to whether latency and subsequent vertical transmission are features of infections with these agents.

Salmonid herpesviruses - The first observations of a herpesvirus (salmonid herpesvirus type 1) from salmonid fish in North America were made by Wolf and Taylor (1975) from rainbow trout in Washington state. Destruction of the affected broodstock and sanitation of the hatchery were successful in eradicating the virus as no subsequent isolations were made from the facility. After a period of 15 years of acquiescence however, herpesviruses were isolated from both anadromous and freshwater stocks of steelhead and rainbow trout in California (Hedrick et al. 1987a). Further studies by Eaton (1988) showed that adult carrier rates were up to 5.8% in one hatchery where 255 steelhead were examined. He further showed that rainbow trout and chinook salmon (Oncorhynchus tshawytscha) exposed to the virus by waterborne routes resulted in subclinical infections. Virus replication was demonstrated for periods up to 35 wks following exposure. Brown trout (Salmo trutta) and coho salmon were found to be resistant to infections by the same routes.

In a second group of experiments Eaton (1988) showed that the viruses isolated from California salmonids were closely related but distinguishable from the virus isolated by Wolf and Taylor (1975). Furthermore a comparison of the serological (Hedrick et al. 1987a) and biochemical (Eaton 1988) properties of the viruses from North American and Japanese salmonids showed they were clearly separable into two groups. Those from North America referred to as salmonid herpesvirus type 1 and those from Japan as salmonid herpesvirus type 2. In agreement with the current taxonomic trends for the herpesviruses, we suggest adoption of this format for nomenclature for future members of the family herpesviridae isolated from fishes.

Ictalurid herpesvirus - Studies examining the presence or absence of a latent stage of the virus in channel catfish (Ictalurus punctatus) were conducted by Wise et al. (1985). They found that a cloned fragment of the DNA from the ictalurid herpesvirus genome hybridized to DNA from selected adult channel catfish. Positive hybridizations were found among groups of fish with and without detectable neutralizing antibodies in their sera. These studies suggest that portions of the viral genome can exist in catfish following initial infections. Uncertainty regarding reactivation of the virus however, remains.

The humoral response of catfish to the virus was examined further by Bowser and Munson (1986) who found seasonal correlations to the levels of neutralizing antibodies present in the sera. Hedrick et al. (1987b) however, were unable to show that increased water temperature significantly influenced the level of antibodies in survivors of experimental infections. Re-exposure of adult catfish exposed to the virus as juveniles did not result in re-infection with virus nor was there any apparent effect on the levels of circulating antivirus antibody (Hedrick et al. 1987b). In those studies however, it was clearly shown that control adult fish never previously exposed to the virus could be infected and succumb to virus infections via the waterborne route. Bowser et al. (1985) were the first to detect virus in pond-reared adult channel catfish and the frequency of recovery of the virus was enhanced by injections with corticosteroids.

In a study examining the significance of circulating antibody in adult catfish Hedrick and McDowell (1987) showed that passive transfer of serum from adults with positive neutralizing activity to juveniles conferred protection to subsequent challenge with the virus. Although these studies demonstrated that neutralizing antibodies are present for extended periods following first exposure to the virus and that they can confer protection to virus challenge, yet unanswered is whether they are indicators of latently infected fish. Catfish examined by both Bowser and Munson (1986) and Hedrick et al. (1987b) were survivors of infections reared subsequently in virus-free well water yet showed positive and occasionally rising titers of neutralizing antibodies. This would suggest that antibodies are produced to endogenous sources of antigens, presumably coded by viral genes perhaps those detected in DNA-DNA hybridization studies by Wise et al. (1985).

Control

Two types of investigations have been pursued recently with hopes of controlling infections by the fish herpesviruses. Kimura et al. (1983) tested the efficacy of the acyclovir (9-[2-hydroxyethoxymethyl] guanine) with the salmonid herpesvirus type 1 and 2 and the ictalurid herpesvirus. In an examination of the DNA polymerase of the salmonid herpesvirus type 2 (OMV), Suzuki et al. (1986) found it to have an optimum activity at 25 C and to be more resistant to phosphonacetic acid and aphidicolin than the human cytomegalovirus polymerase.

Awad et al. (1988) tested the immunogenicity of the membrane proteins of the ictalurid herpesvirus and found some protection could be afforded to juvenile catfish later challenged with the virus.

Conclusions

At least 16 apparently distinct herpesviruses have been isolated or identified by electron microscopy among fishes. A majority of the agents are associated with proliferative conditions of the epidermis. A seasonal and transient effect is common but with some of the salmonid herpesviruses progressive neoplastic growths result from initial infections. Four of the viruses (NeVTA, CCV, lake trout and the flounder herpesviruses) are associated with epizootics among young fish. Evidence suggests latent infections with CCV are possible but this has yet to be proven. Further studies on the molecular biology of the agents in their hosts are warranted as the fish herpesviruses may be useful tools in understanding the diverse outcomes of infection with members of the family herpesviridae in all animals.

References

Anders, K. and H. Möller. 1985. Spawning papillomatosis of smelt, Osmerus eperlanus L., from the Elbe estuary. J. Fish Dis. 8:233-235.

Awad, M. A. and Y. J. Brady. 1988. Preliminary studies on a newly developed subunit vaccine for channel catfish virus. Abstract from the American Fisheries Society, Fish Health Section International Fish Health Conference, Vancouver, British Columbia, Canada. July 19-21. pp. 105.

Békési, L., E. Kovacs-Gayer, F. Ratz and O. Turkovics. 1981. Skin infections on the sheatfish (Silurus glanis) caused by a herpesvirus.In Proceedings of an International Seminar on Fish, Pathogens and the Environment in European Polyculture. Ed. J. Olah, K. Molnar and Z. Jeney. Fisheries Research Institute, Szarvas, Hungary, June 23-27, 1981. pp. 59-69.

Békési, L, I. Horvath, E. Kovacs-Gayer and G. Csaba. 1986. Demonstration of herpesvirus-like particles in skin lesions of the European eel (Anguilla anguilla). J. Appl. Icthyol. 4:190-192.

Boon, J. H., T. McDowell and R. P. Hedrick. 1988. Resistance of the African (Clarias gariepinus) and the Asian catfish (Clarias batrachus) to channel catfish virus. Aquaculture (in press).

Bowser, P. R. and A. D. Munson. 1986. Seasonal variation in channel catfish virus antibody titers in adult catfish. Prog. Fish. Cult. 48:198-199.

Bowser, P. R., A. D. Munson, H. H. Harboe, R. Francis-Floyd and P. R. Waterstrat. 1985. Isolation of channel catfish virus from channel catfish (Ictalurus punctatus) Rafinesque, broodstock. J. Fish Dis. 8:557-561.

Bradley, T., P. Chang, D. Medina and J. McClain. 1988. Viral epizootic epitheliotropic disease of lake trout (Salvelinus namaycush). Fish Health Section, American Fisheries Society, Newsletter 16(2):5.

Buchanan, J. S. and C. R. Madeley. Studies on Herpesvirus scophthalmi infection of turbot Scophthalmus maximus (L.) ultrastructural observations. J. Fish Dis. 1:283-295.

Chumnongsitathum, B, J. A. Plumb and V. Hilge. 1988. Histopathology, electron microscopy and isolation of channel catfish virus in experimentally infected European catfish, Silurus glanis L. J. Fish Dis. 11:351-358.

Eaton, W. D. 1988. Characteristics of the steelhead herpesvirus. PhD thesis, Microbiology Graduate Group, University of California, Davis, California. 98 pp.

Fijan, N.N., T. L. Wellborne Jr. and J. P. Naftel. 1970. An acute viral disease of channel catfish. U. S. Sport Fisheries and Wildlife Technical Paper No. 43, 11 pp.

Hedrick, R. P. and T. McDowell. 1987. Passive transfer of sera with antivirus neutralizing activity from adult channel catfish protects juveniles from channel catfish virus disease. Trans. Am. Fish. Soc. 116:277-281.

Hedrick, R. P., T. McDowell, W. D. Eaton, T. Kimura and T. Sano. 1987a. Serological relationships of five herpesviruses isolated from salmonid fishes. J. Appl. Ichthyol. 3:87-92.

Hedrick, R. P., J. M. Groff and T. McDowell. 1987b. Response of adult channel catfish to waterborne exposures to channel catfish virus. Prog. Fish Cult. 49:181-187.

Iida, Y., K. Masumura, T. Nakai, M. Sorimachi and H. Matsuda. 1988. A viral disease occurred in larvae and juveniles of Japanese flounder, Paralichthys olivaceus. Abstract from the American Fisheries Society, Fish Health Section International Fish Health Conference, Vancouver, British Columbia, Canada. July 19-21. pp. 159.

Kelly, R. K., O. Nielsen, S. C. Mitchell and T. Yamamoto. 1983. Characterization of Herpesvirus vitreum isolated from hyperplastic epidermal tissue of walleye, Stizostedion vitreum vitreum (Mitchill). J. Fish Dis. 6:249-260.

Kimura, T., M. Yoshimizu, M. Tanaka and H. Sannohe. 1981a. Studies on a new virus (OMV) from Oncorhynchus masou - I. Characteristics and pathogenicity. Fish Pathol. 15:143-147.

Kimura, T., M. Yoshimizu and M. Tanaka. 1981b. Studies on a new virus (OMV) from Oncorhynchus masou - II. Characteristics and pathogenicity. Fish Pathol. 15:149-153.

Kimura, T., S. Susuki and M. Yoshimizu. 1983. In vitro antiviral effect of 9-(2-hydroxyethoxymethyl) guanine on the fish herpesvirus, Oncorhynchus masou (OMV). Antiviral Research 3:93-101.

Leibovitz, L. and S. S. Lebouitz. 1985. A viral dermatitis of the smooth dogfish, Mustelus canis (Mitchill). J. Fish Dis. 8:273-279.

McAllister, P. E., B. C. Lidgerding, R. L. Herman, L. C. Hoyer and J. Hankins. 1985. Viral diseases of fish: First report of carp pox in golden ide (Leuciscus idus) in North America. J. Wild. Dis. 21:199-204.

McArn, G. E., B. McCain and S. R. Wellings. 1978. Skin lesions and associated virus in Pacific cod (Gadus macrocephalus) in the Bering Sea. Fed. Proc. 37:937.

Plumb, J. A., V. Hilge and E. E. Quinlan. 1985. Resistance of the European catfish (Silurus glanis) to channel catfish virus. J. Appl. Ichthyol. 2:87-89.

Sano, T. 1976. Viral diseases of cultured fishes in Japan. Fish Pathol. 10:221-226.

Sano, T. 1988. Characterization, pathogenicity and oncogenicity of herpesviruses in fish. 1988. Abstract from the American Fisheries Society, Fish Health Section International Fish Health Conference, Vancouver, British Columbia, Canada. July 19-21. pp. 157.

Sano, T., H. Fukuda, N. Okamoto and F. Kaneko. 1983. Yamame tumor virus: Lethality and oncogenicity. Bull. Jap. Soc. Sci. Fish. 49:1159-1163.

Sano, T., H. Fukuda, M. Furukawa, H. Hosoya and Y. Moriya. 1985a. A herpesvirus isolated from carp papilloma in Japan. Fish and Shellfish Pathology. Ed. A. E. Ellis. Academic Press, London. pp. 307-311.

Sano, T., H. Fukuda and M. Furukawa. 1985b. Herpesvirus cyprini: Biological and oncogenic properties. Fish Pathol. 20:381-388.

Sano, T., M. Sano and H. Fukuda. 1988. Characterization of a newly isolated herpesvirus from eel. Abstract Third International Colloquim on Pathology in Marine Aquaculture. Gloucester Point, Virginia. October 2-6, 1988. pp. 11.

Schubert, G. H. 1966. The infective agent in carp pox. Bull. Off. Int. Epizooties 65:1011-1022.

Suzuki, S., T. Kimura and M. Saneyoshi. 1986. Characteristics of DNA polymerase induced by salmon herpesvirus, Oncorhynchus masou virus. J. Gen. Virol. 67:405-408.

Wise, J. A., P. R. Bowser and J. A. Boyle. 1985. Detection of channel catfish virus in asymptomatic adult channel catfish, Ictalurus punctatus (Rafinesque). J. Fish Dis. 8:485-493.

Wolf, K. and W. G. Taylor. 1975. Salmonid viruses: A syncytium-forming agent from rainbow trout. Fish Health News 4:3.

Yamamoto, T., R. K. Kelley and O. Nielsen. 1983. Epidermal hyperplasia of northern pike (Esox lucius) associated with herpesviruses and c-type particles. Arch. Virol. 97:255-272.

Salmon Herpesvirus: OMV, *Oncorhynchus Masou* Virus

T. Kimura and M. Yoshimizu
Laboratory of Microbiology
Faculty of Fisheries
Hokkaido University
Minato 3-1-1
Hakodate 041
Japan

Introduction

Oncorhynchus masou virus (OMV) is a fish herpesvirus discovered from a population of landlocked masu salmon (*Oncorhynchus masou*) in Hokkaido, Japan. Members of the genus *Oncorhynchus* belong to the Pacific salmon and are mainly anadromous, but some small populations have been isolated and landlocked. This virus is pathogenic and more significantly oncogenic for the young masu salmon and several other salmonid fish. OMV is found to be widely distributed in the northern part of Japan.

Initially in September 1978, 60 ovary fluid samples were collected from 800 brood fish of apparently normal landlocked masu salmon during the course of a virological survey at the Otobe Salmon Hatchery, Hokkaido. Population history of the hatchery showed a pattern of low survival among the progeny fry. Cytopathic effects (CPE) were observed in 4 samples on RTG-2 cells. In the following years, until summer 1984, all fish were killed to exterminate the OMV infection, thus OMV was isolated annually from the ovary fluid and tumor tissues of masu salmon cultured in this hatchery (Yoshimizu et al., 1988b).

Biophysical and Biochemical Properties of OMV

At a near optimal incubation of 15°C, OMV shows distinctive CPE within 5 to 7 days, rounded cells followed by syncytium formation, and eventual lysis of RTG-2 and other salmonid cell lines. The cells derived from non-salmonid species are refractory (Yoshimizu et al., 1988a). The maximum titer of culture-grown virus is about 10^6 $TCID_{50}$/ml with some variation depending on the cell lines (Kimura et al., 1980a).

The OMV is heat-, ether-, and acid(pH 3)-labile and does not hemagglutinate human O-cells. It is completely inactivated by ultraviolet (UV) irradiation with 3.0×10^3 μW·sec/cm^2. In the presence of 50 μg/ml of the pyrimidine analogue, 5-iododeoxyuridine (IUdR), replication is inhibited. Replication of OMV is also inhibited by anti-herpesvirus agents such as phosphonoacetate (PA), acyclovir (ACV), (E)-5-(2-bromovinyl)-2'-deoxyuridine (BVdU), and 1-β-D-arabinofuranosylcytosine (Ara-C), which is caused by the inhibition of DNA polymerase induced by OMV (Kimura et al., 1981a, 1983b,c; Kimura et al., 1988; Suzuki et al., 1986).

Electron microscopy of infected cells reveals that the intranuclear hexagonal capsids have a diameter of 115 nm. Abundance of budding, enveloped virions, 200 x 240 nm in diameter, was also observed on the surface and inside cytoplasmic vesicles (Figs. 1 and 2). The calculated number of capsomears of negatively stained virions is 162 (Fig. 3). These features of OMV confirm the characteristics of the herpesvirus group (Matthews, 1982). Optimal growth temperature of OMV is considered to be 15°C. Viral growth was not observed at 20°C or higher, although replication was recognized at 18°C. This psychrophilic nature of OMV highly differs from channel cat fish herpesvirus (CCV, Wolf and Darlington, 1971) and herpesvirus found in amphibians. OMV is now provisionally designated as salmonid herpesvirus 2 (Roizman et al., 1981; Matthews, 1982; Roizman, 1982; Wolf, 1983).

The general properties of OMV are similar to those of *Herpesvirus salmonis*, salmonid herpesvirus 1 (Wolf et al., 1978); however, OMV differs in the virion size and in the optimal growth temperature. Furthermore, OMV is different from other known fish herpesviruses with respect to the virus-induced polypeptide patterns; 34 polypeptides appearing in OMV-infected cells are designated as virus-specific. These polypeptides possess molecular weights ranging from 19,000 to 227,000. However, *H. salmonis* induced 25 polypeptides with molecular weights ranging from 19,500 to 250,000 (Fig. 4). CCV induced 32 polypeptides (Dixon and Farber, 1980), and the polypeptide patterns of CCV are also distinct from those of OMV. Differences in the electrophoretic mi-

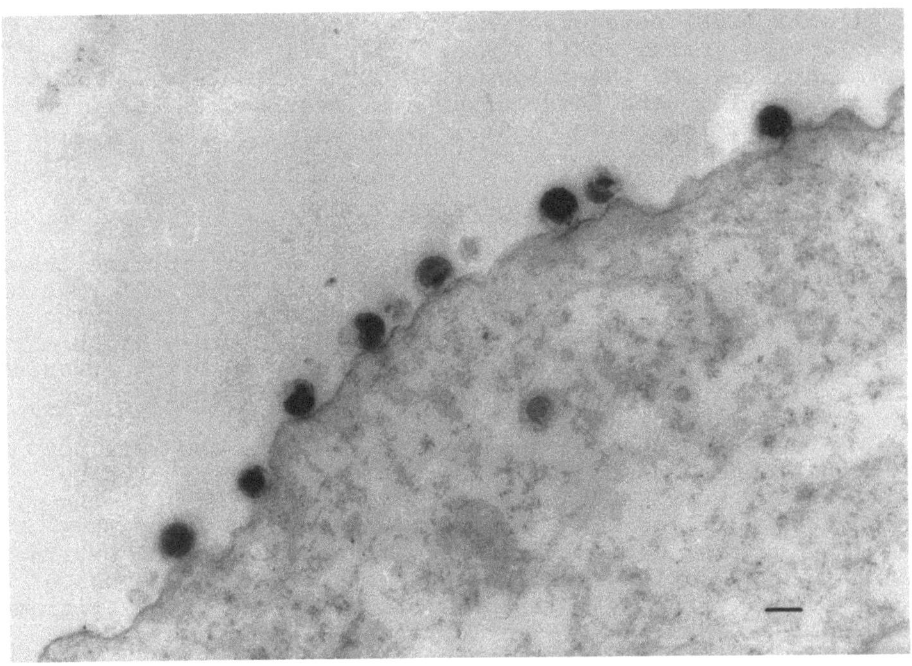

Fig. 1. Electron micrograph of enveloped virions of OMV in the process of budding from the cell membrane of an RTG-2 cell (—— 200 nm)

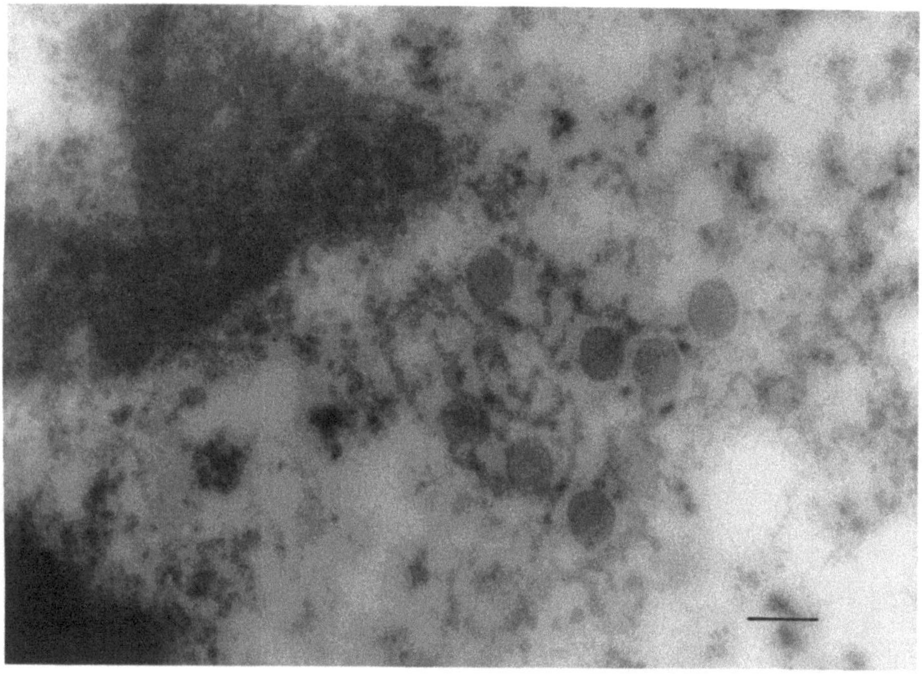

Fig. 2. Electron micrograph of intranuclear hexagonal capsids of OMV in an RTG-2 cell (—— 100 nm)

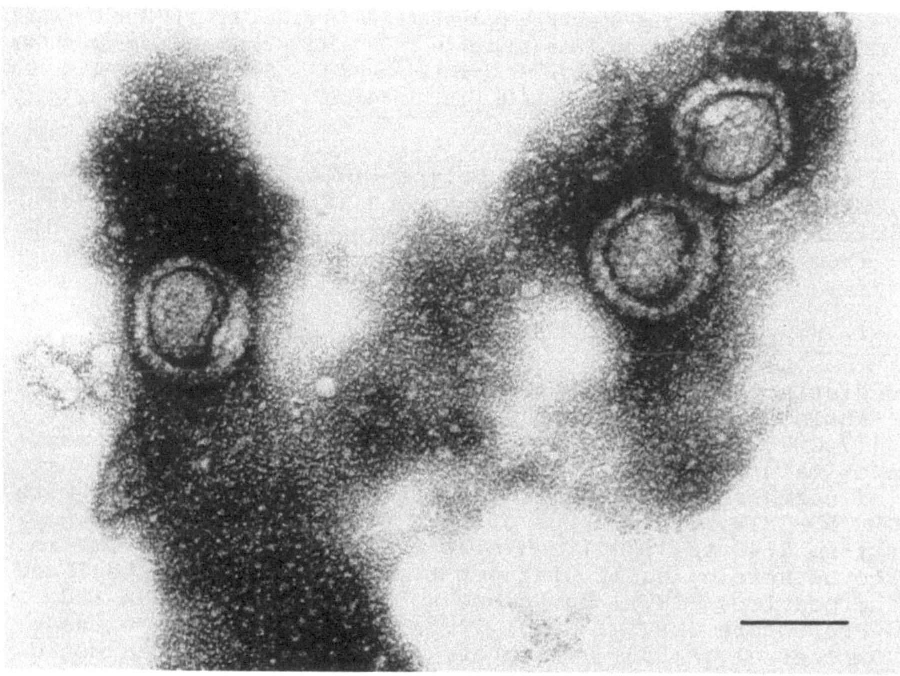

Fig. 3. Electron micrograph of negatively stained (phosphonotungstic acid) virions of OMV. (—— 100 nm)

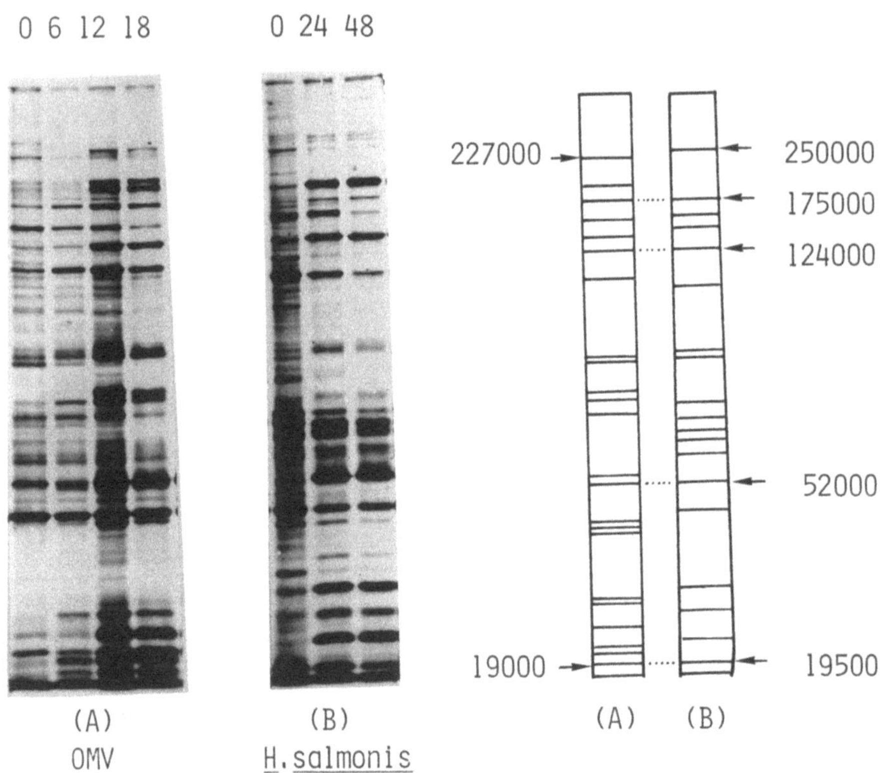

Fig. 4. Comparison of virus-specific polypeptides induced by OMV and *H. salmonis*. The numbers above each lane indicate radioactive pulse. Diagram shows the polypeptides profile of OMV from 18 to 24 h post infection in KO-6 cells (*left*) and *H. salmonis* from 48 to 72 h post infection in RTG-2 cells (*right*)

gration of 2 of 34 OMV-specific polypeptides led to the classification of 12 OMV strains into 6 groups (Kimura et al., 1984). Restriction endonuclease cleavage patterns of OMV DNAs were different from those of *H. salmonis* (Hayashi et al., 1986, 1987).

Antigenic Properties

OMV is distinctively neutralized by homologous antiserum but not by the antiserum of the other salmonid viruses. From 1978 to 1986, 177 OMV strains were isolated from ovary fluid or tumor tissue of mature masu salmon in hatcheries located in various parts of northern Japan. All these strains are neutralized with anti-OMV OO-7812, a prototype of OMV, rabbit serum; the ND_{50} ranges from 1:40 to 1:80 (Yoshimizu et al., 1988b). The OMV is found to be serologically distinct from *H. salmonis* and steelhead trout herpesvirus (SHV), but close to nerka virus Towada and Aomori Prefecture (NeVTA; Sano, 1976) and yamame tumor virus (YTV; Sano et al., 1983), which are all isolated from salmonid in Japan (Hedrick et al., 1987).

Pathogenicity

Susceptibility of several salmonid fry to OMV has been studied experimentally by immersion in water containing 100 $TCID_{50}$/ml OMV at 10°C for 1 h. Compared to the five different salmonid fry, at the age of 1 month, kokanee salmon (*O. nerka*) exhibited the greatest sensitivity and 100% of them died. Masu and chum salmon (*O. keta*) also exhibited high sensitivity at 87% and 83%, respectively. Coho salmon (*O. kisutch*) and rainbow trout (*Salmo gairdneri*) were shown to be less sensitive to OMV infection at 39% and 29% mortality, respectively. Thus the host range of OMV should be wide in salmonid species (Kimura et al., 1983a). Several age groups of chum salmon, 0, 1, 2, 3, 4, 5, 6, and 7 months old, were immersed under the same conditions. The cumulative mortality of just hatching chum salmon, observed in ensuing 4 months, was 35%, but between 1-month and 5-month-old fry, it was more than 80%. In particular, at 3 months, the fry exhibited 98% mortality, which should be the greatest sensitivity. At 6 and 7 months, the fry's susceptibility was reduced and only 7% and 2% fish had succumbed. There were no deaths among 8-month-old fingerlings which were immersed in a suspension of virus and additionally injected intraperitoneally with 200 $TCID_{50}$/fish. On the other hand, 1-month-old masu salmon fry was most sensitive and the cumulative mortality reached 87%. In 3- to 5-month-old fry, cumulative mortality decreased from 65% to 24% (Kimura et al., 1980b, 1983a).

Fish-to-fish transmission was effected by holding 5-month-old fry with fry that had been infected by immersion; the resulting mortality was similar to that observed as a result of original infection by immersion (Kimura et al., 1981a). From the evidence obtained, OMV is considered to be a virulent salmonid virus, and its influence on salmon culture should not be neglected.

Disease Signs

Affected fish are inappetent and exophthalmic, they show petechiae on the body surface, especially beneath the lower jaw (Fig. 5). Agonal or abnormal swimming behavior has not been observed. Internally, livers are noted with white spots, and in advanced cases the whole liver becomes pearly white. In some cases the spleen is found to be swollen. The digestive tract is devoid of food (Kimura et al., 1981a, 1983a).

Histopathology

Histopathologically the kidney of OMV-infected 1- and 3-month-old masu salmon, 1-month-old coho salmon and 2-month-old chum salmon is found to be the principal target organ for the virus as judged by the severity of histopathological changes found in infected 1-month-old masu salmon. Necrosis of epithelial cells of the mouth, jaw, operculum and head is also observed in the early moribund specimens while partial necrosis of the liver, spleen and pancreas is found in later moribund specimens from this group.

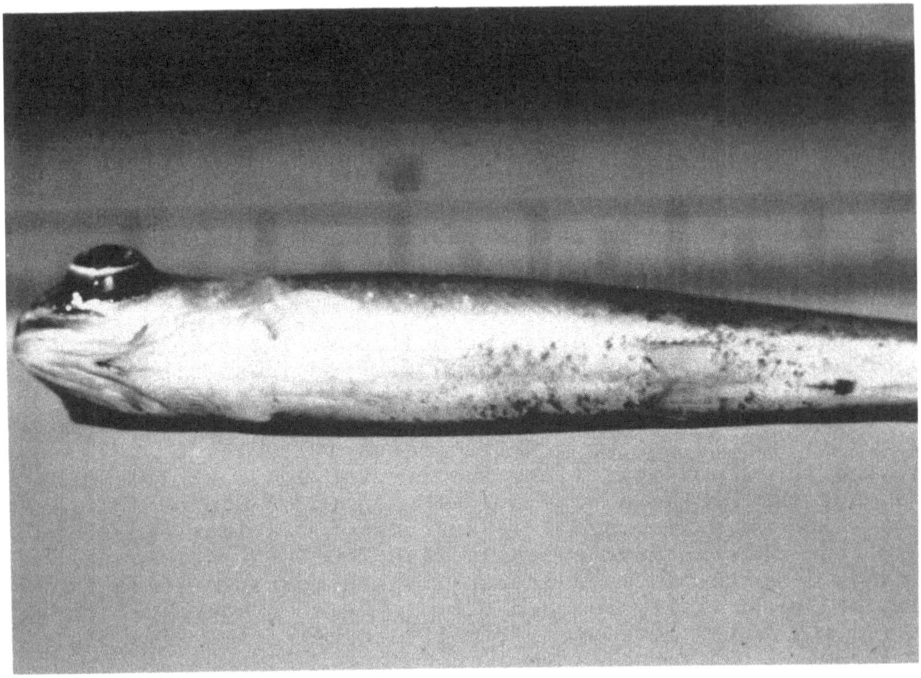

Fig. 5. OMV-infected chum salmon. Exphthalmia and petechiae are observed

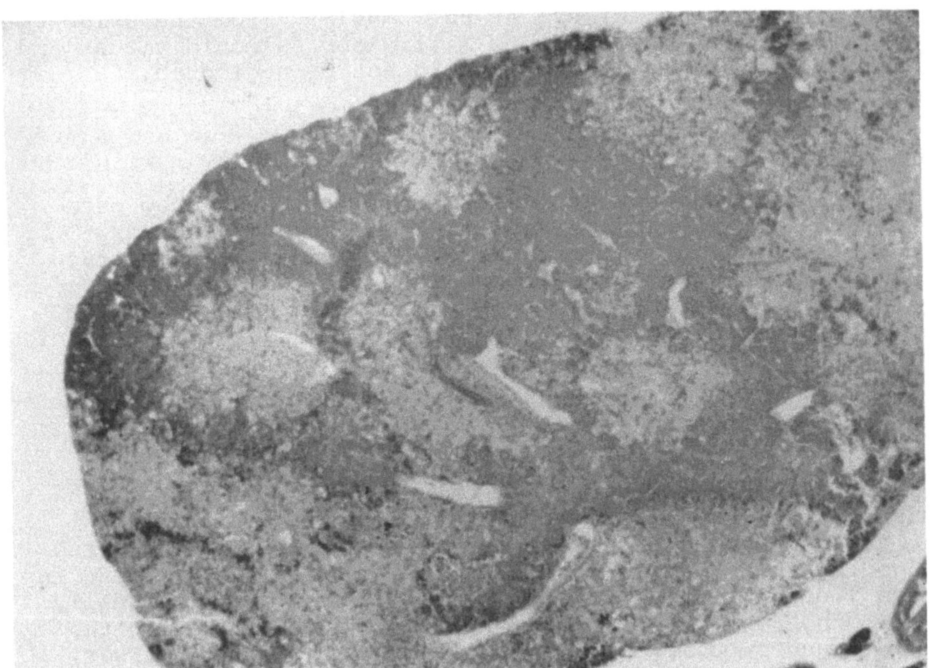

Fig. 6. Multiple foci of severe necrosis in a liver section from moribund chum salmon fry infected with OMV. Methylene blue stain

Necrosis of the kidney hematopoietic tissue has been observed in infected 3-month-old masu salmon, while hyperplasia of the hematopoietic cell was also observed simultaneously. Thus, the kidney of 3-month-old infected masu salmon gradually becomes resistant to OMV infection. From these results, the principal target organ of the virus changed from the kidney to the liver, with marked histopathological changes in the liver observed at the later stage (Fig. 6). Foci of necrosis in the liver tended to become more severe during longer incubation periods. Hepatocytes showing margination of chromatin were present. Cell degeneration in the spleen, pancreas, cardiac muscle and brain was also observed. Histopathological changes observed in coho salmon and chum salmon are the same as those of masu salmon.

In the case of 3-month-old chum salmon, marked histopathological changes were observed in liver sections. These are multiple foci of severe necrosis and syncytia formation. Partial necrosis was present in the spleen sections, and cardiac muscle was edematous in some fish. Pancreatic and kidney tissues were essentially normal, however, young fish, such as 1-month-old, showed evidence of systemic infection with necrosis in each vital organ (Kimura et al., 1981a; Tanaka et al., 1984).

Ultrastructure of Infected Tissue

Electron micrographs of heptatocytes of 80-day-old, 3-month-old, and 4-month-old chum salmon infected with OMV revealed that at first nuclei become marginated with subsequent aggregation of chromatin and syncytia. The nuclear membranes become smooth and virus replication has been observed in the nucleus of those cells. Mature virus particles were observed in the vesicle and around the cell membrane, and enveloped virus particles were seen in the focus of the hepatocyte (Fig. 7). The size of enveloped virus particles is approximately 200 x 240 nm and that of the nucleocapsid is 115 nm in diameter. Fibril strands reported in *H. salmonis* infected cells could not be found (Tanaka et al., 1987).

Tumor Induction by Clonic Infection

Beginning at about 4 months and persisting at least 1 year post-infection, 12% to 100% of surviving chum, coho and masu salmon and rainbow trout, developed tumors of differentiated epithelioma. Tumors in chum and coho salmon were first observed at 120 days post-infection, and after 200 days 40% to 60% of chum and 35% of the coho salmon are found to be affected. The rate of tumor induction was not influenced by the age of the fish at the time of infection. Tumors of rainbow trout and masu salmon were not present at 200 days post-infection but appeared after 240 and 270 days, and the rate of tumor induction reached 12% for rainbow trout and almost 100% for masu salmon after 365 days (Yoshimizu et al., 1987).

Tumors occurred mainly around the mouth (Fig. 8) and maxillary and mandibular regions; in decreasing order of frequency, tumors are also found on the caudal fin, gill-cover, body sur-

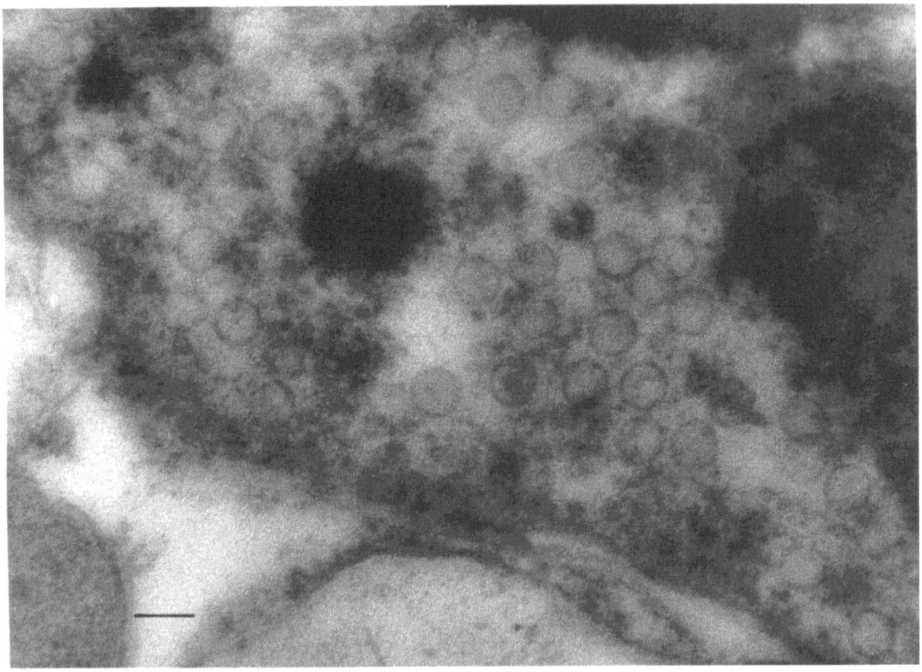

Fig. 7. Electron micrograph showing large numbers of empty capsids and nucleocapsids in cell nuclei of a hepatocyte from a moribund chum salmon fry infected with OMV (—— 100 nm)

face, corneas of the eye and in the kidney. The reisolation of OMV from tumor tissues was succeeded by the primary culture of tumor cells (Kimura et al., 1981b,c; Yoshimizu et al., 1987).

Serum samples from survivors by experimental infection were assayed with anti-OMV neutralizing activity. Sera of surviving fish, with or without tumors, showed a higher titer of neutralizing antibody than the uninfected control fish (Kimura et al., 1981b,c).

Histopathology of Tumor Tissues

The neoplasms are characterized as papillomatous consisting of abnormally proliferating epithelial cells. There are several layers of squamous epithelial cells in a papillomatous array and supported by fine connective tissue stroma. Abundant mitotic figures suggested a highly proliferative nature (Fig. 9). Tumors appearing on the caudal fin, gill-cover, body surface, corneas of the eye and kidney showed characteristics similar to those of the mouth (Kimura et al., 1981b; Yoshimizu et al., 1987).

Electron microscopy revealed that the tumor cells had a typical neoplastic feature of variability in nuclear size, and loose intracellular connection. However, OMV particles have not been found in the nuclei or the cytoplasm of the tumor cells (Kimura et al., 1981b,c; Yoshimizu et al., 1987).

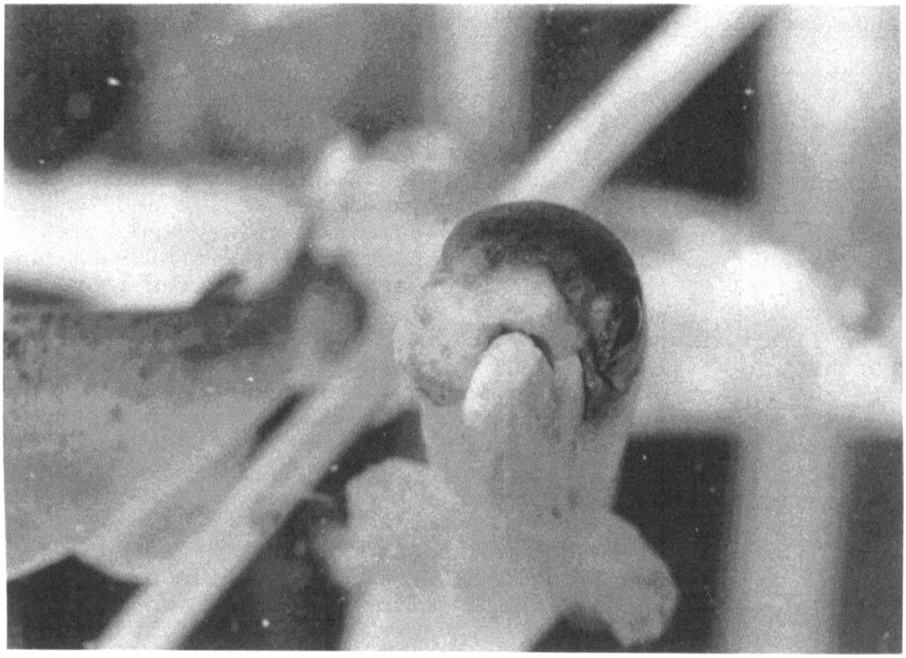

Fig. 8. Tumor developing on the jaw of chum salmon fingerling induced by clonic infection of OMV

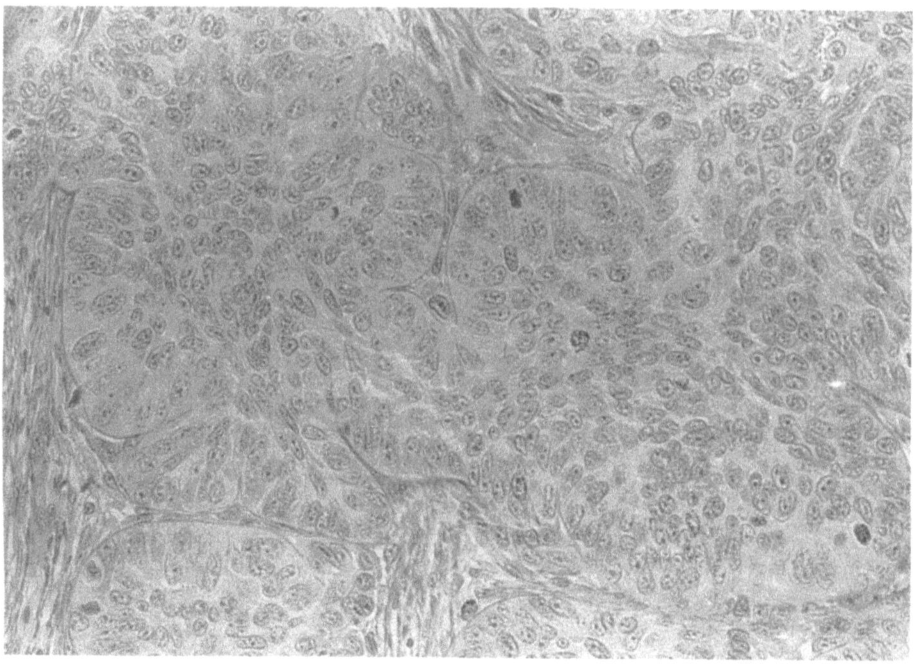

Fig. 9. Histological section of tumor tissue from the jaw of a chum salmon fingerling. High magnification of a tumor tissue section characterized abnormal proliferation of epithelial cells

Host Range and Geographic Distribution

From September 1976 to November 1986, 5058 ovarian fluid specimens from 5 species of 10,028 females and 21 seminal specimens from 2 species of 155 males of mature salmonid fish, including masu, chum, pink (*O. gorbuscha*), kokanee salmon, and rainbow trout were collected from 25 catching stations located in Hokkaido and 5 in Honshu, 6 hatcheries in Hokkaido and 5 in Honshu and 10 fish farms and 1 lake in Hokkaido. Also, 161 mixed kidney and spleen specimens taken from 805 of these fish and 139 tumor tissues observed among 4288 of these fish were employed for virus inspection.

OMV was first discovered in 1978 from masu salmon at Otobe Salmo Hatchery. At all sites permitting collection of 60 specimens with the exception of one hatchery, OMV has subsequently been isolated from ovary fluid or from epithelial tumor tissue around the mouth of mature masu salmon. It has not been found in the other species of fish. Incidence of OMV has decreased since we suggested iodophore treatment at the eyed stage (Yoshimizu et al., 1988b).

However, recently OMV was isolated from the tumor tissue of pen cultured coho salmon (Yoshimizu et al., unpublished data) and from fresh-water and sea-water cultured coho salmon (Horiuchi et al., 1988). Several hatcheries, where OMV was isolated, had a poor annual survival rate; OMV could not be isolated from the fry of those hatcheries. Although OMV is distributed widely in the northern part of Japan, no OMV isolation has been reported from outside Japan.

Diagnosis

The agent's infectivity remains unchanged for at least 2 weeks at 0°C to 5°C, but at -20°C, 99.9% of the infectivity is lost within 17 days. Virus isolation should be done using the fish, transported with ice to the laboratory. For filtration of the OMV, a 0.40 µm nucleopore filter is better than a 0.45 µm cellulose acetate membrane filter, because 99.99% of the infectivity is reduced using the latter.

For the purpose of a virological survey of mature salmonid, ovary fluid is collected by the method described by Yoshimizu et al. (1985), with the addition of the same volume of antibiotic solution (Amos, K.H., 1985) and reacted at 5°C, overnight. In the case of the tumor tissue, tissue is cut and disinfected with iodophore (50 ppm, 15 min), then washed with Hank's BSS and transported with antibiotic solution to the laboratory. Tumor tissue must be prepared for the primary culture or co-culture with RTG-2 cells. After the one transplantation of primary culture cells, the virus inspection of the culture medium should be carried out. Usually RTG-2 cells are harvested and inoculated, suitable incubation temperature is 15°C.

Antiviral Chemotherapy

In vitro, ACV and BVdU showed high efficacy against the OMV, *H. salmonis* and channel catfish virus. The cytopathic effect induced by 100 $TCID_{50}$/ml OMV in RTG-2 cells was inhibited by 1.25 µg/ml ACV and BVdU. ACV and BVdU are more effective than other compounds such as 9-β-D-arabinofuranosyladenine (Ara-A), IUdR and PA. Growth of RTG-2 cells is considerably inhibited by 25 µg/ml ACV or BVdU, but no morphological changes were observed in the cells (Kimura et al., 1983b, 1988).

In vivo, daily immersion of chum salmon into ACV solution (25 µg/ml, 30 min/day, 15 times) reduced the mortality of the infected fish. Oral administration of the drug did not affect the survival of the chum salmon. On the contrary, the group administered with IUdR by the oral route showed a higher survival than the ACV-administered group. Daily immersion of infected fish into ACV solution (25 µg/ml, 30 min/day, 60 times) considerably suppressed the development of tumors induced by OMV (Kimura et al., 1983c).

Control and Prevention of OMV Infection

The agents are sensitive to ultraviolet (UV) irradiation, ID_{99} is $1.0 - 2.0 \times 10^3$ µw·sec/cm^2 and completely inactivated by UV irradiation with 3.0×10^3 µw·sec/cm^2 (Yoshimizu et al., 1986), and also inactivated by iodophore treatment at 50 ppm, 20 min (Kimura et al., unpublished data). The contaminated eggs with OMV must be treated and disinfected with iodophore, then eggs must be incubated using the UV-treated water for supply. The fry susceptible to OMV should also be kept in UV-treated water.

References

Amos KH (ed) (1985) Procedures for the detection and identification of certain fish pathogens. Fish Health Section, AFS, 114 p

Dixson R, Farber F (1980) Channel catfish virus: Physiochemical properties of the viral genome and identification of viral polypeptides. Virol 103, 267-278

Hayashi Y, Kodama H, Ishigaki K, Mikami T, Izawa H, Sakai DK (1986) Characteristics of a new herpesviral isolate from salmonid fish. Jpn J Vet Sci 48, 915-924

Hayashi Y, Kodama H, Mikami T, Izawa H (1987) Analysis of three salmonid herpesvirus DNAs by restriction endonuclease cleavage patterns. Jpn J Vet Sci 49, 251-260

Hedrick RP, McDowell T, Eaton WD, Kimura T, Sano T (1987) Serological relationships of five herpesviruses isolated from salmonid fishes. J Ichthyol 3, 87-92

Horiuchi M, Miyazawa M, Nakata M, Iida K, Nishimura S (1988) Herpesvirus infection of cultured coho salmon in fresh water. Proceeding of annual meeting of Jap. Fish Pathology, pp 21

Kimura T, Yoshimizu M, Tanaka M (1980a) Salmonid viruses: a syncytium forming herpesvirus from landlocked *Oncorhynchus masou*. Fish Health News 9(1), iii

Kimura T, Yoshimizu M, Tanaka M (1980b) Salmonid viruses: Effect of *Oncorhynchus masou* virus (OMV) in fry of chum salmon (*Oncorhynchus keta*). Fish Health News 9(1), ii-iii

Kimura T, Yoshimizu M, Tanaka M, Sannohe H (1981a) Studies on a new virus (OMV) from *Oncorhynchus masou*-I. Characteristics and pathogenicity. Fish Pathology 15, 143-147
Kimura T, Yoshimizu M, Tanaka M (1981b) Studies on a new virus (OMV) from *Oncorhynchus masou*-II. Oncogenic nature. Fish Pathology 15, 149-153
Kimura T, Yoshimizu M, Tanaka M (1981c) Fish viruses: Tumor induction in *Oncorhynchus keta* by the herpesvirus. In: CJ Dawe et al (ed) Phyletic approaches to cancer. Japan Sci Soc Press, Tokyo, pp 59-68
Kimura T, Yoshimizu M, Tanaka M (1983a) Susceptibility of different fry stages of representative salmonid species to *Oncorhynchus masou* virus (OMV). Fish Pathology 17, 251-258
Kimura T, Suzuki S, Yoshimizu M (1983b) In vitro antiviral effect of 9-(2-hydroxyethoxymethyl)guanine on the fish herpesvirus, *Oncorhynchus masou* virus (OMV). Antiviral Research 3, 93-101
Kimura T, Suzuki S, Yoshimizu M (1983c) In vivo antiviral effect of 9-(2-hydroxyethoxymethyl)guanine on experimental infection of chum salmon (*Oncorhynchus keta*) fry with *Oncorhynchus masou* virus (OMV). Antiviral Research 3, 103-108
Kimura T, Yoshimizu M, Yano Y (1984) Comparison of virus induced polypeptides from fish herpesvirus — OMV, *H. salmonis* and CCV. In: Abstracts of sixth international congress of virology, pp 347
Kimura T, Nishizawa T, Yoshimizu M, De Clercq E (1988) Inhibitory activity of (E)-5-(2-bromovinyl)-2'-deoxyuridine on the salmonid herpesvirus, *Oncorhynchus masou* virus (OMV) and *Herpesvirus salmonis*. Microbiol Immunol 32, 57-65
Matthews REF (1982) Classification and nomenclature of viruses (Book). Karger, 199 p
Roizman B, Carmichael LE, Deinhardt F, De Nahmias AJ, Prowright W, Rapp F, Sheldrick P, Takahashi M, Wolf K (1981) Herpesviridae, definition, provisional nomenclature, and taxonomy. Intervirology 16, 201-217
Roizman B (ed) (1982) The Herpesviruses, Vol 1 (Book), Plenum Press, 445 p
Sano T (1976) Viral disease of cultured fishes in Japan. Fish Pathology 10, 223-226
Sano T, Fukuda H, Okamoto N, Kaneko F (1983) Yamame tumor virus: Lethality and oncogenicity. Bull Jap Soc Sci Fish 49, 1159-1163
Suzuki S, Kimura T, Saneyoshi M (1986) Characterization of DNA polymerase induced by salmon herpesvirus, *Oncorhynchus masou* virus. J Gen Virol 67, 405-408
Tanaka M, Yoshimizu M, Kimura T (1984) *Oncorhynchus masou* virus: Pathological changes in masu salmon (*Oncorhynchus masou*), chum salmon (*O. keta*) and coho salmon (*O. kisutch*) fry infected with OMV by immersion method. Bull Jap Soc Sci Fish 50, 431-437
Tanaka M, Yoshimizu M, Kimura T (1987) *Oncorhynchus masou* virus (OMV): Ultrastructure of OMV infected RTG-2 cell and hepatocyte of chum salmon (*O. keta*). Bull Jap Soc Sci Fish 53, 47-55
Yoshimizu M, Kimura T, Winton JR (1985) An improved technique for collecting reproductive fluid samples from salmonid fishes. Prog Fish Cult 47, 199-200
Yoshimizu M, Takizawa H, Kimura T (1986) UV susceptibility of some fish pathogenic viruses. Fish Pathology 21, 47-52
Yoshimizu M, Tanaka M, Kimura T (1987) *Oncorhynchus masou* virus (OMV): Incidence of tumor development among experimentally infected representative salmonid species. Fish Pathology 22, 7-10
Yoshimizu M, Kamei M, Dirakbusrakom S, Kimura T (1988a) Fish cell lines: Susceptibility to salmonid viruses. Invertebrate and fish tissue culture. Kuroda Y, Kurustak E, Maramorosh K (eds) Jap Sci Soc Press, Tokyo/Springer-Verlag Berlin, 207-210

Yoshimizu M, Nomura T, Awakura T, Kimura T (1988b) Incidence of fish pathogenic viruses among anadromous salmonid in northern part of Japan (1976-1986). Sci Rep Hokkaido Salmon Hatchery 42, 1-20

Wolf K, Darlington RW (1971) Channel catfish virus: A new herpesvirus of ictalurid fish. J Virol 8, 525-533

Wolf K, Darlington RW, Taylor WG, Quimby MC, Nagabayashi Y (1978) *Herpesvirus salmonis*: Characterization of a new pathogen of rainbow trout. J Virol 27, 659-666

Wolf K (1983) Biology and properties of fish and repellian herpesviruses. In: Roizman B (ed) The Herpesviruses. Vol 2, 319-366

A Herpesvirus Associated with an Epizootic Epidermal Papillomatosis in European Smelt *(Osmerus Eperlanus)*

K. Anders
Institut für Meereskunde
Düsternbrooker Weg 20
2300 Kiel 1
West-Germany

During a survey on fish stocks and fish diseases in the lower Elbe River, European smelt, Osmerus eperlanus L., was the most abundant species, as well as that with the largest variety of externally visible disease signs (Möller, 1984a). Among them was a hitherto unrecorded type of benign epidermal tumor which occurred at high prevalences in adult smelt only during and shortly before spawning season in spring. Because of its histopathology and peculiar seasonality the condition was named "spawning papillomatosis".

Spawning papillomatosis of European smelt is characterized macroscopically by smooth whitish tumors. Two types of tumors could be differentiated. The hemispherical "fin tumor" was found mainly on the fins and occasionally also on the head as well as in the buccal and nasal cavities Fig. 1A). The less frequently occurring flat "trunk tumors" were located mainly on the body surface, rarely on fins and gills as well (Fig. 1B). While fin tumors grew no bigger than 5 mm in diameter, trunk tumors sometimes covered as flat layers large areas of the body surface giving the fish a "chalky" appearance which is a typical sign for severe cases of "fish pox". This condition is a well-known papillomatosis occurring in many fish species, especially in cyprinids of Europe (Schubert, 1964, Andrews and Bucke, 1982), Japan (Sano et al., 1985a,b), and North America (Nigrelli, 1948, McAllister et al., 1985).

Histologically, fin tumors are characterized as nodular proliferations of epithelial cells which are "encapsulated" by epidermal layers (Fig. 1C). These peculiar hyperplastic growths may arise from subepidermal tissues but this remains to be proven. As the lesions develop they turn into globular forms being attached to the fish's body only by a small tissue stalk (Fig. 1D). Eventually, the epidermal capsule bursts and the softened

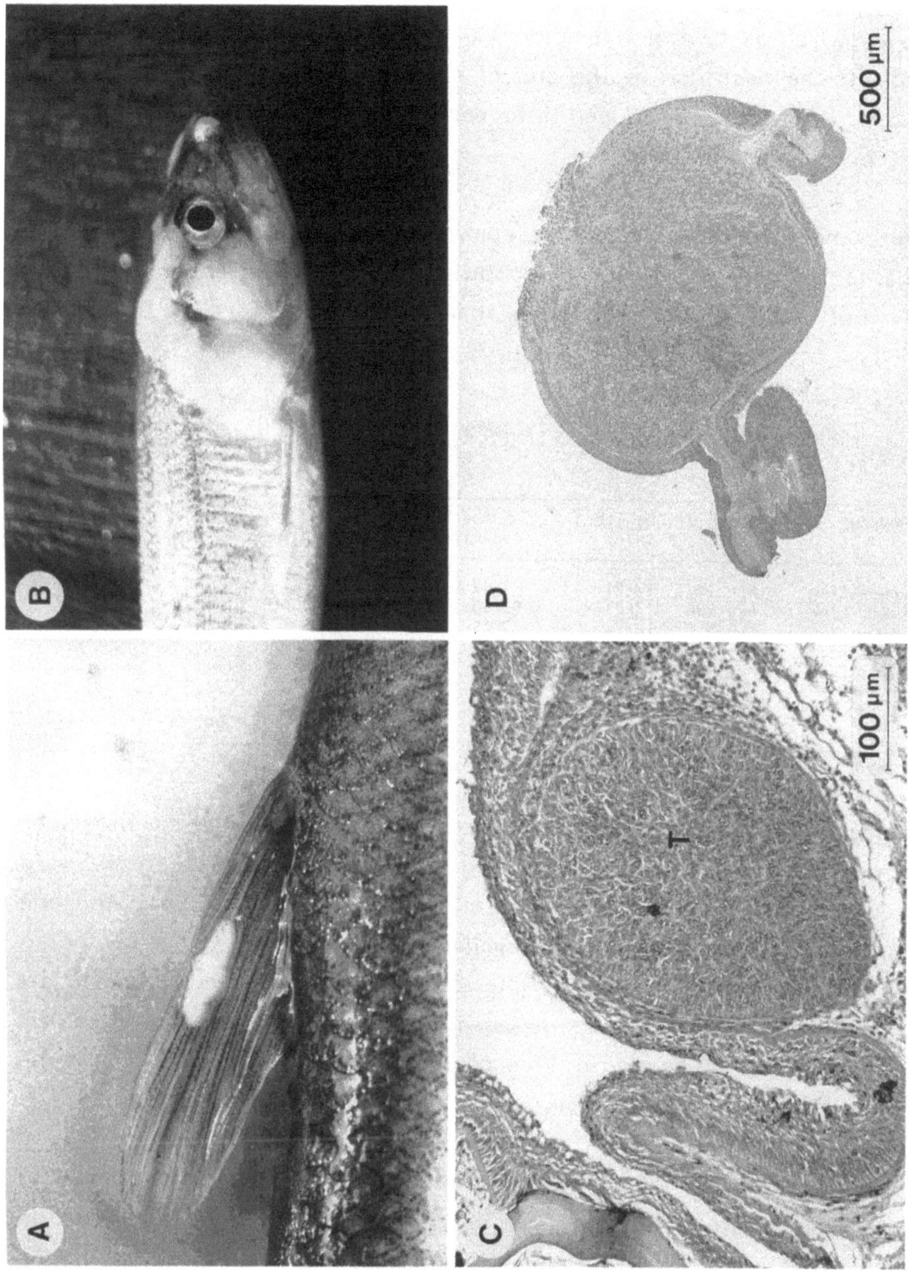

Fig. 1: Spawning papillomatosis of European smelt from the lower Elbe River. A: Typical roundish fin tumor on the dorsal fin of an adult specimen (Anders and Möller, 1985); B: Typical flat trunk tumor on the body surface of an adult specimen (Anders, 1988); C: Early stage of a fin tumor (T) "encapsulated" by epidermal layers, December 1985, H&E stain (Anders, 1988); D: "Ripe" fin tumor stage, March 1983, H&E stain (Anders, 1988).

epithelial tumor tissue is released into the environment. In contrast, trunk tumors appeared as typical flat epidermal hyperplasias and papillomas which also fall off the host fish at the end of spawning season. In one case only a benign growth had developed into early stages of malignancy (Harshbarger, pers comm, 1987).

However, the debilitating effect of spawning papillomatosis on its host is thought to be generally low, as condition factors of diseased fish were slightly but not significantly lower than those of tumor-free specimens (Tab. 1).

Tab. 1: Comparison of condition factors of tumor-bearing and tumor-free Elbe smelt.

Source	Month of capture	Length in cm	Number examined	Condition factor tumor-free	Condition factor with tumors
Lower Elbe	Apr 1984	17-21	77	0.469	
Lower Elbe	Apr 1984	17-21	35		0.464
Lower Elbe	May 1984	19-21	59	0.428	
Lower Elbe	May 1984	19-21	10		0.411
Eider	Apr 1984	16-21	103	0.467	
Eider	Apr 1984	16-21	33		0.462

Spawning papillomatosis has been reported before two times. Breslauer (1916) described a frequently occurring papillomatous growth on the skin of European smelt from coastal waters of the southern Baltic Sea. And one smelt specimen with macroscopically similar lesions, caught at the beginning of this century in British coastal waters, is included in James Johnstone's collection of fish parasites and diseases which is presently placed in the Merseyside County Museums in Liverpool (England). The disease was no more recorded in the following 60 years.

Many epidemiological findings favour the assumption that an outbreak of spawning papillomatosis in European smelt is connected closely with its spawning cycle.

(1) There was a clear seasonal fluctuation in disease prevalence (Tab. 2). The first tiny small tumor stages started to appear on the fins of Elbe smelt in October. This coincides with the beginning maturation of gonads (Lillelund, 1961). Prevalences and tumor sizes reached maximum values of

Tab. 2: Seasonal fluctuations of prevalences of spawning papillomas and skin ulcerations in Elbe smelt (21-23 cm, summary of all stations).

	1984 Nov	Dec	1985 Mar	May	Jun	Jul	Aug	Sep	Oct	Nov	Dec	1986 Jan	Mar	May	Jun	Jul
n smelt investigated	292	260	440	162	18	75	45	74	244	49	183	221	375	98	65	89
% with skin ulcerations	0	0,4	0	14,8	5,6	0	0	0	0,4	0	0	0	0,3	7,1	4,6	0
% with fin tumors	20,9	22,7	39,3	3,7	0	0	0	0	27,9	20,4	33,9	33,0	40,3	10,2	0	0
% with body tumors	0	0	13,6	0	0	0	0	0	0	2,0	0	3,2	7,2	0	0	0

40% and 2.6 mm in diameter, respectively, in March when spawning was at its peak. In May, at the end of spawning season, tumors became soft and could easily be shed by handling. Tumors could never been demonstrated on smelt caught from June to September.

There was a striking parallel development of the prevalences of trunk tumors and skin ulcerations, with the latter disease occurring one month later than the tumors (Fig. 2). It is assumed that skin abrasions left behind by shedding processes of trunk tumors favour the settlement of pathogenic bacteria there which are known to cause ulcerative changes of

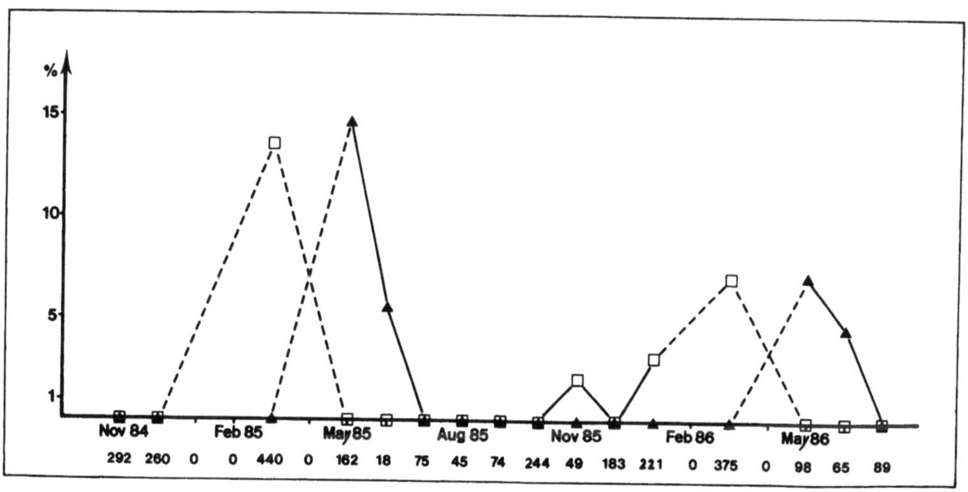

Fig. 2: Prevalences of spawning papillomas located on the body surface (squares) and skin ulcerations (triangles) in Elbe smelt (21-23 cm, summary of all stations) in relation to season.

the fish's skin. Furthermore, smelt are in weak body condition at the end of spawning (Möller, 1984b) and are therefore more susceptible to detrimental effects of pathogenic infectious agents. Similar successions of infections have been described in Baltic cod (Gadus morhua) from Danish coastal waters (Jensen et al., 1979) and in turbot (Scophthalmus maximus) from a brackishwater aquaculture facility in northern Germany (Quantz, pers comm, 1985).

(2) The occurrence of both smelt tumor types is related to fish length (Fig. 3). The smallest smelt affected measured 13 cm in total body length. This coincides with the beginning "puberty". Then the prevalence of trunk tumors is steadily increasing with increasing fish length, while that of fin tumors reaches its maximum at 22 cm and decreases thereafter. A similar tendency as for trunk tumor prevalences has been described repeatedly for epidermal papillomas of dab (Limanda limanda) from the southeastern North Sea (Dethlefsen, 1984, Mellergaard and Nielsen, 1985, Wolthaus, 1984a).

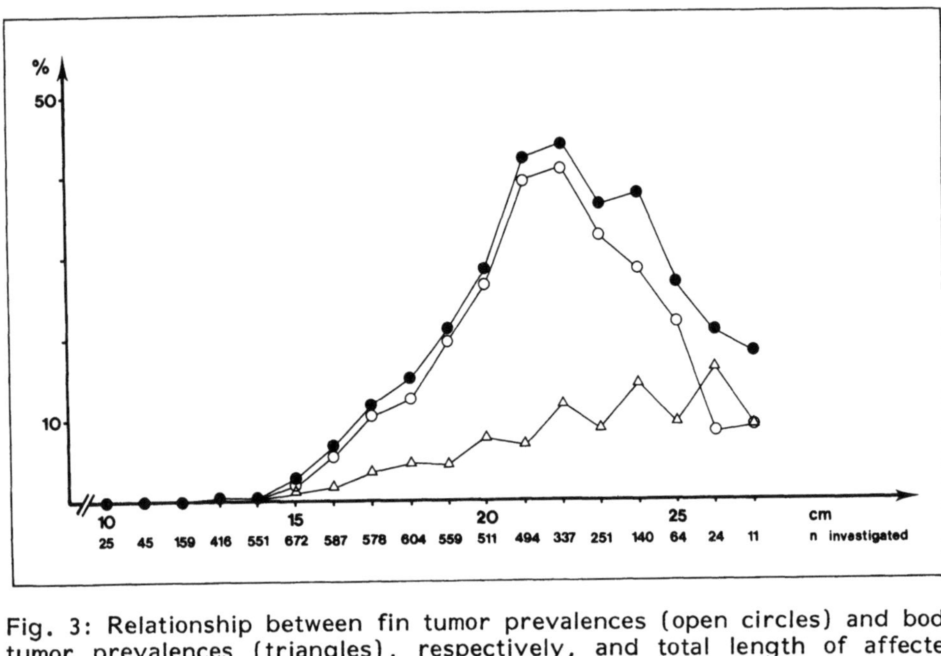

Fig. 3: Relationship between fin tumor prevalences (open circles) and body tumor prevalences (triangles), respectively, and total length of affected Elbe smelt (filled circles: total prevalence). Summary of all stations sampled in Febr-Apr 1984, March 1985 and Jan-Mar 1986. Below: number of specimens investigated.

(3) Spawning tumors of smelt have never been found on specimens which were immature.

(4) There were no differences in tumor prevalences when male and female specimens of the same length and caught at the same place/time (Elbe River/March 1985) were compared with each other.

(5) In recent years, the disease has been observed in spawning European smelt from several German tributary rivers to the North Sea (Ems, Weser, Elbe, and Eider River) and from the Schlei Fjord which is connected with the Baltic Sea. Furthermore, typical nodular fin tumors could frequently been found in adult European smelt from the Bothnian Sea near Oulu (Finland) in June 1988. Compared with German coastal waters, in this northern part of the Baltic Sea spawning season of European smelt is shifted towards the summertime and so is obviously the occurrence of spawning papillomatosis.

Interestingly, no tumors of any kind were seen in three Japanese smelt species from Hokkaido coastal and inland waters investigated at their peak spawning times in spring 1988. Altogether 87 rainbow smelt (Osmerus eperlanus mordax) (203-287 mm total body length), 111 "Chika" smelt (Hypomesus pretiosus japonicus) (113-180 mm), and 199 pond smelt (Hypomesus transpacificus nipponensis) have been carefully checked for externally visible lesions.

In summary of all these findings it seems evident that physiological changes accompanying maturation of gonads and spawning may act as releasing factors for the activation of an infectious agent.

There are convincing indications that a herpesvirus is the causative agent of spawning papillomatosis of European smelt although River's postulates could not be fulfilled until now.

Herpesvirus particles were found in nuclei, cytoplasm and intercellular spaces of both fin and trunk tumors (Fig. 4A). Virus particles in nuclei measured 85-95 nm in diameter in ultrathin sections, those in cytoplasm and intercellular spaces 110-135 nm. Variations in particle size were found to be largely due to preparation procedures. Many of the enveloped intracytoplasmic and intercellular particles had a peculiar appearance: it seemed as if they weared "tails" which gave them a comet-like outer shape.

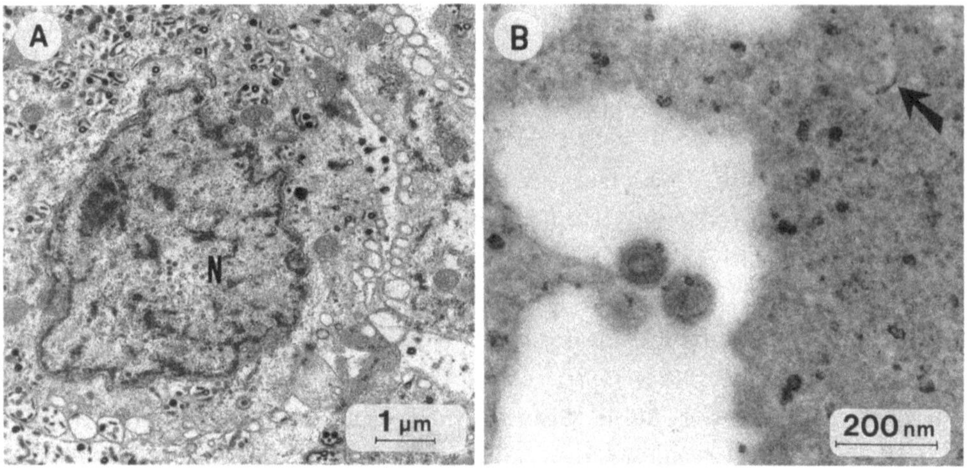

Fig. 4: Virus particles associated with spawning papillomatosis of European smelt. A: Herpesvirus particles in nuclei (N) and cytoplasm of an epithelial tumor cell from Elbe smelt; B: Retrovirus-like particles in intercellular space of tumor tissue from Elbe smelt, single suspected virus particle in the cytoplasm of a tumor cell (arrow) (Anders, 1988).

There was a clear seasonality in the mean diameter of fin tumors and the occurrence of virus particles in the tumor tissue (Tab. 3, Anders, 1988). Maximum values were found in March when the mean tumor size reached 2.6 mm in diameter and the percentage of fin tumors showing typical herpesvirus particles in TEM preparations was 100%. Intensities of occurrence of virus particles in tumor tissue varied between single to numerous,

Tab. 3: Frequency and estimated intensity of occurrence of herpesvirus particles in nuclei and cytoplasm of malpighian fin tumor cells in relation to season and tumor size (tumor development). Data are basing on transmission electron microscopical investigations. Summary of all data from November 1984 to May 1986. Data for fish/ tumor/ tissue blocks respectively.

month of catch (tumor size range)	n tumors investigated	n tumors with virus particles	% with virus particles	intensity of virus occurrence
Oct-Nov 1984 Oct-Nov 1985 (0,4-1,25 mm)	5/5/5	0/0/0	0/0/0	-
Dec-Jan 1984/85 Dec-Jan 1985/86 (1-3 mm)	8/15/16	1/4/4	13/27/25	single to frequent
Mar 1986 (1,5-5 mm)	5/5/10	5/5/9	100/100/90	single to mass occurrence
May 1985 May 1986 (2-6 mm)	6/6/8	2/2/2	33/33/25	single, extracellular only

even between neighbouring cells. At that "highest point" of tumor development in March typical viral inclusion bodies of Cowdry A type could be demonstrated in the nuclei of tumor cells.

Herpesvirus particles were not only found in tumors of smelt from German coastal waters but also in smelt from the Bothnian Sea. First results showed that the "Finnish smelt virus" had exactly the same morphology than the "German" one.

First attempts to isolate the virus in tissue culture failed so far. This is supposed to be due to the circumstance that no cell lines from European smelt were available. However, work on this aspect is promising and will be continued (Yoshimizu, pers comm, 1988).

Nothing is known about serological, biochemical and physico-chemical properties of the virus. At present, studies are under way to compare the antigenic relationship between carp pox herpesvirus and smelt herpesvirus (Hill, pers comm, 1988).

In addition to the herpesvirus particles, in approximately 10% of the fin tumors investigated by electron microscopy, numerous retrovirus-like particles could be demonstrated in intercellular spaces (Fig. 4B). Two different types of particles occurred, one measuring 88-101 nm, the other one measuring 55-76 nm in diameter. The role of these retrovirus-like particles in tumor formation and development is unclear.

Until now 11 cases have been reported where herpesviruses were found to be associated with epidermal hyperplasias and papillomas of teleost fish (Tab. 4). Also adenoviruses and retrovirus-like particles, as well as perhaps rhabdoviruses and papovaviruses seem to be involved. By reviewing the literature on epidermal tumors of fish it became evident that all retroviruses and almost all herpes- and adenoviruses found so far in fish were tumor-associated. For mammalian species it is well-known that numerous members of all these virus groups are able to cause the formation of tumors.

For fish, so far three herpesviruses were proven experimentally to be oncogenic: Oncorhynchus masou virus (OMV), Yamame tumor virus (YTV),

Tab. 4: Viruses and virus-like particles of fish associated with epidermal hyperplasias and papillomas. *: Proven oncogenicity.

Host fish species	Tumor type Hyperplasia Papilloma	Associated virus; Virus-Like-Particle	Evidence by Isolation TransmissElMi.	Reference
Anguilla anguilla	P	rhabdovirus	I	AHNE & al. (1987)
	P	papovavirus(?)	T	SAUER (pers. comm.)
	H	herpesvirus	T	BEKESI & al. (1986)
Oncorhynchus masou	P	herpesvirus OMV*	I	KIMURA & al. (1981a, b)
	P	herpesvirus YTV*	I	SANO & al. (1983)
Salmo salar	H P	VLP	T	CARLISLE (1975, 1977)
Osmerus eperlanus	H	herpesvirus	T	ANDERS & MÖLLER (1985)
	H	retro-VLP	T	own observations
Esox lucius	H	herpesvirus	T	YAMAMOTO & al. (1984)
	H	retro-VLP	T	WINQVIST & al. (1968) YAMAMOTO & al. (1984)
Cyprinus carpio	P	herpesvirus*	I	SANO & al. (1985a, b)
Leuciscus idus	H P	herpesvirus	T	McALLISTER & al. (1985)
Catostomus commersoni	P	retro-VLP	T	SONSTEGARD (1977)
Ictalurus nebulosus	P	VLP	T	EDWARDS & SAMSONOFF (1977)
Silurus glanis	H P	herpesvirus	T	BEKESI & al. (1981)
Gadus morhua	H	adenovirus	T	JENSEN & BLOCH (1980)
Merlangius merlangus	H	VLP	T	own observations
Stizostedion vitreum	H	herpesvirus	I	KELLY & al. (1983) YAMAMOTO & al. (1985b)
	H	retro-VLP	T	WALKER (1969a), YAMAMOTO & al. (1976, 1985b)
Sparus auratus	P	VLP	T	GUTIERREZ & al. (1977)
Paralichthys olivaceus	H	herpesvirus	T	IIDA & al. (1988)
	P	herpesvirus	T	MIYAZAKI & al. (1988)
Limanda limanda	H P	adenovirus	T	BLOCH & al. (1986)
Pseudopleuronectes americanus	H P	VLP	T	EMERSON & al. (1985)

and Herpesvirus cyprini (Tab. 5). In the case of OMV experimental tumors developed into malignant epitheliomas (Yoshimizu et al., 1988).

As seen from the literature in most fish species which are known to be susceptible to "fish pox"-like papillomatoses highest prevalences of epidermal tumors occurred at the respective spawning times in late winter, spring and early summer (Tab. 6). In Atlantic salmon (Salmo salar) maximum values were noted during the smoltification period when strong physiological alterations take place in fish. In European eel (Anguilla anguilla) from the Elbe River maximum rates occurred at a time when total body length reached 16-20 cm. This is the moment when the gonads start to

Tab. 5: Epidermal tumors of fish caused experimentally by infection with herpesviruses (preliminary nomenclature of viruses).

Host fish species	Tumor localisation	Oncogenic herpesvirus (hv)	References
Salmo gairdneri	mouth, fins, trunk, opercula, corneae	Salmonid hv3 (OMV)	YOSHIMIZU & al. (1987)
Oncorhynchus keta	lips, caudal fin, opercula, eyes, kidney	Salmonid hv3	KIMURA & al. (1981a, b, 1983)
Oncorhynchus keta	lips	Salmonid hv4 (YTV)	SANO & al. (1983)
Oncorhynchus keta + Oncorhynchus masou + Oncorhynchus kisutch	mouth, fins, trunk, opercula, corneae	Salmonid hv3	YOSHIMIZU & al. (1987)
Oncorhynchus masou	edges of lower jaw	Salmonid hv4	SANO & al. (1983)
Cyprinus carpio	trunk, fins	Cyprinid hv1	SANO & al. (1985a, b)

Tab. 6: Seasons and periods of fish life with maximum prevalences of epidermal hyperplasias and papillomas.

Host fish species	Season with max. prevalence Spring Summer Autumn Winter	Period of life with max. prev. Young fish Adult fish; Wild catch Aquaculture		References
Anguilla anguilla	Late-Su	Y	W	MÖLLER (1987), PETERS (1975, 1977)
Salmo salar	Late-Su	Y	WA	BYLUND & al. (1980), CARLISLE (1975)
Osmerus eperlanus	W Sp	f	W	ANDERS & MÖLLER (1985), BRESLAUER (1916)
Esox lucius	Sp	?	W	YAMAMOTO & al. (1984), BYLUND (pers. comm.)
Barbus fluviatilis	Sp Early-Su	?	W	KEYSSELITZ (1908)
Cyprinus carpio	W Sp (Su?)	f	A	HAEMPEL (1929), HINES & al. (1974)
Leuciscus idus	Sp	?	A	McALLISTER & al. (1985)
Rutilus rutilus	Sp	f	W	ANDREWS & BUCKE (1982), VALTONEN (pers. comm.)
Catostomus commersoni	?	f	W	SONSTEGARD (1977)
Ictalurus nebulosus	?	f	W	LUCKE & SCHLUMBERGER (1941)
Merlangius merlangus	W?	?	W	WATERMANN & DETHLEFSEN (1985)
Lepomis macrochirus	Sp Early-Su	?	W	NIGRELLI (1948)
Perca flavescens	W?	f	W	WALKER (1969b)
Stizostedion vitreum	W Sp	?	W	YAMAMOTO & al. (1985a, b)
Genyonemus lineatus	?	f?	W	MEARNS & SHERWOOD (1977)
Limanda limanda	Sp	f	W	DETHLEFSEN (1984), WOLTHAUS (1984a, b)

mature (Tesch, 1983). In all cases the highest tumor prevalences occurred at "critical times" in the fish's life cycle when fundamental hormonal changes take place. These changes may result in a lowering of the immune protection. Furthermore, this could mean that viruses which are already present in a host organism could then be activated to induce a transformation of normal cells into tumor cells.

For fish this is still a speculation but it is a well-proven fact for mammalian species (Nienhaus, 1985).

References

Ahne W, Schwanz-Pfitzner I, Thomsen I (1987) Serological identification of 9 viral isolates from European eels (Anguilla anguilla) with stomatopapilloma by means of neutralization tests. J appl Ichthyol 3:30-32

Anders K (1988) Biologie von Tumor- und tumorähnlichen Krankheiten der Elbfische. Möller Kiel 173pp

Anders K, Möller H (1985) Spawning papillomatosis of smelt, Osmerus eperlanus L., from the Elbe estuary. J Fish Dis 8:233-235

Andrews C, Bucke D (1982) Epidermal hyperplasia in a coarse fish population. Bull Eur Ass Fish Pathol 2:29-31

Bekesi L, Horvath I, Kovacs-Gayer E, Csaba G (1986) Demonstration of herpesvirus-like particles in skin lesions of European eel (Anguilla anguilla). J appl Ichthyol 2:190-192

Bekesi L, Kovacs-Gayer E, Ratz F, Turkovics O (1981) Skin infection of the sheatfish (Silurus glanis L.) caused by a herpesvirus. In: Olah J, Molnar K, Jeney Z (eds) Fish, pathogens, and environment in European polyculture. Fish Res Inst Szarvas 59-69

Bloch B, Mellergaard S, Nielsen E (1986) Adenovirus-like particles associated with epithelial hyperplasias in dab, Limanda limanda (L.). J Fish Dis 9:281-285

Breslauer T (1916) Zur Kenntnis der Epidermoidalgeschwülste von Kaltblütern. Histologische Veränderungen des Integuments und der Mundschleimhaut beim Stint (Osmerus eperlanus L.). Arch mikrosk Anat 87:200-263

Bylund G (pers comm 1988) Åbo Akademi Finland

Bylund G, Valtonen ET, Niemelä E (1980) Observations on epidermal papillomata in wild and cultured Atlantic salmon Salmo salar L. in Finland. J Fish Dis 3:525-528

Carlisle JC (1975) An epidermal papilloma of the Atlantic salmon (Salmo salar). A study of its epizootiology, pathology, and immunology. MSc thesis University of Stirling Scotland

Carlisle JC (1977) An epidermal papilloma of the Atlantic salmon. II: Ultrastructure and etiology. J Wildl Dis 13:235-239

Dethlefsen V (1984) Diseases in North Sea fishes. Helgol Meeresunters 37:353-374

Edwards MR, Samsonoff WA (1977) Electron microscopic observations on virus-like particles of a catfish papilloma. Proc ann Meet electronmicr Soc Am 35:394-395

Emerson CJ, Payne JF, Bal AK (1985) Evidence for the presence of a viral non-lymphocystis type disease in winter flounder, Pseudopleuronectes americanus (Walbaum), from the north-west Atlantic. J Fish Dis 8:91-102

Gutierrez M, Perez Crespo J, Arias A (1977) Particulas virus-like en un tumor en boca de dorada, Sparus aurata L. Invest Pesq 41:331-336

Haempel O (1929) Die Pockenkrankheit des Karpfens. Oesterr FischZtg 26:34-36, 55-56

Harshbarger JC (pers comm 1987) RTLA, Mus Nat Hist Smithsonian Inst Washington DC (USA)

Hill BJ (pers comm 1988) MAFF Fish Dis Lab Weymouth (England)

Hines RS, Wohlfahrt GW, Moav R, Hulata G (1974) Genetic differences in susceptibility to two diseases among strains of the common carp. Aquaculture 3:187-197

Iida Y, Masumura K, Nakai T, Sorimachi M, Matsuda H (1988) A viral disease occurred in larvae and juveniles of Japanese flounder, Paralichthys olivaceus. Abstr Int Fish Health Conf Vancouver Canada July 19-21 1988:159

Jensen NJ, Bloch B (1980) Adenovirus-like particles associated with epidermal hyperplasia in cod (Gadus morhua). Nord VetMed 32:173-175
Jensen NJ, Bloch B, Larsen JL (1979) The ulcus-syndrome in cod (Gadus morhua). III. A preliminary virological report. Nord VetMed 31:436-442
Kelly RK, Nielsen O, Mitchell SC, Yamamoto T (1983) Characterization of Herpesvirus vitreum isolated from hyperplastic epidermal tissue of walleye, Stizostedion vitreum vitreum (Mitchill). J Fish Dis 6:249-260
Keysselitz G (1908) Über ein Epithelioma der Barben. Arch Protistenk 11:326-333
Kimura T, Suzuki S, Yoshimizu M (1983) In vivo antiviral effect of 9-(2-hydroxyethoxymethyl) guanine on experimental infection of chum salmon (Oncorhynchus keta) fry with Oncorhynchus masou virus (OMV). Antivir Res 3:103-108
Kimura T, Yoshimizu M, Tanaka M (1981a) Studies on a new virus (OMV) from Oncorhynchus masou.-II. Oncogenic nature. Fish Pathol 15:149-153
Kimura T, Yoshimizu M, Tanaka M (1981b) Fish viruses: tumor induction in Oncorhynchus keta by the herpesvirus. In: Dawe CJ, Harshbarger JC, Kondo S, Sugimura T, Takayama S (eds) Phyletic approaches to cancer. Jap Sci Soc Press Tokyo 59-68
Lillelund K (1961) Untersuchungen über die Biologie und Populationsdynamik des Stintes Osmerus eperlanus (Linnaeus 1758), der Elbe. Arch FischWiss 12 Beiheft:1-128
Lucke B, Schlumberger HG (1941) Transplantable epitheliomas of the lip and mouth of catfish. I. Pathology. Transplantation to anterior chamber of eye and into cornea. J exp Med 74:397-408
McAllister PE, Lidgerding BC, Herman RL, Hoyer LC, Hankins J (1985) Viral diseases of fish: First report of carp pox in golden ide (Leuciscus idus) in North America. J Wildl Dis 21:199-204
Mearns AJ, Sherwood MJ (1977) Distribution of neoplasms and other diseases in marine fishes relative to the discharge of waste water. Ann NY Acad Sci 298:210-224
Mellergaard S, Nielsen E (1985) Fish diseases in the eastern North Sea dab (Limanda limanda) population with special reference to the epidemiology of epidermal hyperplasias/papillomas. Int Counc Explor Sea CM E:14
Miyazaki T, Fujiwara K, Kobara J, Matsumoto N (1988) Histopathological studies on two viral diseases of larvae and juvenile; epidermal necrosis of Japanese flounder and epithelial necrosis of black sea bream. Abstr Int Fish Health Conf Vancouver Canada July 19-21 1988:162
Möller H (1984a) Dynamics of fish diseases in the lower Elbe River. Helgol Meeresunters 37:389-413
Möller H (1984b) Daten zur Biologie der Elbfische. Möller Kiel 217pp
Möller H (1987) Fischbestände und Fischkrankheiten in der Unterelbe 1984-1986. Möller Kiel 344pp
Nienhaus F (1985) Viren, Mykoplasmen, Rickettsien - Parasiten an der Schwelle des Lebendigen. Ulmer Stuttgart 264pp
Nigrelli RF (1948) Hyperplastic epidermal disease in the bluegill sunfish, Lepomis macrochirus Rafinesque. Zoologica 33:133-137
Peters G (1975) Seasonal fluctuations in the incidence of epidermal papillomas of the European eel Anguilla anguilla L. J Fish Biol 7:415-422
Peters G (1977) The papillomatosis of the European eel (Anguilla anguilla L.): Analysis of seasonal fluctuations in the tumor incidence. Arch FischWiss 27:251-263
Quantz G (pers comm 1985) Fa. BUTT Strande near Kiel (West-Germany)
Sano T, Fukuda H, Furukawa M (1985a) Herpesvirus cyprini: Biological and oncogenic properties. Fish Pathol 20:381-388

Sano T, Fukuda H, Furukawa M, Hosoya H, Moriya Y (1985b) A herpesvirus isolated from carp papilloma in Japan. In: Ellis AE (ed) Fish and shellfish pathology. Academic Press London: 307-311

Sano T, Fukuda H, Okamoto N, Kaneko F (1983) Yamame tumor virus: Lethality and oncogenicity. Bull Jap Soc Sci Fish 49:1159-1163

Sauer G (pers comm 1985) German Center of Cancer Research Heidelberg

Schubert G (1964) Elektronenmikroskopische Untersuchungen zur Pockenkrankheit des Karpfens. Z Naturforsch 19b:675-682

Sonstegard RA (1977) Environmental carcinogenesis studies in fishes of the Great Lakes of North America. Ann NY Acad Sci 298:261-269

Tesch FW (1983) Der Aal: Biologie und Fischerei. Hamburg Berlin Parey 340pp

Valtonen ET (pers comm 1988) University of Jyväskylä (Finland)

Walker R (1969a) Virus associated with epidermal hyperplasia in fish. Natl Canc Inst Monogr 31:195-207

Walker R (1969b) Epidermal hyperplasia in fish: Two types without visible virus. Natl Canc Inst Monogr 31:209-213

Watermann B, Dethlefsen V (1985) Epidermal hyperplasia and dermal degenerative changes as cell damage effects in gadoid skin. Arch FischWiss 35:205-221

Winqvist G, Ljungberg O, Hellstroem B (1968) Skin tumours of northern pike (Esox lucius L.). II. Viral particles in epidermal proliferations. Bull Off int Epiz 69:1023-1031

Wolthaus BG (1984a) Populationsdynamik der Kliesche (Limanda limanda L.) und jahreszeitlicher Verlauf von Krankheitsbefallsraten mit Lymphocystis, epidermalen Papillomen, Geschwüren und Flossenfäule in der Deutschen Bucht. PhD thesis University of Bremen West-Germany 211pp

Wolthaus BG (1984b) Seasonal changes in frequency of diseases in dab, Limanda limanda, from the southern North Sea. Helgol Meeresunters 37:375-387

Yamamoto T, Kelly RK, Nielsen O (1984) Epidermal hyperplasias of northern pike (Esox lucius) associated with herpesvirus and C-type particles. Archs Virol 79:255-272

Yamamoto T, Kelly RK, Nielsen O (1985a) Epidermal hyperplasia of walleye, Stizostedion vitreum vitreum (Mitchill), associated with retrovirus-like type-C particles: prevalence, histologic and electron microscopic observations. J Fish Dis 8:425-436

Yamamoto T, Kelly RK, Nielsen O (1985b) Morphological differentiation of virus-associated skin tumors of walleye (Stizostedion vitreum vitreum). Fish Pathol 20:361-372

Yamamoto T, MacDonald RE, Gillespie DC, Kelly RK (1976) Viruses associated with lymphocystis disease and the dermal sarcoma of walleye (Stizostedion vitreum). J Fish Res Bd Can 33:2408-2419

Yoshimizu M (pers comm 1988) Fac Fish Hokkaido Univ Hakodate (Japan)

Yoshimizu M, Tanaka M, Kimura T (1987) Oncorhynchus masou virus (OMV): Incidence of tumor development among experimentally infected representative salmonid species. Fish Pathol 22:7-10

Yoshimizu M, Tanaka M, Kimura T (1988) Histopathological study of tumors induced by Oncorhynchus masou virus (OMV) infection. Fish Pathol 23:133-138

Channel Catfish Herpesvirus

J.A. Plumb
Department of Fisheries and Allied Aquacultures
and Alabama Agricultural Experiment Station
Auburn University
Alabama 36849
USA

Introduction

Herpesvirus of channel catfish causes a serious problem in channel catfish fingerlings in the Southern United States and other areas where this species is cultured. The causative agent of channel catfish virus disease (CCVD) is also known as *Herpesvirus ictaluri* and commonly called channel catfish virus (CCV) (Wolf and Darlington, 1971; Fijan, 1968). The following review of herpesvirus of channel catfish included discussion of its epizootiology, clinical signs, pathogenicity, pathology and other aspects of CCVD.

Epizootiology

Channel catfish (*Ictalurus punctatus*) is the principal species affected by herpesvirus of channel catfish. Young-of-the-year channel catfish are the most susceptible, with the disease occurring in fish from 2- to 3-week-old fry up to fingerlings 6 months old. CCV has occasionally been diagnosed in fish that were 1 year old. However, size has an effect on susceptibility as natural infections are confined to fish less than 10 g. Experimental infection can be initiated by virus injection in channel catfish up to 50 g (Plumb, 1971a).

Experimental infection was induced in fingerling blue catfish (*I. furcatus*) and in channel catfish x blue catfish hybrids by injection, but could not be induced by oral administration or by cohabitation with virus-infected channel catfish fingerlings (Plumb and Chappell, 1978); in

1987 however, CCV was isolated from diseased blue catfish fingerlings (Plumb, unpublished data). Brown bullheads (I. nebulosus) or yellow bullheads (I. natalis) could not be infected by injecting or feeding virus (Plumb, unpublished data). Plumb et al. (1985) demonstrated that the European catfish (Silurus glanis) is resistant to CCV. Feeding virus to different strains of channel catfish fry showed variation in susceptibility among the different strains (Plumb et al., 1975). Young interstrain-hybrid channel catfish were significantly (P = 0.05) more resistant to water borne exposure to CCV than pure strains (Figure 1). McGlamery and Gratzek (1974) showed that the growth of survivors of CCV epizootics may be stunted, however, in commercial production operations, survivors of CCV epizootics have been grown to a satisfactory size and marketed.

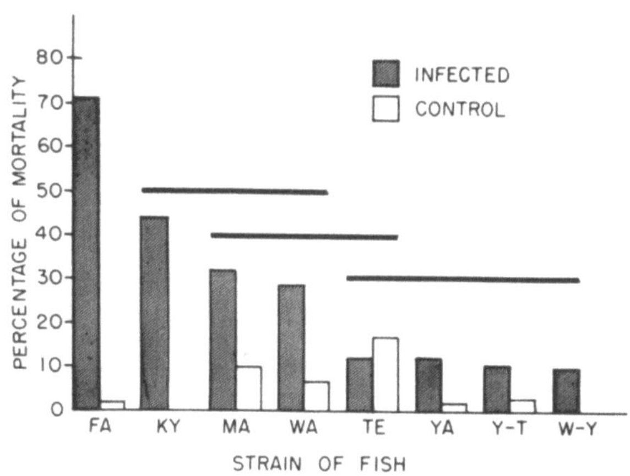

Figure 1. Mortality of different strains of channel catfish when infected with channel catfish virus at 1, 2, and 3 months of age. The strains infected were: Falcon (FA), Kentucky (KY), Marion (MA), Warrior (WA), Tennessee (TF), Yazoo (YA), Yazoo X Tennessee (Y-T) and Warrior x Yazoo (W-Y). The means of any two strains that do not fall under the same line are significantly different (p = 0.05). (From Plumb et al., 1975).

Channel catfish virus has been reported from channel catfish in most Southern States of the United States and from culture systems in most other areas of the United States where channel catfish are grown. The virus was also isolated from a group of fry shipped from the United States to Honduras, Central America, and Dr. Y. Temnihanov (personal communication) reported channel catfish exhibiting clinical signs of CCVD in Russia, and the isolation of a viral agent from these fish. Although cytopathic effect (CPE) was similar to that of CCV, virus specific antisera were not available for CCV confirmation. CCV isolation from wild channel catfish has not been reported.

Incubation period for CCV in fish is inversely related to temperature. Experimental infection at 30°C is followed by clinical signs in 32 to 72 h, and the first deaths occur several hours later (Plumb, 1973). At 20°C the incubation period is 10 days. Under field conditions at 25 to 30°C, healthy channel catfish fingerlings develop the disease 72 to 78 h after exposure, and up to 100% die within 6 days. In another instance, a group of naturally infected fry held at 28°C developed clinical signs of CCVD when 21 days old and 72 h later all were dead.

Cases of CCVD have been diagnosed from May to September (Plumb, 1971a). Epizootics on catfish farms occur more frequently during years of high water temperature and in densely stocked fingerling ponds. Although Bowser et al. (1985) isolated CCV from adult catfish during January, the virus was not considered detrimental to the fish.

Channel catfish virus has been isolated from channel catfish fry and fingerlings exhibiting clinical signs of the disease during epizootics; therefore, these fish serve as a source of infection. Adult channel catfish have been considered a source of CCV in that it is believed they can transmit the virus to their offspring via reproductive products. However, vertical transmission has not been conclusively demonstrated, although circumstantial evidence is strong. Research efforts to demonstrate vertical transmission have failed, as have numerous attempts to isolate CCV from adult fish that produced fry and fingerlings that contracted the disease (Plumb and Jezek, 1983). Plumb et al. (1981) demonstrated the presence of CCV antigen in the ovarian tissues of female channel catfish by an indirect fluorescent antibody technique, but CCV was not isolated from these tissues. Wise and Boyle (1985) and Colyer et al. (1986) utilized a radio-labeled nucleic acid probe to detect CCV genome in tissues of channel catfish, thus demonstrating that herpesvirus of channel catfish can persist in a dormant state after infection without causing clinical disease. Bird et al (In press) utilized monoclonal antibody to detect CCV mRNA in tissues of CCV injected catfish and clinically normal fingerlings from two fish farms and adult fish. Bowser et al. (1985) successfully isolated CCV from adult channel catfish kidneys during January when water temperature was 8°C. Following the initial isolation from a single fish, 21 additional broodfish from the same population were examined virologically. Seven of these fish were immunosuppressed by dexamethazone injection prior to virus sampling. Of 14 non-immunosuppressed injected fish, three were positive for CCV from either kidneys, gonads, or leukocytes cocultivated with channel catfish ovary (CCO) cells, and all others were negative, while all seven of the leukocyte cocultivations with CCO cells from the dexamethazone-injected fish were positive, but their kidneys and gonads were negative for CCV. While these studies do not prove that CCV is vertically transmitted, they conclusively demonstrate that viral latency can occur in adult fish.

Channel catfish virus is usually a primary pathogen in diseased fingerling channel catfish, although poor environmental conditions, elevated water temperature, and handling may predispose fish to disease. During the latter stages of CCV infection, saprophytic bacteria, particularly

Flexibacter columnaris or *Aeromonas hydrophila*, may also infect the fish. When this occurs, mortality may continue at a prolonged low rate.

Clinical Signs

A sudden increase in morbidity, followed by a rapidly increasing mortality, in young-of-the-year channel catfish is the first indication of CCV infection. Infected fish swim erratically or convulsively, sometimes rotating about their longitudinal axis. Victims sink to the bottom, become quiescent, respire weakly but rapidly and then die. At the time of death, the fish may hang head-up in the water (Fijan, 1968; Fijan et al., 1970), but caution must be used in depending on the "head-up" posture as a clinical sign because many abnormalities will cause small catfish to hang "head-up." Clinical signs of diseased fish vary; some or all of the following may be present: distension of the abdomen due to the accumulation of a clear straw-colored fluid in the peritoneal cavity; exophthalmia; pale or hemorrhagic gills; hemorrhagic areas at the base of fins and throughout the skin, particularly on the ventral surface (Figure 2). A general hyperemia occurs throughout the visceral cavity, although the liver and kidneys may be pale. The spleen is generally congested and enlarged, while the stomach and the intestine are devoid of food but filled with a mucoid secretion.

Pathogenesis and Pathology

Channel catfish virus disease is easily transmitted to susceptible channel catfish by water-borne exposure, swabbing the gills with the virus, cohabitation of non-infected with infected fish, feeding CCV contaminated feed, and by intraperitoneal or intramuscular injection (Fijan et al., 1970; Plumb, 1972, Nusbaum, and Grizzle, 1987).

CCV is believed to penetrate the host through the gills and other epithelial tissues. Nusbaum and Grizzle (1987), utilizing radiolabeled CCV, demonstrated that the virus does enter the fish through gill tissues and the gut and is then concentrated in the liver. Intraperitoneal injection of CCV

Figure 2. Gross clinical signs of channel catfish virus disease. Note the exophthalmia, ascites in abdomen (A) and dark hemorrhage on ventral surface.

into 5-g fish held at 28°C resulted in a generalized viremia 72 h post injection (Plumb, 1971b; Plumb and Gaines, 1975). The trunk kidney is the first organ from which CCV can be isolated, followed by liver, intestine, brain, spleen and blood. Virus titers may reach 1 x $10^{5.5}$ to 1 x $10^{7.5}$ $TCID_{50}$/0.1 g of tissue. The muscle has a low virus titer of 1 x $10^{2.5}$ $TCID_{50}$/0.1 g of tissue, and it was not positive until 168 h post infection. In contrast to this, Nusbaum and Grizzle (1987) reported high concentrations of CCV in the tail of infected catfish.

Method of shedding CCV by infected fish is not clear. However, since infection can be acquired by cohabitation, fish with acute infections do shed virus into the water. The virus does not survive well in feces, but it survives for days in channel catfish urine (Plumb, unpublished data) and this would be a likely method of shedding because the trunk kidney is a primary organ affected by the virus. Other possible shedding sites are the gills and skin. Following a CCVD epizootic, it may not be possible to isolate CCV from survivors, typical of herpesvirus infections. Presumably the virus enters a latent stage in which isolation is not possible, but as previously discussed CCV DNA does remain in catfish tissues in a nonactive state.

Histopathologic changes are similar in natural and experimental CCV infections (Wolf et al., 1972; Plumb et al. 1974; Plumb and Gaines, 1975; Major et al., 1975). Renal hematopoietic tissue is edematous and has extensive areas of necrosis and cellular dissolution coupled with an increase in macrophages. The liver develops regional edema, necrosis and hemorrhage, and hepatic cells have eosinophilic cytoplasmic inclusions. Pancreatic acinar cells are necrotic. The submucosa of the digestive tract is edematous and has dense focal areas of macrophages and hemorrhage. The spleen becomes congested with erythrocytes and this change is coupled with extensive reduction of lymphoid tissue. Virus particles have been seen in electron micrographs of the liver, kidneys, and spleen of infected fish (Figure 3) (Plumb et al., 1974).

Etiology and Virus Properties

The etiological agent of Herpesvirus of channel catfish is channel catfish virus (CCV) (Fijan, 1968; Fijan et al., 1970). Wolf and Darlington (1971) characterized the virus as herpes-like, but it is now considered a true herpesvirus (Robin and Rodrique, 1980) (Figure 3) and is often referred to as *Herpesvirus ictaluri*. The unenveloped icosahedral virion has a diameter of 95 to 105 nm, and with the envelope it measures 175 to 200 nm. Infectivity is inactivated by 20% ether and by treatment with 5% chloroform. CCV is heat-labile at 60°C for 1 h, is unstable in seawater, and is sensitive to ultraviolet light, requiring 20 to 40 min for inactivation (Robin and Rodrique, 1980). It survives for less than 24 h on dried

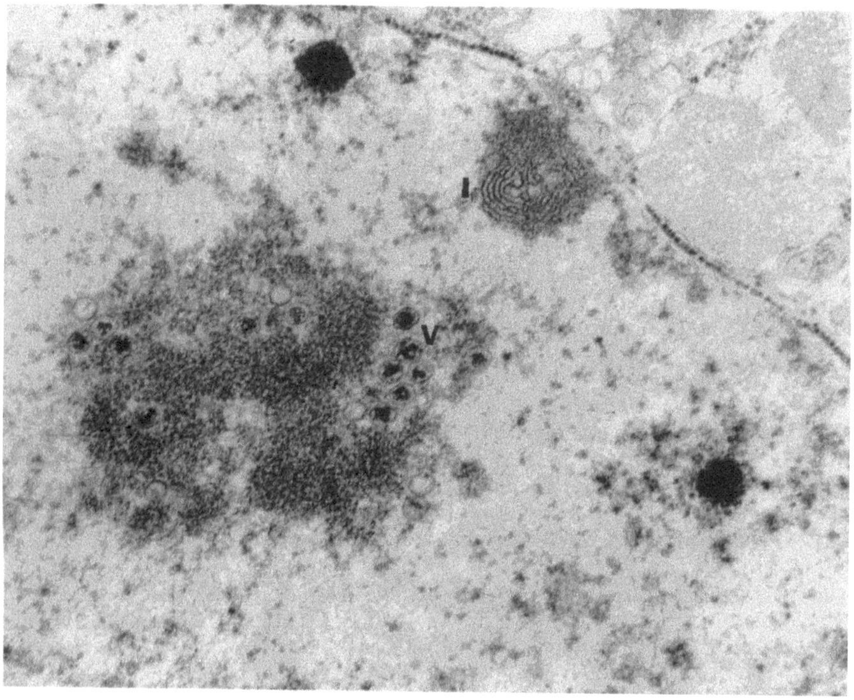

Figure 3. Intranuclear channel catfish virus (V) in a spleen cell of channel catfish. Intranuclear lamellar-like inclusions (I) are also present. X 22,000.

concrete chips and less than 48 h on dried fish netting or glass cover slips (Plumb, 1974). It retains infectivity in pond water for about 2 days at 25°C, but for 28 days at 4°C. In dechlorinated tap water, infectivity is retained for 11 days at 25°C and for over 2 months at 4°C. Under experimental conditions, infectivity is immediately lost in pond bottom mud. Infectious virus could not be isolated from decomposing fish at 22°C, 48 h after death; however, it was recoverable for up to 14 days from iced fish, for 162 days from fish frozen at -20°C, and for 210 days from fish frozen at -80°C (Plumb et al., 1973).

Immunity

The immune response of channel catfish to channel catfish virus was investigated by Plumb (1973), Heartwell (1975), Noga and Hartman (1981),

Amend and McDowell (1983; 1984) and Awad and Brady (1988). Plumb (1973) immunized 3-year-old channel catfish with viable CCV and held the fish at 24°C. The antibody titer was measured by the neutralization index at 30-day intervals and it was determined that peak titers occurred at approximately 60 days post injection. Following a booster injection, the CCV antibody response increased to 1 log higher than the primary response but declined rapidly. These observations were similar to the CCV antibody titer kinetics reported by Heartwell (1975) in channel catfish. CCV antibody in fish serum was reportedly stable at 56°C for 30 min (Plumb, unpublished data), however, Heartwell (1975) found a 20% reduction in neutralization capability of serum heated at 50°C for 30 min and suggested that 45°C incubation would be better. Injection of heat-killed CCV failed to elicit an immune response in 200-g channel catfish (Plumb, 1973).

Heartwell (1975) determined that CCV antibody activity is present in the macroglobulin fraction of serum, and the molecular weight of partially purified globulin was 800,000 to 900,000 daltons. Using an antigen concentrated to about 1 x 10^{10} PFU/ml (ether-tween extraction method) in rabbits, a precipitin reaction in 0.5% agar gells was shown. Although the reaction was faint using serum taken 2 to 3 weeks after inoculation, it was strong at 5 weeks after injection. The electrophoretic mobility of the anti-CCV macroglobulins was similar to that of human IgM and most other fish immunoglobulins.

The immune response of channel catfish to CCV has been used in attempts to identify CCV carrier fish. Plumb (1973) examined the sera from channel catfish broodstock that had produced CCV infected fingerlings for 2 consecutive years. The neutralization indices did not change when assayed three times during a 12-month period. These data led to the use of CCV antibody detection in sera to differentiate adult catfish populations that had been exposed to CCV (with CCV-specific antibodies) from those that had not (without CCV-specific antibodies). Although individual CCV antibody-positive fish could be identified, the procedure should be applied on a population basis rather than to individual fish. Amend and McDowell (1984) applied the serum neutralization test to 17 major brood channel catfish populations in California and found that 7 gave positive specific antibody responses. CCV was subsequently isolated from fingerlings produced by two of the populations that gave a positive response.

Plumb et al. (1981) tested sera from broodfish at a known CCV-exposed farm and broodfish from a farm with no known history of CCV for 10 years. During this period, 99% of the fish from the CCV-negative farm had no CCV neutralization, whereas sera of fish from the CCV-positive farm had 62% of the titers below 1:20 and 38% were above 1:20 (titers determined against 100 $TCID_{50}$/0.1 ml at CCV). Indications were that the fish which produced virus infected fingerlings had higher antibody titers and had higher percentage of positive sera than CCV-negative fish. However, since 1% of the fish from the CCV-negative farm were positive for CCV antibody, there is still some question concerning the reliability of serum neutralization tests.

There has been little work done on vaccination of channel catfish for protective immunity. Noga and Hartmann (1981) and Walczak et al. (1981) reported on the use of attenuated CCV as an immunogen. Channel catfish virus was grown in kidney cells (KIK) of the walking catfish (*Clarias batrachus*). The virus was passed 60 times in KIK cells to obtain attenuation. Channel catfish fingerlings were vaccinated by a hyperosmotic bath in the attenuated virus and by paritoneal injection. A single vaccination by injection provided excellent protection to challenge as did vaccination by bathing, with a booster vaccination 3 weeks after the initial vaccination (Table 1) (Walczak et at., 1981). However, a single bath vaccination provided little protection. These data indicate that channel catfish do respond immunologically to bath exposure and a memory response

Table 1. Survival of channel catfish vaccinated with an attenuated strain of CCV developed in KIK (walking catfish kidney) cells. (From Walczak et al., 1981).

	Survival		
Treatment	Vaccination with booster	Single vaccination	Non-vaccinated
Peritoneal injection	---	91%	0%
Bath immunization	97%	30%	20%

does occur and could be used advantageously in a vaccination program, utilizing an attenuated virus preparation. Awad and Brady (1988) utilized the viral envelope from herpesvirus of channel catfish to immunize eggs and swim-up fry of channel catfish by immersion and challenged them 8 weeks post-vaccination. They reported that CCV challenged fingerlings from vaccinated eggs had 31% survival and single vaccinated fry had 69% survival compared to 0% and 44% survival for non-vaccinates respectively. Fingerlings from vaccinated eggs or swim-up stage and then boosted 2 weeks later had 81% and 80% survival when challenged 8 weeks post-vaccination.

Diagnosis

Diagnosis of CCV is done by observation of the clinical signs found in the field and isolation of the virus in cell cultures. Fish that show clinical signs of CCV infection should be selected and virologically processed as soon as possible. If processing is delayed for 24 h, the fish may be refrigerated, but if they cannot be processed in this time frame, they should be frozen in water (Plumb et al., 1973).

Channel catfish ovary (CCO) cells (Bowser and Plumb, 1980a, 1980b) are the cells of choice for CCV isolation. Brown bullhead (BB) cells may also be used, but they are approximately 1 log less sensitive than CCO cells at 25°C to 33°C (Bowser and Plumb, 1980c). Although cell systems from the walking catfish do support CCV replication, they are not as widely used and are probably less desirable when attempting to diagnose the disease. Most other known poikilothermic cell systems are refractive to CCV.

Cytopathic effect (CPE) is evidenced in CCO cells in 10-12 hours when incubated at 30°C; CPE develops more slowly in BB cells. Focal CPE is characterized by the cells becoming granular followed by enlargement (Figure 4). As the focal area becomes more distinct, a multinucleated syncytium forms in the center and this large cell is connected to normal cells by cytoplasmic strands. The entire cell sheet becomes involved and multiple syncytia will be connected to each other by cytoplasmic strands as CPE progresses. The cell sheet eventually becomes detached from the cell culture vessel surface.

Virus titers in CCV-infected fish may reach 1×10^5 to 1×10^6 $TCID_{50}$ per gram of tissue. However, the titer can be lower during early stages of epizootics or as the epizootic declines. Positive identification of CCV is obtained by neutralization techniques, using CCV specific antiserum made in rabbits, goats, or fish. CCV carrier populations are difficult to detect, however virus specific radiolabeled nucleic acid probes may be used to detect viral antigen (Wise and Boyle, 1985, and Wise et al. (1985).

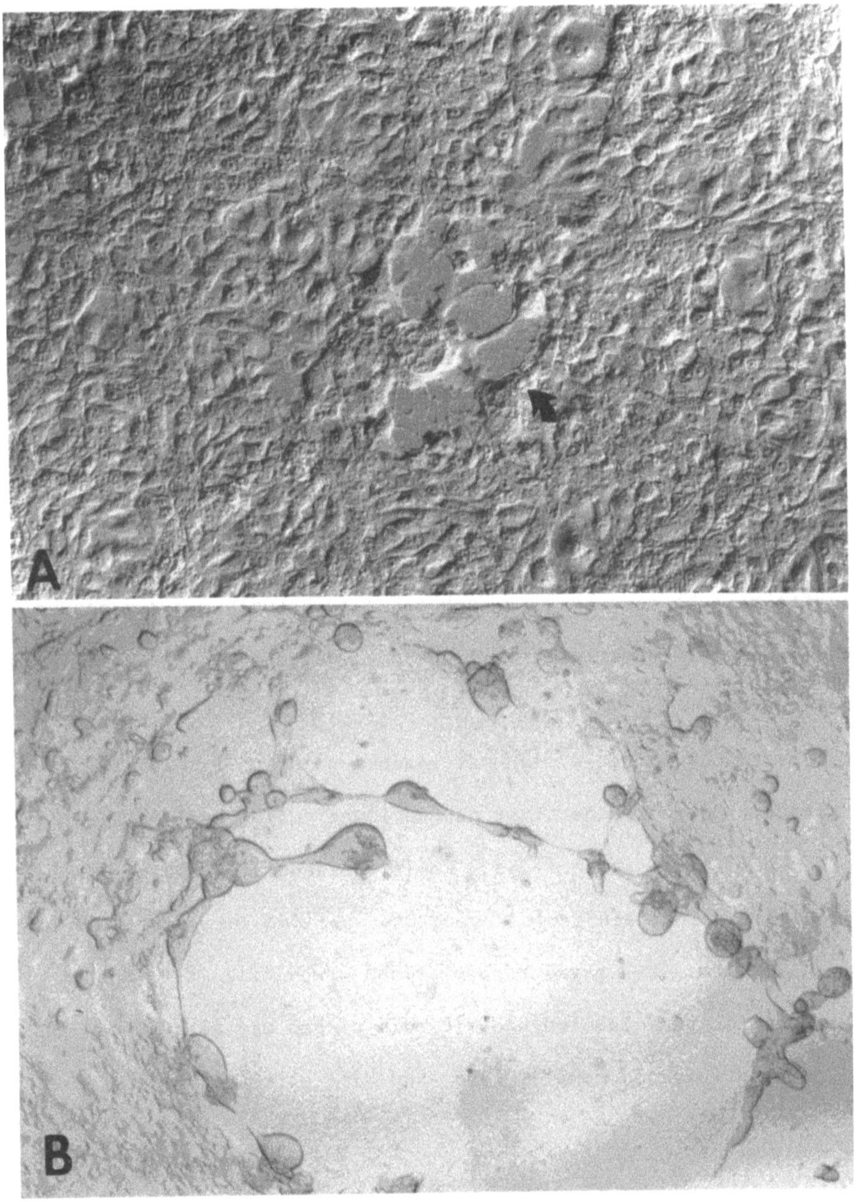

Figure 4. Cytopathic effect (CPE) of CCV in channel catfish ovary cells. (A) uninfected and focal CPE (arrow) at 20 h, (B) extensive CPE at 40 h. 33X. Photos taken with Numarsky system.

Control

The only practical control measures for CCVD are avoidance, quarantine of infected populations, sanitation, and maintaining optimum environmental conditions. Because the incidence of CCVD is closely related to temperature, Plumb (1972) demonstrated that when water temperature was reduced from 28°C to 19°C, or less, 24 h after infection, mortality was reduced from 94% to 24%. Reduction of temperature after onset of disease reduced mortality, but the reduction was not significant. Therefore, reduction of water temperature may be beneficial, but not always practical.

Surveying broodstock for CCV neutralizing antibody and/or testing stocks of fish during winter may be feasible procedures to avoid CCV by not using antibody-CCV positive stocks as spawners. However, these procedures are not generally available to the catfish industry at the present time and are not well refined.

Currently, CCV infections cannot be controlled with chemotherapeutics nor can the virus be prevented through immunization programs that are practical on a commercial scale. When outbreaks occur, ponds from which diseased fish are removed should be drained or disinfected with 40 mg/l of chlorine. Survivors of CCV epizootics may be grown to a marketable size providing the fish are held in ponds that are separated from healthy channel catfish. Under no circumstances should survivors of CCV epizootics be stocked in uninfected waters, nor should they be used as broodstock. Vaccination of channel catfish fry and fingerlings against CCV has shown potential, but whether it can be done economically and efficiently is still unknown. Another hinderance is convincing the catfish industry and potential commercial vaccine producers that the disease is serious enough to warrant a vaccination program.

References

Amend, D. F. and McDowell, T. 1983. Current problems in the control of channel catfish virus. J. World Maric. Soc. 14:261-267.

Amend, D. F. and McDowell, T. 1984. Comparison of various procedures to detect neutralizing antibody to the channel catfish virus in California brood channel catfish. Prog. Fish-Cult. 46(1):6-12.

Awad, M. A. and Y. J. Brady. 1988. Preliminary studies on a newly developed subunit vaccine for channel catfish virus disease. Proceed. Fish Health Sect./American Fisheries Society International Fish Health Conference, Vancouver, B.C. Canada. pp. 105.

Bird, R. C., K. E. Nusbaum, E. A. Scnews, R. R. Young-White, J. M. Girzzle and M. Toivio-Kinnucan. In Press. Molecular cloning of fragments of the channel catfish virus (Herpesviridae) genome and expression of the encoded m RNAs during infection. Am. J. Vet. Res.

Bowser, P. R. and Plumb, J. A. 1980a. Fish cell lines: establishment of a line from ovaries of channel catfish. In Vitro 16(5); 365-368.

Bowser, P. R. and Plumb, J. A. 1980b. Growth rates of a new cell line from channel catfish ovary and channel catfish virus replication at different temperatures. Can. J. Fish. Aquat. Sci. 37(5):871-873.

Bowser, P. R. and Plumb, J. A. 1980c. Channel catfish virus; comparative replication and sensitivity of cell lines from channel catfish ovary and the brown bullhead. J. Wildl. Dis. 16(3):451-454.

Bowser, P. R., Munson, A. D., Jarboe, H. H., Francis-Floyd, R. and Waterstrat, R. P. 1985. Isolation of channel catfish virus from channel catfish Ictalurus punctatus (Rafinesque) broodstock. J. Fish. Dis. 8:537-561.

Colyer, T. E., P. R. Bowser, J. Doyle and J. A. Boyle. 1986. Channel catfish virus: use of nucleic acids in studying viral relationships. Am. J. Vet. Res. 47:2007-2011.

Fijan, N. N. 1968. Progress report on acute mortality of channel catfish fingerlings caused by a virus. Bull. Off. Intl Epiz 69(7-8):1167-1168.

Fijan, N. N., Wellborn, T. J. and Naftel, J. P. 1970. An acute viral disease of channel catfish. Tech. Pap. Bur. Sport Fish. Wildl. 43. 11 pp.

Heartwell, C. M., III. 1975. Immune response and antibody characterization of the channel catfish (Ictalurus punctatus) to a naturally pathogenic bacterium and virus. Tech. Pap. Bur. Sport Fish Wild. 85. 34 pp.

Major, R. D., McCraren, J. R. and Smith, C. E. 1975. Histopathological changes in channel catfish (Ictalurus punctatus) experimentally and naturally infected with channel catfish virus disease. J. Fish Res. Board Can. 32(4):563-567.

McGlamery, M. H., Jr., and Gratzek, J. B. 1974. Stunting syndrome associated with young channel catfish that survived exposure to channel catfish virus. Prog. Fish-Cult. 36(1):38-41.

Noga, E. J. and Hartmann, J. X. 1981. Establishment of walking catfish (Clarias batrachus) cell lines and development of a channel catfish (Ictalurus punctatus) virus vaccine. Can. J. Fish. Aquat. Sci. 38 925-930.

Nusbaum, K. E. and J. M. Grizzle. 1987. Uptake of channel catfish virus from water by channel catfish and bluegills. Am. J. Vet. Res. 48:375-377.

Plumb, J. A. 1971a. Channel catfish virus disease in southern United States. Proc. Conf. Southeast. Assoc. Game and Fish Comm. 25:489-493.

Plumb, J. A. 1971b. Tissue distribution of channel catfish virus. J. Wildl. Dis. 7:213-216.

Plumb, J. A. 1972. Effects of temperature on mortality of fingerling channel catfish (Ictalurus punctatus) experimentally infected with channel catfish virus. J. Fish. Res. Board Can. 30(4):568-570.

Plumb, J. A. 1973. Neutralization of channel catfish virus by serum of channel catfish. J. Wildl. Dis. 9(4):324-330.

Plumb, J. A. 1974. Viral diseases of fishes of the Gulf of Mexico region. In: R. L. Amborski, M. A. Hood and R. R. Miller (Editors) Gulf Coast Regional Symposium on Diseases of Aquatic Animals. LSU-SG-74-05. 55-75.

Plumb, J. A. and J. Chappell. 1978. Susceptibility of blue catfish to channel catfish virus. Proc. Conf. Southeast. Assoc. Game and Fish Comm. Agencies. 32:680-685.

Plumb, J. A. and J. L. Gaines, Jr. 1975. Channel catfish virus disease. In: W. E. Ribelin and G. Migaki (Editors), The Pathology of Fishes, Wisconsin Press, Madison, pp. 287-301.

Plumb, J. A. and Jezek, D. A. 1983. Channel catfish virus disease. In: D. P. Anderson, M. Dorson, and Ph. Dubourget (Editors), Antigens of Fish Pathogens: Development and Production for Vaccines and Serodiagnostics. Collection Marcel Merieux, Lyon, France. pp. 33-50.

Plumb, J. A., Hilge, V., and Quinlan, E. E. 1985. Resistance of the European catfish (Silurus glanis) to channel catfish virus. J. Applied Icthylogy 1(2):87-89.

Plumb, J. A., Thune, R. L., and Klesius, P. H. 1981. Detection of channel catfish virus in adult fish. Intern. Symp. Fish Biol: Serodiag. and Vaccines Develop. Biol. Stand. 49:29-34.

Plumb, J. A., Wright, L. D., and Jones, V. L. 1973. Survival of channel catfish virus in chilled, frozen, and decomposing channel catfish. The Prog. Fish-Cult. 35(3):170-172.

Plumb, J. A., Gaines, J. L., Mora, E. C., and Bradley, G. G. 1974. Histopathology and electron microscopy of channel catfish virus in infected channel catfish, (Ictalurus punctatus (Rafinesque). J. Fish. Biol. 6:661-644.

Plumb, J. A., Green, O. L., Smitherman, R. O., and Pardue, G. B. 1975. Channel catfish virus experiments with different strains of channel catfish. Trans. Am. Fish. Soc. 104(1):140-143.

Robin, J. and Rodrique, A. 1980. Resistance of herpes channel catfish virus (HCCV) to temperature, pH, salinity and ultraviolet irradiation. Rev. Can. Biol. 39(3):153-156.

Walczak, E. E., Noga, E. J., and Hartmann, J. X. 1981. Properties of a vaccine for channel catfish virus disease and a method of administration. Intrn. Symp. Fish. Biol.: Serodiag. and Vaccines Develop. Biol. Stand. 49:419-429.

Wise, J. A. and Boyle, J. A. 1985. Detection of channel catfish virus in channel catfish, Ictalurus punctatus (Rafinesque): use of a nucleic acid probe. J. Fish. Dis. 8:417-424.

Wise, J. A., Bowser, P. R., and Boyle, J. A. 1985. Detection of channel catfish virus in asymptomatic adult catfish, Ictalurus punctatus (Rafinesque). J. Fish. Dis. 8:485-493.

Wolf, K., and Darlington, R. W. 1971. Channel catfish virus: a new herpesvirus of ictalurid fish. J. Virol. 8(4):525-533.

Wolf, K., Herman, R. L., and Carlson, C. P. 1972. Fish viruses; histopathologic changes associated with experimental channel catfish virus disease. J. Fish. Res. Board Can. 29(2):149-150.

Picornaviruses and Reoviruses of Fishes

J. R. Winton
National Fisheries Research Center
Bldg. 204, Naval Station
Seattle, WA 98115 USA

INTRODUCTION

The number of fish viruses isolated in cell culture or observed by electron microscopy continues to increase rapidly. Until recently, most viruses that were isolated from finfish and characterized were found to be members of the Rhabdoviridae, Iridoviridae, or Herpesviridae (Wolf and Mann 1980). In a comprehensive review of fish viruses published in 1984, there were no picornaviruses and only two reoviruses listed (Wolf 1984). The expansion of aquaculture into the rearing of new species at high density in different geographic areas, and the use of improved methods of detection that include newly developed cell lines and increased sampling effort, have led to the discovery of fish viruses representing nearly all families of animal viruses. Among the newest additions, are a member of the family Picornaviridae and several new viruses that belong within the Reoviridae.

PICORNAVIRUSES

Members of the Picornaviridae are small (22-30 nm), non-enveloped, RNA viruses with icosahedral symmetry. The virions are composed of 60 subunits (capsomeres) in a T=1 arrangement. The genome is a 7.5-8.5 kb linear molecule of single-stranded RNA, polyadenylated at 3′ end and having a small protein, VPg, at the 5′ end. The RNA genome is (+) sense, infectious, and serves as messenger RNA for the translation of a large polyprotein that is cleaved to form the four major structural proteins and several proteins involved in replication. Virus replication occurs in the cytoplasm of infected cells. There are four recognized genera within the Picornaviridae that include several important human and animal pathogens (poliovirus, foot and mouth disease; Fenner et al 1987). Currently, one virus has been isolated from finfish that appears to be a member of the Picornaviridae.

Smelt Virus

The smelt virus was isolated in August 1986 from rainbow smelt (Osmerus mordax) in New Brunswick, Canada following a large mortality among landlocked fish (Moore et al 1988). The fish showed no clinical signs of disease, but when examined, a virus was recovered from pools of viscera using the chinook salmon (Oncorhynchus tshawytscha) embryo cell line (CHSE-214) incubated at 15 C. No virus was recovered using rainbow trout (Salmo gairdneri) gonad (RTG-2) or fathead minnow (Pimephales promelas; FHM) cell lines. Cytopathic effect (CPE) was characterized by initial syncytia leading to the formation of plaques with well defined margins and clean interiors. Later, vacuolated floating aggregates were seen. Thin sections of infected cells showed 20-30 nm nonenveloped virions in cytoplasm. The virus was resistant to ether, pH 3, and heating to 37 C for 24 hr, but inactivated after 6 hr at 56 C. The virus was not neutralized by polyvalent antiserum to infectious pancreatic necrosis virus (IPNV). The smelt virus was not pathogenic for brook trout by intraperitoneal injection, but was recovered from all injected fish 1 month post-injection.

REOVIRUSES

Members of the family Reoviridae are nonenveloped, spherical particles, 60-80 nm in diameter, composed of two capsid shells with icosahedral symmetry. The genome is 10, 11, or 12 segments of double-stranded RNA (dsRNA) with a total molecular weight (MW) of 12-20 x 10^6 (Matthews 1982). Five to ten major structural proteins make up the virion. Viral replication occurs in the cytoplasm. There are six recognized genera that include viruses from animals, insects and plants (Joklik 1983). While several members are "orphan" viruses, the family includes some important pathogens (bluetongue virus, rotaviruses; Fenner et al 1987).

In the last few years, a surprising number of reoviruses have been isolated from fish and shellfish in different environments and geographic areas. This increase in aquatic animal reoviruses, probably reflects the examination of new species of fish and shellfish including animals without clinical signs as part of a more intensive surveillance effort.

In general, reoviruses isolated from fish and shellfish have several common properties: the ability to form plaque-like areas or syncytia in a variety of fish cell monolayers, a typical reovirus-like morphology with a double capsid shell, a buoyant density in CsCl of 1.34-1.36 g/ml, a genome composed of 11 segments of dsRNA distributed as three large, three medium, and five small segments with a total of approximately 15 x 10^6 MW, and a virion containing five major structural proteins that range from 137,000 to 32,000 MW (Winton et al 1987). Table 1 lists the reoviruses isolated from finfish or that replicate in fish causing disease. Other reoviruses from shellfish and those from fish which are incompletely characterized are mentioned briefly in the text.

Table 1. Reoviruses of finfish.

Virus	Isolated From	Location	Ref.
Golden Shiner Virus	N. crysoleucas	Southern USA	1
13p2 Reovirus	C. virginica	Eastern USA	2
Chum Salmon Virus	O. keta	Northern Japan	3
Catfish Reovirus	I. punctatus	Western USA	4
Grass Carp Reovirus	C. idella	Mainland China	5
Tench Reovirus	T. tinca	Central Europe	6
Chub Reovirus	L. cephalus	Central Europe	6
Turbot Reovirus	P. maxima	Northern Spain	7
Coho Salmon Reovirus	O. kisutch	Western USA	8

1. Plumb et al (1979) J Fish Res Board Can 36:1390-1394
2. Meyers (1979) J Gen Virol 43:203-212
3. Winton et al (1981) Fish Pathol 15:155-162
4. Amend et al (1984) Can J Fish Aquat Sci 41:807-811
5. Chen and Jiang (1984) Kexue Tongbao 29:832-835
6. Ahne and Kölbl (1987) J Appl Ichtyol 3:139-141
7. Lupiane et al (1988) Abstr, Int Fish Health Conf p 158
8. Winton et al (1988) This conference

Golden shiner virus

The golden shiner (Notemigonus crysoleucas) virus (GSV) was isolated in 1977 from bait fish in the southern USA. Affected fish were lethargic, swam near the surface, and had petechial hemorrhages in the muscles, visceral fat, intestinal mucosa, and cornea. The virus produced plaque-like CPE in the FHM cell line but not in the brown bullhead (Ictalurus nebulosus; BB) or channel catfish (Ictalurus punctatus) ovary (CCO) lines. The virus was resistant to ether and heat, stable at pH 3-10, and infectious after treatment with nucleases. Electron microscopy showed icosahedral virions approximately 70 nm in diameter in the cytoplasm of infected cells (Plumb et al 1979). Infectivity was not neutralized by polyvalent anti-IPNV serum (Schwedler and Plumb 1980). These same workers reported that GSV had an optimal replication temperature of 30 C where the latent period was 8 hr and the yield was 113 infectious units per cell. The virus was very stable at 4 and -70 C (Schwedler & Plumb 1982a). Winton et al (1987) determined that the genome of GSV was composed of eleven segments of dsRNA distributed in three size classes and virions contained five major structural proteins.

Schwedler and Plumb (1982b) reported that mortality caused by the virus was typically low, but higher fish densities resulted in greater infection rates and higher viral titers in infected fish. The virus now appears to be widely distributed among golden shiners in the southern United States.

13p2 Reovirus

Meyers (1979) isolated the 13p2 reovirus from homogenates of juvenile American oysters (*Crassostrea virginica*) collected from Long Island Sound, New York, USA. The virus replicated in the bluegill (*Lepomis macrochirus*) fry (BF-2) cell line where it produced plaques of large syncytia. The virus also replicated in the BB, Atlantic salmon (*Salmo salar*), guppy (*Poecillia reticulata*) embryo, and walleye (*Stizostedion vitreum*) cell lines, but not in the FHM, rainbow trout spleen or RTG-2 cell lines. Viral replication occurred at 15 and 23 C but not at 30 C. The virus was stable at pH 3-9, was not affected by chloroform, did not hemagglutinate human O erythrocytes, and was not neutralized by antiserum against mammalian or avian reoviruses or IPNV. Negatively stained virions were 79 nm in diameter with a double capsid shell. Rotational enhancement suggested a T=3 symmetry (Meyers and Hirai 1980). The genome of the 13p2 reovirus was shown to have eleven segments of dsRNA and the virions contained five structural proteins (Winton et al 1987).

No viral replication and no lesions were observed in oysters exposed to the 13p2 reovirus, but inoculation of bluegills resulted in significant mortality, increasing viral titer, and focal necrotic lesions in the liver (Meyers 1980). No mortality occurred following artificial infection of rainbow trout with 13p2 virus, but a multifocal reticulo-endothelial granulomatous hepatitis and viral replication were observed (Meyers 1983). These results suggested that the 13p2 virus was of marine fish origin and may have been present in oysters as a result of filter feeding.

Chum Salmon Virus

The chum salmon (Oncorhynchus keta) virus (CSV) was first recovered in 1978 from adult chum salmon in Hokkaido, Japan. The virus replicated in CHSE-214 cell monolayers producing foci of syncytia. Optimal replication occurred between 15-20 C. The virus did not grow in BB, FHM or BF-2 cells. Electron microscopy revealed icosahedral particles 75 nm in diameter with a double capsid. The virus was not inactivated by chloroform, was unstable at 56 C, did not hemagglutinate human O erythrocytes, and had a density in CsCl gradients of 1.33 g/ml. Antiserum to IPNV or to mammalian reovirus types 1, 2 or 3 did not neutralize infectivity (Winton 1981). The genome had eleven segments of dsRNA of three size classes with a range of 0.37-2.5 x 10^6 MW. Five major structural proteins and several minor proteins were detected in the virion (Winton et al 1983).

No mortality followed injection of the virus into chinook, kokanee (Oncorhynchus nerka), chum salmon, or rainbow trout fry, but viral replication occurred in each species and focal necrotic lesions were seen in the livers of the chum and chinook fry (Winton 1981). In 1986, a virus very similar to CSV was recovered from masu salmon (Oncorhynchus masou) fry suffering a high mortality at a hatchery in Hokkaido. Since then, the virus has been recovered from returning adult masu salmon at several locations, but recent experiments with three to five month old masu salmon show the isolate is not highly pathogenic for fish of this species (Yoshimizu 1988).

Channel Catfish Reovirus

Channel catfish reovirus (CRV) was isolated from catfish in California, USA suffering a chronic mortality (Amend et al 1984). The virus produced syncytia in the CCO, BB, and CHSE-214 cell lines but did not grow in FHM cells. The optimal temperature for replication was 25 C. Negative stain showed reovirus-like particles with a mean diameter of 75 nm and a double capsid. The virus had low virulence for channel catfish in infectivity trials.

The virus was further characterized by Hedrick et al (1984) who reported the virions had an outer capsid diameter of 75 nm and an inner capsid of 55 nm. Complete virions had a density in CsCl of 1.36 g/ml, empty particles 1.29 g/ml, and cores 1.38 g/ml. The virus genome was composed of eleven segments of dsRNA ranging in size from 0.4-2.5 x 10^6 MW with three large, three medium and five small segments. Five major and two minor polypeptides were detected. The proteins of the virion ranged from 132,000 to 32,000 MW. Neutralization studies using CRV, GSV, CSV, and homologous antisera showed each virus was serologically distinct although some cross-reactions were observed. Since its initial discovery, the virus has been isolated regularly from catfish at locations in California (R. Hedrick personal communication).

Grass Carp Reovirus

A severe hemorrhagic disease of grass carp (*Ctenopharyngodon idella*) cultured in ponds on the Chinese mainland has been reported to be caused by a reovirus (Chen and Jiang 1984a). Virions purified from infected fish were 70 nm icosahedral particles with a double capsid layer. Infectivity of the virus was not affected by heat, pH 3 or chloroform. Unlike other reoviruses of fish, the genome of the grass carp reovirus is reported to contain only 10 segments of dsRNA (Chen and Jiang 1984b). Grass carp injected with the virus had a 70-80% mortality and showed typical signs associated with the disease. Because the disease is widespread in China causing serious losses, an inactivated vaccine was prepared from infected grass carp and tested by injection. Fish receiving the preparation were protected against challenge (Nie and Pan 1985). Sera from immunized grass carp were used for immune electron microscopy (Chen and Jiang 1984a). Additional serological and biochemical studies are required to determine the relationship of the grass carp reovirus to other fish or mammalian reoviruses.

Tench and Chub Reoviruses

Two reoviruses have been recovered from the visceral organs of European tench (*Tinca tinca*) and chub (*Leuciscus cephalus*) (Ahne and Kölbl 1987). The isolates replicated well at 20 C in EPC and FHM cells forming syncytia and causing lysis of the cells. Electron microscopy showed the virions were icosahedral particles, 70-75 nm in diameter, with a double capsid shell. The isolates were resistant to chloroform and were not inhibited by 5-iodo-2-deoxyuridine. Electrophoresis using 9% polyacrylamide gels showed both genomes had 11 segments of dsRNA distributed as three large, three medium, and five small segments. While the electropherotypes of the isolates were generally similar to each other and to CSV and GSV run in the same gel, the chub isolate was more like that of CSV and the tench isolate RNA more like that of GSV. Antiserum prepared against the tench isolate neutralized the tench and chub reoviruses at dilutions of 1:4096, while GSV was neutralized at 1:2048, and CSV at 1:128 showing the two isolates are antigenically related to each other and to GSV. No neutralization was observed with CRV. These new isolates are clearly members of the group of aquatic reoviruses, but further work will have to be done to determine their exact relationship to the other viruses of the group.

Turbot Reovirus

A reovirus was isolated in 1987 from juvenile turbot (Psetta maxima) cultured in Northwestern Spain (Lupiani et al 1988). The turbot reovirus (TRV) and a species of Vibrio were recovered from visceral organs of fish suffering chronic mortality and showing abdominal distension and hemorrhage at the anus and fins. Internal signs were a swollen stomach, an intestine filled with fluid, and hemorrhage of the peritoneal wall. The virus produced syncytia in monolayers of CHSE-214 cells incubated at 15 C. The agent replicated in BF-2 and EPC cells, but did not grow in RTG-2, BB, or FHM cell lines, nor when cultures were incubated at 4C or 25 C. Virions were icosahedral particles, 70 nm in diameter, with a double capsid shell. The virus was not affected by treatment with chloroform, ether, pH 2, 3, or 9, and infectivity was not reduced following heating for 30 min at 50 C or by incubation in the presence of 5-bromo-deoxyuridine. The virus was not neutralized by antisera to IPNV strains VR-299, Sp and Ab. The role of the virus in causing disease and its relationship to other aquatic reoviruses are areas requiring further study.

Coho Salmon Reovirus

Isolates of a virus termed coho salmon (Oncorhynchus kisutch) reovirus (CSR) were recovered from normal adult salmon returning to three locations in the state of Oregon, USA (Winton et al this conference). The virus was characterized by formation of syncytia in cultures of salmonid fish cells and a non-enveloped, icosahedral virion approximately 70 nm in diameter with a double capsid. The virus was not neutralized by antiserum against CSV, the only other reovirus isolated from salmonid fish. Analysis of the genome segments by electrophoresis showed each isolate had 11 segments of dsRNA that fell into similar, but not identical, patterns suggesting the isolates are related to each other and to CSV and 13p2 which were used as controls. Intraperitoneal injection of CSR into rainbow trout produced no mortality and no external signs of disease during an 11 week period, but the virus replicated in these fish producing focal necrotic lesions in the liver of infected fish that persisted during the entire experiment. Studies to determine the relationship of the three isolates to each other and to other reoviruses isolated from aquatic organisms will be required.

Other Reoviruses from Aquatic Animals

A reovirus was recovered from the spleen and kidney of two year old fancy carp in Japan. The characterization of this virus is in progress, but the virus replicates in FHM cells and the electrophoretic mobilities of structural proteins and genome segments were similar to other aquatic reoviruses (T. Sano personal communication).

Examination of oysters (*Crassostrea gigas*) in Japan produced a reovirus isolate termed JOV-1 (Nagabayashi and Mori 1983). The virus replicated in BF-2 and CHSE-214 cell lines at 10 and 15 C causing syncytia. Electron microscopy showed arrays of icosahedral particles 70 nm in diameter. The virus was not neutralized by anti-IPNV serum. Oysters exposed to JOV-1 were reported to experience significant mortality and poor growth. Ten week old rainbow trout infected by intraperitoneal injection with the virus suffered a 40% mortality, but 15 week old rainbow trout exhibited no loss. The virus was recovered from all exposed groups.

A reo-like virus was observed in cells of the spleen of a male puffer (*Fugu pardale*) in Japan during an electron microscopic examination of the hematopoietic tissues (Watanabe and Yamamori 1983). The particles were 60-77 nm in diameter and similar to reoviruses in morphology. The virus was associated with small (20-40 um) necrotic foci in the spleen. As the virus was not isolated in culture, other characteristics were not determined.

Hsu et al (1988) reported the isolation of a reovirus (HCV) from hard clams (*Meretrix lusoria*) in Taiwan. The isolate was a 78 nm icosahedral particle with a double capsid. The virus replicated at 20-27 C in BF-2, CCO, BB, and CHSE-214 cell lines producing plaque-like CPE. The genome was composed of 11 segments of dsRNA distributed among three size classes that ranged from 0.3-2.5 x 10^6 MW. There were five structural proteins in the complete virion. Experimental infection of clams showed 10% mortality and necrosis of the hepatopancreas of infected clams.

DISCUSSION

Increased attention to the health status of lower vertebrates has resulted in the isolation of new members of the Picornaviridae and Reoviridae. The recent detection of these viruses is probably a result of significant improvements in sampling methods and virus isolation techniques (Amos 1985). Whether the reoviruses recovered from shellfish are actually vertebrate fish viruses concentrated by filter-feeding, or if invertebrates are important hosts for aquatic reoviruses will be interesting areas for further research.

As more viruses are isolated from fish and other aquatic species, it will become increasingly important to study the serological and biochemical characteristics of the viruses and to compare them with viruses from higher vertebrates. Presently, there are a sufficient number of aquatic reoviruses to justify a search for type-specific and group antigens by neutralization, compliment fixation, and fluorescent antibody analysis. It would also be interesting to use nucleic acid hybridization to determine the extent of genetic relatedness among these isolates. The taxonomy of some animal virus families may need to be modified to include these aquatic animal isolates.

REFERENCES

Ahne W, Kölbl O (1987) Occurrence of reoviruses in European cyprinid fishes (Tinca tinca Lin.; Leuciscus cephalus Lin.). J Appl Ichthyol 3:139-141

Amend DF, McDowell T, Hedrick RP (1984) Characteristics of a previously unidentified virus from channel catfish (Ictalurus punctatus). Can J Fish Aquat Sci 41:807-811

Amos KH (ed) (1985) Procedures for the detection and identification of certain fish pathogens, 3rd ed. American Fisheries Society, Bethesda

Chen Y-X, Jiang Y-L (1984a) Morphological and physico-chemical characterization of the hemorrhagic virus of grass carp. Kexue Tongbao 29:832-835

Chen Y-X, Jiang Y-L (1984b) Morphological and physico-chemical characterization of the hemorrhagic virus of grass carp. In: Abstracts, Sixth International Congress of Virology, Sendai, Japan, p 373

Fenner F, Bachmann PA, Gibbs EPJ, Murphy FA, Studdert MJ, White DO (1987) Veterinary virology. Academic Press, Orlando

Hedrick RP, Rosemark R, Aronstein D, Winton JR, McDowell T, Amend DF (1984) Characteristics of a new reovirus from channel catfish (Ictalurus punctatus). J Gen Virol 65:1527-1534

Hsu Y-L, Chen B-S, Wu J-L (1988) Biological and biochemical characteristics of a reovirus isolated from hard clam (Meretrix lusoria). In: Abstracts, International Conference on Fish Health, Vancouver, Canada, p 166

Joklik WK (1983) The members of the family Reoviridae. In: Joklik WK (ed) The Reoviridae, Plenum, New York

Lupiani B, Dopazo CP, Ledo A, Barja JL, Hetrick FM, Toranzo AE (1988) A new syndrome of suspected viral etiology in cultured juvenile turbot. In: Abstracts, International Conference on Fish Health, Vancouver, Canada p 158

Matthews REF (1982) Classification and nomenclature of viruses: Fourth report of the international committee on taxonomy of viruses. Intervirol 17:1-199

Meyers TR (1979) A reo-like virus isolated from juvenile American oysters (Crassostrea virginica). J Gen Virol 43:203-212

Meyers TR (1980) Experimental pathogenicity of reovirus 13p2 for juvenile American oysters Crassostrea virginica (Gmelin) and bluegill fingerlings Lepomis macrochirus (Rafinesque). J Fish Dis 3:187-201

Meyers TR (1983) Serological and histopathological responses of rainbow trout, Salmo gairdneri Richardson, to experimental infection with the 13p2 reovirus. J Fish Dis 6:277-292

Meyers TR, Hirai K (1980) Morphology of a reo-like virus isolated from juvenile American oysters (Crassostrea virginica). J Gen Virol 46:249-253

Moore AR, Li MF, McMenemy, M (1988) Isolation of a picorna-like virus from smelt, Osmerus mordax (Mitchill. J Fish Dis 11:179-184

Nagabayashi T, Mori S (1983) Characterization of a virus (JOV-1) isolated from the Japanese oysters (Crassostrea gigas). In: Abstr Soc Invert Path XVI Annual Meeting. Cornell University

Nie D-S, Pan J-P (1985) Diseases of grass carp (Ctenopharyngodon idellus Valenciennes, 1844) in China, a review from 1953 to 1983. Fish Pathol 20: 323-330.

Plumb JA, Bowser PR, Grizzle JM, Mitchell AJ (1979) Fish viruses: A double-stranded RNA icosahedral virus from a North American cyprinid. J Fish Res Board Can 36:1390-1394

Schwedler TE, Plumb JA (1980) Fish viruses: Serological comparison of golden shiner and infectious pancreatic necrosis viruses. J Wildl Dis 16:597-599

Schwedler TE, Plumb JA (1982a) In vitro growth kinetics and thermostability of the golden shiner virus. J Wildl Dis 18:441-446

Schwedler TE, Plumb JA (1982b) Golden shiner virus: Effects of stocking density on incidence of viral infection. Prog. Fish Cult 44:151-152

Watanabe T and Yamamori K 1983 Virus-like particles in the spleen hematopoietic tissue of puffer Fugu pardale. Bul Jap Soc Sci Fish 49:307

Winton JR (1981) Isolation and characterization of a new reovirus from chum salmon (Oncorhynchus keta). Ph.D. thesis, Oregon State University

Winton JR, Lannan CN, Fryer JL, Kimura T (1981) Isolation of a new reovirus from chum salmon in Japan. Fish Pathol 15:155-162

Winton JR, Lannan CN, Fryer JL (1983) Further characterization of a new reovirus of poikilothermic vertebrates. In: Compans RW, Bishop DHL (eds) Double-stranded RNA Viruses. Elsevier, New York

Winton JR, Lannan CL, Fryer JL, Hedrick RP, Meyers TR, Plumb JA, Yamamoto T (1987) Morphological and biochemical properties of four members of a novel group of reoviruses isolated from aquatic animals. J Gen Virol 68:353-364

Wolf K (1984) Fish viruses: Their biology, classification, hosts, pathology and control. In: Kurstak E, Marusyk RG (eds) Control of Virus Diseases, Marcel Dekker, New York

Wolf K, Mann JA (1980) Poikilotherm vertebrate cell lines and viruses: A current listing for fishes. In Vitro 16:168-179

Yoshimizu M (1988) Chum salmon virus isolated from masu salmon. In: Fish and Eggs, Technical Reports of the Hokkaido Salmon Hatchery 157:36-38.

Some Properties of the Etiological Agent of the Hemorrhagic Disease of Grass Carp and Black Carp

Y. Jiang and W. Ahne*
Instititute of Hydrobiology
Academia Sinica
Wuhan, Hubei, P.R. China

Introduction

Aquaculture in fresh water has shown recent expansion in the P.R. China. The production of grass carp (Ctenopharyngodon idella) and black carp (Mylopharyngodon piceus), representing important protein sources, has reached 300.000 tonnes annually. However, Chinese fish culturists are confronted with fish diseases and declines in the productivity.

In the 70's of the 20 century a previously not known hemorrhagic disease of grass carp and black carp was observed in China. The disease proved to be infectious and more then 80% of reared fingerlings and yearlings of grass carp and black carp produced were killed each year. The epizootics are usually recognized in the summer time when the temperatures of water reache above 25°C. The infected fish are exophthalmic and severe hemorrhages occur in fin bases, gills, musculature and intestinal tract.

First observations made by fish pathologists at the Institute of Hydrobiology, Wuhan, P.R. China, lead to the assumption that a viral agent could be involved in the disease (Chen and Jiang, 1984). In this paper we describe some properties of the virus isolated from grass carp and black carp causing the hemorrhagic disease.

---

*Institute of Zoology and Hydrobiology
of the University of Munich,
Kaulbachstr. 37
D-8000 München 22, Fed. Rep. Germany

Material and methods

Isolation of virus

Fingerlings of grass carp and black carp with hemorrhages originated from hatcheries in the Hubei province, P.R. China, were used for the isolation trials. Isolation of virus was done according standard procedures using the adsorption, trypsination and co-cultivation methodes. Infected cell cultures incubated at 15, 20, 25 and 28°C were daily checked for cytopathic effects (CPE). Seven days after infection flasks were frozen at -20°C, thawed and infectious tissue culture medium served as material for the virological investigations.

Cell cultures

The cell cultures used were grown in plastic flasks (75 cm^2) with TC 199 medium supplemented with 10% fetal calf serum and antibiotics (penicillin 100 I.U./ml; streptomycin 100 ug/ml) at 20, 25 and 28°C. 14 different cell lines were used for isolation trials and propagation of virus:

Centrarchidae
BF-2 cells (caudal trunk of *Lepomis macrochirus*)

Cyprinidae:
CAR-cells (blastoderm of *Carassius auratus*)
CF-cells (fin of *Ctenopharyngodon idella*)
CK-cells (kidney of *Ctenopharyngodon idella*)
CO-cells (ovaries of *Ctenopharyngodon idella*)
CLC-cells (leucocytes of *Cyprinus carpio*)
FHM-cells (caudal trunk of *Pimephales promelas*)

Esocidae:
PG-cells (gonads of *Esox lucius*)

Ictaluridae:
BB-cells (caudal trunk of Ictalurus nebulosus)

Salmonidae:
CHSE-214-cells (embryo of Oncorhynchus tshawytscha)
R1-cells (liver of Salmo gairdneri)
RTG-2-cells (gonads of Salmo gairdneri)

Lacertilia:
IGH2-cells (heart of Iguana iguana)

Rodentia:
Mouse cells (Mus musculus)

Titration of infectious material

Since the virus did not produce any cytopathogenic effects in all cell cultures used, the infectivity was tested in vivo by using the LD_{50} (Reed and Muench, 1938).

Electron microscopy

Tissue culture medium of infected cells were clarified by low speed centrifugation (3000xg, 15 min.) and the supernatant was then centrifuged at 85000x g for 2 hours. The pellet was resuspended in 0,1 ml of 0,1 M tris buffer. Preparations applied to carbon coated formvar grids were negatively stained with neutralized phosphotungstic acid and examined with a Zeiss EM 109.

For immuno electron microscopy, purified virus suspension was mixed with serum from grass carp which had survived the infection. After 24 hours at 20°C the mixture was centrifuged at 3000xg for 15 minutes, the preparations were negatively stained and then examined.

Treatment with chloroform, heat, acid and IUdR

Infectious tissue culture medium of CO and CF cells were treated with chloroform (56%), heat (56° C), and acid (pH3). Inhibition of virus multiplication in CO cells using IUdR (50 ug/ml) was carried out in vitro. All tests were done according methods described by Rovozzo and Burke (1973).

Treatment of the virus with alpha-chymotrypsin

The ability of alpha-chymotrypsin to remove the outer capsid layer of the virion was tested by the method as described by Joklik (1972). Purified virions in PBS were digested with alpha-chymotrypsin (100 ug/ml) for 60 minutes at 37° C. Afterwards, particles pelleted by centrifugation (85000xg, 2 hours) were negatively stained and examined in the electron microscope.

SDS PAGE of viral nucleic acid

Virus isolated from grass carp was used for SDS PAGE of viral nucleic acid. Partially purified virus by ultracentrifugation was treated with Genetron (1,1,2-trichloro-1,2,2-trifluoroethane) and after centrifugation for 10 minutes at 1000xg, the aqueous phase was removed, placed onto a continuous sucrose gradient (10 to 50%) and centrifuged for 2 hours at 100.000xg. The viral band of the gradient was removed and pelleted by centrifugation for 1 hour at 100.000xg. The viral nucleic acid was extracted with an equal volume of phenol saturated with buffer. After mixing for 5 minutes the homogenate was centrifuged at 2000 x g for 10 minutes and the aqueous phase was removed. The nucleic acid was precipitated with cold ethanol, collected by centrifugation at 10.000xg for 15 minutes, suspended in buffer and applied to 9% polyacrylamide gels (Laemmli,1970). After electrophoresis for 20 hours at 9 mA, the RNA segments were stained with silver.

Induction of interferon in vitro

CO cells and CF cells were infected with the isolated virus of the 11th passage. After incubation at 28°C for 7 days, the medium of infected tissue cultures was harvested. FHM and PG cells grown in microtiter plates were exposed for 24 hours to serial 2-fold dilutions of the medium. Afterwards, the inoculum was removed and the cells were washed twice with PBS before they were challenged with 100 $TCID_{50}$ of frog virus 3 (DNA virus) and the virus of hemorrhagic septicemia of trout (VHS), (RNA virus). After 7 days of incubation at 15°C the cell cultures were stained with crystal violet and the titer of inhibitor was determined.

Hemagglutination test

To determine the hemagglutinating activity of the virus, infectious tissue culture medium free of cells was diluted 2-fold and it was incubated with human type 0 red blood cells at 20°C as described by Rovozzo and Burke (1973). A myxovirus (Ahne et al. 1987) served as the positive control.

Infection experiments

Cross infection trials were carried out by water bath and by injection method using fingerlings of grass carp and black carp. In each group 38 fish were intraperitoneally infected with 0,1 ml of infectious tissue culture medium.

The bath infection was done by keeping the fish for 60 minutes in 10 liter of of 1:100 diluted organ suspension of grass carp and black carp with hemorrhagic disease. The temperatures of water were 25°C during the infection.

Infected fish and the controls were kept 100 liter aquaria with through running water (25-28°C).

RESULTS

Isolation of virus

All cell lines used did not show any cytopathic effects (CPE) after infection with organ suspensions from sick grass carp and black carp. 30 blind passages were carried out with each cell line but no alterations of cells like inclusion bodies, syncytia or lysis were observed. Infected cells treated with Giemsa stain did not show any differences to the non infected cells. This applied also to the use of different methods and different temperatures. As demonstrated by electronmicroscopy only cells originated from grass carp were susceptible to the virus but the infection remained latent during all passages. CF and CO cells multiplied the virus without showing any CPE. After eleven blind passages of virus in CF and CO cells numerous virus particles could be detected in the supernatant of cell culture medium by electron microscopy.

In negatively stained preparations, spherical particles approximately 65-75 nm in diameter, with a double capsid shell could be observed. Both, complete and incomplete (coreless) particles were present (Fig.1)

Particles viewed along the fivefold axis of symmetry exhibited 20 capsomeres (Fig.2).

Immuno electron microscopy

The identity of the agents isolated from grass carp and from black carp was demonstrated by immuno electron microscopy. Virus preparations from both fish clearly formed antigen-antibody-complexes after reaction with convalscence serum from grass carp and from black carp, respectively.

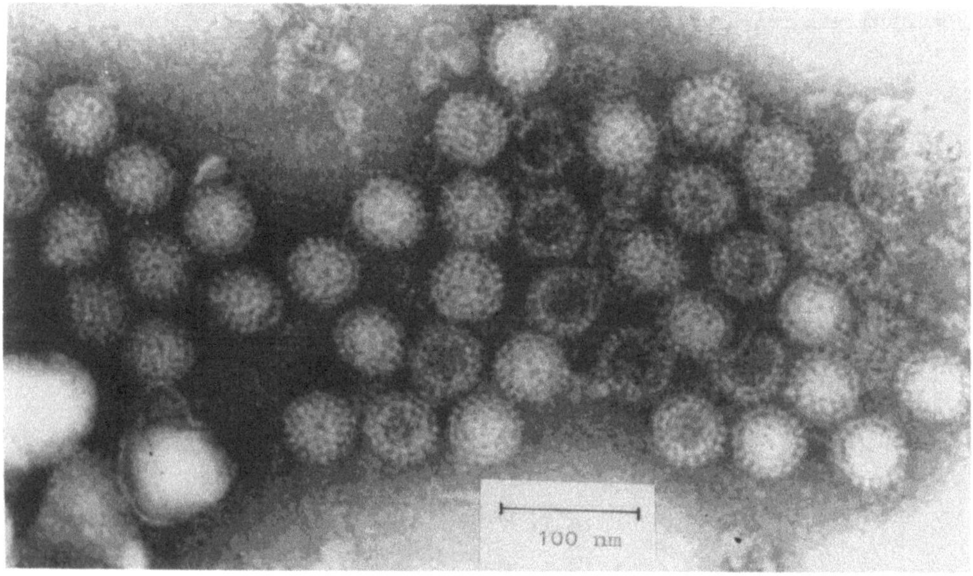

Fig.1 Electron micrographs of negatively stained virions isolated from grass carp and from black carp showing hemorrhagic disease

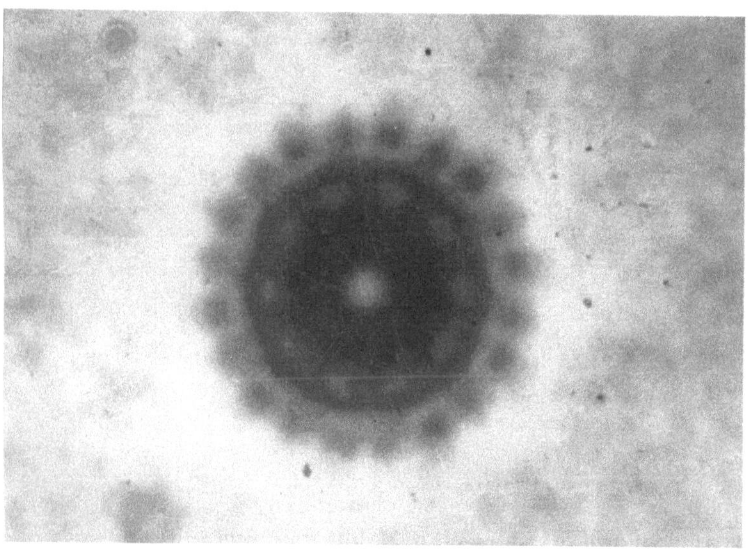

Fig.2 A negatively stained intact virus particle isolated from grass carp. The surface of which is covered by 20 distinct capsomeres (about x 750.000)

Treatment of virus with heat, pH3, chloroform and JUdR

The virus proved to be stabel to treatments with heat (56° C), pH_3 and chloroform. JUdr did nor have effects on multiplication of virus. The infectivity of virus before and after treatment was determined by infection experiments of grass carp (Table 1).

Table 1 Infectivity (LD_{50}) of virus before and after treatment with heat, pH3 ,chloroform and JUdr

		Infectivity (log10 infection dose) before	after
56° C	(30 min.)	4,6	4,5
pH_3	(60 min.)	6,2	6,7
$CHCl_3$	(12 hours)	5,4	5,5
JUdr	(50 ug/ml)	5,2	5,0

Hemagglutination

The isolated viruses did not hemagglutinate human type 0 red blood cells whereas a control orthomyxovirus (Ahne et al. 1987) showed hemagglutination titers up to 1:128.

Treatment of virus with alpha-chymotrypsin

The double capsid structure of the virus became evident by particles where virions in various stages of digestion following treatment with alpha-chymotrypsin were seen with the electron microscope.

Polacrylamide gel electrophoresis of nucleic acid

Electrophoresis of viral nucleic acid using SDS-polyacrylamide gels revealed groups of large, medium and small segments. The genome of the virus exhibited 10-11 segments of double-stranded RNA.

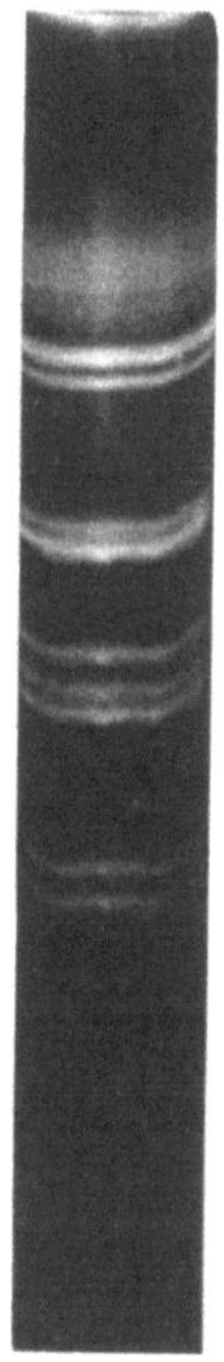

Fig.3 Analysis of virion RNS in 9% polyacrylamide slab gel

Production of antiviral substances by the virus in vitro

Infected CF and CO cells produced an interferon like substance which proved to be stable to a treatment with pH_3. The antiviral effects was defined as the reciprocal of the dilution of the tissue culture medium which inhibited the CPE of frog virus and VHSV completely (Table 2).

Table 2. Inhibitory effect of tissue culture fluid of CO cells and CF cells infected with the reovirus from grass carp and black carp on amphibian (Frog virus 3) and fish viruses (VHS)

Virus	Cell cultures	Inhibitory effect of tissue culture fluid from cells infected with the isolate from grass carp/	black carp
Frog virus 3	FHM	200*	100
VHSV	FMH	3000	1000
VHSV	PG	0	0

*reciprocal of dilution giving protection of 100%

Infection experiments

The virus isolated from sick grass carp and black carp produced the hemorrhagic disease in grass carp and in black carp. Typical signs of the disease such as bleedings in the skin, in the fin bases and in the eyes occurred within 7-10 days after infection at 25-28°C. The mortality rates of infected fish were between 50-95% whereas the control group did not show any symptoms of the disease and no losses occurred. The virus isolated from grass carp and from black carp proved to be infectious to both species of fish (Table 3) and the agent could be reisolated from the animals infected.

Table 3 Mortality rates of grass carp and black carp infected intraperitoneally and by bath method with the virus isolated from grass carp and black carp

Virus	Infection of fish with virus isolated from grass carp				black carp			
Fisch	GC		BC		GC		BC	
Method of infection	inj.	bath	inj.	bath	inj.	bath	inj.	bath
Mortality rates (%)	80	70	8o	70	67	50	95	70

(GC=grass carp, BC=black carp)

Discussion

Disease and parasitism are most prevalent in cultured fish populations over the whole world. Viruses, fungi and bacteria cause considerable debilitation and mortality in fish farms in the P.R. China. Although the P.R. China represents one of the largest aquaculture system , hitherto only slight attention has been paid to the diseases of fish. However, the increasing intensification of aquaculture has focussed attention on infectious diseases of fish, and their etiology today is the concern of many institutes in the P.R. China.

The literature on fish virology shows that there are about 50 fish pathogenic viruses known (Ahne, 1985).The agents belonging to different virus families cause more or less severe epizootics in cultured fish. The channel catfish virus (Herpesviriridae), the spring viremia of carp virus, the virus of infectious hematopoietic necrosis , the virus of hemorrhagic septicemia (Rhabdoviridae), and the virus of infectious pancreatic necrosis of trout (Birnaviridae) are the most important viral agents causing heavy losses in fish farms in different countries.

Two viruses have been isoltated from grass carp so far e.g. a rhabdovirus which is serologically undistinguishable from the Pike Fry Rhabdovirus, PFR, (Ahne, 1975) and the described reovirus. Whereas the pathogenicity of the rhabdovirus to grass carp is low, the virulence of the reovirus is very high. The high virulence of this reovirus is in contrast to the other reoviruses found in teleost fish so far.

Five viruses classified as Reoviridae have been described in fish e.g. the golden shiner virus from golden shiner (Notemogonus crysoleucas), (Plumb et al. 1979); the chum salmon reovirus from chum salmon (Oncorhynchus keta), (Winton et al., 1981); the channel catfish reovirus from channel catfish (Ictalurus punctatus), (Amend et. al., 1984); the reoviruses from tench (Tinca tinca) and from chub (Leuciscus cephalus),(Ahne and Kölbl, 1987).

The virus isolated from grass carp and black carp possess characteristics of the Reoviridae (Matthews, 1982). The agent has an icosahedral morphology and its genome contains 10-11 segments of dsRNA. It is clearly composed of double capsid having 20 peripheral capsomeres most like the members of the genus Reovirus. Winton et al. (1987) have compared certain morphological and biochemical features of reoviruses isolated from fish and shellfish so far, but the viruses were unlike described members of the family Reoviridae. The authors suggested to establish a novel group of Reoviridae consisting of reoviruses of aquatic animals. The reovirus isolated from grass carp and black carp could be a member of the novel group of Reoviridae.

In contrast to the aquatic reoviruses the isolate from grass carp and black carp did not induce cytopathic effects (syncytia formation, lysis of the cells). That fact complicated the detection and characterization of the agent. We were not able to confirm results concerning cytopathic effects of the reovirus isolated from grass carp as described by several Chinese scientists (Chen and Jiang, 1984; Deng et al.,1985; Zhang and Yang, 1985). However, viral multiplication in cells originated from grass carp could be demonstratet by electron microscopy and by infection experiments.

The virus was able to produce a viral inhibitory substance in cultured cells obtained from grass carp and from black carp. The interferon like substance proved to be stable at pH_3 and it showed to be of species specificity (e.g. the CPE was inhibited in cells from cyprinids only) but it was not of virus specificity. The viral inhibitory substance induced by the reovirus from grass carp and blackcarp may be similar to the interferon described in mammals and birds (Isaacs, 1963).

Acknowledgements

This work was supported by the Commission of the European Communities, Contract No. TSDA 290 B.

References

Ahne, W.(1975) A rhabdovirus isolated from grass carp. Arch. Virol. 48: 181-185

Ahne, W.(1985) Viral infections in fishes: Etiology, diagnosis and control. J.Vet.Med. 32:237-264

Ahne, W., Kölbl,O.(1987) Occurrence of reoviruses in European cyprinid fishes (Tinca tinca Lin.; Leuciscus cephalus Lin.) J.Appl.Ichthyol. 3:139-141

Ahne, W., Neubert, W.J., Thomsen, I. (1987) Reptilian viruses: isolation of myxovirus-like particles from the snake Elaphe oxycephala. J.Vet.Med. 34:607-612

Amend, P.F., McDowell, T.; Hedrick, R.P. (1984) Characteristics of a previously unidentified virus from channel catfish (Ictalurus punctatus). Can.J.Fish.Aquat.Sci. 41:8o7-811

Chen,Y.; Jiang, Y. (1984) Morphological and physico-chemical characterization of the hemorrhagic virus of grass carp. Kexue-Tongbao, 29:832-835

Deng, C., Yang, X., Chen, H. (1985) The susceptibility to grass carp reovirus of grass carp kidney cell line and its clones. Abstr. Asia Symp. Freshwater Fish Cult. Beijing

Isaacs, A. (1963) Interferon, Adv.Virus.Res. 10:1-35

Joklik, W.K. (1972) Studies on the effect of chymotrypsin on reoviruses. Virol.49:700-715

Matthews, R.E.F. (1982) Classification and Nomenclature of Viruses. Intervirol. 17 (1-3):199 p.

Plumb, J.A., Bowser, P.R., Grizzle, J.M., Mitchell, A.J. (1979) Fish viruses: a double-stranded RNA icosahedral virus from a North American cyprinid. J.Fish.Res.Bd.Can. 36:1390-1394

Reed, L.J., Muench, H. (1938) A simple method of estimating fifty percent end points. Am.J.Hyg. 27:493-497

Rovozzo, G.C., Burke, C.N. (1973) A manual of basic virological techniques. Inc.Englewood Cliffs,N.J.

Winton, J.R., Lannan, C.N., Fryer, J.L., Kimura, T. (1981) Isolation of a new reovirus from chum salmon in Japan. Fish Pathol. 15:155-162

Winton, J.R., Lannan, C.N., Fryer, J.L., Hedrick, R.P., Meyers T.R., Plumb, J.A., Yamamoto, T. (1987) Morphological and biochemical properties of four members of a novel group of reoviruses isolated from aquatic animals. J.gen.Virol. 68:353-363

Zhang, N., Yang, G. (1985) The establishment and characteristics of cell line ZC-7901 from the snout tissue cells of grass carp. Abstr. Asia Symp. Freshwater Fish Cult.Beijing

Characteristics of Reoviruses Isolated from Cyprinid Fishes in California, USA

R.P. Hedrick, J.M. Groff, T. McDowell, and W.H. Wingfield*
Aquaculture and Fisheries Program
Department of Medicine
School of Veterinary Medicine
University of California
Davis, California, USA 95616

Introduction

The movement of pathogens with shipments of live fish or eggs has always been a concern in the spread of these agents to new geographical locations. A majority of this concern however has dealt with salmonid fish and appropriate regulations governing fish movements have been established at the state and federal level in the USA. Similar regulations exist for several other countries.

The detection of serious pathogens among non-salmonid fishes has received considerably less attention eventhough these fish may be placed into waters with salmonid fishes or commericially and ecologically important species of non-salmonid fishes. Little studied are the effects of some of these agents on the non-salmonid fish themselves.

Sano et al. (1977) found that European eels imported to Japan were infected with several viruses one of which was later found to be pathogenic to rainbow trout. Adair and Ferguson (1981) isolated infectious pancreatic necrosis virus (IPNV) from ornamental fish and recently Michel et al. (1986) have shown that fathead minnows (Pimephales promelas) can suffer from Yersinia ruckerii infections and the bacterium was also pathogenic to salmonid fish.

In the present study we have examined grass carp (Ctenopharyngodon idella) and golden shiners (Notemigonus crysoleucas) for the presence of viral agents. These fish were bound for import or already present in California waters.

---

*Department of Fish and Game
Fish Disease Laboratory
2111 Nimbus Road
Rancho Cordova, California, USA 95670

Methods and Results

Virus Isolations - The spleen and kidney from five-fish pools were processed by standard methods and inoculated onto monolayers of the EPC cell line from common carp (*Cyprinus carpio*). Viruses were isolated from five different groups of fish examined over the period (Table 1). There were no signs of disease eventhough numerous individuals harbored viruses.

Two types of cytopathic effects (CPE) were observed in EPC cells. One was typical of a recently described group of reoviruses from aquatic animals (Winton et al. 1987) and was characterized by massive syncytia. The two members of this group of viruses which are known from fish in North America are the golden shiner virus (GSV) and catfish reovirus (CRV). All five viruses isolated with this type of CPE during the course

Table 1. Isolates of Reoviruses From Cyprinid Fishes in California and Reference Viruses

Designation GSV Serum	Host	Neutralization by anti-GSV-Ser
86-61	Grass Carp	Yes
86-77	Grass Carp	Yes
86-86A	Golden Shiners	Yes
86-86B	Golden Shiners	No
86-89	Golden Shiners	Yes
86-93	Golden Shiners	Yes
87-6	Golden Shiners	Yes
87-13	Golden Shiners	Yes
87-20	Golden Shiners	Yes
Control Viruses		
CSV	Chum Salmon	No
GSV	Golden Shiners	Yes
13p2	Juvenile American Oysters	No

of our study were strongly neutralized by hyperimmune rabbit serum prepared to GSV (Table 1). In contrast, the agent causing a different type of CPE characterized by cell rounding was not neutralized by immune serums prepared against GSV, CRV, IPNV or infectious hematopoietic necrosis viruses.

Infectivity Trials - To determine the virulence of the viruses from cyprinid fishes, infectivity trials were conducted in alevin grass carp (mean wt. 0.05 g) and golden shiners (mean wt. 0.21 g) by immersion exposure to the virus isolated from juvenile grass carp. Four replicate 20 L aquaria containing 50 alevins of each species were drained to contain 1 L of well water (25 C). The culture fluids from infected EPC cells were then added to two aquaria containing grass carp and two containing golden shiners so that they received $10^{5.3}$ $TCID_{50}$/ml of virus. A parallel set of control fish were treated in the same manner but received only culture fluids from uninfected EPC cells. After 30 min of exposure the water flow was resumed to all aquaria (25 $\pm$ 4 C). Both species were fed a commercial larval fish diet by automatic feeders.

Ten fish from each aquaria were examined for virus at each sample period. An additional five fish were also placed into fixative for later histological examinations. There were no mortalities attributed to the virus nor any indications of infection over a 28 d period (Table 2). Virus was however, recovered from pooled samples of kidney, liver and spleen from five fish of both species 7 d following exposure. Virus was detected in three of four pooled samples of grass carp (mean concentration $10^{3.2}$ $TCID_{50}$ /g) and from all four pooled samples from golden shiners (mean concentration $10^{5.4}$ $TCID_{50}$/g). No virus was detected in either control group of golden shiners or grass carp. A subsequent examination of 20 fish of both species at 4 wks showed virus in only one pooled sample from golden shiners but at low concentrations. There was no evidence of viral induced lesions among 10 fish examined from each group at both sampling periods following microscopic examinations of saggital sections of whole fish.

In a second study with the reovirus from grass carp, 30 juvenile grass carp (mean wt. 148.3 g) were injected intraperitoneally with $10^{7.3}$ $TCID_{50}$ of GSV propagated in EPC cells. A parallel control group received injections with culture fluids from uninfected EPC cells. All fish remained healthy in appearance throughout the study. After 3 wk, at a water temperature of 25 C and then again at 3 mo., fish were examined for the presence of virus (Table 3). No virus was detected in controls but

Table 2. Response of Grass Carp and Golden Shiner Fry to Waterborne Exposures to Golden Shiner Virus (GSV)

	Fish			
	Grass Carp		Golden Shiners	
Day	No. positive pools / No. pools	Mean Virus Conc. ($TCID_{50}$/g)	No. positive pools / No. pools	Mean Virus Conc. ($TCID_{50}$/g)
0	0/4	-	0/4	-
7	3/4	$10^{3.2}$	4/4	$10^{5.3}$
28	0/4	-	1/4	$10^{2.2}$

Four replicate groups of 50 fish each were exposed to GSV at $10^{5.3}$ $TCID_{50}$/ml for 30 min before 26 C flowing water was resumed to the aquaria. Control groups received MEM only.

Grass carp and golden shiner fry weighed 0.05 and 0.21 g respectively at the beginning of the study.

No mortalities nor virus were observed among control fish.

Table 3. Experimental infections of juvenile grass carp with golden shiner virus (GSV)

Day ($TCID_{50}$/ml	Virus Recovered	Mean Concentration
0	0/10	-
21	2/10	$10^{3.5}$
90	12/16	$10^{3.3}$

Grass carp juveniles (mean wt. = 148.3 g) received intraperitoneal injections of $10^{7.2}$ $TCID_{50}$ of GSV from EPC cells. Controls received MEM only.

Pooled samples of kidney, liver and spleen were titrated on EPC cells by $TCID_{50}$ analyses.

virus was found in pooled samples of the liver, kidney and spleen of two of 10 fish at 3 wk ($10^{3.5}$ $TCID_{50}/g$). At 3 mo. post injection the remaining fish were examined and virus was detected in 12/16 (mean concentration was $10^{3.3}$ $TCID_{50}/g$) experimental fish but in none of the controls (0/17).

A new agent not neutralized by anti-GSV serum was isolated from one group of golden shiners. This agent caused CPE characterized by foci of rounded cells that gradually increased in size to involve the entire monolayer. There is no indication of syncytium formation seen with the GSV-like isolates. In addition the CPE develops slowly at 25 C (7 versus 1 - 2 d for GSV). Numerous virus particles were seen in the pellet from the supernatant following centrifugation at 90,000 x g for 1 h of the supernatant from infected cells. Similar particles were also seen within membrane bound vacuoles in sections made from infected cells. The virions were round to hexagonal with a mean diameter of 81 nm (n=20) and possessed an electron dense center. The virus lost infectivity following chloroform treatments which suggested the presence of essential lipids although no envelope was detected by electron microscopy. The virus was only partially neutralized by antiserum prepared to GSV (Table 4).

Table 4. Neutralization of a virus from golden shiners (86-86B) with anti-GSV serum

Virus	Anti-GSV Serum	Control Serum
86-86B	2,000	20
GSV	192,000	40

Both GSV and 86-86B were adjusted so that 100 $TCID_{50}$ of virus was present in each test well.

In initial trials similar to those describe above, there was no evidence for replication of the agent in golden shiners tested after water expousure to the agent.

Discussion

Reoviruses have now been isolated from eight species of fish and one specie of shellfish, in both fresh and marine water habitats (Winton et al. 1987; Winton et al. 1989; Ahne and Kölbl 1988; Lupiani et al. 1989).

Most of these viruses are not associated with overt disease although experimental trials with certain isolates have demonstrated their ability to induce disease and mortality (Meyers 1980). With perhaps the exception of the hemorrhagic virus of grass carp (Chen and Jiang 1984), the viruses isolated from cyprinid fishes although replicating to significant titers in the fish's tissue, fail to cause disease signs or mortality. The most studied cyprinid reovirus is golden shiner virus (GSV).

Initial descriptions of GSV were made by Mitchell et al. (1978), Plumb et al. (1979) and Schwedler and Plumb (1980). Mitchell found that GSV could induce mortality in 10 cm golden shiners following intraperitoneal injections of the virus but details on the concentrations of virus recovered and pathological changes in experimentally and naturally-infected fish were not given. Later studies by Brady (1985) however, failed to demonstrate mortality or histological evidence of infection in golden shiners (6 cm length) inoculated with GSV although virus was recovered from fish at 10, 20 and 30 d post intraperitoneal injection. In our study, we were also unable to attribute any disease or histological evidence for tissue changes associated with virus replication in golden shiners or grass carp. From our brief survey however, it is apparent that GSV and related agents are widespread in at least two species of cyprinid fishes (Table 1). All but one of the viruses isolated from the cyprinids examined in our study were neutralized by antiserum made to GSV and furthermore showed similar RNA segment patterns in polyacrylamide gels (data not shown).

Ahne and Kölbl (1988) were also able to recover a virus similar to GSV from apparently healthy tench (*Tinca tinca*) and chub (*Leuciscus cephalus*) in Germany. Although direct comparisons of the viruses from European, North American and Chinese cyprinids are pending, it is apparent that cyprinids are appropriate hosts to these reoviruses. Movement of these cyprinids might therefore be responsible for the appearance of similar viruses in cyprinids from distant geographical regions. Fortunately, the agents may not be catastrophic pathogens as demonstrated in our studies with alevin and juvenile grass carp and alevin golden shiners (Table 2 and 3). It is clear however, that the host can carry the viruses for long periods of time without any signs of infection.

Our viral examinations of cyprinids in California also resulted in the isolation of a previously unknown agent from golden shiners. This virus is not neutralized by anti-GSV serum and preliminary studies have shown that the genome does not consist of ds RNA. No evidence for replication of the virus in golden shiners could be demonstrated in infectivity trials which would suggest that the agent is not a potential pathogen of cyprinid fishes. Further studies to characterize the agent are underway.

In conclusion, it is evident that the cyprinid fishes from California examined in this study are host to a least two types of viruses the most prominent being agents similar to GSV. The effects of GSV on the host however, are unclear and recent studies suggest that it is not a serious pathogen under controlled laboratory conditions. The more subtle effects of the agent on growth, reproduction and resistance to other pathogens is not known. Caution should be taken during movements of these infected fishes as contact with new species in other geographical regions may result in a more severe host-pathogen relationship with potentially disastrous and unforseen results.

References

Adair, B. M. and H. W. Ferguson. 1981. Isolation of infectious pancreatic necrosis (IPN) virus from non-salmonid fish. J. Fish Dis. 4:69-76.

Ahne, W. and O. Kölbl. 1988. Occurrence of reoviruses in European cyprinid fishes. J. Appl. Ichthyol. 3:139-141.

Brady, Y. J. 1985. Comparative serological and histopatholoy of golden shiner virus, chum salmon virus, reovirus 13P2 and channel catfish reovirus infection in golden shiners (Notemigonus crysoleucas Mitchell). PhD Thesis, Auburn University, Auburn, Alabama.

Chen, Y. and Y. Jiang. 1984. Morphological and physico-chemical characteriziation of the hemorrhagic virus of grass carp. Kexue Tongbao 29:832-835.

Meyers, T. R. 1980. Experimental pathogenicity of reovirus 13p2 for juvenile American oysters Crassostrea gigas (Gmelin) and bluegill fingerlings Lepomis macrochirus (Rafinesque). J. Fish Dis. 3:187-201.

Michel, C., B. Faivere and P. De Kinkelin. 1986. A clinical case of enteric redmouth in minnows (Pimephales promelas) imported in Europe as bait-fish. Bull. Eur. Ass. Fish Pathol. 6:97-99.

Mitchell, A. J., J. A. Plumb and P. R. Bowser. 1978. Field and laboratory findings on a viral disease of Golden Shiners. Proceedings of the Joint Third Biennial Fish Health Section and the Ninth Annual Midwest Fish Disease Workshops. Kansas City, Missouri. p. 34.

Lupiani, B., C. P. Dopazo, A. Ledo, B. Fouz, J. L. Barja, F. M. Hetrick and A.E. Toranzo. A new syndrome of suspected viral etiology in cultured turbot (Scophthalmus maximus L.). Journal of Aqua. Animal Health (in press).

Plumb, J. A., P. R. Bowser, J. M. Grizzle and A. J. Mitchell. 1979. Fish viruses: A double-stranded RNA icosahedral virus from a North American cyprinid. J. Fish. Res. Board. Can. 36:1390-1394.
Sano, T., T. Nishimura, N. Okamoto and H. Fukuda. 1977. Studies on viral diseases of Japanese fishes - VII. A rhabdovirus isolated from European eel, Anguilla anguilla. Fish Path. 43:491-495.
Schwedler, T. E. and J. A. Plumb. 1980. Fish viruses: Serological comparison of the golden shiner virus and infectious pancreatic necrosis viruses. J. Wild. Dis. 16:597-599.
Winton, J. R., C. N. Lannan, J. L. Fryer, R. P. Hedrick, T. R. Meyers, J. A. Plumb and T. Yamamoto. 1987. Morphological and biochemical properties of four members of a novel group of reoviruses isolated from aquatic animals. J. Gen. Virol. 68:353-364.
Winton, J. R., C. K. Arakawa, C. N. Lannan and J. L. Fryer. 1989. Isolation of a reovirus from coho salmon (Oncorhynchus kisutch) in Oregon, USA. Picornaviruses and reoviruses of fishes. Proceedings of the International Symposium on Viruses of Lower Vertebrates. University of Munich, Federal Republic of Germany. August 22 - 25, 1988.

Some Properties of Reoviruses Isolated from Tench (*Tinca tinca*) and Chub (*Leuciscus cephalus*)

W. Ahne and O. Kölbl
Institute of Zoology and Hydrobiology of the University of Munich
Kaulbachstr. 37
D-8000 München 22
Fed. Rep. Germany

Introduction

Double stranded RNA viruses (Birnaviridae, Reoviridae) are found to be widely distributed in the nature. Birnaviridae consisting of 2 and Reoviridae consisting of 10-12 segments of dsRNA are described in aquatic and in land animals. Reoviruses obtained from fish e.g. golden shiner (Notemigonus crysoleucas), chum salmon (Oncorhynchus keta), channel catfish (Ictalurus punctatus) were found to be unlike members of the established genera of Reoviridae (Winton et al. 1987). In this paper we describe some properties of reovirus-like particles isolated from chub (Leuciscus cephalus) and tench (Tinca tinca).

Materials and Methods

The viruses used in this study were isolated from healthy appearing tench and chub (Ahne and Kölbl, 1987).
For viral susceptibility different fish cells (FHM, EPC, RTG-2, CHSE-214, CHH1, BB, BF2) were used. Monolayer of cell cultures infected with MOI:1 of each virus were incubated at 20°C for 7 days. Titrations of viruses were done in microtiter plates by TCID50 assay. Endpoints were calculated by the method of REED and MUENCH (1938).
Antisera against the isolate from tench was prepared in rabbits as described elsewhere (Ahne, 1981).
Neutralization of infectivity of the viruses were carried out in microtiter-systems.

*Lannergasse 42
A-3001 Mauerbach
Austria

Effects of chloroform, acid (pH3) and 5-iodo-2-deoxyuridine (IUdR) to the isolates were determined as described by ROVOZZO and BURKE (1973).
For electron microscopy the viruses were sedimented from the infectious tissue culture fluid by ultracentrifugation (85000xg, 3 hours). The negatively stained preparations were examined in a Zeiss EM 109.
From purified virus (sucrose gradient 50-10%, treatment with freon) the nucleic acid was extracted by use of phenol and ethanol. Polyacrylamide gel electrophoresis (9% acrylamide SDS-PAGE vertical slab gels) was used for determination of the nucleic acid segments as described by WINTON et al. (1987).
Testing of the infectivity of the isolates for young cyprinid fishes was done by using carp (Cyprinus carpio), roach (Rutilus rutilus) and redfin (Scardinius erythrophthalmus). Fish infected intraperitoneally with 10^5TCD50 of each virus were kept at a water temperature of 10°C and infectivity of organs was determined 4 weeks later.

Results

The viruses replicated best in FHM cells and to a lower extent in the other cells used except in BF2 cells which proved to be refractory to the viruses. Highest titers up to $10^{6,5}$ TCID50/ml were obtained in FHM cells, lowest titer was found in BB cells (Table 1).

Table 1 **Susceptibility of different fish cells to the cyprinid reoviruses (20°C)**

Cells	Reovirus from Tench	Reovirus from Chub
	$\log_{10}$ TCID50/ml	
FHM	5,2	6,5
EPC	3,5	3,2
RTG-2	3,5	3,5
CHSE-214	2,5	3,5
CHH1	1,8	2,5
BB	1,2	1,2
BF2	0,0	0,0

In Giemsa stained infected FHM cells syncytia formations were evident 24-48 hours after infection (Fig. 1).

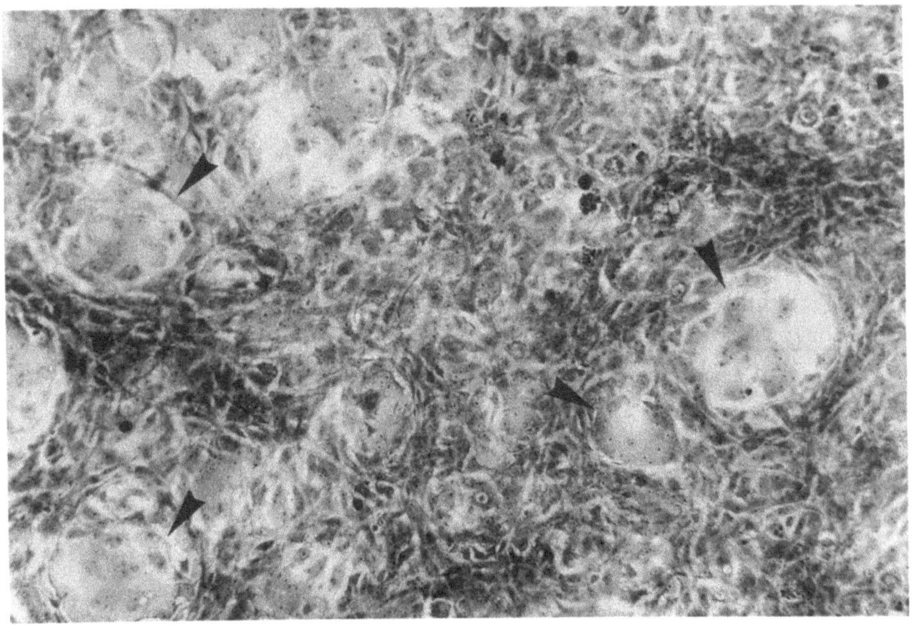

Fig. 1. FHM cells infected with the isolate from tench showing syncytia 72 hours after infection. (magnification 50x)

Treatment of the isolates with chloroform, acid, IUdR and freezing/thawing had no effect on the infectivity of both viruses (Table 2).

Table 2 Treatment of the cyprinid reoviruses with chloroform, acid, IUdR and freezing/thawing

Treatment	Reovirus from Tench, log_{10} TCID50/ml: treated	Reovirus from Tench, log_{10} TCID50/ml: control	Reovirus from Chub, log_{10} TCID50/ml: treated	Reovirus from Chub, log_{10} TCID50/ml: control
Chloroform	6,5	6,2	7,5	7,8
IUdR	6,2	6,2	7,2	7,5
pH3	6,2	6,2	7,2	7,5
freezing/thawing	6,5	6,2	7,2	7,2

Electron microscopy of negatively stained virions revealed spherical to icosahedral particles with a double capsid layer and a diameter of 70-75 nm (Fig.2).

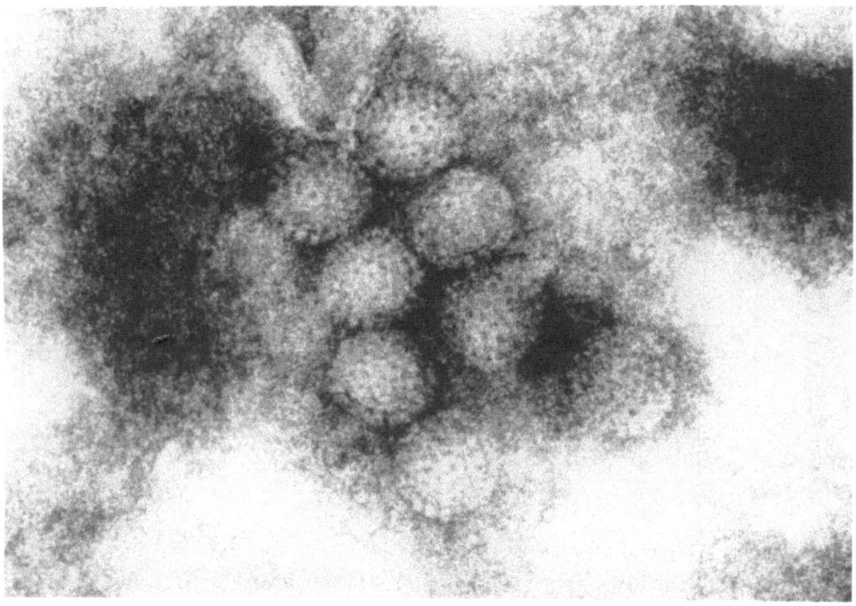

Fig. 2 Electron micrographs of negatively stained virions. The particles show typical reovirus-like morphology. (magnification about 200.000x)

Polyacrylamide gel electrophoresis of viral genome exhibited 11 segments dsRNA. Three large, three medium and five small segments showed similar migration patterns of the isolates, CSV and GSV run in the same gel. Differences were seen between the RNA of the isolates in the migration distances of the three medium and the five small RNA segments. Similar but not identical migration patterns were found between the reference viruses CSV, GSV and the isolates from chub and from tench (Fig. 3).

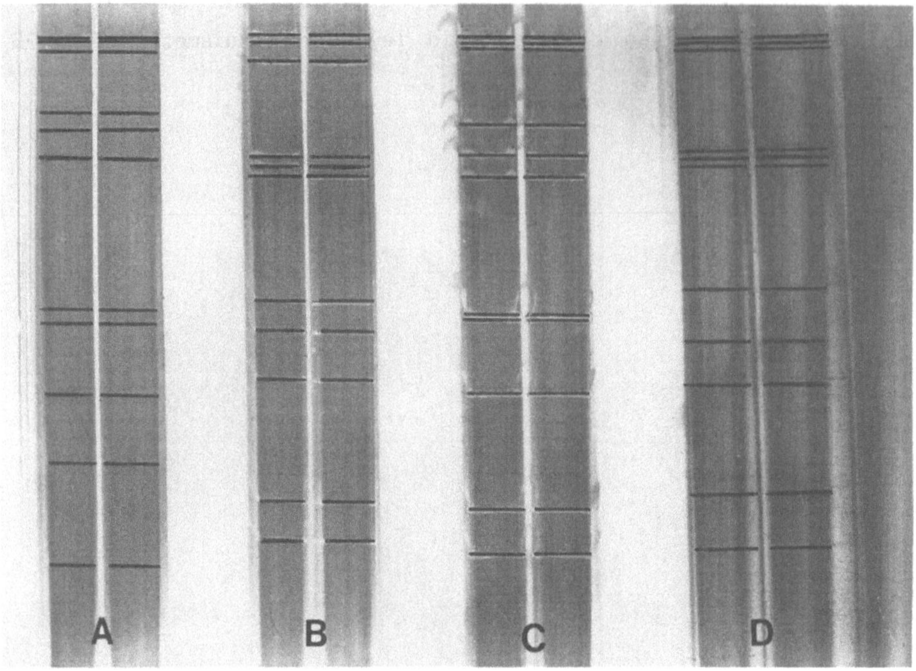

Fig. 3 SDS-PAGE of RNA of CSV(A), GSV(B), reovirus from tench (C) and from chub (D)

Antiserum against the isolate from tench neutralized the tench reovirus and the chub reovirus in dilutions of 1:4096, whereas GSV and CSV were neutralized by an antiserum dilution of 1:2048 and 1:128 respectively. In contrast, the CRV was not neutralized by the antiserum used (Table 3).

Table 3 Neutralization of the isolates and GSV, CSV, CRV with antiserum against the tench reovirus

Virus	Neutralization
tench isolate	1:4096
chub isolate	1:4096
GSV	1:2048
CSV	1:128
CRV	0

The cyprinid fish infected with the reoviruses from tench and chub did not show any signs of diseases four weeks after infection. The viruses could even not be reisolated from the infected fish.

Discussion

Viral diseases of fishes play an important role as limiting factors in aquaculture. Among the more than 50 viruses discovered in teleost fishes there exist members of the 12 established virus families. Recently, a new genus of Reoviridae covering aquatic reoviruses has been suggested (Winton et al. 1987). The aquatic reoviruses possess unique characteristics but they do not fit exactly in any of the genera of Reoviridae e.g. Reovirus, Rotavirus, Orbivirus, Fijivirus, Phytoreovirus, Cypovirus. The viruses isolated from chub and tench exhibited characteristics of Reoviridae e.g. icosahedral morphology, double protein layer, dsRNA and a size of 70-75 nm. Several fish cells replicated the viruses inducing syncytia formation followed by lysis of the cells. The virions composed of 11 segments of dsRNA were much like the reoviruses of aquatic animals described by Winton et al. (1987). Serologically, the viruses from European cyprinids shared common antigens and they showed a serological relationship to the golden shiner virus (GSV) (Plumb et al., 1979) and to the chum salmon reovirus (Winton et al., 1981) but not to the channel catfish reovirus (CRV), Amend et al., 1984). On the basis of the mobilities of the viral RNA segments the reoviruses from chub and tench could be distinguished from each other and from GSV and CSV. However, similarities in the three large segments of RNA of GSV, CSV and the cyprinid reoviruses were evident. The three medium sized segments were similar between GSV and the isolate from chub and between CSV and the isolate from tench respectively. The five small segments differed slightly. Similar results obtained with CRV, CSV, GSV were described by Winton et al. (1987). The viruses obtained from tench and from chub proved to be not pathogenic to carp, roach and redfin. Winton et al. (1987) pointed out that the pathogenicity of reoviruses found in golden shiner, chum salmon and channel catfish is very low for that fish. This observation is in accordance with the mammalian and avian reoviruses which generally do not appear to be important cause of diseases.

Acknowledgement

We would like to thank Dr. J.R. Winton, National Fishery Research Center, Seattle, USA, for providing the reference reovirus.

References

Ahne, W., (1981) Serological techniques currently used in fish virology. Develop.biol. Standard 49, 3-27.

Ahne, W., (1985) Virusinfektionen bei Fischen: Ätiologie, Diagnose und Bekämpfung. Zbl. Vet. Med. B, 32, 237-264.

Ahne, W., Kölbl, O. (1987) Occurrence of reoviruses in European cyprinid fishes (Tinca tinca Lin.; Leuciscus cephalus Lin.)

Amend, D.F., McDowell, T., Hedrick, R.P., (1984) Characteristics of a previously unidentified virus from channel catfish (Ictalurus punctatus). Can. J. Fish. Aquat. Sci. 41, 807-811.

Plumb, J.A., Bowser, P.R., Grizzle, J.M., Mitchell, A.J., (1979) Fish viruses: a double-stranded RNA icosahedral virus from a North American cyprinid. J. Fish Res. Bd. Can. 36, 1390-1394.

Reed, L.J., Muench, H., (1938) A simple method for estimating fifty percent end points. Americ. J. Hyg. 27, 493-497.

Rovozzo, G.C., Burke, C.N., (1973) A manual of basic virological techniques. Prentice-Hall, Inc., Englewood Cliffs, N.J., USA.

Winton, J.R., Lannan, C.N., Fryer, J.L. Kimura, T., (1981) Isolation of a a new reovirus from chum salmon in Japan. Fish Pathol. 15, 155-162.

Winton, J.R., Lannan, C.N., Fryer, J.L., Hedrick, R.P., Meyers, T.R.,Plumb, J.A., Yamamoto, T., (1987) Morphological and biochemical properties of four members of a novel group of reoviruses isolated from aquatic animals. J. gen. Virol. 68, 353-364.

Isolation of a Reovirus from Coho Salmon (*Oncorhynchus kisutch*) in Oregon, USA

J.R. Winton, C.K. Arakawa, C.N. Lannan, and J.L. Fryer
Department of Microbiology
Nash Hall 220
Oregon State University
Corvallis Oregon 97331-3804 USA

INTRODUCTION

Reoviruses isolated from aquatic animals share certain common characteristics: (1) a typical reovirus-like morphology which shows an icosahedral particle with a double capsid that is approximately 75 nm in diameter; (2) a genome with eleven segments of double-stranded RNA (dsRNA) distributed as three large, three medium and five small segments with a total molecular weight of approximately 15 x 10^6; (3) a virion composed of five major and several minor structural proteins that range in molecular weight from 32,000 to 137,000; and (4) form plaque-like syncytia in monolayer cultures of fish cells. Intact virus particles have buoyant densities in CsCl of 1.34 to 1.36 g/ml. The viruses have been isolated from fish and shellfish collected in both the marine and freshwater environments and will replicate in several fish cell lines (Plumb *et al.*, 1979; Meyers and Hirai, 1980; Winton *et al.*, 1981; Nagabayashi and Mori, 1983; Hedrick *et al.*, 1984; Chen and Jiang, 1984). The original four aquatic reovirus isolates have been compared by Winton *et al.*, 1987.

This report describes the isolation of a reovirus from coho salmon (*Oncorhynchus kisutch*). The virus was obtained

---

*Present Address:

U.S. Fish and Wildlife Service
National Fisheries Research Center
Building 204 Naval Station
Seattle Washington 98115 USA

from sexually mature fish returning to the Coquille River located on the central coast of Oregon USA. Partial characterization of the isolated agents indicate they are reoviruses. The exact relationship of the coho salmon reovirus to the other aquatic reoviruses is not known, but it is believed to be a new agent.

MATERIALS AND METHODS

Virus Isolation: This agent was isolated as part of a routine examination of adult fish for virus. Kidney, spleen and ovarian fluid samples were obtained from sexually mature adult coho salmon returning to the Coquille River and the adjacent Eel Lake. Tissues and fluids were collected in five fish pools and prepared for viral examination in accordance with the American Fisheries Society Fish Health Section Blue Book (Amos, 1985). Tissue extracts were inoculated onto monolayer cultures of chinook salmon embryo, *Oncorhynchus tshawytscha*, cells (CHSE-214 ATCC CRL 1681) (Lannan *et al.*, 1984) and incubated at 15°C. After 14 days incubation 0.5 ml of the culture fluid was inoculated onto fresh cells and incubated for an additional 14 days.

Virus Purification: When cytopathogenic effect (CPE) was extensive in cultures inoculated with virus, any remaining cells were scraped from the surface of the flask into the growth medium and the culture fluid was centrifuged for 1 h at 100,000 g. The pellet was resuspended in 2 ml SSC buffer (0.15 M sodium chloride, 0.015 M sodium citrate, pH 7.4) and homogenized with 20 strokes of a Ten Broek homogenizer. Genetron (1,1,2-trichloro-1,2,2-tri-fluoroethane; Aldrich Chemicals) was added to the homogenate and mixed for 5 min. After centrifugation for 10 min at 1000 g, the aqueous phase was removed, placed onto a continuous sucrose gradient (15 to 50% in SSC) and centrifuged for 2 h at 100,000 g. The viral band in the center of this gradient was removed and pelleted by centrifugation for 1 h at 115,000 g. The pellet of partially purified virus was resuspended in 0.5 ml of SSC, layered onto a three-step CsCl gradient (2.0 ml 40%, 1.5 ml 30% and 1.0 ml 20% CsCl in SSC) and centrifuged for 16 h at 115,000 g. The band of purified virus was removed from the gradient, diluted with SSC and pelleted for 1 h at 115,000 g.

Electron Microscopy: Purified virus was resuspended in distilled water and a drop placed onto a Formvar (Electron Microscopy Services, Fort Washington, PA, USA) coated grid. A drop of unbuffered 2% phosphotungstic acid (adjusted with NaOH to pH7) was added for 1 min, the excess removed with a filter paper wedge and the virions examined with a Philips EM300 electron microscope.

Sensitivity of Cell Lines: Eight established fish cell lines were tested for their ability to detect and replicate the coho salmon reovirus. The salmonid lines were: CHSE-214; CHH-1 (ATCC CRL 1680) from chum salmon, *Oncorhynchus keta*; CSE-119 derived from coho salmon, *O. kisutch*; K06 from kokanee salmon, *Oncorhynchus nerka*; and STE-137 from steelhead trout, *Salmo gairdneri*. All these cell lines have been described by Lannan *et al.*, 1984. In addition, the EPC line from the common carp, *Cyprinus carpio* (Fijan *et al.*, 1983); FHM (ATCC CCL 42) from the fathead minnow, *Pimephales promelas* (Gravell and Malsburger, 1965); and PHE, an uncharacterized cell line from the Pacific herring, *Clupea harengus pallasi*, were used. Cells were propagated in Eagle's minimum essential medium (MEM), supplemented with Earle's salts, fetal bovine serum (5%), and 0.2 M L-glutamine. Test cultures were incubated at 18°C.

Pairs of 25 cm^2 flasks of each cell line were inoculated at a multiplicity of infection (MOI) of ~0.1. Culture medium from duplicate flasks was pooled and titered at 7 and 14 days by end point dilution assay ($TCID_{50}$) on CHH-1 cells. Results were computed by the method of Reed and Muench, 1938.

Serology: To prepare antiserum against the coho reovirus, a New Zealand white rabbit was injected intramuscularly (IM) with ~$10^{8.5}$ $TCID_{50}$ partially-purified virus mixed with Freund's complete adjuvant. After four weeks, the rabbit was given a second IM injection with ~10^{9} $TCID_{50}$ purified virus mixed with Freund's incomplete adjuvant. Eight weeks after the initial injection, the rabbit was injected in the ear vein with ~$10^{8.5}$ $TCID_{50}$ purified virus suspended in balanced salt solution (BSS). The rabbit was bled from the ear vein and the serum titered at 2, 4, 8, and 12 weeks. A serum neutralization assay using the alpha procedure of Rovozzo and Burke, 1973, was performed with the coho reovirus and a polyclonal rabbit antiserum specific for the chum salmon virus (Winton *et al.*, 1981).

Effect of Temperature on Viral Replication: Replicate 75 cm^2 flasks of CHH-1 cells were inoculated with coho reovirus at $10^{2.7}$ $TCID_{50}$/ml, and incubated at 9, 12, 15, 18, 21 and 24°C. At 2 day intervals for 14 days 0.5 ml of culture medium was harvested from each flask. Paired samples were pooled and titered on CHH-1 cells.

Pathology: Rainbow trout, *Salmo gairdneri*, and coho salmon fingerlings weighing 1-2 g were injected intraperitoneally with 10^4 $TCID_{50}$ of the coho reovirus suspended in BSS. On the first day and each week for eleven weeks, five fish were removed for histological examination. Control fish were injected with BSS and five fish harvested at day 0, and at weeks 6 and 10. All fish were fixed in Bouin's solution, embedded in paraffin, sectioned, and stained with hematoxalin and eosin.

Nucleic Acid Analysis: The number and molecular weights of genomic RNA segments were determined by polyacrylamide gel electrophoresis, using 9% gels (Laemmli, 1970). The dsRNA was extracted from purified virus using an equal volume of phenol saturated with SSC. After mixing for 5 min, the homogenate was centrifuged at 2000 g for 10 min, the aqueous phase removed and the extraction repeated twice. The RNA was precipitated with cold ethanol, collected by centrifugation at 10,000 g for 15 min, suspended in sample buffer and applied to the gel. After electrophoresis for 20 h at 10 mA, the RNA segments were stained with silver (Merrill et al., 1981).

RESULTS

Virus Isolation: After 14 days incubation cell cultures inoculated with samples from Coquille River fish appeared normal, and 0.5 ml of culture medium was transferred to fresh cells for a second, or blind pass, incubation period. Cells were examined every other day, and cells receiving ovarian fluid samples remained normal. However, on the 14th day of the second incubation period plaque-like syncytial areas were observed in one culture inoculated with pooled kidney and spleen tissues (Fig. 1). Cells exposed to tissue homogenates from Eel Lake fish showed CPE 6 days after initial inoculation. In subsequent passages the time required by either isolate for the appearance of CPE was reduced to approximately 3-5 days. Cytopathogenic effect generated by both isolates was identical to that described by Winton et al., 1981, for the chum salmon reovirus.

Electron Microscopy: Examination of virus taken from CHSE-214 cell cultures revealed icosohedral to spherical particles approximately 75 nm in diameter and frequently with a double capsid. The appearance of the virus was similar to other reoviruses isolated from fish (Winton, et al., 1987) (Fig. 2a and 2b).

Sensitivity of Cell Lines: Cytopathogenic effects were produced in CHSE-214, CHH-1, STE-137, KO-6, and PHE cells. Characteristic CPE was not observed in EPC, FHM or CSE-119

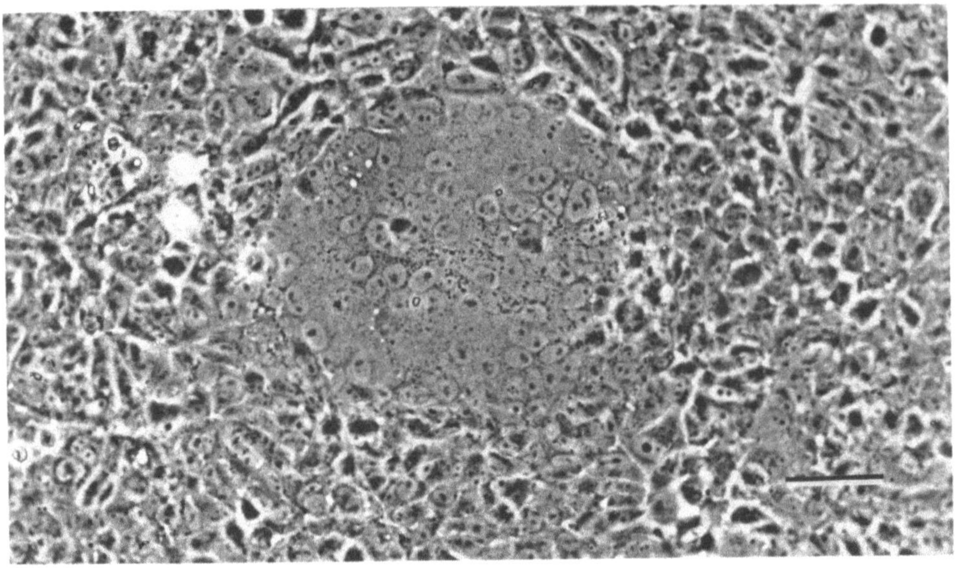

Figure 1. Cytopathogenic effect produced by the coho salmon reovirus in CHH-1 cells incubated at 18°C for 96 hours. Phase contrast microscopy.

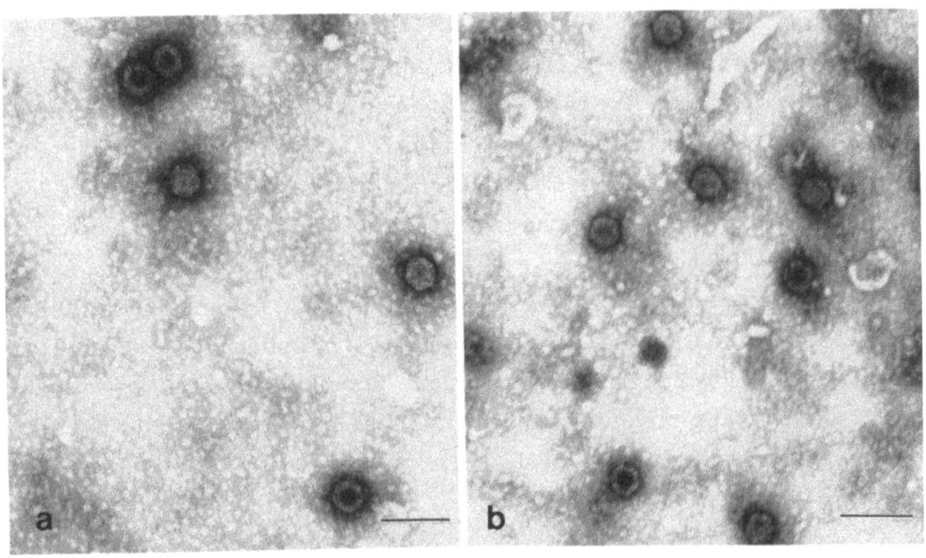

Figure 2a and b. Electron micrographs of virions of the coho salmon reovirus stained with phosphotungstic acid (2a Eel Lake isolate and 2b Coquille isolate).

cultures, although alterations in the appearance of some cells were noted. We did not feel this constituted the typical CPE observed in the sensitive cell lines. The virus replicated in three of the cell lines tested. The CHH-1 cell line produced a titer of $10^{6.3}$ $TCID_{50}$/ml while CHSE-214 and KO-6 produced titers of $10^{6.0}$ $TCID_{50}$/ml. Cytopathogenic effect was apparent in the STE-137 and PHE cells, but no new virus was detected beyond the inoculum (Table 1).

Table 1. Ability of eight fish cell lines to detect and replicate the coho salmon reovirus.

Cell Line	Cytopathogenic Effect	Virus Titer ($TCID_{50}$/ml)
CHSE-214	+	$10^{6.0}$
CHH-1	+	$10^{6.3}$
KO-6	+	$10^{6.0}$
CSE-119	-	-[1]
STE-137	+	-
PHE[2]	+	-
FHM	-	-
EPC	-	-

[1]- indicates $<10^1$ increase in titer above original inoculum.
[2]Pacific Herring Embryo.

Serology: Rabbits inoculated with the coho salmon reovirus failed to respond and no neutralizing antibody was obtained. Neutralization tests were conducted using antiserum against the chum salmon reovirus; however, this antiserum did not neutralize the reovirus from coho salmon.

Effect of Temperature on Viral Replication: Viral replication was measured over a 14 day period at six different temperatures (Table 2). The optimum temperature

Table 2. Replication of the coho salmon reovirus at selected temperatures.

Temperature °C	Log_{10} $TCID_{50}$/ml at days post-infection						
	2	4	6	8	10	12	14
9	2.5	2.6	2.5	2.5	2.5	4.4	4.5
12	3.0	2.6	3.3	4.3	5.0	5.4	5.5
15	2.3	2.0	3.6	5.3	5.3	5.7	5.8
18	2.3	2.3	5.0	5.0	5.3	6.0	6.4
21	2.0	nv	2.5	3.6	4.4	4.3	4.5
24	2.3	nv	2.0	nv	nv	nv	nv

nv = no virus detected (less than 10^2 $TCID_{50}$/ml)

for viral replication appeared to be 18°C. However, viral growth occurred over the range of 9 to 21°C. No replication occurred at 24°C. CPE was first noted on the fourth day at 12, 15, 18, and 21°C. and by the sixth day at 9°C.

Pathology: The coho salmon reovirus, like the chum salmon reovirus, produced a focal, necrotizing hepatitis. To demonstrate this, rainbow trout and coho salmon were injected with virus at 10^4 $TCID_{50}$/fish. No mortality occurred throughout the 11 week period in either the experimental or control groups. Fish collected, sectioned and examined showed hepatic lesions by the end of the third week (Fig. 3), and these lesions remained through the 11th week. Control fish showed no such lesions in the liver or other organs.

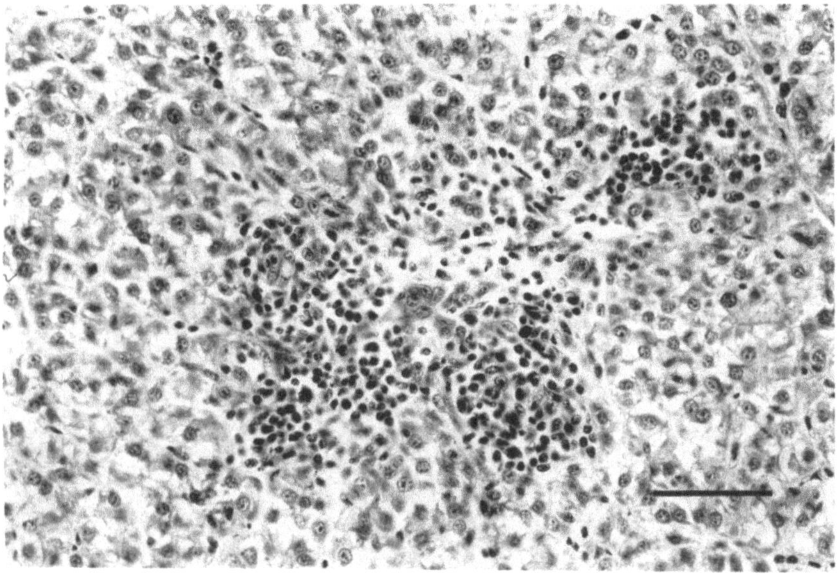

Figure 3. Hepatic lesion in juvenile rainbow trout five weeks post injection with the coho salmon reovirus. Hematoxylin and eosin stain.

The time of appearance and duration of the pathology observed in these experiments was distinctly different from that described for the chum salmon reovirus. Lesions induced by the chum salmon virus appeared earlier and had generally subsided by the end of the sixth week.

Nucleic Acid Analysis: Limited information is available at this time describing the electropherotype of the coho salmon reovirus. It does, however, possess 11 segments of double stranded RNA (Fig. 4). It is distinctly different from the $13p_2$ reovirus described by Meyers from oysters (Meyers, 1979). Note that the patterns shown for the two isolates of reovirus from coho salmon in the Coquille River system are not identical. This indicates differences between two isolates derived from the same host species residing in the same watershed and collected at about the same time.

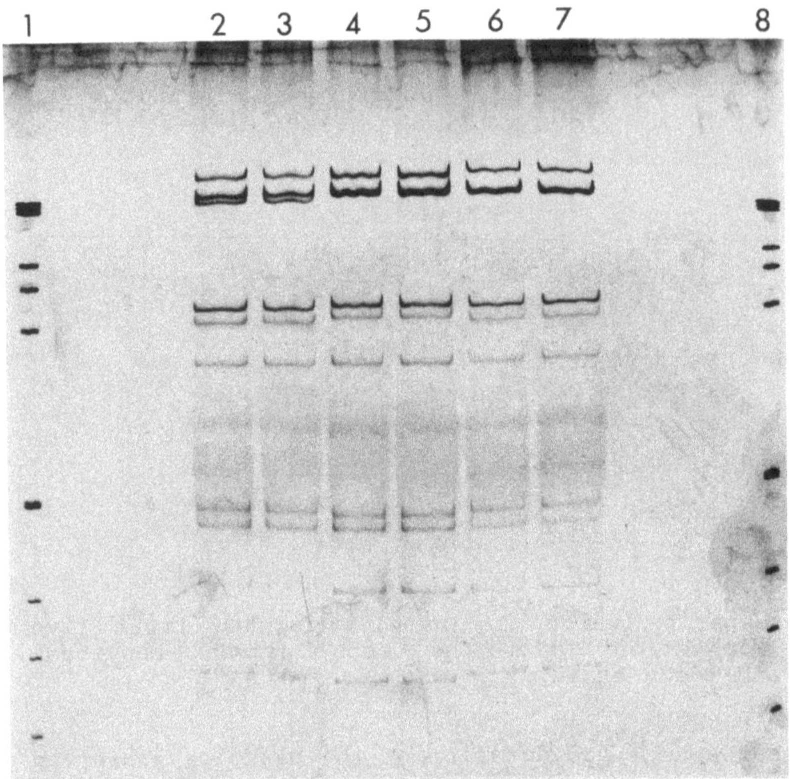

Figure 4. A comparison of the ds RNA genome segments of the chinook salmon reovirus and the two isolates of the coho salmon reovirus on a 9% polyacrylamide slab gel. Electrophoresis was for 20 hours at 10mA. The $13p_2$ reovirus RNA was run in the same gel as a marker. Lanes 1 and 8 are $13p_2$; Lanes 2 and 3 the Coquille coho isolate; Lanes 4 and 5 the Eel Lake coho isolate; Lanes 6 and 7 a chinook salmon isolate.

DISCUSSION

The information contained here represents the first report of a reovirus isolated from cold water fish in North America. Others have been described from warm water species (Plumb, et al., 1970; Hedrick, et al., 1984). It is likely this virus is a member of a newly described group of reoviruses associated with fish and shellfish and referred to as the aquareoviruses. Collectively, the aquareoviruses have the following in common: (1) infect aquatic animals, (2) contain 11 segments of double stranded RNA, (3) exhibit low pathogenicity for their host, and (4) produce a self-limiting disease with what appears to be minor pathology. The pathology produced by the coho salmon reovirus was similar to that described for the chum salmon reovirus. It appeared to be limited to hepatic tissue and there was no associated mortality.

We believe that the coho salmon reovirus is a poor antigen which may make it difficult to obtain significant amounts of high quality antisera against it from rabbits. Similar results have been obtained with $13p_2$ which also appears to be a weak antigen (T. Meyers personal communication, 1988). The genome segments banded differently from the $13p_2$ virus and some differences were noted between the coho isolates described. It is important to do a definitive comparison incorporating all of the reoviruses isolated from fish and perhaps shellfish before any meaningful conclusions can be drawn.

It is interesting to note the virus replicated only in cultures originating from Pacific salmon, but did not replicate or produce CPE in a cell line derived from coho salmon (CSE-119). The origin of the coho salmon reovirus and its role as a pathogen among fish is not well understood. It is possible that it is principally a virus of another host and was only incidentally detected in coho salmon. Since the isolation of the two strains of virus reported here, a third has been obtained at Yaquina Bay near Newport, Oregon from an adult chinook salmon. Only limited information is available about this virus. However, genome segments are very similar to the one obtained from coho salmon at Eel Lake. We believe a variety of reoviruses may exist in fish and shellfish, and in time more of these agents will be revealed with the increased activity directed towards the detection, prevention and control of viral diseases in these populations of animals.

ACKNOWLEDGEMENTS

The authors wish to thank C. Dugan for his excellent help with the photography for this publication and J. Ganzhorn for supplying us with the reovirus isolate from chinook salmon.

This work is a result of research sponsored by the Oregon State University Sea Grant College Program supported by NOAA Office of Sea Grant, U.S. Department of Commerce, under Grant NA85AA-D-SG-095 and is Oregon Agricultural Experiment Station Technical Paper No. 8609.

REFERENCES

Amos KH (ed) (1985) Procedures for the detection and identification of certain fish pathogens. 3d ed Fish Health Section, Am Fish Soc Corvallis Oregon

Chen Y, Jiang Y (1984) Morphological and physico-chemical characterization of the hemorrhagic virus of grass carp. Kexue Tongbao 29:832-835

Fijan N, Sulimanovic D, Bearzotti M, Muzinic D, Zwillenberg LO, Chilmonczyk S, Vautherot JF, de Kineklin P (1983) Some properties of the Epithelioma papulosum cyprini (EPC) cell line from carp Cyprinus carpio. Ann Virol (Inst Pasteur) 134 E:207-220

Gravell M, Malsburger RG (1965) A permanent cell line from the fathead minnow (Pimephales promelas). Ann NY Acad Sci 126:555-565

Hedrick RP, Rosemark R, Aronstein D, Winton JR, McDowell T, Amend DF (1984) Characteristics of a new reovirus from channel catfish (Ictaluras punctatus). J Gen Virol 65:1527-1534

Laemmli UK (1970) Cleavage of the structural proteins during the assembly of the head of bacteriophage T4. Nature, London 227:680-685

Lannan CN, Winton JR, Fryer JL (1984) Fish cell lines: Establishment and characterization of nine cell lines from salmonids. In Vitro 20:671-676

Merrill CR, Goldman D, Sedman SA, Ebert MH (1981) Ultrasensitive stain for proteins in polyacrylamide gels shows regional variation in cerebrospinal fluid proteins. Science 211:143-144

Meyers TR, Hirai K (1980) Morphology of a reo-like virus isolated from juvenile American oysters (Crassostrea virginica). J Gen Virology 46:249-253

Nagabayashi T, Mori S (1983) Characterization of a virus (JOV-1) isolated from the Japanese oysters (Crassostrea gigas). In: Abstracts, Soc Invert Pathol XVI Annual Meeting Cornell Univ NY p56

Plumb JA, Bowser PR, Grizzle JM, Mitchell AJ (1979) Fish viruses: A double-stranded RNA icosahedral virus from a North American cyprinid. J Fish Res Bd Can 36:1390-1394

Reed LJ, Muench H (1938) A simple method for estimating fifty percent end points. Am J Hygiene 27:493-497

Rovozzo GC, Burke CN (1973) A manual of basic virological techniques. Prentice-Hall Inc Englewood Cliffs NJ

Winton JR, Lannan CN, Fryer JL, Kimura T (1981) Isolation of a new reovirus from chum salmon in Japan. Fish Pathol 15:155-162

Winton JR, Lannan CN, Fryer JL, Hedrick, RP, Meyers TR, Plumb JA, Yamamoto T (1987) Morphological and biochemical properties of four members of a novel group of reoviruses isolated from aquatic animals. J Gen Virol 68:353-364

Response of Salmonid Fish to Artificial Infection with Chum Salmon Virus

J.R. Winton[1], C.N. Lannan, M. Yoshimizu[2], and T. Kimura[2]
Oregon State University
Marine Science Center
Newport, Oregon 97365 USA

INTRODUCTION

In the fall of 1978, a reovirus was isolated from normal-appearing adult chum salmon (Oncorhynchus keta) returning to the Tokushibetsu Hatchery in Hokkaido, Japan (Winton et al 1981). The chum salmon virus (CSV) was recovered in the chinook salmon (Oncorhynchus tshawytscha) embryo cell line (CHSE-214) where it replicated at 15-20 C, producing foci of syncytia in the monolayer. Electron microscopy revealed icosahedral particles, 75 nm in diameter, with a double capsid. The virus was not inactivated by chloroform or inhibited by fluorodeoxyuridine. It was unstable at 56 C, did not hemagglutinate human type O erythrocytes, and had a density of 1.33 g/ml in CsCl. The virus was not neutralized by antiserum against infectious pancreatic necrosis virus or mammalian reovirus serotypes 1, 2 or 3 (Winton 1981). Electrophoretic analysis showed the genome was composed of three large, three medium, and five small segments of double stranded RNA (dsRNA) that ranged from 0.37-2.5 x 10^6 molecular weight. The virions contained five major structural proteins and several minor proteins (Winton et al 1983).

After the initial isolation and characterization of CSV, the virus was not recovered again until 1986 despite extensive surveys for fish viruses in Hokkaido (Yoshimizu et al 1988). In 1986, a large mortality among masu salmon (Oncorhynchus masou) fry resulted in the isolation of a virus that was similar, or identical, to CSV. The titer of infectious virus in some

---

[1]U.S. Fish and Wildlife Service
National Fisheries Research Center
Building 204, Naval Station
Seattle, Washington 98115 USA

[2]Laboratory of Microbiology
Hokkaido University
Hakodate, 041 Japan

fish was as high as 10^5/g of internal organ tissue. Since that time, the virus has been recovered from stocks of adult masu salmon at new locations in Hokkaido in 1987 and 1988 (Yoshimizu 1988). This study was designed to examine the pathogenicity of CSV in order to determine if the virus was a threat to these valuable stocks of salmonid fish.

MATERIALS AND METHODS

The study was conducted in two laboratories. The first, at Oregon State University, used rainbow trout (*Salmo gairdneri*) and three species of Pacific salmon. The virus was injected into fry which were examined during the 6 week experiment for mortality, viral titer and pathological changes. Following the 1986 epizootic in masu salmon, a second set of experiments was conducted at Hokkaido University using three and five month old masu salmon challenged by waterborne exposure and intraperitoneal injection. In this experiment, the fish were assayed for mortality and gross external pathology.

Pathogenicity for rainbow trout and chum, chinook and kokanee salmon

The pathogenicity of CSV was tested in chum salmon, chinook salmon, kokanee salmon (*Oncorhynchus nerka*), and rainbow trout fry. Fish weighing 1-2 g were given intraperitoneal (i.p.) injections of 10^4 $TCID_{50}$ of virus in 0.02 ml minimal essential medium (MEM). Two groups of twenty fish were injected with virus and two control groups received 0.02 ml MEM. Fish were held in static aquaria with 12 C fresh water for 42 days and observed daily for mortality. Fry dying during the experiment were frozen at -20 C until they were titered for virus by $TCID_{50}$ assay.

Virus titer in infected fish

The ability of CSV to replicate in chum, chinook, kokanee and rainbow trout fry was tested by injecting 1-2 g fish i.p. with 0.02 ml of MEM containing 10^4 $TCID_{50}$ of virus. At 48 hr intervals for 14 days, then at day 21, 28, and 42, three fish were removed, pooled and frozen at -20 C. At the end of the experiment, fish were homogenized (1:10 w/v) in a balanced salt solution (BSS). The homogenate was centrifuged at 2000 g for 20 min, the supernatant filtered using a 0.45 um pore size membrane filter (Millipore), and the virus titered by $TCID_{50}$ assay.

Histological changes

Histological changes resulting from CSV infection were determined using 1-2 g chum, chinook, and kokanee salmon and rainbow trout fry infected by i.p. injection with 10^4 $TCID_{50}$ of CSV in 0.02 ml MEM. At days 8, 14, 21, 28, and 42, two fish were removed and placed in Bouin's fixative. Tissues were embedded in paraffin, 7 um sections cut of internal organs, stained with hematoxylin and eosin, and examined by light microscopy.

Serial 7 um frozen sections through the visceral portion of infected chum salmon fry were cut using a cryostat (International Equipment Company). The sections were placed on slides and three serial sections were each stained using a different technique.

For acridine orange stain, sections were fixed in Carnoy's fixative, rehydrated in a series of graded alcohols, rinsed in McIlvaine's buffer (pH 3.8), and stained with 0.01% acridine orange in McIlvaine's buffer. The sections were examined using a fluorescence microscope.

For May-Grunwald-Giemsa stain, sections were fixed in absolute methanol for 5 min, stained for 10 min in May-Grunwald, stained for 20 min in Giemsa (1:15), rinsed in distilled water, and examined by light microscopy.

For fluorescent antibody staining, sections were fixed in 100% acetone for 5 min and incubated for 30 min with antiserum against CSV (1:10 in MEM). After incubation, the sections were rinsed with PBS and reacted for 30 min with goat-anti rabbit conjugated with fluorescein isothiocyanate. The slides were rinsed in PBS and examined with a fluorescence microscope.

Experiments with masu salmon

A waterborne challenge with CSV used groups of 50 three month old masu salmon (Oncorhynchus masou) that were exposed for 60 min to the Nakagawa strain of the virus at $10^{2.75}$ $TCID_{50}$/ml or to the Shubuto strain for 60 min at $10^{4.25}$ $TCID_{50}$/ml. Fish were held for 30 days at 10-15 C and examined for mortality and external signs of disease.

An injection challenge used five month old masu salmon that were inoculated i.p. with CSV at $10^{5.25}$ $TCID_{50}$/fish. The fish were held in pathogen-free water and examined for mortality during a 30 day period.

RESULTS

No significant mortality occurred in chum, chinook, kokanee salmon or rainbow trout fry injected with CSV and held for 42 days at 12 C. No gross external signs of infection or gross internal pathology was observed among the few fish dying during the experiment. The $TCID_{50}$ titers of those mortalities in the virus-injected group were not elevated compared to the titers of virus-injected fish that did not die.

The CSV titers of fish were assayed at intervals after infection. Chum salmon virus replicated in the four species tested (Fig. 1). The virus titers increased gradually in the chinook salmon and rainbow trout fry, but replication was more rapid in chum and kokanee salmon fry especially during the second week of infection. The highest titers were recorded in chum salmon and occurred 21 days post-injection. At the end of the 42 day experiment, the titers of all four species were similar.

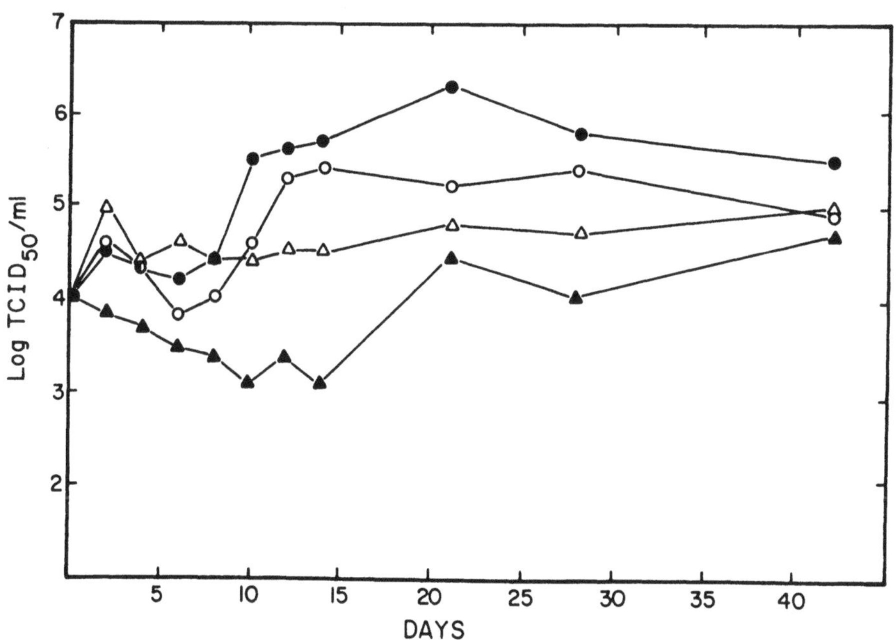

Figure 1. Titer of infectious chum salmon virus in homogenates of whole kokanee (○), chum (●), chinook (△), and rainbow (▲) fry. Fish weighing 1-2 g were injected i.p. with 10^4 $TCID_{50}$ of virus in 0.02 ml MEM and held 42 days at 12 C in 10 L aquaria provided with pathogen-free fresh water.

Histological examination provided evidence of pathology caused by CSV in chum and chinook salmon fry. A focal necrotizing hepatitis was observed in the liver sections from infected fish. In the chum salmon fry, these lesions began as small necrotic foci at 8 days post infection. After 14 days, the lesions were well developed with some syncytia and a coagulation necrosis of the hepatocytes. The lesions began to heal by day 21 as normal cellular architecture began to return (Fig 2). However, at 42 days when the experiment was terminated, the hepatic architecture showed some remaining disorganization. A similar histological picture was seen in the chinook salmon fry, however, the extent and severity of the lesions were less. No pathology was seen in the liver of infected kokanee salmon or rainbow trout. All other internal organs were unremarkable.

A cryostat was used to cut serial frozen sections through the visceral portion of chum salmon that had been infected for 14 days. Acridine orange stain revealed lesions in the liver that were active sites of nucleic acid synthesis. May-Grunwald-Giemsa of the adjacent section showed the lesions were foci of necrosis while the fluorescent antibody stain of a third section showed the presence of high levels of virus antigen in the lesions. No other internal organs or the areas of normal liver surrounding the lesions showed any of these features.

Injection and waterborne exposure of masu salmon to CSV produced no mortality and no external signs of disease. Because the fish were not assayed for virus or histopathological lesions, no estimate of the ability of CSV to replicate in masu salmon can be given.

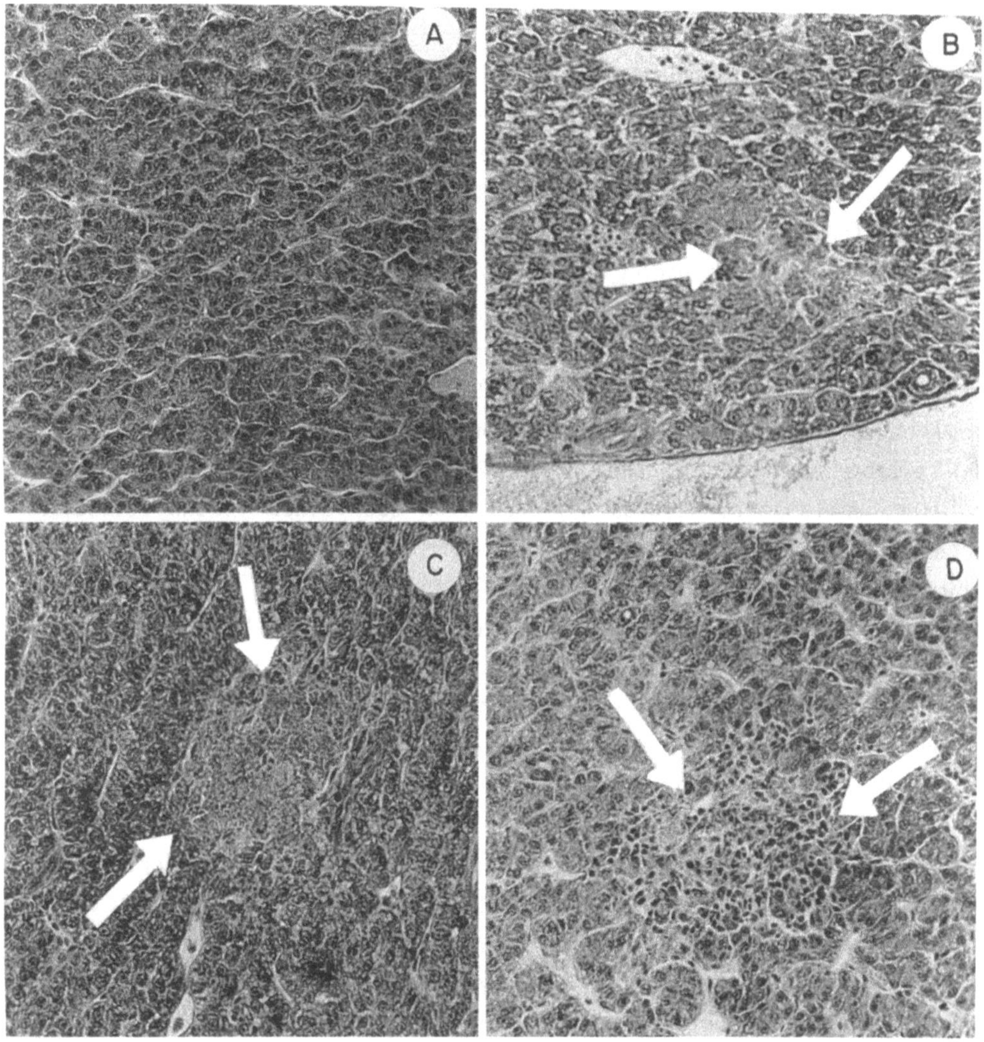

Figure 2. Photomicrographs of hematoxylin and eosin stained sections through the liver of 1-2 g chum salmon fry held at 12 C in 10 L aquaria. Control fish (A) were injected i.p. with 0.02 ml of MEM. Infected fish received 0.02 ml of MEM containing 10^4 $TCID_{50}$ of CSV. Fish were fixed and stained at weekly intervals. The focal hepatic lesions began at day 8 (B), but were most obvious at 14 (C) and 21 (D) days post-infection.

DISCUSSION

The family Reoviridae is composed of non-enveloped viruses, 60-80 nm in diameter with double capsid shells. Members of the family have genomes containing 10-12 segments of dsRNA (Matthews 1982). Although the family contains several important pathogens of vertebrates, many reoviruses have limited pathogenicity for their hosts (Fenner et al 1987). Winton et al (1987) compared the morphology, buoyant density, cell line growth, plaque formation, and molecular weights of genome segments and virion proteins of CSV with those of the golden shiner virus (GSV, Plumb et al 1979), the 13p2 reovirus from oysters (13p2, Meyers 1979), and the catfish reovirus (CRV, Amend et al 1984). The viruses were morphologically similar, had genomes with eleven segments of dsRNA, and virions with five major structural proteins. While these viruses differed from each other in the molecular weight of the structural proteins and genome segments as well as in the optimal temperature and the range of cell lines supporting replication, the four viruses were clearly related. However, they were unlike the members of the six established genera of the Reoviridae (Joklik 1983).

Our results can be compared with those of Meyers (1980) who tested the pathogenicity of the 13p2 virus for bluegill fingerlings. Intraperitoneal injection of bluegills with the 13p2 virus resulted in a significant (44%) mortality. Clinical signs included anorexia, listless behavior and loss of equilibrium. No external lesions were observed. The livers of injected fish were white and a diffuse multifocal hepatitis with large syncytial cells and coagulation necrosis of the hepatocytes was seen. Viral particles were observed in the cytoplasm of hepatocytes of infected fish. Waterborne exposure of bluegills did not result in mortality or in external signs, but the virus was found to replicate within infected fish.

Intraperitoneal injection of adult or fingerling rainbow trout with 13p2 virus produced no mortality and no gross external or internal lesions. Viral replication did occur. Histologically, an extensive, multifocal reticulo-endothelial granulomatous hepatitis was observed. The lesions were most common near the vasculature of the portal areas and the inflammatory areas varied in shape and number. A chronic fibrosing pancreatis was observed in some fish. Sera from infected fish contained neutralizing and precipitating antibodies specific for the virus (Meyers 1983).

Greenfield (1981) reported that the kidney, liver, and spleen tissues of golden shiners (<u>Notemigonus</u> <u>crysoleucas</u>) were able to support GSV replication and appeared to be the target organs of the virus. Brady (1986) compared the pathogenicity of GSV, CRV, 13p2, and CSV by intraperitoneal injection of 10^2 $TCID_{50}$ of each virus into golden shiners. Experimentally infected fish were capable of supporting the replication of each of the viruses at 23 and 28 C. No histopathological changes were seen in the kidney, liver, or spleen of golden shiners infected with GSV. Fish infected with CSV showed lymphocytic infiltration in the liver, but no lesions were seen in spleen or kidney. Fish infected with the 13p2 reovirus showed only pyknotic nuclei in the liver and lesions in the spleen.

Our results show that CSV was not able to cause significant mortality among five species of salmonid fish challenged by artificial exposure. The virus did cause specific pathological lesions in the liver. These lesions may reduce the resistance of the fish to secondary infections or make them more susceptible to predation. Further research on the effect of CSV, and other fish reoviruses, on salmonid fish immunity will be of interest. An examination of various species of marine fish and invertebrates should be conducted to study the pathogenicity of CSV in these species and determine if the host range of the virus extends to other valuable aquatic animals.

ACKNOWLEDGMENTS

This paper is the result research sponsored by the NOAA Sea Grant Office, US Department of Commerce, under Grant No. NA81AA-D-00086 and the National Science Foundation, US-Japan Cooperative Science Program, Grant No. INT 82 10495. The U.S. Government is authorized to produce and distribute reprints for governmental purposes notwithstanding any copyright notation that may appear hereon. Oregon State University Agricultural Experiment Station Technical Paper No. 8691.

REFERENCES

Brady YJ (1986) Comparative serological responses and histopathology of golden shiner virus, chum salmon virus, reovirus 13p2, and channel catfish virus infection in golden shiners (Notemigonus crysoleucas Mitchell). PhD thesis, Auburn University

Fenner F, Bachmann PA, Gibbs EPJ, Murphy FA, Studdert MJ, White DO (1978) Veterinary virology. Academic Press, Orlando

Greenfield DB (1981) Distribution of golden shiner virus in selected tissues and organs of experimentally infected golden shiners (Notemigonus crysoleucas). Ms thesis, Auburn University

Joklik WK (1983) The members of the family Reoviridae. In: Joklik WK (ed) The Reoviridae, Plenum, New York

Matthews REF (1982) Classification and nomenclature of viruses: Fourth report of the international committee on taxonomy of viruses. Intervirol 17:1-199

Meyers TR (1979) A reo-like virus isolated from juvenile American oysters (Crassostrea virginica). J Gen Virol 43:203-212

Meyers TR (1980) Experimental pathogenicity of reovirus 13p2 for juvenile American oysters Crassostrea virginica (Gmelin) and bluegill fingerlings Lepomis macrochirus (Rafinesque). J Fish Dis 3:187-201

Meyers TR (1983) Serological and histopathological responses of rainbow trout, Salmo gairdneri Richardson, to experimental infection with the 13p2 reovirus. J Fish Dis 6:277-292

Winton JR (1981) Isolation and characterization of a new reovirus from chum salmon (Oncorhynchus keta). PhD thesis, Oregon State University

Winton JR, Lannan CN, Fryer JL, Kimura T (1981) Isolation of a new reovirus from chum salmon in Japan. Fish Pathol 15:155-162

Winton JR, Lannan CN, Fryer JL (1983) Further characterization of a new reovirus of poikilothermic vertebrates. In: Compans RW, Bishop DHL (eds) Double-stranded RNA Viruses. Elsevier, New York

Winton JR, Lannan CL, Fryer JL, Hedrick RP, Meyers TR, Plumb JA, Yamamoto T (1987) Morphological and biochemical properties of four members of a novel group of reoviruses isolated from aquatic animals. J Gen Virol 68:353-364

Yoshimizu M (1988) Chum salmon virus isolated from masu salmon. In: Fish and Eggs, Technical Reports of the Hokkaido Salmon Hatchery 157:36-38.

Yoshimizu M, Nomura T, Awakura T, Kimura T (1988) Incidence of fish pathogenic viruses among anadromous salmonid in northern part of Japan. Sci Rep Hokkaido Salmon Hatchery 42:(in press)

A New Serotype of Infectious Pancreatic Necrosis Virus (IPN N 1)

K. E. Christie and L. S. Havarstein
Norbio A/S
Lab. of Biotechnology
University of Bergen
Norway

Infectious pancreatic necrosis virus (IPNV) has been isolated from Norwegian Atlantic salmon during an outbreak with symptoms of IPN and increased mortality. The virus has been compared immunologically and biochemically with the IPNV strains Ab, Sp, Jasper, West Buxton (WB) and VR299, and genetically with the Jasper and Sp strain. The N1 strain differs from all IPNV strains tested and is proposed to represent a new serotype of IPNV.

The IPNV genome contains two segments of ds-RNA (Dobos 1976; Dobos and Rowe 1977; Dobos et al. 1979; Dobos and Roberts 1982; Macdonald and Yamamoto 1977). The A segments encode three polypeptides: preVP2 (precursor protein of VP2), NS (a nonstructural protein) and VP3 (the minor structural protein). The B segment encodes one polypeptide, VP1 (virus RNA polymerase). The major virus capsid polypeptide, VP2, is processed from preVP2. Usually two precursor proteins are detected in IPNV infected cells. VP2 have been shown to contain the antigenic epitopes inducing the

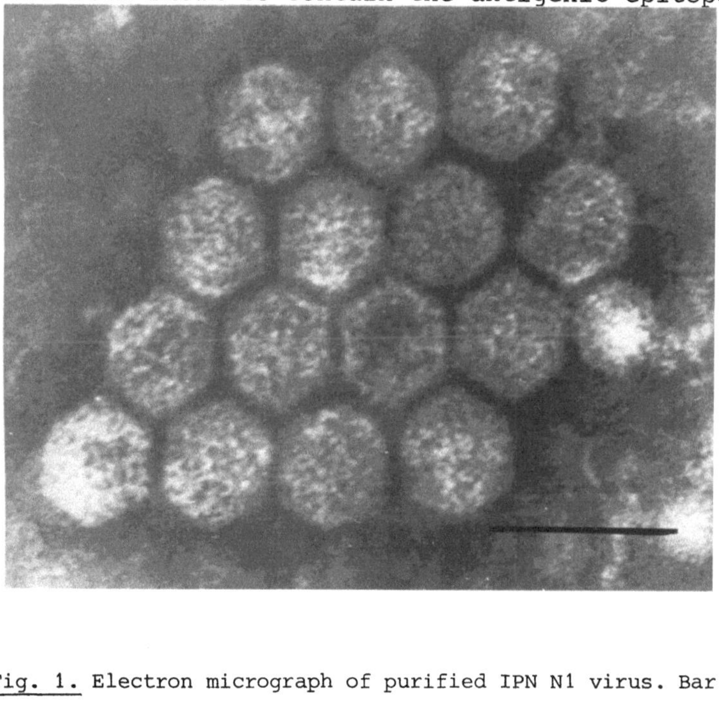

Fig. 1. Electron micrograph of purified IPN N1 virus. Bar = 100 nm

production of neutralizing antibodies against the virus (Caswell-Reno et al. 1986). The molecular weight (MV) of VP2 varies among different IPNV strains from 50 to 54 KD. The MW of VP1, VP3 and NS is about 90, 31 and 29 KD, respectively.

Virus isolated from one year old hatchery reared Atlantic salmon with symptoms of IPN has been compared serologically with the IPNV strains Ab, Sp, Jasper, WB and VR299. Some cross-reactions were observed for all IPNV strains by crossed neutralization assay, especially with antisera raised against the Jasper strain. However, antisera raised against the N1 strain gave a 100 times higher 50% plaque-reduction titer with the N1 virus compared to the titer with heterologous viruses (Christie et al. 1988).

SDS-PAGE analysis of radiolabelled purified N1 virus shows a polypeptide pattern differing considerably from the pattern obtained with the Ab, Sp, Jasper (Figs. 2 and 3) and has also been shown to differ from the pattern of the WB and VR299 strains. The MW of VP2 was 52 KD for the N1 virus compared to 50 KD for the Ab and Sp strain and 54 KD for the Jasper, WB and VR299 strains.

Only one precursor protein with MW of 62 KD was observed with the Jasper strain. Two preVP2 were detected for the other strains and the MW's were about 60 KD and 62 KD for the N1 strain and 54 KD and 55 KD for the Ab and Sp strains. These results may indicate that the major capsid protein of the N1 strain is structurally closer related to the American strains than to the European. The N1 virus did not show the VP4 polypeptide detected for the Jasper, WB and VR299 strains. An additional polypeptide, probably a degradation product of VP3, was observed for all strains. The MW of this polypeptide was about 24 KD of the N1 strain compared to 25 KD of the Jasper and 26 KD of the Ab and Sp strain.

A sandwich ELISA for detection and typing of IPNV in fish has been developed using strain specific antisera as primary and

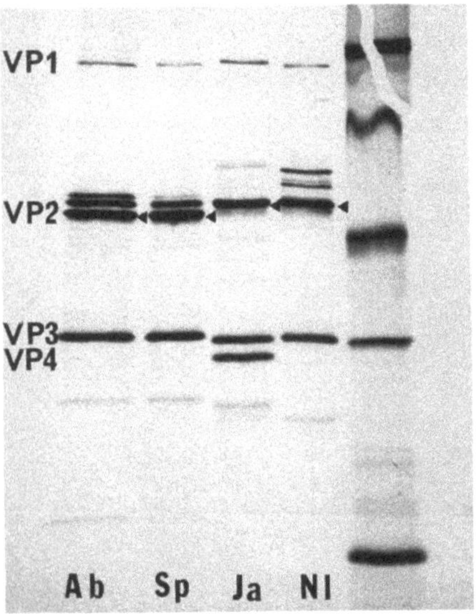

Fig. 2. Autograph of ^{35}-S-methionine labelled polypeptides of purified IPNV strain Ab, Sp, Jasper and N1 analyzed on a 7-20% continuous SDS-PAGE gradient gel together with molecular weight marker proteins of 14, 30, 46, 69 and 92 KD

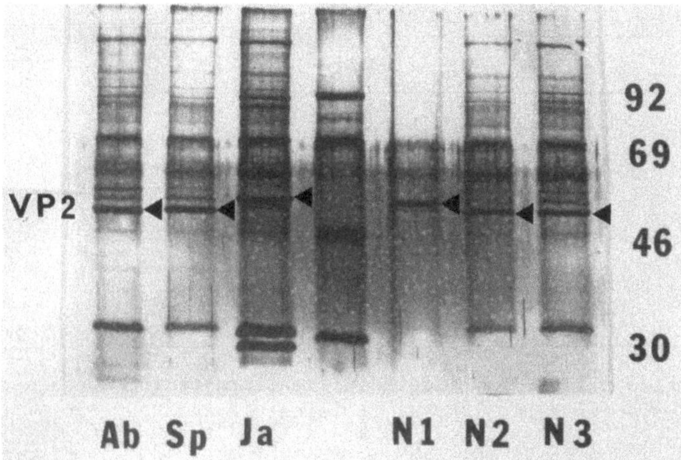

Fig. 3. Silver stained polypeptides of purified IPNV strain Ab, Sp, Ja, and three local IPNV isolates (N1, N2, N3) analyzed on a 12% SDS-PAGE gel together with molecular weight marker proteins of 30, 46, 69 and 92 KD

secondary antibodies. The detection limit of this assay was 10^4PFU/ml, and by this method the N1 strain may be typed.

A library of hybridomas producing antibodies reacting with the N1 virus proteins VP1, VP2 and VP3 has been established. Two monoclonal antibodies against VP2 and two against VP3 have been characterized. The monoclonal antibodies against VP3 did not neutralize the virus but did react as secondary antibodies in a sandwich ELISA with all the IPNV strains Ab, Sp, Jasper, N1, WB and VR299. Monoclonal antibodies against the N1 VP2 protein did neutralize the virus. Preliminary results show that these monoclonal antibodies did neutralize the Jasper, WB and VR299 strains as well as the homologous N1 strain, but did not neutralize the Ab and Sp strains (Fig. 4). These results suggest a relation of the N1 strain with the American and Canadian serotypes.

Molecular cloning and sequencing of the genome segment A have been performed for the Jasper strain (Duncan et al. 1987) and for the Sp strain (Huang et al. 1986). We have made two libraries of the N1 genome in lambda-gt10 and lambda-gt11 vectors. From the lambda-gt10 library the A segment has been cloned in E. coli and a segment of 1800 bp has been sequenced. This segment contains the NS and VP3 genes. The restriction map of this segment, using six different restriction enzymes, is almost identical to the map of the Sp strain. The cDNA base sequence of this segment shows 77% homology with the Jasper strain. The amino acid sequence of the NS gene shows 91% homology with the Sp strain and 80% homology with the Jasper strain.

A new nomenclature of the IPNV has been proposed by Hill and Way (1988) containing two serogroups: A (nine serotypes) and B (one serotype). In addition to the three classical serotypes WB (VR299), Sp and Ab, two European (He and Te) and four Canadian serotypes (Canadian 1, 2, 3) (Macdonald et al. 1983) and Jasper

Neutralization of IPNV strains by monoclonal antibodies		
Virus strain	MAb-VP2	MAb-VP3
VR299	>100	-
WB	>100	-
Sp	-	-
Ab	-	-
Jasper	>100	-
N1	>100	-

Fig. 4. 100% TCID-reduction titer of hybridoma cell culture medium containing monoclonal antibodies against the IPNV N1 proteins VP2 and VP3

have been included. The serotypes are tentatively called A1 - A9. I propose that the N1 strain may represent a new serotype called the A10 serotype.

Conclusion

A new strain of infectious pancreatic necrosis virus has been isolated from Atlantic salmon with IPN during smolt. This strain is regularly detected in fish suffering from Infectious Anemia in Salmon (IAS), a disease of unknown etiology.

The new strain differs serologically and biochemically from the IPNV serotypes WB (VR299), Sp, Ab and Jasper.

Genetic comparison of the NS-VP3 genome suggests a close relationship between the N1 and Sp strains.

The polypeptide pattern of purified virus and neutralization of infectious virus by monoclonal antibodies against the N1 VP2 protein may indicate a closer structural resemblance of N1 to the American/Canadian serotypes than to the European serotypes.

References

Caswell-Reno P, Reno PW, Nicholson BL (1986) Monoclonal antibodies to infectious pancreatic necrosis virus: Analysis of viral epitopes and comparison of different isolates. J Gen Virol 67, 2193-2205

Christie KE, Håvarstein LS, Ness S, Djupvik HO, Endresen C: Characterization of a new serotype of infectious pancreatic necrosis virus isolated from Atlantic salmon. Submitted for publication in Archives of Virology

Dobos P (1976) Size and structure of the genome of infectious pancreatic necrosis virus. Nucl Acid Res 3, 1903-1924

Dobos P, Rowe D (1977) Peptide map comparison of infectious pancreatic necrosis virus-specific polypeptides. J Virol 24, 805-820

Dobos P, Hill BJ, Ross H, Kells DTC, Becht H, Teninges D (1979) Biophysical and biochemical characterization of five animal viruses with bisegmented double-stranded RNA genomes. J Virol 32, 593-605

Dobos P, Roberts TE (1982) The molecular biology of infectious pancreatic necrosis virus: A review. Can J Microbiol 29, 377-384

Duncan R, Nagy E, Krell PJ, Dobos P (1987) Synthesis of the infectious pancreatic necrosis virus polyprotein, detection of a virus-encoded protease, and fine structure mapping of genome segment A coding regions. J Virol 61, 3655-3664

Hill BJ, Way K (1988) Proposed standardization of the serological classification of aquatic Birnaviruses. Presented on the international fish health conference, Vancouver

Huang MTF, Manning DS, Warner M, Stephens EB, Leong JC (1986) A physical map of the viral genome for infectious pancreatic necrosis virus Sp: Analysis of cell-free translation products derived from viral cDNA clones. J Virol 60, 1002-1011

Leong JC, Barrie RC, Engelking JC, Feyereisen-Koener J, Gilmore RD, Huang MTF, Kurath G, Manning DS, Masou C (1988) Development of viral vaccines for fish by molecular cloning. Presented on the international fish health conference, Vancouver

Macdonald RD, Yamamoto T (1977) The structure of infectious pancreatic necrosis virus RNA. J Gen Virol 34, 235-247

Macdonald RD, Moore AR, Souter BW (1983) Three new strains of infectious pancreatic necrosis virus isolated in Canada. Can J Microbiol 29, 137-141

Infectious Pancreatic Necrosis Virus in Norwegian Fish Farms

J. Krogsrud, T. Håstein, and K. Rønningen*

National Veterinary Institute
Pb. 8156 Dep, 0033 Oslo 1
NORWAY

INTRODUCTION

There has been a rapid growth in the industry of sea water fish farming in Norway during the last 20 years. In 1987 the total production of slaughtered fish amounted to 47.000 tons of Atlantic salmon (Salmo salar) and 9.000 tons of rainbow trout (Salmo gairdneri), and the prognosis for 1988 is 80.000 tons of salmon.

The fish farming industry has gradually been more and more specialized, and today the main structure of the breeding programme can be outlined as follows: A large breeding centre supplies 12 broodfish farms with eggs and smolts. These 12 farms furnish some 700 hatcheries with eggs, and the hatcheries provide the smolts needed in the approximately same number of farms producing fish for consumption. Valuable genetic material is exchanged between a separate breeding research station and the large breeding centre. Apart from this cooperative breeding programme there are some other hatcheries depending on their own broodfish.

Infectious pancreatic necrosis virus (IPNV) was isolated for the first time in Norway in 1975 (Håstein and Krogsrud, 1976). The present report gives information on the occurrence of the IPNV infection in Norwegian fish farms, some data about its clinical significance and some preliminary results of a control programme carried out at the breeding centre.

---

* K. Rønningen, Norwegian Fishfarmers' Breeding Centre, 7200 Kyrksaeterøra, Norway

MATERIAL AND METHODS

1. Virus isolation trials

Much of the investigations have been made in connection with the testing for the presence of viral haemorrhagic septicaemia virus in farms exporting fish for consumption to USA. In an increased effort to map the occurence of IPNV samples have been collected also in other farms without apparent disease problems. Diagnostic samples taken in connection with disease outbreaks have been another important part of the material.

Whole fish, kidney, ovarian fluid or milt were sent on ice to the laboratory. As an annual average 5-10 pooled samples from each farm were tested, each sample representing 5 fish. Kidney tissue or fry was homogenized in 9 times its volume of Earle's BSS in a Stomacher 80 (Seward Medical, London). Kidney or fry homogenates and sexual fluids were clarified by sentrifugation and antibiotics (penicillin, streptomycin, kanamycin and amphotericin B) were added.

As a rule the standard methods for virus isolation and serological typing of the isolates were applied as described by Vestergård Jørgensen and Kehlet (1971) and Vestergård Jørgensen (1974). Alternatively, RTG-2 or CHSE-214 cell cultures were used, and the isolates were routinely identified by a constant serum — varying virus neutralization test with antisera against the IPNV types Sp and Ab (Vestergård Jørgensen and Grauballe, 1971).

2. Disease observations

Clinical symptoms and losses during disease outbreaks were recorded by local veterinarians. Routine gross pathological and histological examinations were made on material with a disease history.

3. Control measures

In order to find out whether it is possible to control the IPNV infection in broodfish farms and thus at least to some extent prevent the spread of the virus by transfer of infected material to the hatcheries, a prelimary experiment was initiated along the following lines. In the central breeding centre kidney samples for virus testing were taken from all broodfish intended as origin of new families or in other ways intented for further production at the centre. These samples were taken at slaughter immediately after stripping. The fishes were individually tested. Routine disinfection of eggs with an iodophore (Buffodine^{R}) was carried out to prevent cross contamination between the different hatching trays, and isolation of the individual hatching units was established by physical barriers, separate clothing and equipment and disinfection procedures. Egg and fry batches from parents found infected were removed as soon as possible, and the remaining fry batches were tested for presence of virus at the start feeding stage.

RESULTS

A. Occurrence of infection

As shown in Table 1 there has been a rapid increase in the incidence of IPNV isolations from 1982 and onwards. The infection has been detected in all the broodfish farms, and the prevalence is now very high. The virus has in recent years been frequently found in fry, fingerlings and smolts, as well as in the internal organs and ovarian fluids from adult fish in sea water. By far the most virus strains have been isolated from Atlantic salmon, but the virus has also been occasionally found in rainbow trout, and on single occasions it has been isolated from brown trout (Salmo trutta), pike-perch (Stizostedion lucioperca) and from scallops (Pecten maximus).It has also been found in eels (Anguilla anguilla) imported from the European continent.

TABLE 1. Isolation trials for IPNV in Norwegian fish farms 1975-1987.

Year	Number of farms tested	Number of positive farms	% positive farms
1975	42	4	9.5
1976	37	0	0.0
1977	38	1	2.6
1978	63	4	6.3
1979	55	3	5.5
1980	59	3	5.0
1981	98	4	4.1
1982	296	40	13.5
1983	323	92	28.5
1984	664	197	29.7
1985	548	218	39.8
1986	742	325	43.8
1987	697	401	57.5

B. Clinical appearance and lesions related to serotypes

It was not until 1985 that typical IPN disease was diagnosed in one farm with fry mortality rates of approximately 30 % in two affected tanks. The corresponding rates in the rest of the farm were below 1 %. In 1986-87 clinical outbreaks of IPN were recorded in 11 hatcheries, all with limited losses. In most of these hatcheries the outbreaks were characterized by an increasing mortality for 1-2 weeks. Some whirling movements were observed in the diseased fry. The mortality gradually decreased during the next month. However, some of these outbreaks occurred in connection with smoltification and transfer to sea cages, and after a week with increased losses, the infection seemed not to affect the fish any more.

Both in fry and smolts the gastrointestinal tract contained little or no food, but was more or less extended by mucus. Histological examinations showed an extensive degeneration with necroses in the exocrine pancreas. The acinar tissue was often replaced by a detritus composed of eosinophilic cell fragments, pycnotic nuclei and zymogen granules. Some infiltration with mononuclear leukocytes could also be seen.

All the virus strains isolated in connection with outbreaks of disease have been found serologically related to the Sp strain. Also some strains isolated from fish in farms with no signs of disease have proved Sp-related, while other isolates from similar material have been related to the Ab strain. However, most of the strains have only been identified as IPNV by mixed Sp and Ab antisera.

C. Control of the infection

Table 2 shows the results of the testing at the breeding centre in 1987-1988. Some variations are seen between different strains of fish. Between 8 and 31 % of the broodfish were found virus positive. The corresponding precentages for fry were 8 and 14, respectively. It should be noticed that all the tested fry samples originated from parents found IPNV negative in previous testings.

TABLE 2. Results of the IPNV testing at the breeding centre.

	Number of brood fish		Number of fry batches from parents found IPNV negative	
	Tested	IPNV pos.	Tested	IPNV pos.
Salmon strain 1	160	12	137	18
" " 2	300	11	281	22
Rainbow trout strain 1	75	23	41	4
" " " 2	-*	-*	194	28

* Results not available

There was no significant difference in the mortality rates in the different batches of fry hatched during the winter 1987-88 when compared with the IPNV positive or negative testing results of their parents.

DISCUSSION

This study has shown that IPNV occurs in a very large proportion of all categories of fish farms in Norway. It is reason to believe that hardly any farm is free from the infection. The exceptions might be most of those hatcheries which use eggs from wild fish for the purpose of restocking lakes and rivers with salmon and brown trout.

It is not known from where the IPN viruses were originally brought into Norwegian fish farms. Some of the strains may have been introduced by the import of live fish, for instance rainbow trout or eel. Infection may also have been transmitted from wild fish or lower marine organisms. The extensive trade with eggs and smolts has of course served as an efficient mode of spreading the virus.

According to the Norwegian fish disease legislation IPN is a notifiable disease. During the period of the first virus isolations (1975-1977) infected fish not suitable for transfer to sea water was stamped out. Smolts which could be brought to sea water farms by restricted movements, were allowed raised for human consumption purposes. No fish was allowed sold to other freshwater farms or for restocking purposes. Up to 1984 movement restrictions on fish from farms with IPNV-infection were in force. Faced with the new situation of a large and increasing portion of brood fish farms and hatcheries diagnosed as IPNV-infected and with apparently only minor losses caused by the virus, the veterinary authories decided to alter their policy regarding the restrictions in order not to hamper the whole industry. Therefore, during the last years, restrictions have been enforced only upon farms with clinical outbreaks of the disease and only for the periods with relevant death losses. No transfer of live fish or eggs is allowed from these farms to hatcheries or brood fish farms.

There has been some discussion whether a control programme for IPN should be established or not. Even if the losses have been small up to now, it is difficult to predict how big problems these potentially pathogenic viruses might present in the long run. Especially the fishfarmers and their organizations have argued for a testing/culling/isolation programme in order to try to eliminate the virus from the farms, first of all from the two breeding centres and secondly from the brood fish farms. From a veterinary point of view reservation have been expressed to such a plan, since it might be questionable if it is possible to stamp out the infection by economically acceptable measures. The prevention of reinfection is another matter of serious concern. The water supplied to the breeding stations comes directly from the rivers, and the brood fish is kept in sea water. It is therefore not possible to isolate the fish biologically from possibly infected wild-living fish or other organisms. In the laboratory the sensitivity of the test methods and the limited working capacity are considered the most important success limiting factors.

It is shown that transmission of IPNV from parents to offspring via eggs readily takes place in brook trout (*Salvelinus fontinalis*) and rainbow trout (Wolf et al., 1963; Bullock et al., 1976; Fijan and Giorgetti, 1978). The mechanism of this vertical transmission is not clear, but it is likely that such transmission takes place also in salmon and that it contributes much to spreading of the virus from the broodfish farms to the hatcheries.

Our attempt to reduce the prevalence of IPNV infection by removing batches of eggs from virus positive parents does not seem promising. This might be due either to low sensitivity in the testing programme of broodfish or to cross contamination between individual batches of eggs and fry. Further investigations are necessary in this connection. However, each family group will also be tested at fingerling and smolt stages, even if it will be difficult to keep the different groups sufficiently apart during weighing, sorting, marking and vaccination processes. It is the plan to continue with the testing and culling programme for the coming years, and future results will show if these efforts made within economically reasonable limits are sufficient to keep an IPNV-free stock.

REFERENCES

Bullock GL, Rucker RR, Amend D, Wolf K, Stuckey HM (1976) Infectious pancreatic necrosis: Transmission with iodine-treated and non-treated eggs of brook trout (Salvelinus fontinalis). J Fish Res Bd Can 33:1179-1198.

Fijan NN, Giorgetti G (1978) Infectious pancreatic necrosis: isolation of virus from eyed eggs of rainbow trout, Salmo gairdneri Richardson. J Fish Dis 1:296-270.

Håstein T, Krogsrud J (1976) Infectious pancreatic necrosis. First isolation of virus from fish in Norway. Acta Vet Scand 17:109-111.

Vestergård Jørgensen PE (1974) A study of viral diseases in Danish rainbow trout, their diagnosis and control. Thesis. A/S Carl Fr Mortensen Copenhagen.

Vestergård Jørgensen PE, Grauballe PC (1971) Problems in the serological typing of IPN virus. Acta Vet Scand 12:145-147.

Vestergård Jørgensen PE, Kehlet NP (1971) Infectious pancreatic necrosis (IPN) viruses in Danish rainbow trout. Their serological and pathological properties. Nord Vet-Med 23:568-575.

Wolf K, Quimby MC, Bradford AD (1963) Egg associated transmission of IPN virus of trouts. Virology 21:317-321.

Infectious Pancreatic Necrosis Virus in Atlantic Salmon: Transmission via the Sexual Products?

D.A. Smail and A.L.S. Munro
Department of Agriculture an Fisheries for Scotland
Marine Laboratory
Aberdeen
Scotland

INTRODUCTION

Infectious pancreatic necrosis virus (IPNV) is known to be transmitted via eggs in brook trout (Salvelinus fontinalis), and rainbow trout (Salmo gairdneri) (Wolf et al., 1963; Bullock et al., 1976). The weight of evidence suggests that the virus is largely external to the egg and does not easily penetrate it.

Dorson and Torchy (1985) attempted to infect the gametes of rainbow trout under a variety of conditions ie direct infection of females and males, incubation of ova with virus, insemination in the presence of virus and incubation of milt with virus. They concluded that virus could transmit through the eggs and cause clinical disease in fry only by incubation of milt with virus. Ahne and Negele (1985) showed that eyed eggs of rainbow trout and Arctic char (Salvelinus alpinus) could be infected by IPNV and that infectivity was retained especially by the egg shell until after hatching, providing a source of infection for fry. By contrast, in rainbow trout eggs infected with virus before fertilization, no virus could be recovered after 60 mins in flowing water. These authors showed by scanning electron microscopy that the chorion of unfertilized eggs has a smooth crenellated surface whereas that of the hardened egg has a roughened surface but is lobed and porous, thereby allowing virus uptake.

The purpose of this work was to set up experimental infections of the gametes of Atlantic salmon, to see to what extent previous findings applied to a species which is farmed widely in Scotland. Possible transmission was examined under four main conditions: (a) eggs crossed with sperm, both from fish infected earlier in life, (b) virus-exposed eggs crossed with control sperm, (c) control eggs crossed with virus-adsorbed sperm, (d) eggs from carrier broodfish injected with dexamethasone, as immunosuppressant, crossed with virus negative sperm. Virus isolation from maturing 2+ fish is also reported (Section E).

MATERIALS AND METHODS

1. Source of Fish

2+ broodgrilse with a previous exposure to IPN virus were held at DAFS Experimental Unit, Aultbea and used for experiments A and D. Such carriers were also used as the source of eggs in experiment B: the eggs were sampled for virus at the same time as infection and found negative. Salmon eggs from a commercial farm negative for virus were obtained for experiment C. On-site male grilse with virus negative milt, were used for fertilization and virus adsorption.

2. Sources of Virus Isolates and Growth

The isolates of IPN serotypes Sp and ss (salmon strain = Te in Hill's classification) used are listed in Table 1. Isolates were grown and titrated in AS or CHSE-214 cells and concentrated to $>10^9$ pfu/ml by PEG precipitation by the method of Sano et al. (1981).

TABLE 1. IPN virus isolates used in sections A-E of the Results

Section	Isolates
A	ss/Sal.F (AG.)/9/81
B1	ss/Sal.F(CW)/3/82 Sp/Rt.F(MAFF, 75-13) Sp/Rt/F(KR)/6/77
B2	ss/Sal.F(CW)/3/82 Sp/Rt.F(KR)/6/77
B3	ss/Sal.F(CW)/3/82 Sp/Sal.F(JJ)/2/85
C	Sp/Sal.F(JJ)/2/85
D	ss/Sal.F(CW)/3/82
E	Sp/Sal.F(JJ)/2/85

Table 1 abbreviations etc. The isolate abbreviation denotes: Serotype/Host of origin. Farmed or Wild, (source)/month/year of isolation. Sal. = Atlantic salmon, Rt = Rainbow trout. F = Farmed, W = Wild.

3. Tissue Preparations and Virological Procedures

For virological examination, eggs, organs, sex products and whole fry were processed as follows.

Ovarian fluid was decanted from unfertilized eggs, clarified by sedimentation and processed neat. Unfertilized eggs were stored chilled and decontaminated from bacteria by incubation with Hanks balanced salt solution HBSS(1X) + 10x concentration of antibiotics (penicillin, streptomycin, kanamycin and fungizone) 3 hrs at room temperature, then homogenised in a Stomacher 80 (Seward Medical, London) in 4 vols HBSS(1X). Fertilized eggs were sometimes processed to obtain a separation of the egg contents and the shell as follows. Drained eggs were briefly homogenised, the egg fluid was taken off and diluted in 4 vols HBSS(1X). The remaining shells were weighed and macerated in the Stomacher in up to 9 vols HBSS(1X) plus high antibiotics.

Eyed eggs were washed thoroughly in HBSS(1X) by shaking and surface disinfected by immersion for 15 mins in a 0.3% solution of buffered Wescodyne (Ciba-Geigy) in distilled water, pH 7.2 and rinsed. They were processed in small pools of 5-20 and homogenised in 5-10 vols HBSS(1X) by the Stomacher.

Kidneys were routinely weighed and homogenised in 9 vols HBSS(1X) in the Stomacher. Gonads, especially of immature fish, needed more thorough extraction. Ovaries were finely minced with scalpels, then homogenised by the Stomacher and seminal tubules and testis were homogenised in a Potter hand homogeniser, both tissues being diluted 10x (w/v).

Milt was processed to permit virus isolation from the seminal fluid and the sperm. Milt was clarified and the supernatant used undiluted as inoculum. The sperm cell pellet was diluted in 4 vols HBSS(1x) and sonicated at 180 watts and 20 kHz for 1 min to release virus, then clarified and the supernatant used.

All homogenates were filter sterilized (Millipore HA, 0.45 m) to prepare 0.5 ml virus inocula, which were ovarlaid on drained Atlantic salmon (AS) or Chinook salmon embryo (CHSE-214) cells in No 25 flasks. Viral cytopathic effect was read after seven days incubation at 15°C. Most virus inocula were also titrated by microtitration or plaque assay using agarose.

4. Infection of Broodgrilse

(Expt B1). Three virus strains were grown in AS cells and the supernatants harvested and titrated. One group of three 2+ broodgrilse was injected by the intraperitoneal route with a farm isolate of ss serotype (4 x 10^6 plaque forming units - pfu per kgm) and two groups of three 2+ broodgrilse with two isolates of Sp serotype (1-2 x 10^6 pfu/kgm)

(Expt B3). Two 2+ broodgrilse infected as 0+ fry were selected as ready to spawn. Twenty-four hours before spawning, the fish were injected by the intraperitoneal route with the same serotype of virus at doses >10^9 pfu/kgm body weight (see Table 5). Eggs were fertilized with virus negative milt and egg virus titres determined pre and post fertilization.

5. Infection of Eggs

(Expt B2 and 3). Concentrated virus of Sp and ss isolates was added to freshly spawned eggs in ovarian fluid, gently mixed and incubated at 10°C for 20 mins. The virion to egg ratios were as given in Table 4. The eggs were fertilized with virus negative milt, without prewashing.

6. Infection of Milt (Expt C)

Virus was adsorbed to sperm in virus negative milt by the methods described by Dorson and Torchy (1985).

7. Dexamethasone Administration (Expt D)

To depress humoral immunity in broodgrilse, seven 2+ fish previously infected with ss serotype as 0+ fry were injected i.m. and i.p. with dexamethasone (Sigma Chemicals, 1 mg/ml suspension in PBS) at 1 mg/kgm body weight after the method of Bowser et al. (1985). Fish were injected at weeks 6 and 2 prespawning and a blood sample was taken. Internal tissues were virus tested after spawning and the eggs fertilized with virus negative milt. Three egg lots were incubated at 4-8°C and both dead and live eggs sampled weekly.

8. Serum Neutralization

Fish sera were frozen at -20°C to inactivate complement, filter sterilized and dilutions made from 10x onwards in Eagles maintenance medium. Serum dilutions were incubated one hour with 100 pfu of the type isolate for the serotype infecting fish. The remaining virus was enumerated by plaque assay using agarose and the serum titre reducing 50% of the virus dose determined.

RESULTS

A. Eggs x Sperm, Previous Infection

Virus recovery was followed in a group of mature 2+ fish and their sex products. A stock of 2+ potential grilse was identified as infected with an IPN ss isolate at 30% prevalence three months before spawning.

Nine broodgrilse were spawned separately and each batch of eggs fertilized with pooled milt from three males from the same cage. Kidney virus tests on male and female grilse are shown in Table 2. Virus sampling on egg lots was carried out up to the first-feeding fry stage, at 95% confidence of detecting 3% incidence but no virus was detectable (Table 2).

TABLE 2. Virus status of mature 2+ fish, carriers of IPN/ss/AG/9/81 and virus sampling on the progeny

Fish Nos	Sex	Nos Positive*	Progeny
9	Female	5/9	0/9
3	Male	1/3	

* minimum detectable virus titre = 16 pfu/gm kidney

B. Virus-exposed Eggs x Control Sperm

Three experiments (B1-3) were performed under this heading: (1) low virus dose injection of broodfish, (2) low virus dose bathing of eggs and (3) high virus dose injection of broodfish and higher dose egg bathing.

B1 Low virus dose injection of broodfish

Grilse were spawned of their eggs 26-50 days after the first injection and 7-31 days after the second injection and the eggs fertilized with milt from control male fish from the same original source. The ovarian fluid and kidney from each broodfish were virus titrated and the fertilized eggs sampled for virus up to the first feeding fry stage as nine separate lots (Table 3).

Low levels of virus were detectable in the ovarian fluids of 7/9 grilse and in the kidneys. However no virus was recoverable post fertilization in samples of either dead or living eggs.

B2 Low virus dose bathing of eggs

Virus was not recoverable up to the first-feeding fry stage (Table 4).

TABLE 3. Virus titres of 2+ broodgrilse injected with virus in three groups with three different isolates, and virus status of the males, pre-eyed eggs and alevins (ND = not done)

Virus isolate (dose per injection)	Fish No	Virus titre Ovarian fluid ($TCID_{50}$/ml)	Kidney ($TCID_{50}$/gm)	Virus status Eggs	Alevins
ss/Sal.F.CW/3/82	1	2-40	6.8×10^4	-	-
(4×10^6 pfu/kgm)	2	2-40	>40	-	-
	3	2-40	1.5×10^3	-	-
Sp/Rt.F.MAFF, 75-13	4	<2	ND	-	-
(10^6 pfu/kgm)	5	<2	40-400	-	-
	6	2-40	40-800	-	-
Sp/Rt.F.KR/6/77	7	2-40	40-800	-	-
(2×10^6 pfu/kgm)	8	2-40	6.4×10^3	-	-
	9	2-20	40-400		
Males, 3 pools of 3			<40		

TABLE 4. Infection of eggs at two virus doses, Sp and ss isolates

Isolate	Nos eggs	Virion/egg ratio	Virus recovery (egg to alevin)
ss/Sal.F.CW/3/82	1000	$9.2 \times 10^6/1$	-ve
Sp/Rt.F.KR/6/77	1000	$2.3 \times 10^3/1$	-ve

B3 High virus dose injection of broodfish and egg bathing

Tissue titres in two broodfish injected with isolates of Sp and ss serotypes are shown in Table 5 and recovery of virus in time after high virus dose bathing of eggs.

After injection, virus was detectable in the ovarian fluid and in one lot of eggs at 5 mins after ferilization and water hardening but not by 60 minutes. Virus became associated with the unfertilized egg after virus injection to the body cavity and also with the unfertilized egg by virus bathing.

C. Control eggs x virus-adsorbed sperm

To examine the possibility of virus infection of eggs via virus adsorbed to sperm, a source of IPN negative eggs was chosen and virus adsorbed to virus negative milt at one virion per sperm (Sp/Sal.F.JJ/2/85). This was used to fertilize eggs and the virus levels were followed in both egg shells and egg contents directly after fertilization (Table 6).

TABLE 5. Virus titres in tissues and eggs after either injection of broodfish or virus bathing of eggs (TCID$_{50}$/ml* or TCID$_{50}$/gm^{+})

Virus isolate	K^{+}	OF*	Milt*	Eggs pre-fert	Eggs post-fert 5 mins	1 hr	7 wks
INJECTION 8 x 10^9 pfu/kgm (Sp) 2.4 x 10^9 (ss)							
Sp/Sal.F.JJ/2/85	120	>200	<10	220	10-50	<10	<4
ss/Sal.F.CW/3/82	340	>200	<10	260	<10	<10	<4
DIRECT BATHING 10^7 pfu/egg, 10 mins							
ss/Sal.F.CW/3/82	50	<2	<10	<10	>10^3	<10	<4

Table 6. Virus titres in a broodsalmon, on the milt used for fertilization and in the eggs pre- and post-fertilization. (Titres in pfu/gm-K, ova: orpfu/ml-OF)

Egg lot	K	OF	Virion/sperm	Eggs pre-fert		Post-fert 5 mins	60 mins	2 wks
1	<20	<2	1/1	<10	S	<50	<50	<8
			@ 4.3 x 10^9 sperm		E	1850	<50	<8
2	<20	<2	"	<10	S	<50	<50	<8
					E	266	20	<8

S = shell; E = egg contents; K = kidney; OF = ovarian fluid

Virus was detectable inside the fertilized egg after preincubation of sperm with virus showing that the sperm can cause virus entry into the egg but there was a drop in titre by 60 mins after water hardening.

D. Dexamethasone (DM) injected broodgrilse x virus negative sperm

Results of virus recovery in Table 7 show that of the seven grilse, virus could be isolated from the ovary of fish No 1 (after two passages, fish not spawned) and from a single dead uneyed egg at 200 day-degrees post fertilization, spawned from fish No 5. Interestingly, for both broodgrilse virus could not be isolated from the kidney (MDT 20 pfu/gm). The serum neutralizing titres against the homologous virus serotype for these two fish showed a slight depression from detectable titres after DM injection.

E. Gonadal virus isolation from maturing fish

A group of 2+ potential grilse infected as 0+ fry with an Sp serotype isolate were sampled to assess virus prevalence (Table 8). Isolation was possible from the ovary with or without recovery from the kidney and from immature testis.

TABLE 7. Virus isolation in tissues and eggs spawned from dexamethasone-injected broodgrilse, infected with ss/Sal.F(CW)/3/82 as 0+ fry and serum neutalizing titres of fish nos 1 and 5

Grilse	K	OF	OV	Eggs D A	Serum neut titre* Wk 6 PS	Wk 2 PS	Wk 0 PS
1	-	-	+	NS	1/10	<1/10	<1/10
2	-		-	NS			
3	-		-	- -			
4	+		-	NS			
					Wk 6	Wk 2	Wk 1
5	-	-	-	+ -	1/50	1/10	1/20
6	-		-	NS			
7	+		-	- -			

* serum titre reducing 50% of 100 plaques of IPN/ss. D = dead, A = alive, K = kidney, OF = ovarian fluid, OV = ovary, PS = pre spawning, Wk = week, NS = not spawned.

TABLE 8. Virus recovery from 2+ potential grilse infected as 0+ fry with Sp/Sal.F(JJ)/2/85 (F = female, M = male)

Fish No	Sex	Kidney	Gonad
1	F	+	-
2	M	-	-
3	M	-	-
4	F	-	-
5	F	-	+
6	M	+	-
7	M	+	-
8	F	-	-
9	F	+	+
10	M	+	+
Totals		5/10	3/10

DISCUSSION AND CONCLUSIONS

We have been unable to demonstrate complete vertical transmission of IPN serotypes Sp and ss in Atlantic salmon by artificial gamete infection. However, the occurrence of virus in the gonads of a very few maturing fish has been noted so that there remains the small probability that virus could transmit via the gametes and cause clinical disease in fry. The main reasons for the general result are considered under four headings: (1) virus inactivation within the egg, (2) washing away and dilution of virus, (3) sampling numbers and experiment scale and (4) health conditions of maturing fish, including stressors, immunosuppression and/or secondary infections.

The results for lack of virus recovery after infection of unfertilized eggs fit the pattern that has been described by Dorson and Torchy (1985) and Ahne and Negele (1985) for rainbow trout egg infection ie no detectable virus after 60 minutes of water hardening and no subsequent infection. Wolf et al. (1963) found in carrier brood brook trout that only in fish with ovarian fluid virus titres in the range 10^{4-6} ID_{50}/ml was virus detectable in egg homogenates. Such high titres were certainly not obtained in this study. There is also evidence from several salmonid species that ova contain either lectins (Chinook salmon, Voss et al. 1978) or protectins (Atlantic salmon, Voak et al., 1974) which agglutinate human type B erythrocytes. The lectin has been shown to bind and inhibit the growth of virulent bacteria, probably by virtue of combining with a-galactosyl-like determinants and these molecules may possibly inactivate viruses also. For IHN virus, experimental infection of eggs of masu and chum salmon before fertilization led to a steady decrease in titre thereafter (Yoshimizu et al., 1988) and the anti-viral activity of the egg contents was attributed to a yolk phospholipid.

For virus adsorption to sperm, there was clearly virus entry into the egg via the sperm but virus was undetectable after 60 minutes of water hardening. This is hardly surprising since only one virion was adsorbed to each sperm and it would be interesting to examine the effect of higher virion dose per sperm on virus recovery and embryo development.

The scale of the experiments carried out in no way mimics farm conditions in terms of numbers of broodfish and eggs sampled. Wolf et al. (1968) first pointed out that the number of infected eggs may be very low ie 1 in 1000 and in this respect it is interesting that only one dead egg was found virus positive in the dexamethasone trial.

Another significant effect on vertical transmission may be the general broodfish health. Sauter et al. (1987) reported that surface-disinfected unfertilized eggs from apparently healthy chinook salmon could yield a variety of commonly isolated bacteria. If broodfish were additionally stressed the penetration of unfertilized eggs by bacteria and viruses might thus be increased.

These results raise questions about the exact location of virus in the egg and the precise effect of the input virus dose on the developing embryo, which are subjects for further study.

ACKNOWLEDGEMENTS

We thank Mrs L McFarlane, Miss A Thomson and Miss K Ross for technical assistance.

REFERENCES

Ahne W, Negele RD (1985) IPNV transmission via eyed eggs. p261-9. In 'Fish and Shellfish Pathology' ed A Ellis. Academic Press, 1985

Bowser PR, Munson AD, Jarboe HH, Francis-Floyd R, Waterstrat PR (1985) Isolation of channel catfish virus from channel catfish, Ictalurus punctatus (Rafinesque), broodstock.

Bullock GL, Rucker RR, Amend D, Wolf K, Stuckey HM (1976) Infectious pancreatic necrosis: Transmission with iodine-treated and non-treated eggs of brook trout (Salvelinus fontinalis). J Fish Res Bd Can 33(5): 1197-1198.

Dorson M, Torchy C (1985) Experimental transmission of infectious pancreatic necrosis virus via the sexual products. p251-260. In 'Fish and Shellfish Pathology' ed A Ellis. Academic Press, 1985.

Sano T, Tanaka K, Fukuzaki S (1981) Immune response in adult trout against formalin killed concentrated IPNV. Develop Biol Standard 49: 63-70 (S Karger, Basel 1981).

Sauter RW, Williams C, Meyer EA, Celnik B, Banks VL, Leith DA (1987) A study of bacteria present within unfertilized salmon eggs at the time of spawning and their possible relation to early lifestage disease. J Fish Dis 10: 193-203.

Voak D, Todd GM, Pardoe GI (1974) A study of the serological behaviour and nature of the anti-B/P/P^K activity of salmonidae roe proteins. Vox Sang 26: 176-188.

Voss EW, Fryer VL, Banowetz GM (1978) Isolation, purification and partial characterisation of a lectin from chinook salmon ova. Archiv Biochem Biophys 186(1): 25-34.

Wolf K, Quimby MC, Bradford AD (1963) Egg-associated transmission of IPN virus of trouts. Virology 2(3): 317-321.

Wolk K, Quimby MC, Carlson CP, Bullock GL (1968) Infectious pancreatic necrosis: selection of virus-free stock from a population of carrier trout. J Fish Res Bd Can 25(2): 383-391.

Yoshimizu M, Sami M, Kimura T (1988) Survival of infectious haematopoietic necrosis virus (IHNV) in fertilized eggs of masu (Oncorhynchus masou) and chum salmon (O. keta). Am Fish Soc Fish Health Section Int Fish Health Conf Vancouver Canada July 1988 Abstracts p5.

The Paramyxoviruses and Orthomyxoviruses

J.L. Fryer
Department of Microbiology
Nash Hall 220
Oregon State University
Corvallis Oregon 97331-3804 USA

PARAMYXOVIRIDAE

Paramyxoviruses are responsible for a number of serious diseases of humans and animals (mumps, measles, parainfluenza, Newcastle disease, distemper, rinderpest). Newcastle disease virus (NDV) is found in flocks of birds throughout most of the world. Virulent strains are lethal, especially those that replicate in and destroy the host tissues needed for immunologic response. Other paramyxoviruses are associated primarily with respiratory tract infections in animals. The paramyxoviridae share with the orthomyxoviridae their affinity for the sialic acid of cell surface glycoproteins used as receptors (Fraenkel-Conrat et at., 1988). Table 1 compares the paramyxo- and orthomyxoviruses (Kingsbury, 1985).

The virion's shape is pleomorphic, ranging from spheres to filaments with diameters from 150 to 300 nm. The surface is covered with short spikes (peplomers).

Particularly characteristic of the paramyxoviruses and that which differentiates them from the orthomyxoviruses, is the nature of the RNA and the structure of their nucleocapsid. The nucleocapsid is a helix, 18 nm in diameter and protects the RNA against ribonuclease attack. It is flexible and up to 1000 nm long. The molecular weight of the RNA genome has been estimated to be between 5 and 7 x 10^6. This size is in agreement with the sedimentation rate of the isolated RNA (56 S), and with sequence data.

The lipid-rich envelope consists of two glycoproteins (80 and 65 K, respectively). The larger of these glycoproteins forms spikes that carry both hemagglutinating and, in most paramyxoviruses, neuraminidase activity. This enzyme is able to split off the terminal sialic acid residues from the oligosaccharide chains on the host.

Table 1. Comparison of orthomyxoviruses and paramyxoviruses (Kingsbury, 1985)

Similarities

Single-stranded RNA genome of negative polarity
Helical nucleocapsid
Transcriptase in virion
Lipid-containing virion envelope
Virion buds from cell-surface membrane
Two surface glycoprotein species
  Hemagglutination (some paramyxoviurses)
  Neuraminidase activity (some influenza viruses and paramyxoviruses)
  Fusion of virion envelope with cell-surface membrane requires proteolytic cleavage of a viral glycoprotein
Respiratory pathogens, primarily

Differences

Property	Ortho-myxovirus[a]	Para-myxovirus[b]
Genome organization	Segmented	Nonsegmented
Cellular site of RNA synthesis	Nucleus	Cytoplasm
Primer requirement for transcription	Yes	No
Auxiliary nucleocapsid protein species	Three	Two
Nonstructural proteins	Three	One
Cell membrane at site of envelope fusion	Endosome	Surface
Rate of genetic recombination	High	Nil
Rate of evolution	High	Low

[a]Influenza A viruses.
[b]Sendai virus.

ORTHOMYXOVIRIDAE

This family of animal RNA viruses, called influenza or flu viruses, have unusual features including epidemiology, molecular biology and structure. Type A has been responsible for the most serious flu epidemics. It infects humans and many species of mammals and birds. Type B, and probably C, infect only humans. They all show the same unique structural features of the group. Types A, B, and C can be differentiated by the marked serological differences between their nucleocapsids. Within each type, particularly A, there exist many subtypes and strains, and these vary in their surface proteins. These differences may also be demonstrated by serological tests. All members of one type show the same or similar internal proteins. Flu virus is an enveloped RNA particle consisting of an internal nucleocapsid and an envelope made up of a matrix protein, a lipid bilayer, and external glycoproteins.

PARAMYXOVIRUS IN FISH

Examination of the literature indicates the number of paramyxoviruses isolated from fish is very limited. Miyazaki _et al._, 1988, reported what they believed to be a paramyxovirus from black sea bream (_Acanthopagrus schlegeli_) in Japan. Viral epithelial necrosis occurred in tank-raised larvae of black sea bream in 1987. Observation by light microscopy showed separation of rounded epithelial cells of the epidermis, gills and intestine. These rounded epithelial cells were necrotized and destroyed. Electron microscopy revealed enveloped virions displaying features of paramyxovirus in the cytoplasm of necrotized cells. Water-borne exposure of uninfected black sea bream with the ultrafiltrate of diseased fish tissue could establish the infection.

Perhaps the best characterized paramyxovirus from fish was isolated from chinook salmon (_Oncorhynchus tshawytscha_) in 1982 and 1983 (Winton, _et al._, 1985). The virus was recovered from pooled kidney and spleen tissue using CHSE-214 cells. Cytopathic effect (CPE) required 28 days to develop on primary isolation. The virus replicated in CHSE-214,

CHH-1, KO-6, CSE-119 and FHM cells. It had an optimal growth temperature of 18°C and no virus was produced at 24°C or above.

The virus was stable in minimal essential medium (MEM) at pH 3-11 and replication was not affected in cells incubated with MEM containing 50 ug/ml iododeoxyuridine. Infectious virus had a density of 1.20 g/cc. The agent was sensitive to chloroform and caused hemagglutination of erythrocytes from several species of fish, mammals and birds.

Electron microscopy revealed enveloped particles 125-250 nm in diameter containing a coiled nucleocapsid. Electron micrographs of freon-treated preparations showed the extracted nucleocapsid to be a helix 18 nm in diameter and up to 1000 nm in length. Based on morphology and biochemical features, the virus appeared to be a member of the Paramyxoviridae.

With improvement in diagnostic techniques and intensity of effort, It is likely other viruses will be observed which belong to the paramyxo group. However, at this time it appears they are limited in number and of little importance as a cause of mortality among cultured fish.

ORTHOMYXOVIRUS IN FISH

Perhaps the first isolate believed to be an orthomyxovirus was the bluegill virus (BGV) reported by Hoffman _et al._, 1969. Work by a variety of investigators emphasizing the molecular biology of this agent has proven interesting and informative. This enveloped, ether-sensitive, acid labile virus isolated from _Lepomis macrochirus_, the bluegill, was further described by Beckwith and Malsberger, 1979. Virus replication was limited to teleostean cell lines and achieved highest titres in centrarchid cells. The viral genome was found to be RNA. Bluegill virus had no haemagglutinin or neuraminidase activity nor did it share antigens with prototypes of orthomyxoviruses, paramyxoviruses or arenaviruses. Thin-section electron microscopy revealed a virion 80 to 100 nm in dia. These authors believed the BGV represented a novel, undescribed, teleostean virus.

Robin and Dery (1982) reported the Eastern Fish Disease Laboratory (EFDL) strain of BGV grew in the bluegill fry cell line (BF-2) with medium containing (^{3}H)uridine or (^{32}P)orthophosphate. They reported only 8.5 and 0.5%, respectively, of the host proteins and the host RNA were recovered with the viral band, while recovery of viral infectivity was approximately 20%. The nucleic acid extracted from purified virus had the following properties: (1) it migrated homogeneously during electrophoresis at the rate expected from a single-stranded nucleic acid with a MW of 2.17×10^6, (2) it sedimentated as a single molecular species. (3) it banded in a neutral Cs_2SO_4 gradient at a density of 1.69 g/cm^3 consistent with single-stranded RNA, (4) the base composition was that of a single-stranded RNA molecule. The nucleic acid of BGV is single-stranded RNA (MW 2.17×10^6).

Bluegill virus grown in BF-2 cells was studied by negative staining and ultrathin sections of infected cells (Berthiaume et al., 1982). Although BGV resembled bunyaviruses in physical appearance, it differed in several important aspects. They believed BGV should not be classified as a member of the Bunyaviridae and that it could be a representative of a novel family of enveloped RNA viruses.

Robin and Lariviere-Durand, 1983, have indicated the BGV is a ribovirus with a positive-strand genome polarity. BGV virions lacked detectable virion-associated transcriptase activity while BGV RNA was infectious and had 3'-terminal poly (A) sequences. This suggests BGV contains a genome with a messenger function. They also stated that BGV belonged to class 1 of the positive-strand viruses.

A reclassification of the BGV is almost certain to occur and the concensus among experts would seem to indicate this agent is sufficiently different that a new family should be provided.

In addition to BGV, an orthomyxovirus designated EV-2 was isolated from external tumor tissue and internal organs of European Eels (Anguilla anguilla) with stomatopapilloma (Nagabayashi and Wolf, 1979; Neukirch, 1985). It contained RNA and was ether, acid, and temperature labile above 4°C. Concentrated virus preparations agglutinated chicken and sheep erythrocytes. Buoyant density of the virus was 1.19 g/cm^3. EV-2 was spherical in shape but pleomorphic with a diameter ranging from 80 to 140 nm. The virion had regularly spaced surface projections about 10 nm long. Replication in FHM cells at 15°C indicated new virus appearing at 10 h postinfection and was complete at 20 h. Notable cytopathic effects were cell fusion and syncytia.

As with the paramyxoviruses, the orthomyxoviruses are not well-represented among fish. Their numbers are few and they do not appear to be an important cause of mortality among cultured fish. Improved methodology and more intense effort on the part of fish pathologists may yet reveal additional members of the orthomyxoviruses.

REFERENCES

Beckwith DG, Malsberger RG (1979) Characterization and morphology of the bluegill virus. J Gen Virol 43:489-501

Berthiaume L, Robin J, Alain R (1982) Electron microscopic study of bluegill virus. Can J Microbiol 28(4):398-402

Fraenkel-Conrat H, Kimball PC, Levy JA (1988) Virology Prentice Hall 440 pgs

Hoffman GL, Dunbar CE, Wolf K, Zwillenberg LO (1969) Epitheliocystis, a new infectious disease of the bluegill (Lepomis macrochirus). Antonie van Leeuwenhoek 35:146-158

Kingsbury DW (1985) Orthomyxo- and paramyxoviruses and their replication. In: Virology Fields BN, Editor Raven Press pp 1157-1177

Miyazaki T, Fujiwara K, Kobara J, Matsumoto N (1988) Histopathological studies on two viral diseases of larvae and juvenile; epidermal necrosis of Japanese flounder and epithelial necrosis of Black Sea Bream. Fish Health Section International Fish Health Conference Vancouver B.C. Canada July 19-21 p 162

Nagabayashi T, Wolf K (1979) Characterization of EV-2, a virus isolated from European eels (Anguilla anguilla) with stomatopapilloma. J Virol 30:358-364
Neukirch M (1985) Isolation of a orthomyxovirus-like agent from European eel (Anguilla anguilla). Bul EAFP 5(1):12-13
Robin J, Dery C (1982) The genome of bluegill virus. Can J Microbiol 28(1):58-64
Robin J, Lariviere-Durand C (1983) Bluegill virus is a ribovirus of positive-strand polarity. Arch Virol 77:119-125
Winton JR, Lannan CN, Ranson DP, Fryer JL (1985) Isolation of a new virus from chinook salmon (Oncorhynchus tshawytscha) in Oregon, USA. Fish Pathol 20(2/3):373-380

Persistent Infections of Fish Cell Lines by Paramyxovirus Isolates from Chinook Salmon (*Oncorhynchus tshawytscha*)

C.N. Lannan, C.K. Arakawa*, J.R. Winton*, and J.L. Fryer
Department of Microbiology
Oregon State University
Corvallis, Oregon 97331-3804 USA

INTRODUCTION

We have reported the isolation of a paramyxovirus from stocks of adult chinook salmon (Oncorhynchus tshawytscha) returning to coastal rivers of Oregon, USA (Winton et al 1985). The isolates were obtained from kidney and spleen tissues using the chinook salmon embryo cell line, CHSE-214. Initial cytopathic effect (CPE) was slow to develop, requiring 28 days incubation at 18°C. The virus replicated in CHSE-214, chum heart (CHH-1), kokanee ovary (KO-6), coho salmon embryo (CSE-119), and fathead minnow (FHM) cell lines where it produced a lytic type of CPE.

The virus was stable at pH 3-11 and iodo-deoxyuridine did not inhibit viral replication. Infectivity was lost after treatment with chloroform indicating the presence of essential lipids. The density of virions in CsCl was 1.2 g/ml. The virus hemagglutinated cells of 11 of 14 species of birds, mammals, and fish tested. Electron microscopy of infected cells revealed enveloped particles 125-250 nm in dia. containing coiled nucleocapsids and examination of freon-treated virions showed the nucleocapsid was a helix approximately 18 nm in dia. and > 1000 nm in length (Winton et al 1985).

In addition to causing hemagglutination, members of the Paramyxoviridae are known for the ability to establish persistent infections of cell lines (Choppin and Compans 1975). The purpose of this study was to determine if the paramyxovirus isolates from salmon were able to establish persistent infections in fish cell lines and to study the nature of the infection.

---

*Present address:
U.S. Fish and Wildlife Service
National Fisheries Research Center
Bldg. 204, Naval Station
Seattle, WA 98115, USA

MATERIALS AND METHODS

Cell lines and virus isolates

We used chinook salmon embryo (CHSE-214; ATCC CRL 1681), chum salmon (*Oncorhynchus keta*) heart (CHH-1; ATCC CRL 1680), and coho salmon (*Oncorhynchus kisutch*) embryo (CSE-119, Lannan et al 1984) cell lines for growth and assay of the viruses. Eagle's minimum essential medium (MEM) was supplemented with 10% fetal bovine serum (MEM-10) for cell growth and 5% serum (MEM-5) for virus titration. Methods for culturing cells were essentially those described by Wolf and Quimby (1976). All cells were incubated at 18°C.

Two paramyxovirus isolates were used in this study. The OAF strain was recovered in 1982 from adult chinook salmon returning to the Oregon Aqua Foods salt water recapture site at Newport, Oregon. The KS-8 strain was isolated in 1983 from adult chinook salmon returning to the Trask River Hatchery on the northern Oregon coast. Stocks of the viruses were prepared using CHSE-214 cells.

Assay for infectivity

We used 96-well microplates (Corning) containing monolayer cultures of CHSE-214 cells inoculated with serial ten-fold dilutions of virus to assay viral infectivity. After 14 days incubation, the wells were examined for CPE and the titer determined by the method of Reed and Muench (1938).

Establishment and culture of persistent infections

Persistent infections of CHH-1 and CSE-119 cells were established using monolayer cultures in 25 cm^2 flasks (Corning). The cells were inoculated with virus at selected multiplicities of infection (MOI). When cytopathic effect was complete, dead cells and spent culture fluid were removed and the flasks provided with new medium. The cells were allowed to replicate until the monolayer was reestablished. Thereafter, cells were subcultured at appropriate intervals by standard methods except that an aliquot of spent culture fluid was saved for determination of virus titer.

Comparison of two isolates for initiation of persistent infections

The OAF and KS-8 isolates were compared for the ability to initiate persistent infections of coho salmon embryo cells. Monolayer cultures of

CSE-119 cells in 25 cm^2 flasks were infected with each isolate and the resulting persistently infected cultures were routinely subcultured at approximately weekly intervals for 31 weeks. At each subculture, the titer of the spent culture fluid was determined.

Effect of multiplicity of infection

We compared the abilities of the OAF and KS-8 isolates to initiate persistent infections from different multiplicities of infection. Monolayer cultures of CSE-119 cells in 25 cm^2 flasks were inoculated with $10^{5.5}$ $TCID_{50}$ (high MOI) and $10^{2.5}$ $TCID_{50}$ (low MOI) of each isolate. The titers of each culture were determined at 7 and 14 days post-infection.

In a second experiment, cultures of CHH-1 cells in 25 cm^2 flasks were infected with $10^{4.5}$ $TCID_{50}$ of the OAF isolate (high MOI) or $10^{2.5}$ $TCID_{50}$ of the KS-8 isolate (low MOI). These cultures were propagated for 15 weeks and the titer of the culture medium was determined at selected passages.

Infection of trout and salmon fry

We infected chinook salmon, coho salmon (*Oncorhynchus kisutch*), chum salmon (*Oncorhynchus keta*), and rainbow trout (*Salmo gairdneri*) fry by intraperitoneal injection of 0.01 ml of the OAF isolate ($10^{7.0}$ $TCID_{50}$/ml) or the KS-8 isolate ($10^{6.8}$ $TCID_{50}$/ml). Immediately post-injection and at day 7, 14, 21, 28, and 42, fish were removed, weighed, diluted 1:10 in Hank's balanced salt solution, homogenized, and the titer determined. All fish dying during the experiment were assayed for virus.

RESULTS

Persistent infections

Both the OAF and the KS-8 isolates were able to establish persistently infected cell cultures (Table 1). Following the initial appearance of CPE in the infected cultures, normal appearing cells began to fill the vacant areas of the monolayer. After the monolayer had reformed, it appeared similar to uninfected monolayers and could be subcultured at intervals by standard methods without adverse effect. The virus titer of the culture fluid from persistently infected cultures varied at each passage, but was usually in the range of 10^4 to 10^6 $TCID_{50}$/ml. At most stages of persistent infection, the cell sheet was indistinguishable from that of an uninfected monolayer and low to moderate levels of virus were produced. At intervals, the titer of the persistently infected cultures would rise and CPE would appear in the monolayers. The most extensive CPE was concurrent with highest titer (usually over 10^6 $TCID_{50}$/ml). In some cases, this "lytic crisis" resulted in the nearly complete destruction of the monolayer. At this point, the titers would drop and the persistent state would reestablish itself in a manner similar to that following the initial viral infection. When the experiment was terminated at 31 weeks post-infection, the persistent cultures appeared stable and well established.

Table 1. Titers of two isolates of chinook salmon paramyxovirus from persistently infected coho salmon embryo cell lines.

	$\log_{10}$ $TCID_{50}$/ml at week of culture										
Isolate	1	2	3	4	5	6	7	8	9	10	11
OAF	4.0	6.0	6.0	6.6	6.2	5.6	5.7	6.0	5.8	5.7	4.6
KS-8	4.0	4.7	3.8	4.6	4.8	5.0	5.0	6.0	6.7	6.7	6.3
	12	13	14	15	16	20	21	24	27	31	
OAF	5.0	4.6	6.5	5.7	6.5	5.6	5.0	5.4	5.4	4.3	
KS-8	4.7	6.5	6.5	6.4	6.8	5.6	5.2	4.4	4.8	4.5	

Effect of multiplicity of infection

In this experiment, we compared two paramyxovirus isolates for their ability to initiate persistent infections when inoculated at high and low multiplicities of infection. Both isolates showed similar differences between the titers obtained from the high and low MOI cultures at 7 days (Table 2). After 14 days incubation, the titer of the high MOI cultures remained about the same or declined slightly, but the titer in the low MOI cultures increased significantly, especially for the OAF isolate.

Table 2. Titers of two paramyxovirus isolates from chinook salmon in coho salmon cell line inoculated at high and low multiplicities of infection (MOI).

		$\log_{10}$ $TCID_{50}$/ml	
Isolate	MOI	Day 7	Day 14
OAF	High	6.5	6.6
	Low	4.0	6.0
KS-8	High	6.4	5.6
	Low	4.0	4.7

In a second experiment, CHH-1 cells were inoculated at high or low MOI with the OAF isolate and the culture fluid titer determined at selected passages. In these cultures, a large difference was seen in the titers at week 1 (Table 3). After the persistent infections had become established, the titers of the cultures rose and fell during the 15 week experiment in a manner similar to that of persistently infected CSE-119 cells.

Table 3. Titers of the chinook salmon paramyxovirus (OAF isolate) from persistently infected chum heart cells established from high and low multiplicities of infection (MOI).

	$\log_{10}$ $TCID_{50}$/ml at week of culture					
MOI	1	3	7	9	11	15
High	4.4	5.6	5.8	6.3	4.6	5.5
Low	1.8	6.6	6.0	6.2	5.0	5.5

Infection of salmon and trout fry

Curiously, we were never able to recover either virus isolate from the chinook fry following intraperitoneal injection. We could recover both OAF and KS-8 isolates from coho and chum salmon and rainbow trout fry on the day of injection but not thereafter. These day 0 titers for the coho were $10^{3.6}$/g for the OAF isolate and $10^{4.6}$/g for the KS-8 isolate; for chum the titers were $10^{3.6}$/g for OAF and $10^{4.6}$/g for KS-8; and for rainbow they were $10^{3.8}$/g for OAF and $10^{4.7}$/g for KS-8. We did not recover virus from any fish at day 7, 14, 21, 28 or 42 nor from the few fish dying during the experiment.

DISCUSSION

Persistent infections by RNA viruses have been shown to be maintained by several different mechanisms (Kauffman and Fields 1985). One is through mutation or evolution of the virus during passage in cell culture to create strains with altered biological properties, e.g. slower growth rate, lower virulence, altered host range, or temperature sensitivity (Youngner and Preble 1980). A second method is mediated by formation of large numbers of defective interfering particles (Holland et al 1980). Another mechanism involves the action of external factors e.g. interferon (Joklik 1985) or antibody (Fujinami and Oldstone 1984), while a fourth method that has received little attention is the evolution of populations of cells with lowered sensitivity to virus (Ramig 1985).

Persistent infections of CHSE-214 cell cultures by a fish rhabdovirus, infectious hematopoietic necrosis virus (Engleking and Leong 1980), and a birnavirus, infectious pancreatic necrosis virus (IPNV) have been reported (Ahne 1977, Hedrick et al 1978, Hedrick and Fryer 1981, MacDonald and Kennedy 1979). Kennedy and MacDonald (1982) proposed that IPNV persistence was maintained by defective interfering (DI) particles rather than by interferon although the rate of DI particle production by CHSE-214 cells was low. Hedrick and Fryer (1982) showed that the persistently infected CHSE-214 cell lines were analogous to the carrier state in trout.

In our experiments, we were able to establish persistent infections of salmonid cell lines with relative ease. These cultures typically exhibited limited cytopathic effect and continuously produced infectious virus in a manner similar to vertebrate cell lines persistently infected with other paramyxoviruses (Choppin and Compans 1975, Kingsbury 1977, White and Fenner 1986). The mechanism by which our cultures remained persistently infected was not determined, but others have shown persistent infections of cell cultures with paramyxoviruses are maintained by both evolution of strains and defective interference (Youngner and Preble 1980, Holland et al 1980).

Although we were able to establish lytic and persistent infections of chinook salmon cell cultures, when we injected chinook, coho, and chum salmon, and rainbow trout fry with approximately 105 $TCID_{50}$ of the OAF or the KS-8 isolate, we were only able to recover the virus from chum and coho salmon, and rainbow trout fry immediately post-injection.

It may be difficult to explain the apparent resistance of chinook salmon fry to infection, in light of the isolation of the virus from fish in three separate stocks of adult chinook salmon in several years.

Following initial recovery of the OAF isolate from adult chinook salmon, the progeny of the fish were extensively tested and showed no evidence of infection. One explanation may be that the passage of virus in cell culture resulted in the rapid selection of host range mutants that were not infectious for chinook salmon. A second possibility is that returning fish are hormonally or immunologically altered to make them more susceptible to infection. A third, less likely, explanation is that the virus rapidly established a latent state from which no infectious virus was detected. This hypothesis is supported by the results of Koch et al (1984) who established a lifelong latency in mice with Sendai virus. Although no infectious virus could be recovered 74 days after infection, and no viral protein was expressed, cDNA probes were able to demonstrate virus-specific nucleic acid was present in the brain tissues of the mice.

ACKNOWLEDGEMENTS

This work is a result of research sponsored by NOAA Office of Sea Grant, Department of Commerce, under Grant No. NA81AA-D-00086. The U.S. Government is authorized to produce and distribute reprints for governmental purposes, notwithstanding any copyright notation that may appear hereon. Oregon State University Agricultural Experiment Station Technical Paper No. 8728.

REFERENCES

Ahne W (1977) Persistent infection in CHSE-214 cells with IPN virus isolated from pike (Esox lucius). Bull Off Int Epiz 87:415-416

Choppin PW, Compans RW (1975) Reproduction of paramyxoviruses. In: Fraenkel-Conrat H, Wagner R (eds) Comprehensive Virology, Vol 4. Plenum, New York

Engleking M, Leong J (1980) IHNV persistently infects chinook salmon embryo cells. Virol 109:47-58

Fujinami RS, Oldstone MBA (1984) Antibody initiates viral persistence: Immune modulation and measles virus infection. In: Notkins AL, Oldstone MBA (eds) Concepts in Viral Pathogenesis. Springer-Verlag, New York

Hedrick RP, Fryer JL (1981) Persistent infection of three salmonid cell lines with infectious pancreatic necrosis virus (IPNV). Fish Pathol 15:163-172

Hedrick RP, Fryer JL (1982) Persistent infections of salmonid cell lines with infectious pancreatic necrosis virus (IPNV): A model for the carrier state in trout. Fish Pathol 16:163-172

Hedrick RP, Leong JC, Fryer JL (1978) Persistent infections in salmonid fish cells with infectious pancreatic necrosis virus (IPNV). J Fish Dis 1:297-308

Holland JJ, Kennedy SIT, Semler BL, Jones CL, Roux L, Grabau EA (1980) Defective interfering RNA viruses and the host-cell response. In: Fraenkel-Conrat H, Wagner RR (eds) Comprehensive Virology Vol 16. Plenum, New York

Joklik WK (1985) Interferons. In: Fields BN, Knipe DM, Channock RM, Melnick J, Roizman B, Shope R (eds) Virology. Raven, New York
Kauffman RS and Fields BN (1985) Pathogenesis of viral infections. In: Fields BN, Knipe DM, Channock RM, Melnick J, Roizman, B Shope R (eds) Virology. Raven, New York
Kennedy JC, MacDonald RD (1982) Persistent infection with infectious pancreatic necrosis virus mediated by defective-interfering (DI) particles in a cell line showing strong interference but little DI replication. J Gen Virol 58:361-371
Kingsbury DW (1977) Paramyxoviruses. In: Nayak D (ed) The Molecular Biology of Animal Viruses. Marcel Dekker, New York
Koch EM, Neubert WJ, Hofschneider PH (1984) Lifelong persistence of paramyxovirus Sendai-6/94 in C129 mice: detection of a latent viral RNA by hybridization with a cloned genomic cDNA probe. Virol 136:78-88
Lannan CN, Winton JR, Fryer JL (1984) Fish cell lines: Establishment and characterization of nine cell lines from salmonids. In Vitro 20:671-676
MacDonald RD, Kennedy JC (1979) Infectious pancreatic necrosis virus persistently infects chinook salmon embryo cells independent of interferon. Virol 95:260-264
Ramig RF (1985) Principles of animal virus genetics. In: Fields BN, Knipe DM, Channock RM, Melnick J, Roizman B, Shope R (eds) Virology. Raven, New York
Reed JL, Muench H (1938) A simple method of estimating fifty percent endpoints. Am J Hyg 27:493-497
White DO, Fenner F (1986) Medical Virology. Academic Press, London
Winton JR, Lannan CN, Ransom DP, Fryer JL (1985) Isolation of a new virus from chinook salmon (Oncorhynchus tshawytscha) in Oregon USA. Fish Pathol 20:373-380
Wolf K, Quimby MC (1976) Procedures for subculturing fish cells and propagating fish cell lines. TCA Manual 2:471-474
Youngner JS, Preble OT (1980) Viral persistence: Evolution of viral populations. In: Fraenkel-Conrat H, Wagner RR (eds) Comprehensive Virology Vol 16. Plenum, New York

Rhabdoviruses of Fishes

G.N. Frerichs
Institute of Aquaculture
University of Stirling
Stirling FK9 4LA
Scotland

The rhabdoviruses constitute the largest group of viruses isolated from teleost fish and, collectively, have been associated with major disease epizootics and catastrophic losses of intensively reared juvenile and adult fish throughout the world. To date, 12 or 13 species, depending upon the classification system adopted, have been identified. The chronological sequence of virus isolations is shown in Table I and, not surprisingly, broadly reflects the evolution of intensive fish culture over the past 30 years. The progressive development of rainbow trout culture in the USA and Europe, Pacific salmon farming in the river systems of the western seaboard of North America, carp pond culture in central and eastern Europe, the intensive rearing of eels and other species in Japan and rice-field fish culture in southeast Asia have successively given rise to new disease problems, among which rhabdovirus infections have appeared with disturbing regularity. Moreover, the continuing expansion of both freshwater and marine fish culture, involving an ever increasing variety of fish species, culture systems and environmental conditions may be expected to lead to the identification of yet more virus species and subspecies in the years ahead.

Some of the rhabdoviruses, and the diseases with which they are associated, have been the subject of considerable research effort over many years whereas others have received relatively little attention. It is an unfortunate fact of fish disease research that the extent and depth of any investigative programme is largely determined by the monetary value of the product at risk, and hence the level of financial investment

TABLE I Rhabdoviruses of fishes

Virus	Origin	Reference
Viral haemorrhagic septicaemia virus (VHSV)	Rainbow trout (*Salmo gairdneri*)	Jensen (1963)
Infectious haematopoietic necrosis virus (IHNV)	Sockeye salmon (*Oncorhynchus nerka*) Rainbow trout (*Salmo gairdneri*)	Amend et al (1969)
Spring viraemia of carp virus (SVCV) (*Rhabdovirus carpio*)	Common carp (*Cyprinus carpio*)	Fijan et al (1971)
Pike fry rhabdovirus (PFRV)	Pike (*Esox lucius*)	de Kinkelin et al (1973)
Eel virus A (EVA)	American eel (*Anguilla rostrata*)	Sano (1976)
Eel virus Europe X (EVEX or EVX)	European eel (*Anguilla anguilla*)	Sano et al (1977)
Cod ulcus-syndrome rhabdovirus	Atlantic cod (*Gadus morhua*)	Jensen et al (1979)
Rio Grande perch rhabdovirus	Rio Grande perch (*Cichlasoma cyanoguttatum*)	Malsberger & Lautenslager (1980)
Eel virus B_{12}	European eel (*Anguilla anguilla*)	Castric & Chastel (1980)
Rhabdovirus salmonis	Rainbow trout (*Salmo gairdneri*)	Osadchaya & Nakonechnaya (1981)
Perch rhabdovirus	Perch (*Perca fluviatilis*)	Dorson et al (1984)
Hirame rhabdovirus (HRV) (*Rhabdovirus olivaceus*)	Japanese flounder (*Paralichthys olivaceus*) Ayu (*Plecoglossus altivelis*)	Kimura et al (1986)
Ulcerative disease rhabdovirus (UDRV)	Striped snakehead (*Ophicephalus striatus*) Freshwater eel (*Fluta alba*)	Frerichs et al (1986)

in fish health services engendered by these commercial considerations. It is therefore not surprising to find that it is the rhabdoviruses of salmon, particularly IHNV, which have so far been most thoroughly examined. It must be hoped that similar manpower and material resources will be made available to study the diseases of the non-luxury species of fish which form the principal source of animal protein for many millions of people in the less affluent regions of the world.

Virus subtypes

VHSV is the only rhabdovirus for which more than one serotype is recognised in polyclonal antibody serum neutralization tests. Serotype 1 (Egtved virus strain F-1) is most commonly encountered and has been detected in several European countries. Serotype 2 (strain He) has been isolated in Denmark and serotype 3 (strain 23.75) in France (Ahne, 1981). All three serotypes show a degree of cross-reaction in neutralization tests. Although only one serotype of IHNV has been identified using polyclonal antibodies, an element of antigenic diversity has been found between strains isolated from different host species (McCain et al, 1971). It would now seem likely that the application of monoclonal antibody neutralization procedures will re-define the serotypes of both VHSV and IHNV and also identify serotypes within those rhabdovirus species which have not yet been serologically subdivided.

EVA and EVX were initially considered to be separate species of eel rhabdovirus but are now regarded as antigenic variants of the same virus for which the name *Rhabdovirus anguilla* (strains A and X) has been proposed (Hill et al, 1980). The adoption of this name would certainly help to readily distinguish these two rhabdoviruses from other unrelated eel viruses which have been assigned such confusing and uninformative appellations as EVE and EV-2, but it is not easy to see how this proposed classification could incorporate the more recently identified and antigenically distinct eel virus B_{12}.

Virus composition and structure

All the fish rhabdoviruses appear to possess the typical characteristics of the animal rhabdovirus family. When examined, fish isolates have been shown to contain a single stranded, non-segmented RNA genome with a sedimentation coefficient of 38-40S, and virion associated RNA-dependent RNA polymerase activity capable of transcribing genomic RNA to messenger RNA (mRNA). The viruses have a buoyant density of 1.14-1.20 g/cm^3 and infectivity is lost following treatment with lipid solvents, exposure to pH 3 and heating at 60°C for 15 minutes.

The morphological and structural features of the different species are shown in Table II. All isolates have been found to possess the animal rhabdovirus bullet-shaped morphology although bacillary and pleomorphic forms have also been identified in some preparations. Reported virus sizes vary considerably both between and within species but much of the variation within species can be accounted for by the presence of structurally defective particles and differences in the types of preparation used to determine virion size. Polypeptide analyses have been carried out on some, but not all, virus species and five virus specific proteins have been identified in all cases (McAllister, 1979). The electrophoretic mobility profiles of VHSV, IHNV and eel virus B_{12} polypeptides are similar to those of the lyssavirus (rabies) group whereas the SVCV, PFRV, EVA, and EVX profiles resemble the vesiculovirus (VSV) types, but formal inclusion of these species in the lyssavirus or vesiculovirus genus has not yet been accepted by the International Committee on Taxonomy of Viruses (Matthews, 1982).

TABLE II Rhabdovirus structure

Virus	Shape	Virion size (nm)	Protein type	References
VHSV	Bullet and bacillary	240 x 75 180 x 70 175 x 60	lyssavirus	Zwillenberg et al (1965) Cohen & Lenoir (1974) de Kinkelin & Scherrer (1970)
IHNV	Bullet	188 x 62 120 x 95 160 x 90	lyssavirus	Cohen & Lenoir (1974) Amend & Chambers (1970)
SVCV	Bullet	180 x 80 90-140 x 70 125 x 83	vesiculovirus	Hill et al (1975) Bachmann & Ahne (1973) Cohen & Lenoir (1974)
PFRV	Bullet	115 - 135 x 72-88	vesiculovirus	de Kinkelin et al (1973)
EVA/EVX	Bullet	160 x 75 100 x 80 143 x 84	vesiculovirus	Hill et al (1980) Nishimura et al (1981)
Cod ulcus-syndrome	Bullet	115-195 x 50-80		Jensen et al (1979)
Rio Grande perch	Bullet and bacillary			Malsberger & Lautenslager (1980)
Eel virus B_{12}	Bullet and pleomorphic	130 - 240 x 70-80	lyssavirus	Castric & Chastel (1980) Castric et al (1984)
Rhabdovirus salmonis	Bullet	130-140 x 60-70		Osadchaya & Nakonechnaya (1981)
Perch rhabdovirus	Bullet	200 x 100		Dorson et al (1984)
HRV	Bullet	180-200 x 80		Kimura et al (1986)
UDRV	Bullet	110 - 130 x 75-85		Frerichs et al (1986)

Polypeptide analysis has more recently been used to distinguish between strains of IHNV of different origins. Although no apparent fish species-specific strain has been identified, different electrophoretic mobility patterns of the G and N polypeptides appear to be characteristic for strains in a given geographical area, suggesting that this technique may have particular application in epidemiological studies of IHNV infections (Leong et al, 1983).

Virus stability

The infectivity of all the fish rhabdoviruses can be preserved almost indefinitely by lyophilisation or at -20°C or lower in cell culture medium supplemented with 2-10% serum. Rates of inactivation increase with rising temperatures and, although not identical, most species show a broadly similar heat sensitivity pattern (Table III). A notable exception is UDRV which appears to be markedly more resistant to heat inactivation than the other species. The viruses are generally less stable when stored in serum-free distilled water, tap water or lake water and also appear to be adversely affected by the high level of electrolytes in seawater.

Exposure for 10 minutes to 6 hours to methylene blue, malachite green, benzalkonium chloride, copper sulphate, acriflavine or thiomersal at concentrations similar to those used for the treatment of fish or the disinfection of eggs has been found to have no effect on the infectivity of VHSV, IHNV, SVCV or PFRV (Amend and Pietsch, 1972; Ahne, 1982). Formalin (2%), sodium hydroxide (2%) and sodium hypochlorite (100mg Cl_2/l) destroy these viruses within 10 minutes but iodophore disinfectants are the chemicals of choice for inactivation. It is considered that a 5 minute exposure to an iodophore solution containing 25 ppm available I_2 will destroy all known salmonid viruses (Amend and Pietsch, 1972).

TABLE III Rhabdovirus stability

Virus	≥99% reduction in infectivity: Culture medium	Water	References
VHSV	50°, <10 min 30°, <24 hr 4°, many mnth	Tap 14°, >24 hr Tap 10°, 4 wk	Jorgensen (1974) Ahne (1982)
IHNV	32°, 8 hr 15°, 2 mnth 4°, 5 mnth	Dist. 10°, 7 d Lake 10°, 5 wk Sea 15°, 2 d	Pietsch et al (1977) Wedemeyer et al (1978)
SVCV	45°, 15 mn	Tap 10°, 3 wk	Liversidge & Munro (1978) Ahne (1982)
PFRV	45°, 30 min 14°, 2½ mnth 4°, 5 mnth	Dist. 14°, 2 wk	de Kinkelin et al (1974)
EVA/EVX	50°, 30 min 20°, 2 wk		Sano et al (1977)
<u>Rhabdovirus salmonis</u>	4°, 9 mnth		Osadchaya & Nakonechnaya (1981)
Perch rhabdovirus	37°, 30 min		Dorson et al (1984)
HRV	50°, 2 min		Kimura et al (1986)
UDRV	35°, 7 d		Frerichs et al (1989b)

No information available on the stability of cod ulcus-syndrome Rio Grande perch or eel B_{12} rhabdoviruses.

Cell Culture

In contrast to some of the other groups of fish viruses, particularly the herpesviruses and retroviruses, all the known rhabdoviruses have been isolated in cell cultures (Table IV). Most of the viruses can be readily propagated in a wide range of cells derived both from fish species known to be susceptible to natural infection with the virus and from non-susceptible species. Other isolates appear to have a more limited host cell range, but probably only because the cell susceptibility spectrum has not been evaluated to the same extent. Not all cell lines which support the growth of a given virus are equally susceptible to infection and refractory cell substrates have been recorded for some species. As the number of rhabdovirus isolates increases, so it becomes correspondingly more important for comparative purposes to know the resistant as well as the susceptible cell lines, and it is to be hoped that this aspect of virus characterization will be more frequently recorded in the literature in future.

As with the other fish viruses, the rhabdoviruses are capable of growth over a wide temperature range. The actual rate of virus maturation and release from infected cells is generally temperature dependent, with an optimum range occurring towards the upper growth limit point. For those viruses capable of multiplication in a number of cell lines which in themselves have different optimal growth temperatures, the optimum temperature for virus growth may also vary according to the cell line used.

The cytopathological effects (cpe) produced by the rhabdoviruses in cell cultures are broadly similar for all virus species. Granulation and rounding up of cells, deformation of nuclei and margination of nuclear chromatin, followed by detachment and lysis of cells leading to complete destruction of the monolayer within 2-5 days is the general observation. Intracytoplasmic inclusions may be identified if cells are

TABLE IV Rhabdovirus fish substrates

Virus	Sensitive cells	Refractory cells	Virus growth temp. (°C)	References
VHSV	RTG-2, FHM, EPC CHSE-214, STE-137 RTF-1, RF, PG		6-18 opt. 14	Wolf & Mann (1980) McAllister (1979)
IHNV	RTG-2, FHM, EPC CHSE-214, STE-137 CHSE-114, CSE-119 SSE-5, SSE-30, AS RTF-1, RTH-149 RBS, GE-4		4-20 opt. 13-18	Wolf & Mann (1980) McAllister (1979)
SVCV	FHM, EPC, RTG-2 CHSE-214, BF-2 BB, PG		4-32 opt. varies with cells	Ahne (1981) Wolf & Mann (1980) McAllister (1979)
PFRV	FHM, EPC, RTG-2 BB, PG		10-28 opt. 21-28	de Kinkelin et al (1974) Wolf & Mann (1980)
EVA/EVX	RTG-2, FHM, BF-2 EPC		10-29	Hill et al (1980) Castric & Chastel (1980) Nishimura et al (1981)
Cod ulcus-syndrome	PS	BF-2, FHM RTG-2	15	Jensen et al (1979)
Rio Grande perch	FHM, RTG-2, BF-2		23-33	Malsberger & Lautenslager (1980)
Eel virus B_{12}	EPC	RTG-2	10-20	Castric & Chastel (1980)
<u>Rhabdovirus salmonis</u>	RTG-2, FHM, EPC ASE		18-20	Osadchaya & Nakonechaya (1981)
Perch rhabdovirus	RTG-2	EPC	14	Dorson et al (1984)
HRV	RTG-2, FHM, EPC YNK, BF-2, BB EK-1, STE-137 CCO, HF-1	CHH-1, KO-6 CHSE-214	5-20	Kimura et al (1986)
UDRV	SSN-1, BF-2, AS RSN, CP	CHSE-214, EPC ONP, GCP	15-30 opt. 25-30	Frerichs et al (1989a)

examined in the early stages of the infective cycle.

Defective virions

Defective interfering (DI) particles have either been demonstrated or suspected in preparations of IHNV (McAllister and Pilcher, 1974), SVCV (de Kinkelin and le Berre, 1974), PFRV (de Kinkelin et al, 1974), cod ulcus syndrome rhabdovirus (Jensen et al, 1979) and EVA/EVX (Hill et al, 1980) but not VHSV (McAllister, 1979). The formation of DI particles is favoured when cells are infected under conditions of high input multiplicity and it is now standard practice with all fish rhabdoviruses to dilute inocula to ensure a low multiplicity of infection (MOI) and the avoidance of autointerference.

Persistent cell infection

The phenomenon of regrowth of persistently infected cells which have survived a cytolytic virus infection has been reported for CHSE-214 cells infected with IHNV (Engelking and Leong, 1981) and snakehead cells (SSN-1) infected with UDRV (Frerichs et al, 1989b). In both instances, infected cells have appeared microscopically indistinguishable from normal cells and have continued to produce cytolytic virus throughout a number of serial cell passages. In view of the widely recognised occurrence of latent rhabdovirus infections in fish it would be surprising if similar persistently infected cells did not arise with other rhabdovirus species but this has yet to be established.

Interferon

VHSV and IHNV have been shown to induce interferon synthesis in fish and in RTG-2 and FHM cell cultures. Replication of these viruses is also inhibited by exogenously supplied cell culture or fish derived interferon. In contrast, SVCV and PFRV do not elicit an interferon response and are not sensitive to exogenously supplied interferon (McAllister, 1979). The status of the other rhabdoviruses with respect to interferon synthesis and sensitivity is not known.

Virus identification

Although the type of cpe produced by some of the rhabdoviruses can be differentiated by experienced workers, this cannot be relied upon as more than a presumptive diagnosis and identity should always be confirmed by specific tests. Serum neutralization is the most widely used test but fluorescent antibody, immunoperoxidase, immunoprecipitation and ELISA tests may also be used. Fortunately for the diagnostician, the great majority of the rhabdoviruses are antigenically distinct and a clear-cut identification is normally obtained in serum neutralization tests. One exception to this is in the relationship between SVCV and PFRV where some PFRV isolates from non-pike hosts have shown a degree of cross-neutralization with anti-SVCV serum (Ahne et al, 1982; Ahne and Thomsen, 1986). The main disadvantages of the serum neutralization test are the comparatively lengthy time required to obtain a result and the practical difficulty of obtaining high titre antiserum. Many of the fish rhabdoviruses are weak immunogens which stimulate only low levels of antibody in rabbits used for antiserum production (Hill et al, 1981). However, with the development of hybridoma technology, specific, high titre cell culture or mouse ascitic tumour derived antibody may become more readily available.

Fish hosts

The range of naturally infected fish from which rhabdoviruses have been isolated is shown in Table V. Additionally, various species have been found to be susceptible to experimental

TABLE V Rhabdovirus fish hosts

Virus	Naturally infected fish		References
VHSV	Rainbow trout	(S. gairdneri)	Jensen (1963)
	Brown trout	(S. trutta)	Jorgensen (1980)
	Grayling	(T. thymallus)	Wizigmann et al (1980)
	Pile	(E. lucius)	Meier & Jorgensen (1979)
	White fish	(Coregonus sp)	Ahne & Thomsen (1985)
IHNV	Sockeye salmon	(O. nerka)	Amend et al (1969)
	Kokanee	(O. nerka)	Follett et al (1987)
	Chinook salmon	(O. tschawytscha)	Groberg (1983)
	Chum salmon	(O. keta)	Mulcahy & Wood (1986)
	Rainbow trout	(S. gairdneri)	
	Steelhead trout	(S. gairdneri)	
	Cutthroat trout	(S. clarki)	
	Brook trout	(S. fontinalis)	
	Atlantic salmon	(S. salar)	
SVCV	Common carp	(C. carpio)	Fijan et al (1971)
	Bighead carp	(H. nobilis)	Rudikov et al (1975)
	Crucian carp	(C. carassius)	Kolbl (1975)
	Sheatfish	(S. glanis)	Fijan et al (1984)
PFRV	Pike	(E. lucius)	de Kinkelin et al (1973)
	Grass carp	(C. idella)	Ahne (1975)
	Tench	(T. tinca)	Ahne et al (1982)
	White bream	(B. bjoerkna)	Ahne & Thomsen (1986)
	Brown trout	(S. trutta)	Adair & McLoughlin (1986)
	Pseudorasbora parva		
EVA/EVX	American eel	(A. rostrata)	Sano (1976)
	European eel	(A. anguilla)	Sano et al (1977)
Cod ulcus-syndrome	Atlantic cod	(G. morhua)	
Rio Grande perch	Rio Grande perch	(C. cyanoguttatum)	Malsberger & Lautenslager (1980)
Eel virus B_{12}	European eel	(A. anguilla)	Castric & Chastel (1980)
Rhabdovirus salmonis	Rainbow trout	(S. gairdneri)	Osadchaya & Nakonechnaya (1981)
Perch rhabdovirus	Perch	(P. fluviatilis)	Dorson et al (1984)
HRV	Japanese flounder	(P. olivaceus)	Kimura et al (1986)
	Ayu	(P. altivelis)	
UDRV	Striped snakehead	(O. striatus)	Frerichs et al (1986)
	Green snakehead	(O. punctatus)	Frerichs et al (1989b)
	Freshwater eel	(F. alba)	

infections, suggesting an even greater potential for cross-species disease transfer than is indicated by this listing. For example, rainbow trout have been successfully infected with EVA/EVX and hirame rhabdoviruses (Hill et al, 1980; Kimura et al, 1986) and sea bass and turbot with VHSV (Castric and de Kinkelin, 1984).

Despite the concerted efforts of scientists and government agencies over the past 30 years, no practical or economically feasible means of controlling viral fish diseases has yet been found beyond the establishment and maintenance of disease free stocks supported by controls on the movement of fish and eggs between farms, areas, countries and continents. However, the demand for importation, fed by a desire for exportation, of new genetic strains or species of food fish, and an expanding international trade in exotic ornamental fish of totally unknown disease status, seems likely to lead to an increasing rather than decreasing incidence of all virus diseases for the forseeable future.

References

Adair BM, McLoughlin M (1986) Isolation of pike fry rhabdovirus from brown trout (*Salmo trutta*). Bull Eur Ass Fish Pathol 6: 85-86

Ahne W (1975) A rhabdovirus isolated from grass carp (*Ctenopharyngodon idella* Val). Arch Virol 48: 181-185

Ahne W (1981) European reference strains for fish virology. Bull Eur Ass Fish Pathol 1: 10-12

Ahne W (1982) Vergleichende untersuchungen uber die stabilitat von vier fishpathogenen viren (VHSV, PFR, SVCV, IPNV). Zbl Vet Med B 29: 457-476

Ahne W, Mahnel H, Steinhagen P (1982) Isolation of pike fry rhabdovirus from tench, *Tinca tinca* L., and white bream, *Blicca bjoerkna* (L). J Fish Dis 5: 535-537

Ahne W, Thomsen I (1985) Occurrence of VHS virus in wild white fish (*Coregonus* sp.). Zbl Vet Med B 32: 73-75

Ahne W, Thomsen I (1986) Isolation of pike fry rhabdovirus from *Pseudorasbora parva* (Temminck & Schlegel). J Fish Dis 9: 555-556

Amend DF, Chambers VC (1970) Morphology of certain viruses of salmonid fishes. I. In vitro studies of some viruses causing haematopoietic necrosis. J Fish Res Bd Can 27: 1285-1293

Amend DF, Pietsch JP (1972) Virucidal activity of two iodophores to salmonid viruses. J Fish Res Bd Can 29: 61-65

Amend DF, Yasutake WT, Mead RW (1969) A haematopoietic virus disease of rainbow trout and sockeye salmon. Trans Amer Fish Soc 98: 796-804
Bachmann PA, Ahne W (1973) Isolation and characterization of agent causing swim bladder inflammation in carp. Nature 244: 235-237
Castric J, Chastel C (1980) Isolation and characterization attempts of three viruses from European eel, Anguilla anguilla: preliminary results. Ann Virol (Inst Pasteur) 131E: 435-448
Castric J, de Kinkelin P (1984) Experimental study of the susceptibility of two marine fish species, sea bass (Dicentrachus labrax) and turbot (Scophthalmus maximus) to viral haemorrhagic septicaemia. Aquaculture 41: 203-212
Castric J, Rasschaert D, Bernard J (1984) Evidence of lyssaviruses among rhabdovirus isolates from the European eel Anguilla anguilla. Ann Virol (Inst Pasteur) 135E: 35-55
Cohen J, Lenoir G (1974) Ultrastructure et morphologie de quatre rhabdovirus de poissons. Ann Rech veter 5: 443-450
Dorson M, Torchy C, Chilmonczyk P, de Kinkelin P, Michel C (1984) A rhabdovirus pathogenic for perch, Perca fluviatilis L.: isolation and preliminary study. J Fish Dis 7: 241-245
Engelking HM, Leong JC (1981) IHNV persistently infects chinook salmon embryo cells. Virology 109: 47-58
Fijan N, Matasin Z, Soney Z, Olah S, Zwillenberg LO (1984) Isolation of Rhabdovirus carpio from sheatfish (Siluris glanis) fry. Symposia Biologica Hungarica 23: 17-24
Fijan N, Petrinec Z, Sulimanovic D, Zwillenberg LO (1971) Isolation of the viral causative agent from the acute form of infectious dropsy of carp. Vet Arhiv Zagreb 41: 125-135
Follet JE, Thomas JB, Huck AK (1987) Infectious haematopoietic necrosis virus in moribund and dead juvenile chum, Oncorhynchus keta (Walbaum), and chinook, O. tshawytscha (Walbaum), salmon and spawning adult chum salmon at an Alaskan hatchery. J Fish Dis 10: 309-313
Frerichs GN, Hill BJ, Way K (1989a) Ulcerative disease rhabdovirus: cell-line susceptibility and serological comparison with other fish rhabdoviruses. J Fish Dis (in press)
Frerichs GN, Millar SD, Alexander M (1989b) Rhabdovirus infection of ulcerated fish in south-east Asia. This publication
Frerichs GN, Millar SD, Roberts RJ (1986) Ulcerative rhabdovirus in fish in south-east Asia. Nature 322 (6076): 216
Groberg WJ, Jr (1983) The status of viral fish diseases in the Columbia River basin. In Proceedings: Workshop on Viral Diseases of Salmonid Fishes in the Columbia River Basin. Leong JC, Barila TY (eds) p 1-22. Bonneville Power Admin, Portland, Oregon
Hill BJ, Williams RF, Finlay J (1981) Preparation of antisera against fish virus disease agents. Develop biol Standard 49: 209-218
Hill BJ, Williams RF, Smale CJ, Underwood BO, Brown F (1980) Physicochemical and serological characterization of two

rhabdoviruses isolated from eels. Intervirology 14: 208-212
Jensen MH (1963) Preparation of fish tissues for virus research. Bull Off int Epizoot 59: 131-134
Jensen NJ, Bloch B, Larsen JL (1979) The ulcus-syndrome in cod (Gadus morhua) III A preliminary virological report. Nord Vet-Med 31: 436-442
Jorgensen PEV (1974) A study of viral diseases in Danish rainbow trout, their diagnosis and control. Carl Fr. Mortensen, Copenhagen
Jorgensen PEV (1980) Egtved virus: the susceptibility of brown trout and rainbow trout to eight virus isolates and the significance of the findings for the VHS control. In Fish Diseases 3rd COPRAQ session. Ahne W (ed) p 3-7. Springer-Verlag, Berlin
Kimura T, Yoshimizu M, Gorie S (1986) A new rhabdovirus isolated in Japan from cultured hirame (Japanese flounder) Paralichthys olivaceus and ayu Plecoglossus altivelis. Dis aquat Org 1: 209-217
de Kinkelin P, le Berre M (1974) Rhabdovirus des poissons. II Proprietes in vitro du virus de la viremie printaniere de la carpe. Ann Microbiol (Inst Pasteur) 125A: 113-124
de Kinkelin P, le Berre M, Lenoir G (1974) Rhabdovirus des poissons. I Proprietes in vitro du virus de la maladie rouge de l'alevin de brochet. Ann Microbiol (Inst Pasteur) 125A: 93-111
de Kinkelin P, Galimard B, Bootsma P (1973) Isolation and identification of the causative agent of "red disease" of pike (Esox lucius L. 1766). Nature 241: 465-467
de Kinkelin P, Scherrer R (1970) Le virus d'Egtved. I. Stabilite, developpement et structure du virus de la souche Danoise F_1. Ann Rech veter 1: 17-30
Kolbl O (1975) Neue erkentnisse uber die bauchwassersucht bei karpfen. Osterreichs Fischerei 28: 69-72
Leong JC, Hsu YL, Engelking HM, Fendrick JL, Durrin LK, Kurath G (1983) Methods for diagnosing IHNV infection in fish. In Proceedings: Workshop on Viral Diseases of Salmonid Fishes in the Columbia River Basin. Leong JC, Barila TY (eds) p 23-47. Bonneville Power Admin, Portland, Oregon
Liversidge J, Munro ALS (1978) The virology of teleosts. In Fish Pathology. Roberts RJ (ed) p 114-143. Bailliere Tindall, London
McAllister PE (1979) Fish viruses and viral infections. In Comprehensive Virology vol 14. Fraenkel-Conrat H, Wagner RR (eds) p 401-470. Plenum Press, New York
McAllister PE, Pilcher KS (1974) Autointerference in infectious haematopoietic necrosis virus of salmonid fish. Proc Soc Exp Biol Med 145: 840-845
McCain BB, Fryer JL, Pilcher KS (1971) Antigenic relationships in a group of three viruses of salmonid fish by cross neutralization. Proc Soc Exp Biol Med 137: 1042-1046
Malsberger RG, Lautenslager G (1980) Fish viruses: rhabdovirus isolated from a species of the family Cichlidae. Fish Hlth News 9: i-ii
Matthews REF (1982) Classification and nomenclature of viruses. Fourth Report of the International Committee on Taxonomy of Viruses. Intervirology 17: No 1-3

Meier W, Jorgensen PEV (1979) Egtved virus: characteristics of a virus strain isolated from pike fry (Esox lucius L.). Nord Vet-Med 31: 484-485

Mulcahy D, Wood J (1986) A natural epizootic of infectious haematopoietic necrosis in imported Atlantic salmon, Salmo salar L., reared in the enzootic region. J Fish Dis 9: 173-175

Nishimura T, Toba M, Ban F, Okamoto N, Sano T (1981) Eel rhabdovirus, EVA, EVEX and their infectivity to fishes. Fish Pathol 15: 173-184

Osadchaya YF, Nakonechnaya MG (1981) Rhabdovirus salmonis, the cause of a new disease in rainbow trout, Salmo gairdneri. J Ichthyol 21: 113-121

Pietsch JP, Amend DF, Miller CM (1977) Survival of infectious haematopoietic necrosis virus held under various environmental conditions. J Fish Res Bd Can 34: 1360-1364

Rudikov NI, Griscenko JL, Lobuncov KA (1975) Vesenyaya virusnaya bolezn ryb. Bjul Vses ord Lenina Inst Exp Vet 20: 16-19

Sano T (1976) Viral diseases of cultured fishes in Japan. Fish Pathol 10: 221-226

Sano T, Nishimura T, Okamoto N, Fukuda H (1977) Studies on viral diseases of Japanese fishes - VII A rhabdovirus isolated from European eel, Anguilla anguilla. Bull Jap Soc Sci Fish 43: 491-495

Wedemeyer GA, Nelson NC, Smith CA (1978) Survival of the salmonid viruses infectious haematopoietic necrosis (IHNV) and infectious pancreatic necrosis (IPNV) in ozonated, chlorinated and untreated waters. J Fish Res Bd Can 35: 875-879

Wizigmann G, Baath CH, Hoffmann R (1980) Isolierung des virus der viralen hamorrhagischen septikamie (VHS) aus regenbogenforellen-, hecht- und aschenbrut. Zbl Vet Med B 27: 79-81

Wolf K, Mann JA (1980) Poikilotherm vertebrate cell lines and viruses: a current listing for fishes. In Vitro 16: 168-179

Zwillenberg LO, Jensen MH, Zwillenberg HL (1965) Electron microscopy of the virus of viral haemorrhagic septicaemia of rainbow trout. Arch ges Virusforsch 17: 1-19

Rhabdovirus Carpio in Herpivorous Fishes: Isolation, Pathology and Comparative Susceptibility of Fishes

I.S. Shchelkunov and T.I. Shchelkunova
Laboratory of Fish Pathology
The All-Union Research Institute of Pond Fisheries
Rybnoe
Dmitrov Region
Moscow Province 141821
USSR

INTRODUCTION

In the USSR as well as in other countries, where the Orient herbivorous fishes are reared, some diseases of these fishes associated with a haemorrhagic syndrome have been registered. They have been considered as diseases presumably of infectious nature. In some cases pathogenic bacteria (sp. Aeromonas) were isolated from affected fish. Regular virological examinations of diseased fishes were not carried out.

Meanwhile, a similar condition caused by a reovirus has been revealed in grass carp fingerlings (Ctenopharyngodon idella) in P.R.China (Chen & Jiang, 1983). The rhabdovirus (GRV) was isolated from 3-yr-old grass carp in F.R.Germany (Ahne, 1975). This virus has been shown to be closely related to pike fry rhabdovirus (Clerx & Horzinek, 1978). Under experimental conditions GRV was pathogenic to grass carp, less pathogenic to carp and caused no mortality in silver carp (Ahne, 1979).

Grass carp reovirus and GRV have not been found in the USSR. At the same time Rhabdovirus carpio (SVCV) was isolated recently from yearling silver carp with focal scale protrusion, anaemic liver and blood exsudate in abdominal cavity (Shchelkunov et al., 1984). Earlier N.Rudikov et al.(1975) isolated a rhabdovirus from a bighead carp (Aristichthys nobilis). The virus was considered at first as a strain of Rhabdovirus carpio. It was stated later by means of serum neutralization test that this virus is not related to SVCV (Pichugina, 1983). The virus remained to be identified, its pathogenicity to herbivorous fishes was not studied.

There are few data on susceptibility of herbivorous fishes to SVCV (Fijan, 1978; Ahne, 1979). At the same time importance of this problem is obvious taking into account, that herbivorous fishes are reared today on a large scale in polyculture with carp, Cyprinus carpio.

As a whole, the question about the role of viruses in diseases of herbivorous fishes remains not sufficiently studied (particularly in our country). Results of virological examination of diseased herbivorous fishes and some biological properties of viruses isolated are given in this paper.

MATERIALS AND METHODS

Cell cultures

The fish cell lines FHM (Gravell & Malsberger, 1965) and EPC (Fijan et al., 1983) were used in the study. Cells were grown in Eagle's minimal essential medium containing 10% fetal bovine serum, 2% tryptose phosphate broth (all Gibco), 100 IU/ml penicillin and 100 µg/ml streptomycin (MEM-10-2) at 25°C(EPC) and 28°C(FHM). Inoculated cell cultures were supplemented with MEM-2-2 and incubated at 20-22°C.

Virus isolation

Examination of fish was carried out in 1986 at farms suspected of the presence of carp spring viraemia. Yearlings of silver carp, Hypophthalmichthys molitrix, bighead, Aristichthys nobilis, grass carp, Ctenopharyngodon idella, and farmed carp, Cyprinus carpio, were examined. The fishes were sampled and processed using the standard method(Hill, 1976). Kidney, liver, spleen and separately gills of 6-7 fishes were pooled and homogenated. Suspensions of these organs were decontaminated by antibiotics and inoculated onto FHM and EPC cell cultures. If no cytopathic effect (CPE) appeared after 10-14 days, two "blind passages" were carried out additionally.

The Rhabdovirus carpio strain M2 isolated earlier from yearling silver carp (Shchelkunov et al., 1984) was also used in this study. Virus end-point titration was calculated by the method of Reed and Muench.

Fish

Bighead x silver carp hybrid, grass carp and farmed carp delivered from farms free of viral diseases were used in fish infectivity tests. Moreover the herbivorous fishes of various ages used throughout all the experiments were taken and grown from a single lot of fish delivered in 1986.

Before the beginning of every experiment the fish was held in a running water at 8 - 10°C for several weeks. And only herbivorous fishes from fry to 16-wk-old fingerlings were continually held at 18 - 22°C and fed. Water temperature was adjusted to the desirable level in 1 day after experimental infection of fishes.

Fish infectivity tests

Farmed carp and herbivorous fishes were challenged either with several R.carpio strains (as it was done using fry and 7,5-8-mo-old fingerlings) or with the only strain M2 (16-wk-old fingerlings, yearlings and 17-mo-old fishes). The fish was challenged intra-peritoneally (i.p.) or by water bath method. Viral strains 1, 3 and 5 passaged twice (test in fry) or 5 times (test in 7,5-8-mo-old fishes) in cell culture and strain M2 after 9-13 passages were used for the challenge. When challenging fishes i.p. approximately constant virus dose of $10^{6,4}$-$10^{7,5}$ $TCID_{50}$/g fish body mass was used.

Equal number of fishes of every species were used as a control. These fishes were handled by the same mode using virus-free cell culture fluid instead of virus.

Experiments were carried out in running water 80 l aquaria. Herbivorous fishes fry was fed with zooplankton. 16-wk-old fishes were fed with zooplankton, granular mixed feed RK-S and water plants. Fishes of older ages were not fed during experiments. For virus re-isolation kidney, liver and spleen were sampled from an individual fish, or organs from 3-5 fishes were pooled and processed as mentioned above. Fry fish was processed as a whole after head and tail were removed. Details of experiments are shown in Tables 1 and 2.

In some tests a possibility of cross-infection among herbivorous fishes and carp under conditions of cohabitation was investigated. For this aim groups of healthy herbivorous fishes were introduced into groups of carp that had been challenged with R.carpio 3 - 7 days before, and vice versa. Fish was held in a running water aquaria at 13 - 15°C without feeding. Health condition of introduced fish was continuously monitored; parenchymatous organs of the fish were regularly sampled for virus isolation in cell culture (Details are in Table 6).

Statistics

To evaluate a mortality dynamics of challenged fish we calculated a mean death time ($\bar{T}_d$, calculated as the arithmetic mean of individual death times measured in days post infection) and standard deviation of this value (σ_d). Statistical difference between different fish age groups was determined by means of Student's t-criterion when using the first parameter ($\bar{T}_d$) and by means of Fisher's F-criterion when using the second parameter (σ_d). To evaluate a statistical difference between total mortality levels of different test fish groups the Student's t-criterion was also used.

RESULTS

Isolation of virus

Examination of one of the farms with a few years known history of spring viraemia of carp (SVC) showed the following. Under conditions of net cage polyculture rearing of carp and herbivorous fishes earlier only carp was affected with the disease. At the moment of examination (April 1986) an increased mortality besides farmed carp was observed among bighead and grass carp yearlings. The greatest mortality had been registered at 13 - 17°C.

Clinical and patho-anatomical signs in diseased yearlings of carp were typical of acute SVC. Affected herbivorous fishes looked weak and had necrotic damaged dorsal and caudal fins. Sometimes eye haemorrhages occured.

From bighead, grass carp and farmed carp yearlings taken

from net cages of the farm viral agents were isolated (isolates 1, 3 and 5 from organ pools of bighead, grass carp and carp; isolates 2 and 4 from gills of bighead and grass carp, respectively).

Agents isolated were identified by means of serum neutralization test as the strains of Rhabdovirus carpio. Neutralization indexes of 1800 - 10000 were found for different strains. The test had been carried out in FHM cell culture using guinea pig antiserum to SVCV diluted 1:100. The antiserum was supplied by Dr. N.I. Rudikov (The All-Union Research Institute of Experimental Veterinary, Moscow, USSR).

Experimental study of susceptibility of herbivorous fishes and carp to challenge with isolated strains of Rhabdovirus carpio

Under experimental conditions all tested viral strains caused disease and mortality in herbivorous fishes as well as in carp (Tables 1 and 2; Figures 1 - 4). Clinical and patho-anatomical signs of diseased fishes were typical of SVC. Among herbivorous fishes the fry was affected most severely. Hyperaemia and swelling of abdomen, exophthalmos, haemorrhages in fin tissue, on the body surface, gill opercula and around mouth, hyperaemia and oedema of caudal trunk, skin bubbles filled with exsudate and scale protrusion were observed in the diseased fish.

Organs and tissues of moribund fish had the high virus titers (Tables 3 and 4).

If herbivorous fishes fry was highly susceptible to water bath challenge, this mode of infection did not cause the disease in 16-wk-old herbivorous fishes. At the same time significant mortality was observed in farmed carp yearlings challenged by this method. When challenging i.p. older herbivorous fishes as well as 16-wk-old fishes were taken ill and died.

I.p. inoculation of 16-wk-old grass carp with viral strain M2 produced 100% disease at both water temperatures tested (15 and 20°C), but the disease had more severe course and higher mortality occured at 15° than at 20°C. The similar pattern was observed in bighead x silver carp hybrid with such

Table 1. Susceptibility of herbivorous fishes and farmed carp to water bath infection with Rhabdovirus carpio strain M2

Fish	fry *)		fingerlings				yearlings	
	g.carp	hybrid	g.carp		hybrid		g.carp	carp
Age	19 d	50 d	16 wk				12 mo	
Body mass, g	0,150	0,047	10 - 30		0,5 - 3,0		26	29
Challenge, $log_{10}TCID_{50}$/ml water	5,5 - 6,5; 3 hr		5,5; 3 hr				7,25; 6 hr	
Water temperature, °C	19 - 22		15	20	15	20	13 - 15	
Number of fish	100	100	11	11	28	28	30	30
Total mortality, %	67	24	0	0	0	0	0	43,3
Virus re-isolation	+	+	+	-	-	-	+	n.d.**)

*) Simultaneously under the same experimental conditions the pathogenicity of R.carpio strains 1, 3 and 5 was tested in herbivorous fishes fry. Fish mortality data are shown in Fig. 1.
**) Not done

Table 2. Susceptibility of herbivorous fishes and farmed carp to intra-peritoneal inoculation of Rhabdovirus carpio strain M2

Fish	fingerlings		fingerlings *)			yearlings		2-summer-old fish		
	g.carp	hybrid	g.carp	hybrid	carp	g.carp	carp	g.carp	hybrid	carp
Age	16 wk		7,5 mo		8 mo	12 mo		17 mo		
Body mass, g	10-30	0,5-3,0	18,5	13	40	26	29	150	260	270
Challenge dose, lgTCID$_{50}$/g body mass	6,5	6,9	6,9			7,5		6,4		
Water temperature, °C	15 20	15 20	13 - 15			13 - 15		13 - 15		
Number of fish	11 10	41 41	16	18	25	30	30	25	25	25
Total mortality, %	54,5 20	17,1 0	62,5	0	100	6,7	43,3	0	0	0

*) Simultaneously under the same experimental conditions the pathogenicity of R.carpio strains 1, 3 and 5 was tested in fingerlings of herbivorous fishes and carp. Fish mortality data are shown in Fig. 3.

a difference that at 15° they were feeling better than grass carp and practically looked healthy at 20°.

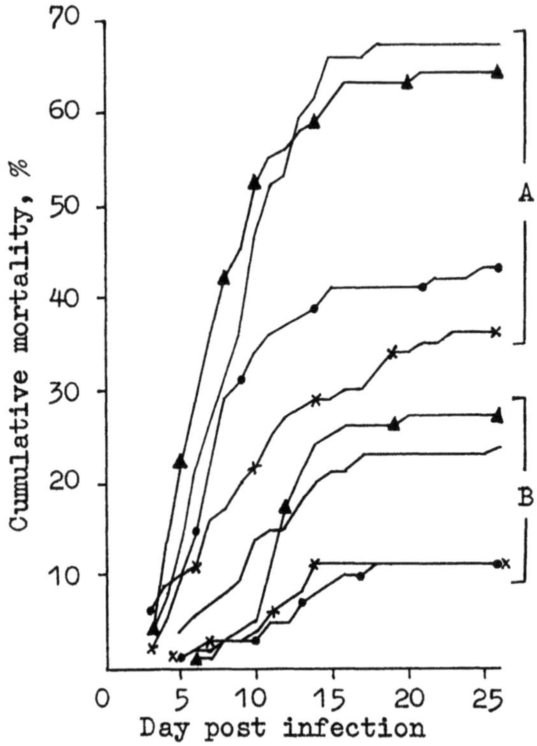

Fig. 1

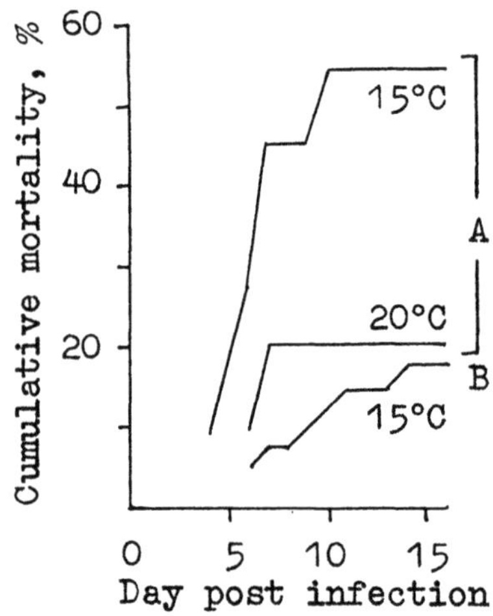

Fig. 2

Fig. 1. Cumulative mortality in grass carp (A) and bighead x silver carp hybrid (B) fry at 19 - 22°C following experimental water bath infection with different strains of Rhabdovirus carpio

—•— strain 1, —▲— strain 3, —×— strain 5, —— strain M2
Design of the experiment see in Table 1.

Fig. 2. Cumulative mortality in 16-wk-old grass carp (A) and bighead x silver carp hybrid (B) at different water temperatures following intra-peritoneal inoculation of Rhabdovirus carpio strain M2

Design of the experiment see in Table 2.

In this test healthy grass carp and grass carp, taken ill after i.p. inoculation of M2 viral strain, were examined bacteriologically. For this aim whole kidneys of 4 control and 4 diseased fishes (two from each aquaria with water temperature 15 and 20°) were sampled individually, weighed and homogenized

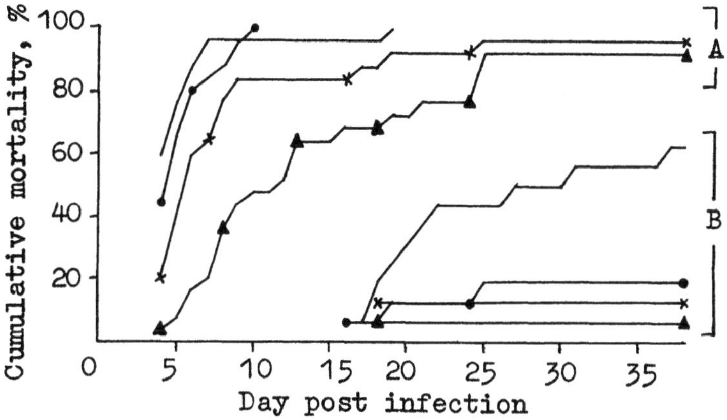

Fig. 3. Cumulative mortality in 7,5-8-mo-old farmed carp (A) and grass carp (B) at 13 - 15°C following intra-peritoneal inoculation of different Rhabdovirus carpio strains

—●— strain 1, —▲— strain 3, —×— strain 5, —— strain M2
Design of the experiment see in Table 2

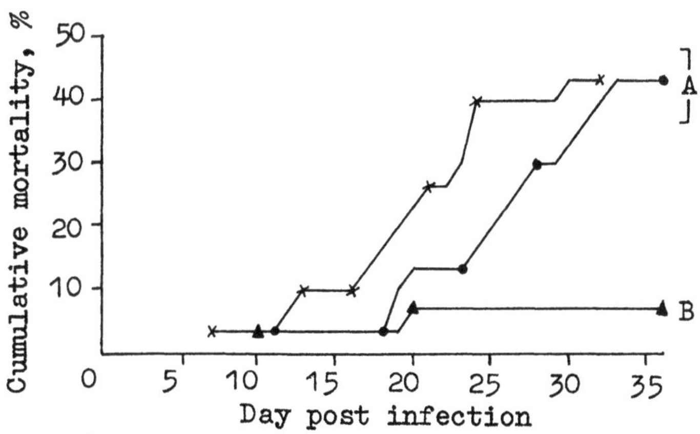

Fig. 4. Cumulative mortality in yearlings of farmed carp (A) and grass carp (B) at 13 - 15°C following experimental infection with Rhabdovirus carpio strain M2

×, ▲ - intra-peritoneal inoculation
● - water bath infection
Design of the experiment see in Tables 1&2

with a mortar and pestle. 0,1 ml aliquots of 10-fold dilutions, that had been prepared in Eagle's basal medium (BME), were inoculated into Petri dishes with erythritol - agar and incubated

Table 3. Virus infectivity titer in organ - tissue pool of herbivorous fishes fry infected by water bath method

Strain of R. carpio	Virus titer, $\log_{10}TCID_{50}/g$	
	grass carp*)	bighead x silver carp hybrid**)
1	7,5	6,75
3	4,25	8,25
5	7,5	7,75
M2	5,75	7,75

*) Day 11 post infection
**) Day 13 post infection

Table 4. Virus infectivity titer in organ - tissue pool of 16-wk-old herbivorous fishes fingerlings at day 13 post intra-peritoneal inoculation of Rhabdovirus carpio strain M2

Water temperature, °C	Virus titer, $\log_{10}TCID_{50}/g$	
	grass carp	bighead x silver carp hybrid
15	7,75	6,75
20	4,5	3,5

at 28°C. No bacteria were found in kidneys.

No mortality was registered in control groups of fishes throughout all the experiments.

Comparative susceptibility of fishes to challenge with Rhabdovirus carpio

To evaluate fish susceptibility to R.carpio we chose the strain M2 as one retaining its high virulence in fishes of different species and ages. Data on total fish mortality when i.p. inoculation of this strain was used are shown in Fig. 5.

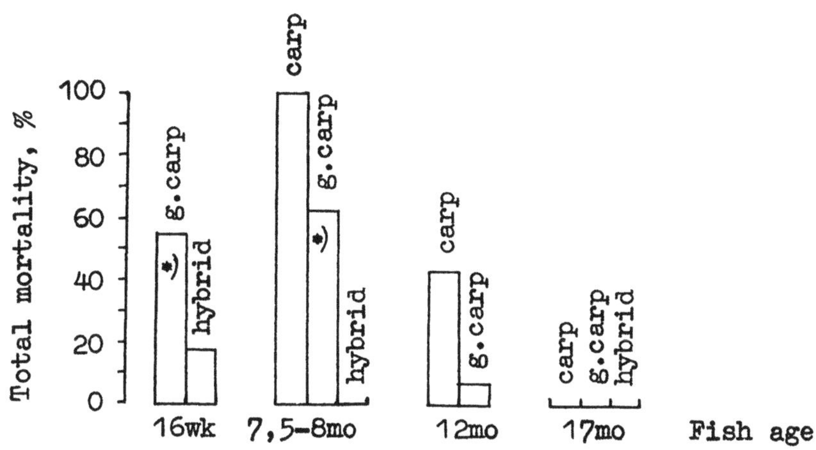

Fig. 5. Total mortality in herbivorous fishes and farmed carp of different ages following intra-peritoneal inoculation of Rhabdovirus carpio strain M2 in the dose of $10^{6,4}$– $10^{7,5}$ $TCID_{50}$/g body mass
*) no statistical difference ($p > 0,5$)

Age-related increase of fish resistance to the virus was not limited by the decrease of total fish mortality. The mortality dynamics itself changed. The older was the fish the later mortality appeared and continued for a longer period of time (mortality curves move to the right and diminish their slope).

Such a tendency is well seen when comparing mortality dynamics of 16-wk-old and 7,5-mo-old grass carps (Figs.2 and 3), and 8-mo-old and yearling farmed carps also (Figs. 3 and 4) at 15°C when i.p. inoculation of M2 viral strain was used.

We expressed both the processes in terms of changing of: 1) mean death time ($\bar{T}_d$) and 2) standard deviation of this value (σ_d). Statistical analysis showed high reliability of changings registered (Table 5).

Cross-infection among herbivorous fishes and carp under conditions of cohabitation

16 tests on cohabitation of healthy and SVCV-challenged fishes were carried out. The results showed that herbivorous fishes of various ages as well as carp shed the SVC virus into environment that leads to infection of cohabitating healthy fish (Table 6). SVCV was isolated from introduced fish in 7 of 16

Table 5. Age-related changings of mortality parameters in fishes intra-peritoneally inoculated with Rhabdovirus carpio (strain M2, 13 - 15°C)

Fish	Farmed carp		Grass carp	
Age	8 mo	12 mo	16 wk	7,5 mo
Mean death time ($\bar{T}_d$), days	5,24[1)]	19,38[1)]	6,5[2)]	22,9[2)]
Standard deviation (σ_d), days	3,03[3)]	6,12[3)]	2,07[4)]	6,71[4)]

$p < 0{,}001$ for 1) and 2); $p < 0{,}01$ for 3) and 4)

tests, moreover in one of the tests the acute disease and mortality with signs of SVC occured. It is interesting that cohabitation infection was achieved more often when using bath challenge (4 of 5 occasions) than i.p. inoculation (3 of 11 occasions).

DISCUSSION

It has been established that during the course of spring viraemia of carp in polyculture conditions infection also extends to herbivorous fishes. Just in a such a situation occured at one of the fish farms we succeeded in isolating of SVC virus from carp, grass carp and bighead carp yearlings simultaneously. Fishes of the two latter species have been first found to harbor this virus. In our opinion, SVCV-infection could not be the direct cause of mortality among herbivorous fishes at the examined farm because the very low pathogenicity of the virus to herbivorous fishes of this age was revealed in experiments.

Herbivorous fishes are susceptible to the challenge with SVCV by bath method and especially by i.p. inoculation. Clinical and patho-anatomical symptoms of the disease appearing in fishes following challenge are similar to that in the affected carp (typical of SVCV-infection). Herbivorous fishes finger-

Table 6. Cross - infection of herbivorous fishes and farmed carp following introduction of non - infected fishes to Rhabdovirus carpio - infected fishes

R.carpio-infected fish			Introduced fish	
Mode of infection	Fish species	Number and age of fish	Species, number and age of fish	Virus isolation
Intra-peritoneal inoculation	carp	10/25*); 0+	hybrid, 7/grass carp, 21**); 0+	+/+
		30; 1	grass carp, 16; 1	–
		25; 1+	grass carp, 25; 0+	–
	grass carp	40; 0+	carp, 12; 0+	–
		30; 1	carp, 15; 1	–
		25; 1+	carp, 30; 0+	–
	bighead x silver carp hybrid	22/55; 0+	carp, 30/carp, 12; 0+	–/–
		10; 1	carp, 20; 1	–
		25; 1+	carp, 30; 0+	+
Water bath	carp	25; 0+	grass carp, 21; 0+	+
		30; 1	grass carp, 16; 1	–
	grass carp	40; 0+	carp, 12; 0+	+
		30; 1	carp, 15; 1	+
	bighead x silver carp hybrid	55; 0+	carp, 12; 0+	+***)

*) Two tests with different number of fishes in each one.
**) Two tests: one with hybrid, another with grass carp.
***) Death with symptoms of SVC.

lings inoculated i.p. with the virus have a more severe course of the disease at 15°C than at 20°C. Bath challenged herbivorous fishes fry was severely ill at 19 - 22° too. The onsets of SVCV-infection at unusually high temperatures (up to 23°C) were described earlier among sheatfish, Silurus glanis, fry and farmed carp fry (Fijan et al., 1981). Apparently they were possible thanks to functional immaturity of immune system in the fish of this age.

It was found experimentally that bacteria did not play any role in the development of the disease in SVCV challenged herbivorous fishes at 20°C as well as at 15°C. Infected herbivorous fishes shed SVCV as carp does.

Cohabitation infection was achieved easier when using bath challenge than i.p. inoculation. This curious incident undoubtedly should be taken into account in future work of this kind.

Organs and tissues of challenged herbivorous fishes had the high virus titer (more than 10^8 $TCID_{50}$/g body mass). At the same time, titrating suspensions of fish tissues, we were faced more than once with such a phenomenon when there were no signs of cytopathic effect (CPE) in the cell cultures inoculated with the first (low) dilutions of suspension, while the next (high) dilutions produced the well developed virus-specific CPE. Probably it may be due to the presence of some inhibitors of virus infectivity in fish tissues homogenates as it is suggested for infectious pancreatic necrosis virus (IPNV) of salmonids (Dixon, 1987). With this in mind, it should be considered as a rule (especially when sampling asymptomatic virus carriers) when isolating SVCV in cell culture one should use for inoculation not less than 2 - 3 consecutive 10-fold dilutions of initial 10% suspension of tissue homogenate.

Our experimental data showed that resistance to SVCV increased in rows: 1) farmed carp - grass carp - bighead x silver carp hybrid, and 2) fry - fingerling - yearling - 1,5-yr-old fish. Additional study of young fishes is essential to determine the most susceptible to SVCV fish age.

We studied the fish mortality dynamics in conditions of long-term cumulative mortality tests. Apparently that kind of dynamics resembles to some extent a nature course of the disease. Such a dynamics can be characterized at least by three parameters: 1) mean death time; 2) standard deviation of this value (this parameter is related to the duration of the disease course in a group of fish); 3) total mortality level.

Fish susceptibility to virus modifies as fish becomes older. Evidently, this modification will manifest in changing of every of above mentioned parameters. We believe that total mortality level will be maximal in fish of the most susceptible age. Simultaneously, the two other parameters will have their minimal values. When moving away from this age to younger or to older ages, parameters mentioned will change in reverse way.

We have revealed the high reliability of age-related changings of mortality dynamics parameters in grass carp and carp fingerlings and yearlings. The changings registered meet the hypothesis propounded above. Additional studies are essential to verify it.

Apparently, the same pattern of virus-host interaction will occur in the event when a susceptible host of some age is faced with various strains of some virus, differing in virulence (for illustration see data on grass carp in Fig. 1 and data on carp in Fig. 3).

Of four tested R.carpio strains strain M2 appeared as the most virulent one in fishes of various species and ages. The viral strain 3 had the virulence as high as M2 did when testing them in herbivorous fishes fry (Fig. 1). But at the same time this strain appeared as the less virulent when testing in fingerlings (particularly in grass carp fingerlings; Fig.3). It is the well known fact that virus strains can change their virulence following passaging, storage etc. Our data add one more fact to this one: virulence of any strain of some virus in relation to its host species is not a constant property but changes in a different way in fishes of various ages. Apparently, this property may be the marker sign of a virus strain.

ACKNOWLEDGMENTS

We thank Dr. N.I. Rudikov, Moscow, for supplying the SVCV-antiserum, A.G. Novikova for technical assistance and V.M. Kozlova for photography.

LITERATURE

Ahne W (1975) A rhabdovirus isolated from grass carp (Ctenopharyngodon idella Val). Archives of Virology 48: 181-185

Ahne W (1979) Untersuchungen über die akute Form der infektiösen Bauchwassersucht bei Cypriniden (Cyprinus carpio, Ctenopharyngodon idella). München: 1-112

Chen YX, Jiang YL (1983) Morphological and physico-chemical characterization of reovirus from haemorrhage of grass carp. A Monthly J. Sci. 18: 1138-1140

Clerx J, Horzinek M (1978) Comparative protein analysis of nonsalmonid fish rhabdoviruses. J. gen. Virology 40: 287-295

Dixon PF (1987) Inhibition of infectious pancreatic necrosis virus infectivity by extracts of rainbow trout, Salmo gairdneri Richardson, tissue. J.Fish Dis. 10: 371-378

Fijan N (1978) Frühlingsvirämie der Karpfen. Ätiologie,Vorkommen und Bekämpfung. Tagungsbericht des Seminars der DVG "Diagnose und Behandlungen von Fischkrankheiten" München

Fijan N, Matasin Z, Jeney Zs, Olah J, Zwillenberg LO (1981) Isolation of Rhabdovirus carpio from sheatfish, Silurus glanis, fry. Proc. Int. Seminar "Fish, Pathogens and Environment in European Polyculture" Szarvas, Hungary : 48-58

Fijan N, Sulimanović D, Bearzotti M, Muzinić D, Zwillenberg L, Chilmonczyk S, Vautherot J, de Kinkelin P (1983) Some properties of the Epithelioma Papulosum Cyprini (EPC) cell line from carp Cyprinus carpio. Ann. Virol.(Inst. Pasteur)134E: 207-220

Gravell M, Malsberger R (1965) A permanent cell line from the fathead minnow (Pimephales promelas). Ann. N.Y. Acad. Sci. 126: 555-565

Hill B (1976) Procedures for the isolation and identification of IPN, VHS, IHN and SVC viruses from diseased fish. Fisheries research technical report, Lowestoft 27: 1-16

Pichugina TD (1983) Comparative study of isolates causing spring viral disease of fish by means of serum neutralization test. Proc. of the All-Union Res. Inst.Exp. Veter., Moscow 57: 141-143 (In Russian)

Rudikov NI, Grishchenko LI, Lobuntsov KA (1975) Spring viral disease of fish. Bul. of the All-Union Res. Inst. Exp. Veter., Moscow 20: 16-19 (In Russian)

Shchelkunov IS, Yukhimenko LN, Shchelkunova TI, Trombitsky ID, Manchu AP (1984) Isolation of a virus from silver carp with infectious dropsy syndrome. Fishery employment of inland water reservoirs, Express-information CNIITEIRCH, Moscow 4: 3-7 (In Russian)

SVCV and PFR Viruses: Serological Examination of 22 Isolates Indicates Close Relationship Between the Two Fish Rhabdoviruses

P.E. Vestergård Jorgensen, N.J. Olesen, W. Ahne, and N Lorenzen
National Veterinary Laboratory, Hangøvej 2, 82oo Arhus N, Denmark.

Introduction

The aim of the present study was to elucidate the antigenic variation and cross reactivity inside the groups of fish viruses represented by spring viraemia of carp virus (SVCV) (Fijan et al., 1971) and pike fry rhabdovirus (PFR) (deKinkelin et al., 1973), respectively, viruses which belong to the so-called vesiculo-group of fish rhabdoviruses (Lenoir and deKinkelin, 1975).

It was hoped that the results would provide a well documented basis for serological identification of virus isolates belonging to either of the groups, a basis which has been inadequately elaborated so far. Such virus identification procedures are necessary tools in connection with diagnosis and control of viral diseases and with health certification procedures.

A more theoretical goal was to provide information about the taxonomic distance between SVCV and PFR by means of immunochemical and biological techniques.

The viruses both induce systemic infections with generalized viraemia and haemorrhages in viscera and muscles, but in part have different host spectrum.

* Institut für Zoologie und Hydrobiologie, Universität München, Kaulbachstrasse 37, 8ooo München 22, W.Germany.

In addition to the SVCV and PFR reference strains the study included a PFR-related rhabdovirus isolated from grass carp (Ahne, 1975) and 19 field isolates from 7 fish species representing the Cyprinidae, the Siluridae and the Esocidae.

Materials and methods.

Cell lines:
FHM cells (Gravell and Malsberger, 1965), EPC cells (Fijan et al. 1983) and CLC cells (Faisal and Ahne, in press) were grown at temperatures between 25° and 30°C in Eagle's MEM supplemented with 10% foetal bovine serum and antibiotics in standard concentrations. Medium for cells which were grown in closed units (bottles and tubes) were buffered with bicarbonate. Medium for other purposes was buffered with Tris or Hepes.

Following infection with virus the cell cultures were incubated at 15°C to 25°C as specified later.

Virus:
The following virus isolates were used in the study:
The reference strains of spring viraemia of carp virus (SVCV) (Fijan et al. 1971) and pike fry rhabdovirus (PFR) (deKinkelin et al., 1973), originally obtained from Dr. P. deKinkelin, Grignon, France, 8 field isolates from Hungary (a gift from Dr. E. Kovács-Gayer, Budapest), 2 field isolates from Denmark originating from fish in an import quarantine station, and finally 10 isolates from W. Germany, among these the grass carp rhabdovirus (GRV) previously described by Ahne (1975).

Data on the fish species origin of the various virus isolates are given in Table 1 together with the identification of the isolates based on 50% plaque neutralization tests carried out as a part of the present work.

Passage levels of the field isolates were between 3 and 8 and of the reference strains between 12 and 15.

The majority of the field isolates had not been serologically examined previously. They were all known to be ether sensitive and most of them to be of rhabdovirus morphology as revealed by electron microscopy.

In some tests Egtved virus (VHSV), the virus of viral haemorrhagic septicaemia (VHS) (Jensen, 1965) was used for control purposes.

Preparation of viral antisera:

The reference strains of SVCV, and PFR, respectively, following multiplication in FHM cells at 25°C, were gradient purified on 10-50% sucrose gradient in Beckmann SW 28 or SW 50.1 rotors as previously described (Ahne, 1978). Purified virus suspension with titres of 5 x 10^8 to 5 x 10^9 $TCID_{50}$ per ml were mixed with Freund's complete adjuvant and injected intramuscularly into rabbits on day 0 in amounts of 1 ml per rabbit. Booster injections of 0.5 ml without adjuvant were given on days 14, 21 and 28 after the first injection (Procedure A (PFR)).

Another group of rabbits were given injections at weeks 0, 2, 4, 6, 8, 16, 19 and 22, with the inocula at weeks 0, 2, 4, 16 and 19 containing Freund's incomplete adjuvant, the others being without adjuvant. Inocula with adjuvant were given subcutaneously, inocula without adjuvant were given intraperitoneally (Procedure B (SVCV and PFR)).

Bleeding was performed 6 days after the last injection in both immunization programs.

PFR antisera produced according to procedure A were used only in the neutralization tests. SVCV and PFR antisera produced according to procedure B were used in all the techniques applied.

Before being stored at -8o°C the antisera were heat inactivated at 56°C for 3o minutes except for small aliquots kept untreated for reference purposes.

For use in ELISA immunoglobulin G (IgG) was purified by affinity chromatography on Protein A Sepharose according to standard procedures. Elution was performed by means of 1 M acetic acid, pH 3.o and protein concentration determined by means of spectrophotometry at 28o nm.

Enzyme-linked immunosorbent assay (ELISA):
A technique based on the modified double antibody sandwich ELISA of Voller et al. (1979) was developed for the purpose of comparing SVCV and PFR isolates.

Microtiter plates (Nunc-immuno plate, Maxisorp, Nunc, Denmark) were incubated over night at 4°C with 5o ul per well of purified IgG (15-3o ng) from SVCV- or PFR antisera diluted in o.1 M $NaHCO_3$, pH 9.6. The plates were washed three times with phosphate buffered saline (PBS) containing o.o5% Tween 2o (PBS-T) and incubated for one hour at 37°C with 5o ul of serial four-fold dilutions of virus in the form of membrane filtered (o.45 um) supernatant from cell cultures showing complete cytopathic effect (CPE). For each virus six dilutions starting with 1:4 were used. Each dilution was applied to two wells. The diluent was PBS-T with 1% bovine serum albumin (Sigma) (PBS-T-BSA). The washing procedure was repeated and the plates incubated for one hour at room temperature with 5o ul of either biotinylated anti-SVCV Ig or anti-PFR Ig. After washing the plates were incubated for one hour at room temperature with horse raddish peroxidase (HRP)-conjugated Streptavidin (Dakopatts, Copenhagen) diluted 1:5ooo in PBS-T-BSA. After a last rinse the bound enzyme was visualized using H_2O_2 - o-phenylenediamine (OPD) substrate. The colour reaction was stopped with 1 M H_2SO_4 and the reactions read spectrophotometrically at 492 nm (Titertek Multiscan). All virus isolates were multiplied at 15°C.

Biotinylation of antibodies:
Biotinylation was carried out as described by Tejssen (1987). Briefly, Biotin N-Hydroxysuccinimide-esther (BNHS), dissolved in N,N-Dimethylformamide, was reacted with IgG at pH 8.o at room temperature for two hours with a ratio of 1 between BNHS and free aminogroups.

Indirect fluorescent antibody technique (IFAT):
EPC, FHM or CLC cells were seeded on 9 x 22 mm coverglasses in glass Petri dishes (diameter 1o cm) and cultivated for 24 hours before being infected with virus. Each of the 22 virus

isolates tested were inoculated into coverglass cell cultures at two different levels of infectivity by adding 1oo ul and 1o ul, respectively, of membrane filtered medium from infected cell cultures (showing complete CPE) into a Petri dish. After incubation for 2o to 24 hours at 25°C, medium was removed, the coverglasses rinsed twice in Eagle's MEM without serum and the cells subsequently fixed in acetone for 1o minutes or in 3% paraformaldehyde in o.1 M PBS, pH 7.6 for 3o minutes.

Staining was performed with dilutions of rabbit anti-SVCV or -PFR, respectively, as first antibody layer, followed by fluorescein isothiocyanate (FITC) or tetramethylrhodamine isothiocyanate (Rhodamine) conjugated swine antibodies to rabbit Ig (Dakopatts, Copenhagen) as second antibody layer. Rinsing and mounting was according to standard procedures as previously described (Jørgensen, 1974a). Examination was carried out in an Olympus Vanox epifluorescence microscope supplied with a 2oo W mercury lamp, interference filters (Rygård and Olsen, 1969) and fluorescence oil-objectives x 4o and x 1oo.

Sodium dodecyl sulfate polyacrylamide gel electrophoresis (SDS-PAGE):

SDS-PAGE was performed according to the method of Laemmli (197o) in 1 mm thick slab gels in a vertical slab gel system. The stacking gel contained 3.5% acrylamide / o.11% bisacrylamide, the separating gel 1o% / o.31%. The gels were loaded with gradient purified virus of the SVCV, PFR or Egtved reference strains. After electrophoresis the gels were used for immunoblotting.

Immunoblotting:

The protein bands were transferred from polyacrylamide gels to polyvinylidene difluoride membranes (Immobilon, Millipore) according to procedures described by Kyhse-Andersen (1984). Free binding sites were blocked by incubation in PBS with 5% BSA for one hour at 37°C. After washing in PBS-T the membranes were cut into 4 mm wide strips and incubated over night with dilutions of rabbit anti-SVCV or -PFR, respectively, previously absorbed twice in dilution 1:5o on monolayers of EPC cells. Following rinsing as described under ELISA, HRP conjugated swine antibodies to rabbit Ig (Dakopatts, Copenhagen) were added at dilution 1:1ooo. Visualization of bound conjugate and total protein was performed as previously described (Lorenzen et al., 1988).

5o% plaque neutralization tests (5o% PNT):
5o% plaque neutralization tests in EPC cells were carried out with all the 22 virus isolates under study against selected SVCV- and PFR antisera of rabbit origin using a micro-technique previously described (Olesen and Jørgensen, 1986). Complement, when used, was in the form of guinea pig serum. Fixation and staining was carried out after 5 days of incubation at 15°C under a semisolid medium containing o.5% methyl cellulose.

Plaque titration:
Plaque titration was carried out in EPC cell using microtiter plates (Nunc, Denmark) and methyl cellulose overlay medium. Cell cultures were fixed with 1o% Formalin and stained with crystal violet after incubation for 7 days at 15°C.

Heat inactivation:
Samples of membrane filtered virus in Eagle's MEM with 1o% serum were placed in a waterbath at 45°C or at room temperature, respectively, for 3o minutes. After rapid cooling in an ice water bath plaque titrations were performed and the pre- and post heating titres compared. The inactivation rate was calculated as log. 1o of the preheating titre divided by the postheating titre.

Comparison of plaque size:
Each of the 22 isolates were plaqued in EPC cells at 21°C under an overlay medium containing o.5% agarose and fixed after 48 hours as described by deKinkelin et al., 1974. The diameters of ten plaques for each virus isolate were measured to the nearest o.1 mm and the average diameters calculated.

Results.

The IFAT results showed that both anti SVCV and anti PFR (but not anti Egtved) gave a positive reaction with the SVCV and PFR reference strains as well as all of the 2o test isolates in acetone fixed cells. In contrast anti SVCV and anti PFR did not stain Egtved virus antigen (Table 2).

Table 2. IFAT with SVCV and PFR isolates.

	Anti SVCV	Anti PFR	Anti Egtved
SVCV reference strain	+	+	-
PFR reference strain	+	+	-
20 test isolates	+	+	-
Egtved virus, F 1	-	-	+

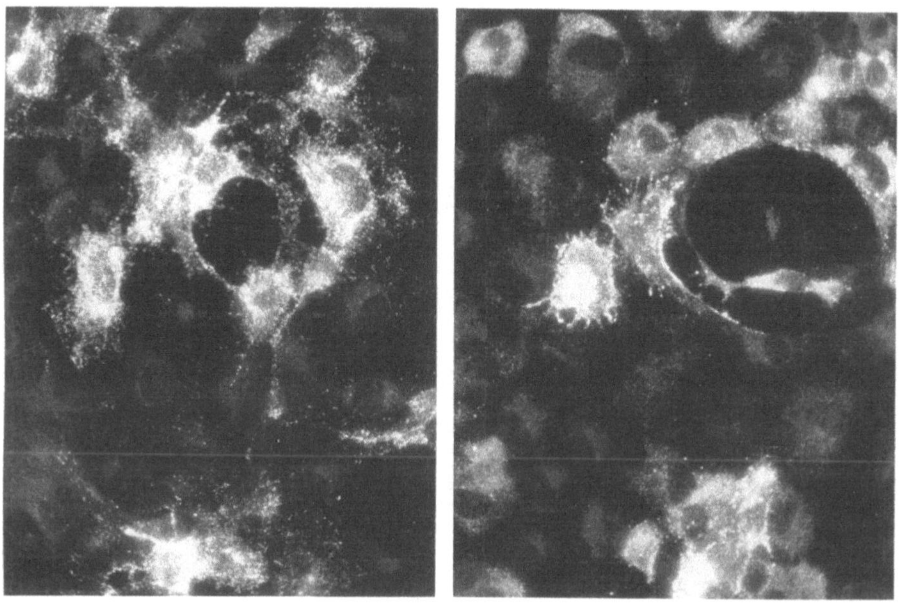

1 2

Figure 1 & 2: EPC cells fixed with acetone 2o to 24 hours after infection with SVCV (Fig. 1) or PFR (Fig. 2) and stained with rabbit antiserum to SVCV (dilution 1:8oo) and rhodamine conjugated swine antibodies to rabbit Ig (dilution 1:8o).

TABLE 1: Serological and biological characteristics of 22 virus isolates from Cyprinidae, Siluridae, and Esocidae

A: SVCV-like virus isolates

Designation of isolate	Fish species of origin	Titre with anti SVCV 1)	Titre with anti PFR 1)	Plaque size (mm) 2)	Log_{10} of heat reduction factor
SVCV ref.str.	*Cyprinus carpio L.* common carp	1280	< 40	1.2 (0.2)	1.8
12840/7	- " -	640	< 40	1.4 (0.2)	5.0
17314/5	- " -	160	< 40	1.9 (0.5)	3.6
17312/5	- " -	640	< 40	1.6 (0.5)	4.7
287	- " -	160	< 40	1.3 (0.2)	3.4
450	- " -	640	< 40	2.1 (0.1)	1.7
455	- " -	320	< 40	1.8 (0.4)	1.5
10/6	- " -	160	< 40	1.0 (0.3)	3.3
65/82	- " -	320	< 40	1.1 (0.3)	1.7
435	- " -	80	< 40	1.3 (0.2)	4.4
14286/3	*Silurus glanis L.* sheatfish	1280	< 40	1.3 (0.2)	0
17417/3	- " -	640	< 40	1.4 (0.4)	5.9
3587	*Carassius auratus L.* goldfish	640	< 40	1.6 (0.4)	3.4

B: PFR-like virus isolates

Designation of isolate	Fish species of origin	Titre with anti SVCV 1)	Titre with anti PFR 1)	Plaque size (mm) 2)	Log_{10} of heat reduction factor
PFR ref.str.	<u>Esox lucius L</u> pike	< 40	2560	0.7 (0.2)	0.8
"Hecht"	- " -	< 40	80	1.3 (0.3	5.7
V 76	<u>Ctenopharyngodon idella V</u> grass carp	< 40	2560	1.3 (0.2)	3.3
3605	- " -	< 40	2560	1.7 (0.4	1.8
GRV	- " -	< 40	5120	1.0 (0.4)	3.3
14241/6	<u>Siluris glanis L.</u> sheatfish	< 40	80	1.1 (0.2)	0.8
14249/5	- " -	< 40	80	1.2 (0.3)	1.4
S 64	<u>Tinca tinca L.</u> tench	< 40	5120	0.8 (0.1)	3.7
332	<u>Blicca bjoerkna L.</u> white bream	< 40	5120	1.8 (0.4)	3.7

1) 5o% Plaque neutralization titre (reciprocal value)

2) Standard deviation in brackets

The positive reaction with both antisera consisted of an intense staining of granular material in the cytoplasm of infected cells. There was no obvious difference in staining intensity when a homologous reaction was compared with a heterologous one (Figures 1 and 2).

Cross reaction was also observed when fixation was carried out with paraformaldehyde, a compound which leaves cell membranes impermeable to immunoglobulin molecules so that no antigenic material inside the membrane of the cell is stained by IFAT (Biberfeld et al., 1974). Under such conditions probably only the transmembrane G protein of a rhabdovirus would be stained. In the case of Egtved virus, another fish rhabdovirus, this was confirmed by the use of protein specific monoclonal antibodies (MAbs). Only MAbs specific for the viral G protein were able to detect virus infected cells in cultures fixed with paraformaldehyde (to be published elsewhere).

The ELISA results (Figs. 3 and 4) similarly showed that both anti-SVCV and anti-PFR gave positive reactions with all the virus isolates tested. However, the reaction of either of the two antisera was strongest with virus isolates which according to neutralization tests were homologous to the antiserum used. In contrast, Egtved virus gave rise to low absorption values whether one or the other antiserum was used.

The mobility in SDS-PAGE of G, N and M proteins in SVCV and PFR were very much alike (Fig. 5). The protein suspected to be NS, however, in the case of PFR moved more slowly than in that of SVCV.

Both antisera reacted strongly with G and N protein in homologous as well as heterologous virus. Anti-PFR stained the M protein of both SVCV and PFR, whereas anti-SVCV stained only homologous M protein. Both antisera stained homologous, but not heterologous NS protein.

5o% plaque neutralization tests with selected antisera (Table 1) allowed identification of all the 22 isolates as either SVCV or PFR. Only one out of five PFR antisera neutralized PFR specifically, i.e. without neutralizing SVCV as well. This antiserum in addition to two others was

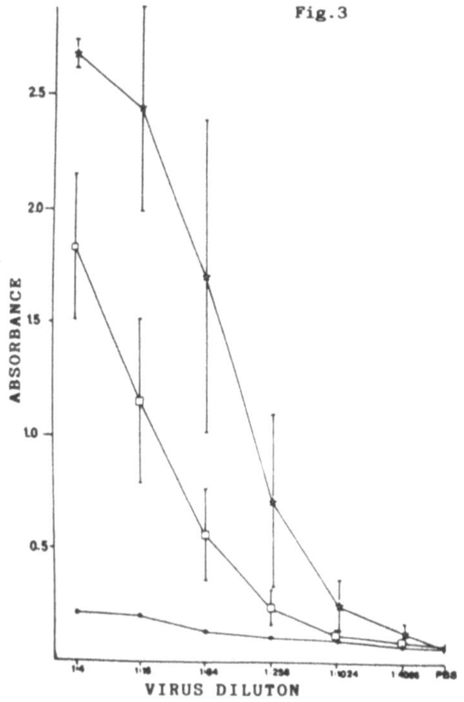

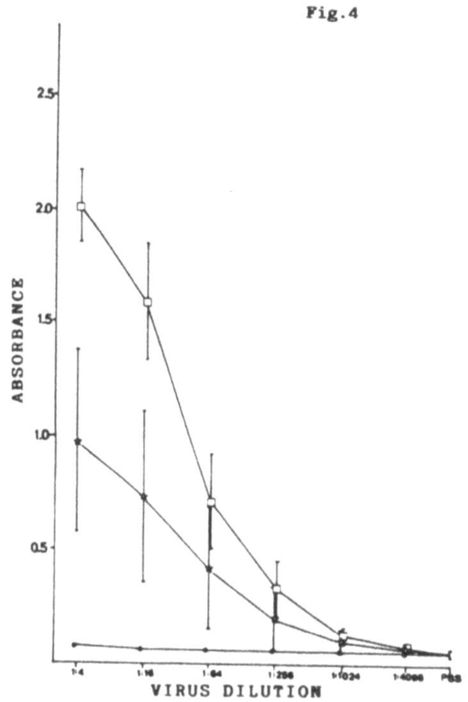

Figure 3 & 4: ELISA. ☆—☆ 13 SVCV isolates, □—□ 9 PFR isolates and ●—● 1 Egtved virus isolate examined against anti SVCV (Fig. 3) and anti PFR (Fig. 4). Vertical bars: Standard deviation.

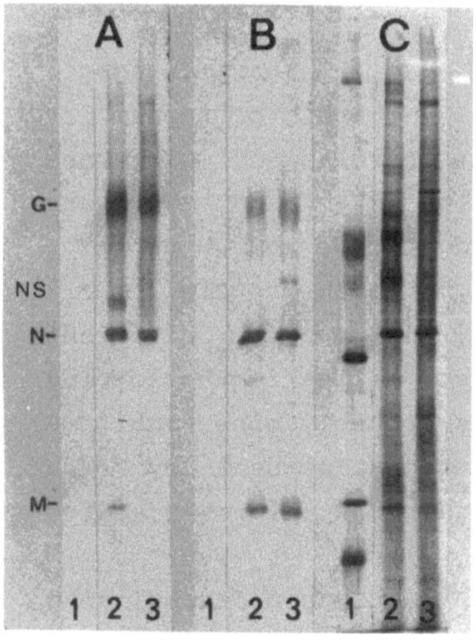

Figure 5: Immunoblotting with VHSV (1), SVCV (2), and PFR (3) antigens. Viral proteins were separated in a polyacrylamide gel and transferred to a membrane matrix. Membrane strips were either stained by immunological means (anti SVCV (A) and anti PFR (B)) or by colloidal gold (C).

prepared according to procedure A. In contrast both of two antisera to SVCV prepared according to procedure B neutralized SVCV specifically. However, if antisera were not heat inactivated or if complement was added to the reaction mixtures, the difference between homologous and heterologous titre of a given antiserum was reduced to an insignificant level (data not shown).

As shown in Table 1 the various virus isolates produced plaques of very varying size. Plaque diameters ranged from o.7 mm to 2.1 mm.

Resistance to heat also varied greatly between isolates (Table 1).

For both heat resistance and plaque size the differences between the isolates in each of the virus groups were more pronounced than the differences between the reference strains. Considering the reference strains separately it was verified that SVCV produce plaques which are bigger than those of PFR and that PFR is more resistant to heat than SVCV as reported by deKinkelin et al. (1974).

Discussion.

The present work shows that SVCV and PFR share antigenic determinants to a large extent on the G, N and M proteins.
This finding is in contrast with the results of Clerx and Horzinek (1978) according to which PFR and SVCV are antigenically unrelated with regard to G and M proteins although slightly related with regard to the N protein.
The difference in results probably is due to different sensitivity of the techniques applied. The immunoblotting technique used in the present work is probably more sensitive than the rocket immunoelectrophoresis technique used by Clerx and Horzinek (1978) and in addition allows a safer identification of protein bands.

Due to the occurrence of common antigenic determinants distinction between isolates belonging to the SVCV or PFR group, respectively, could not be achieved by means of IFAT.

In ELISA, the differences observed between homologous and heterologous absorption values of either antiserum were large enough to allow such a distinction provided that each virus isolate was tested against both SVCV and PFR antiserum and with proper controls included in the test. This was the case even though the virus concentrations were not adjusted to a uniform level prior to the testing. Performing such adjustments probably would decrease the standard deviations observed.

The suspected NS proteins of the two reference strains did not share antigenic determinants which could be revealed by the antisera used. Distinction between the reference strains therefore also was possible by means of immunoblotting in that the NS proteins of both viruses were stained only by homologous antiserum. This feature, however, was not studied with any of the field isolates.

A more detailed knowledge of the structure of the NS molecule in rhabdoviruses would make it easier to evaluate whether major antigenic differences would be likely to occur between different isolates of one virus. The observed differences between the reference strains concerning the mobility of the NS proteins in SDS PAGE are probably not valuable taxonomic characteristics either, since this parameter seems to be strongly influenced by minor variations in amino acid sequence leading to differences in intramolecular phosphodiesther bindings (Evans et al., 1979). Perhaps in accordance with this the NS mobility in SDS PAGE seems to vary from one isolate to another of one and the same virus e.g. PFR (Clerx and Horzinek, 1978).

In the case of vesicular stomatitis virus (VSV) it has been described that the NS proteins of virus strains belonging to two different serotypes (New Jersey and Indiana) in some cases show very little sequence homology (20%) (Banerjee, 1987). When this is the case it appears likely that also antigenic characteristics might be so different that antisera would not crossreact. In other words, failure to detect common antigens at the NS protein level may not necessarily indicate a "species"-difference.

The mobility of the G, N, NS and M proteins of SVCV and PFR in SDS-PAGE/Immunoblotting observed in the present work corresponded very well with the results of Clerx and Horzinek (1978). These authors found a slight mobility-difference between the respective NS proteins, indicating that PFR NS protein had a slightly higher molecular weight than the NS of SVCV. The other proteins had almost identical molecular weights as also reported by Lenoir and deKinkelin (1975).

The reason why PFR upon its isolation was considered to be a new virus was lack of neutralization by SVCV antiserum, in combination with different temperature resistance and plaque size when compared with the SVCV (deKinkelin et al., 1974). With regard to the neutralization results lack of cross reaction between SVCV and PFR was later confirmed by Hill et al. (1975), and Ahne (1986), both using heat inactivated immune sera from rabbits. However, it appears from the data of the above authors that the neutralizing titres of PFR-antisera were low compared to those of anti-SVCV, and this may explain the inability of the antisera to cross react.

Loss of specificity of SVCV and PFR antisera after addition of complement has previously been observed (Clerx et al., 1978).

It appears from the above results that SVCV and PFR can indeed be distinguished from each other in neutralization tests.

However, presumably only a small part of the G protein of rhabdoviruses is involved in the complement independent neutralization process, and differences in the ability of rhabdoviruses to become neutralized therefore may reflect only very minor biochemical differences (Wunner et al., 1985). These may very well be connected with serotype differences rather than "species" differences.

If the above pattern of serological cross reaction observed between SVCV and PFR is compared to that described for the viruses belonging to the lyssa-group of fish rhabdoviruses, i.e. the Egtved (VHSV)-IHN virus group (Lenoir and deKinkelin, 1975) a striking difference becomes apparent.

Serological cross reactions between VHSV and IHN do not occur neither in neutralization tests nor in IFAT (Jørgensen, 1972, 1974b, McAllister et al. 1974). It therefore appears logical to consider VHSV and IHN as separate viruses. Inside the VHSV group subtypes (sometimes designated serotypes) exist, which cross react very strongly in IFAT but not always in neutralization tests (Meier and Jørgensen, 1979, Le Berre et al., 1972). The differences observed between SVCV and PFR clearly resemble the differences between VHS subtypes more than the difference between VHS and IHN.

On the above overall background it appears that it may be more appropriate to consider SVCV and PFR as representatives of two serotypes of one virus species rather than as two separate virus-"species". This view seems to gain support from the finding that differences with regard to heat resistance and plaque size according to the present work do not correspond with neutralization characteristics, a finding which tends to minimize the taxonomic significance of these biological characteristics.

As shown in Table 1 two out of four isolates from sheatfish (Silurus glanis) were identified as PFR and two as SVCV isolates. This finding which reveals sheatfish as a natural host for both of the viruses seems to be important from an epizootiological point of view and hopefully will stimulate more research in the area. Isolation of SVCV from sheatfish has previously been reported (Fijan et al., 1984).

Among the isolates which were identified as SVCV 1o out of 13, originated from common carp (Cyprinus carpio). In contrast none of the PFR isolates were of that origin.

Further studies concerning the host range of the viruses, the possible influence of the host on the neutralization characteristics of a given virus isolate and the reaction in immunoblotting of more virus isolates have to be carried out before final conclusions can be drawn with regard to the taxonomic status of SVCV and PFR. In addition it may become necessary to include RNA homology studies in the work.

Based on the results of the present work it can be concluded that 5o% plaque neutralization tests with carefully selected antisera will allow distinction between virus isolates belonging to the SVCV group and isolates belonging to the PFR group. Such a distinction can also be based on ELISA. Considering the difficulties encountered in the production of antibodies which neutralize PFR specifically ELISA may be the method of choice for identification purposes in many laboratories.

Acknowledgement:

The present work was supported by the Commission of the European Communities (Grant ST 2J -o1o8-2-DK).

References.

Ahne W (1975) A rhabdovirus isolated from grass carp (Ctenopharyngodon idella Val). Arch Virol 48: 181-185

Ahne W (1978) Laboratoriumsdiagnostik fischpathogener Viren. Tierärzliche Umschau 33: 584-594

Ahne W (1986) Unterschiedliche biologische Eigenschaften 4 cyprinidenpathogener Rhabdovirusisolate. J Vet Med B 33:253-259

Banerjee AK (1987) Transcription and replication of rhabdoviruses. Microbiological Reviews 51:66-87

Biberfeld P, Biberfeld G, Molnar Z, and Fagraeus A (1974) Fixation of cell-bound antibody in the membrane immunofluorescence test. J Immunol Methods 4:135-148

Clerx JPM, and Horzinek MC (1978) Comparative protein analysis of non-salmonid fish rhabdoviruses. J gen Virol 4o:287-295

Clerx JPM, Horzinek MC, and Osterhaus ADME (1978) Neutralization and enhancement of infectivity of non-salmonid fish rhabdoviruses by rabbit and pike immune sera. J gen Virol 4o:297-3o8

Evans D, Pringle CR, and Szilàgyi JF (1979) Temperature sensitive mutants of complementation group E of vesicular stomatitis virus New Jersey serotype possess altered NS polypeptides. J Virol 31:325-333

Fijan N, Petrinec Z, Sulimanovic D, and Zwillenberg LO (1971) Isolation of the causative agent from the acute form of infectious dropsy of carp. Vet Arhiv 41:125-138

Fijan N, Sulimanovic D, Bearzotti M, Muzinic D, Zwillenberg LO, Chilmonczyk S, Vauthorst JF, and deKinkelin P (1983) Some properties of the epithelioma papulosum cyprini (EPC) cell line from carp (Cyprinus carpio). Ann Inst Pasteur/Virol 134 E: 2o7-22o

Fijan N, Matasin Z, Jeney Z, Olah J, and Zwillenberg LO (1984) Isolation of Rhabdovirus carpio from sheatfish (Silurus glanis) fry. Symp Biol Hung 23:17-24

Faisal M, and Ahne W (1988) A cell line of adherent peripheral blood mononuclear cells of normal common carp (Cyprinus carpio). Developmental and Comparative Immunology (in press)

Gravell M, and Malsberger RG (1965) A permanent cell line from the fathead minnow (Pimephales promelas). Ann N Y Acad Sci 126:555-565

Hill BJ, Underwood BO, Smale CJ, and Brown F (1975) Physico-chemical and serological characterization of five rhabdoviruses infecting fish. J gen Virol 27: 369-378

Jensen MH (1965) Research on the virus of Egtved disease. Ann N Y Acad Sci 126:422-426

Jørgensen PEV (1972) Antigenic variation in 76 virus isolates examined in neutralization tests and by means of the fluorescent antibody technique. Symp Zool Soc London 3o:333-34o Mawdsley-Thomas L (ed) Academic Press London and New York

Jørgensen PEV (1974a) Indirect fluorescent antibody techniques for demonstration of trout viruses and corresponding antibody. Acta Vet Scand 15:198-2o5

Jørgensen PEV (1974b) A study of viral diseases in Danish rainbow trout. Their diagnosis and control. Thesis commissioned by CF Mortensen Bülowsvej 5 Copenhagen V

deKinkelin P, Galimard B, and Bootsma R (1973) Isolation and identification of the causative agent of "red-disease" of pike (Esox lucius L 1766). Nature Lond 241:465-467

deKinkelin P et Le Berre M (1974) Rhabdovirus des Poissons II Propriétés in vitro du virus de la virémie printanière de la Carpe. Ann Microbiol (Inst Pasteur) 125 A:113-124

deKinkelin P, Le Berre M et Lenoir G (1974) Rhabdovirus des Poissons I Propriétés in vitro du virus de la maladie rouge de l'alvin de Brochet. Ann Microbiol (Inst. Pasteur) 125 A:93-111

Kyhse-Andersen J (1984) Electroblotting of multiple gels: a simple apparatus without buffer tank for rapid transfer of proteins from polyacrylamide to nitrocellulose. J Biochem Biophys Methods 1o:2o3-2o9

Laemmli UK (197o) Cleavage of structural proteins during assembly of the head of bacteriophage T4. Nature Lond 227:68o-685

LeBerre M, deKinkelin P, and Metzger A (1977) Identification sérologique des Rhabdovirus des Salmonidés. Bull Off int Epiz 87: 391-393

Lenoir G, and deKinkelin P (1975) Fish rhabdoviruses Comparative study of protein structure. J Virol 16: 259-262

Lorenzen N, Olesen NJ, and Jørgensen PEV (1988) Production and characterization of monoclonal antibodies to four Egtved virus structural proteins. Dis Aquat Org 4: 35-42

McAllister PE, Fryer JL, and Pilcher KS (1974) An antigenic comparison between infectious hematopoietic virus (OSV strain) and the virus of haemorrhagic septicemia of rainbow trout (Salmo gairdneri). J Wildl Dis 1o:1o1-1o3

Meier W, and Jørgensen PEV (1979) Egtved virus: Characteristics of a virus strain isolated from pike fry (Esox lucius L). Nord Vet -Med 31: 484-485

Olesen NJ, and Jørgensen PEV (1986) Detection of neutralizing antibodies to Egtved virus in rainbow trout (Salmo gairdneri) by plaque neutralization test with complement addition. J Appl Ichtyol 2:33-41

Rygaard J, and Olsen V (1969) Interference filters for improved immunofluorescence microscopy. Acta Path Microbiol Scand 76:146-148

Tejssen P (1987) Practice and theory of enzyme immunoassays. Elsevier Amsterdam

Wunner WH, Dietzschold B, and Wiktor TJ (1985) Antigenic structure of rhabdoviruses. In Immunochemistry of viruses The basis for serodiagnosis and vaccines. (Eds van Regenmortel MHV and Neurath AR) Elsevier Amsterdam New York Oxford

The Significance of Rainbow Trout Brain and Excretory Kidney for the Propagation of Viral Haemorrhagic Septicaemia (VHS) Virus*

P. Kruse and M. Neukirch
Fachgebiet Fischkrankheiten und Fischhaltung
Tierärztliche Hochschule Hannover
Bünteweg 17
3000 Hannover 71
Federal Republic of Germany

INTRODUCTION

Viral haemorrhagic septicaemia (VHS) is one of the most common virus diseases in European fish farms mainly affecting rainbow trout. Besides the triphasic pattern of this disease - the acute, the chronic, and the nervous form (Ghittino, 1965) - the development of a persistent infection without showing any clinical signs but combined with potential virus excretion has been demonstrated in surviving trout (Köbl, 1978; Vestergård Jφrgensen, 1982a, 1982b; Neukirch & Glass, 1984; Neukirch 1986). Especially the brain and the kidney were presumed to be the favoured organs for harbouring persistent VHS virus.
The multiplication of VHS virus in the brain of experimentally infected rainbow trout has already been demonstrated (Neukirch, 1984). Furthermore, VHS virus was reisolated from the brain of apparently healthy trout more than 400 days after experimental infection without detecting the virus in the blood or in other internal organs (Neukirch, 1986). On the other hand, virus excretion has also been demonstrated in the urine of catheterized trout in spite of the fact that in some cases

* This study was supported by a grant of the DFG (Deutsche Forschungsgemeinschaft).

the reisolation of the virus from the blood completely failed (Neukirch & Glass, 1984). So, virus multiplication was supposed to occure in both the cells of the nervous tissue as well as the epithelia of the renal excretory system. The aim of the investigations presented here was, therefore, to obtain more informations about the location in the brain and the excretory kidney where VHS virus can multiply and often persist.

MATERIAL AND METHODS

Rainbow trout (Salmo gaidneri Rich.) of length 10-15 cm were used for the infection trials. The trout were maintained for 2 h at 9°C in an aqueous suspension of VHS virus with an infectivity titre of nearly $10^{4.0}$ $CCID_{50}$/ml. After exposure the fish were replaced in 70-litre tanks with running tap water and kept at temperatures between 9°-10°C.

For virus demonstration the organs of the infected fish were removed at predefined intervalls and bisected lengthwise. One half of each sample was homogenized and investigated for VHS virus infectivity using RTG cells (Wolf & Quimby, 1962). The corresponding halves were embedded in paraffin after they had been fixed in buffered formalin (10%) immediately after collection. The specimens were then cut in 5-µm sections and stained by the indirect immunoperoxidase technique[2] (Sternberger, 1979; Bourne, 1983) for demonstration of virus antigen. Counterstaining was performed with Mayer's hemalum solution.

[2] VHS-Antiserum was kindly provided by Dr. Enzmann, Tübingen.

RESULTS AND DISCUSSION

In this experiment 29 rainbow trout were waterborne infected. During the first two weeks the brain and the kidney of one infected fish were collected daily. The next three weeks the samples were collected twice a week and thereafter once a week. The results of virus demonstration are summerized in table 1.

VHS virus could be detected in 11 of 29 fish (about 38%). Seven trout showed VHS virus in both the brain and the kidney. In two trout VHSV was demonstrated only in the kidney, and in further two trout only in the brain. No external signs distinguished the latter two fish from healthy trout except nervous symptoms like looping swimming behaviour which, however, was not always manifest.

Table 1. Demonstration of VHS virus in rainbow trout brain and kidney using the indirect immunoperoxidase method (IP) and the cell culture technique (CC).

	BRAIN		KIDNEY		
D.p.i.	CC	IP	CC	IP	VHS Symptoms
1	-	-	-	-	-
2	-	-	-	-	-
3	-	-	-	-	-
4	-	-	-	-	-
5	+	+	+	+	-
6	-	-	-	-	-
7	+	-	+	-	Petechia
8	+	-	+	-	-
9	+	+	+	+	Petechia
10	+	-	+	+	Petechia, Exophthalmus
11	-	-	+	+	Dark colouring
12	-	-	-	-	-
13	-	-	+	+	-
16	-	+	+	+	Dark colouring
20	+	+	+	-	Dark colouring, Petechia
23	-	-	-	-	-
27	-	-	-	-	-
30	-	-	-	-	-
34	+	+	-	-	Nervous signs
43	-	-	-	-	-
52	+	+	-	-	Nervous signs
62	-	-	-	-	-
69	-	-	-	-	-
76	-	-	-	-	-
83	-	-	-	-	-
90	-	-	-	-	-
97	-	-	-	-	-
105	-	-	-	-	-
112	-	-	-	-	-
Totals	8	6	9	6	
	9		9		

Excretory Kidney

In this attempt the first virus antigen was detected 5 days p.i. in the endothelia of the renal blood vessels manifested by few weakly marked antigen spots. Number and size of marked spots increased during the following days. The virus antigen was often associated with the basal membrane of the tubulus cells (Fig. 1). In the epithelia of the excretory kidney virus antigen occured on day 8 p.i. the first time. The prefered location where antigen generally became evident was found to be the cells of the proximal nephron (Fig. 2). Reduced amounts of virus antigen were observed in cells of the distal nephron and only very few renal corpuscles seem to be involved in the process of virus shedding (Fig. 3). Virus antigen that was supposed to be excreted via urine could be demonstrated in the cytoplasma of tubulus cells and in the lumina of the distal segment of the proximal nephron (Fig. 4).

These results do not agree with experiments of Hoffmann and Wizigmann (1984) who failed in detecting VHS virus in the excretory system of the kidney. However, only the virus multiplication in tubulus cells of the excretory system, demonstrated in our experiments, can explain the phenomenon of virus shedding via urine in spite of the failure of virus demonstration in the serum or leucocytes of the infected fish. Sometimes VHS virus is presumed to penetrate into the lumina of the tubuli by the intercellular way (Fig. 5). The virus, therefore, is suggested to enter the tubulus lumina in two ways, first after VHS virus had been multiplied in the epithelial cells of the renal tubuli and, second, passively by the intercellular way from endothelia of the renal vascular system, especially during the acute form of VHS disease with generally high virus titres in the blood and also in the urine.

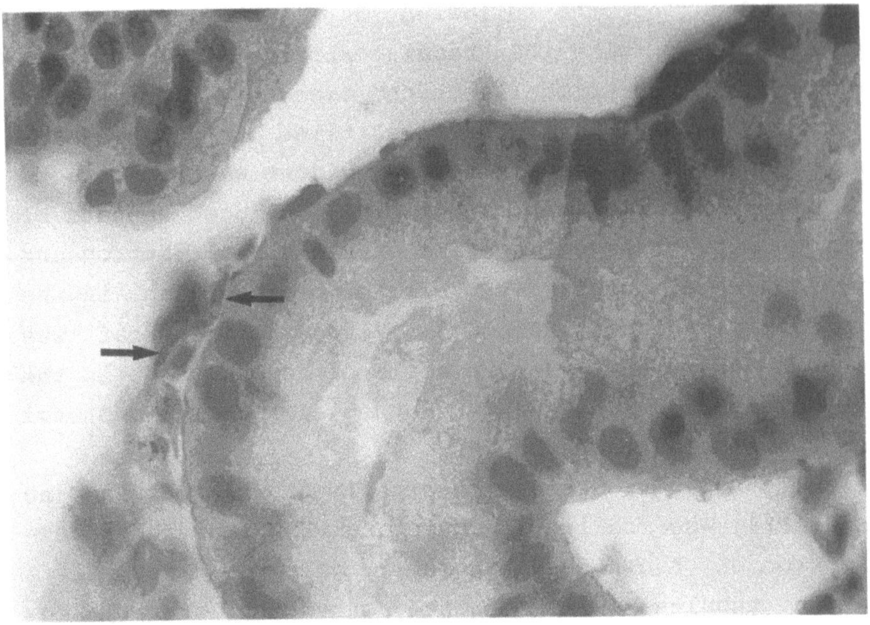

Fig. 1. Excretory kidney, X 788. VHS virus in endothelia associated with the basal membran of tubulus cells. In this and the following pictures arrows indicate virus antigen.

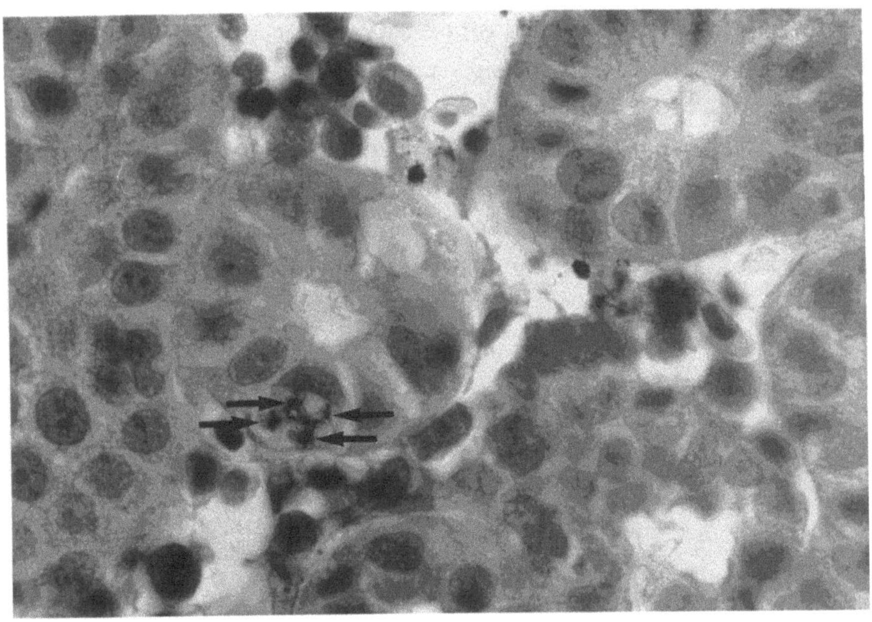

Fig. 2. Excretory kidney, X 788. VHS virus located in a distal nephron cell.

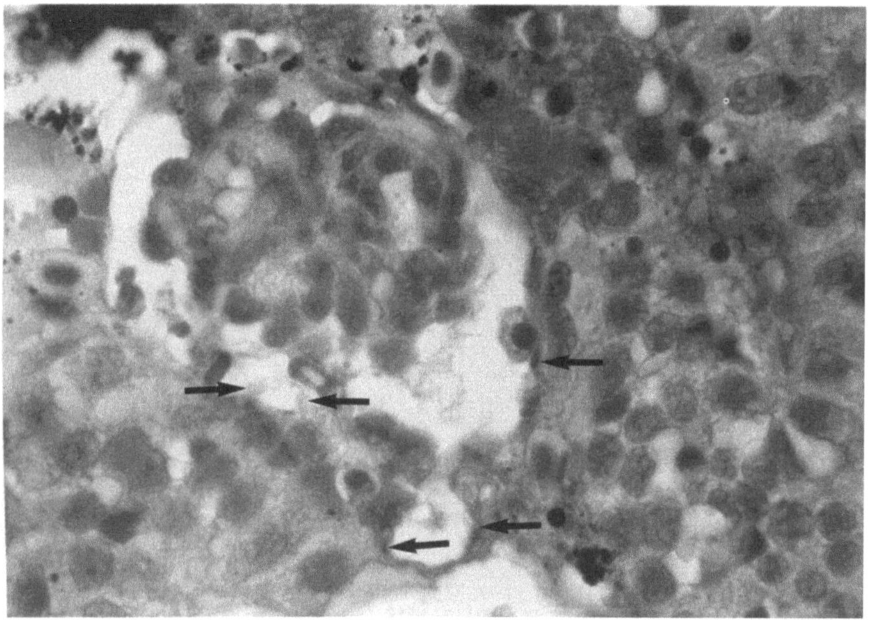

Fig. 3. Excretory kidney, X 788. Glomerulus affected with VHS virus. Virus antigen in mesangial cells and endothelia of tubulus neck segment.

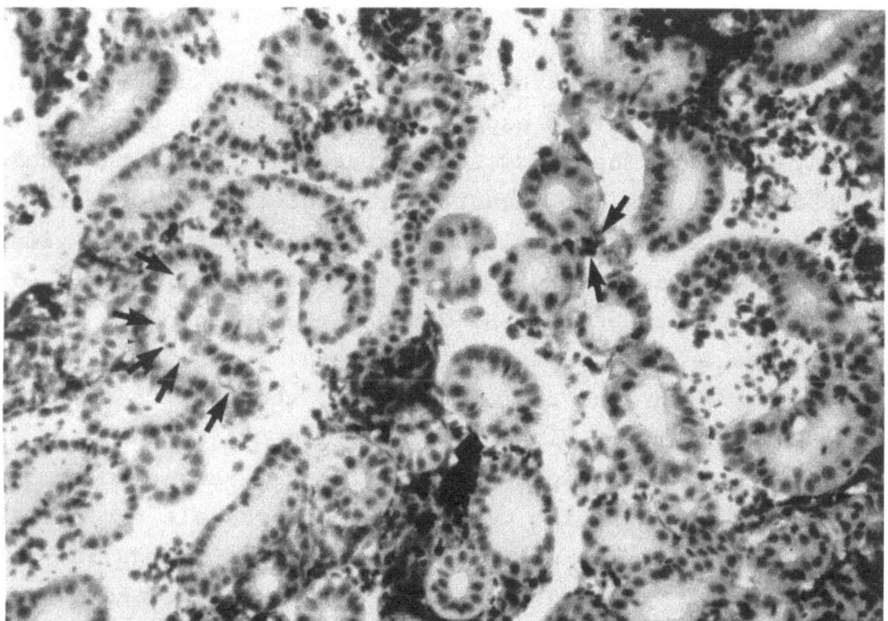

Fig. 4. Excretory kidney, X 200. VHS virus located in tubulus lumina.

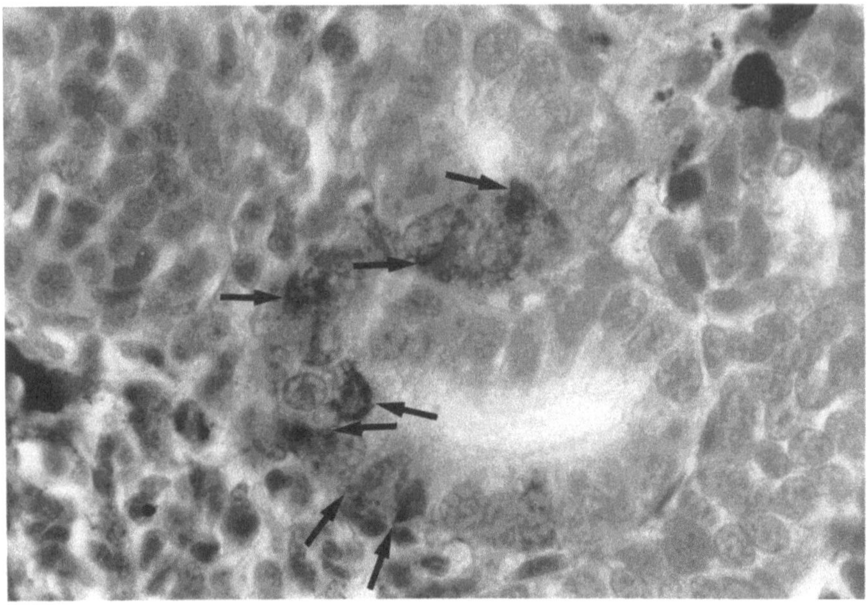

Fig. 5. Excretory kidney, X 788. VHS virus entering the tubulus lumen by the intercellular way.

Brain

Comparing the brain with the kidney no delay in demonstrating VHS virus resulted in the experiment. Antigen was also detected on day 5 p.i. The process of virus spread into the brain tissue probably seems to emanate from infected endothelia of brain vessels. During the initial phase small amounts of virus were found mainly associated with the leptomeninx (Fig. 6). In more progressive stages of VHS disease the antigen was disseminated in nearly all parts of the brain. The nervous tissue of the diencephalon and the optic lobe, however, seem to be favoured in harbouring virus antigen. These findings are in contrast to those of Wizigmann and Hoffmann (1982), who demonstrated VHS virus only in cerebral blood vessels but not in the nervous tissue. The multiplication of the virus in the nervous tissue of the brain explain former experimental results (Neukirch, 1986) concerning the reisolation of VHS virus from the brain without detecting virus in other organs.

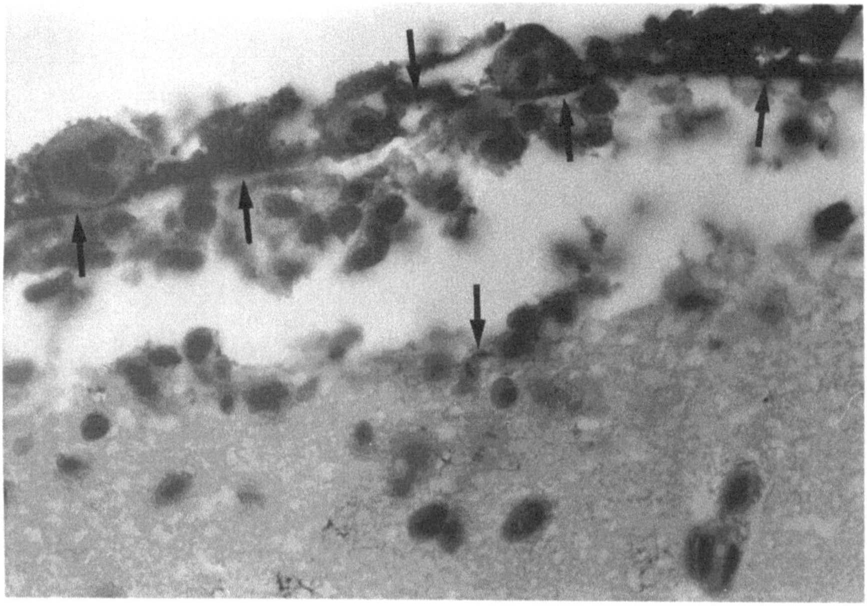

Fig. 6. Central nervous system, optic lobe, X 788. Virus antigen associated with the leptomeninx.

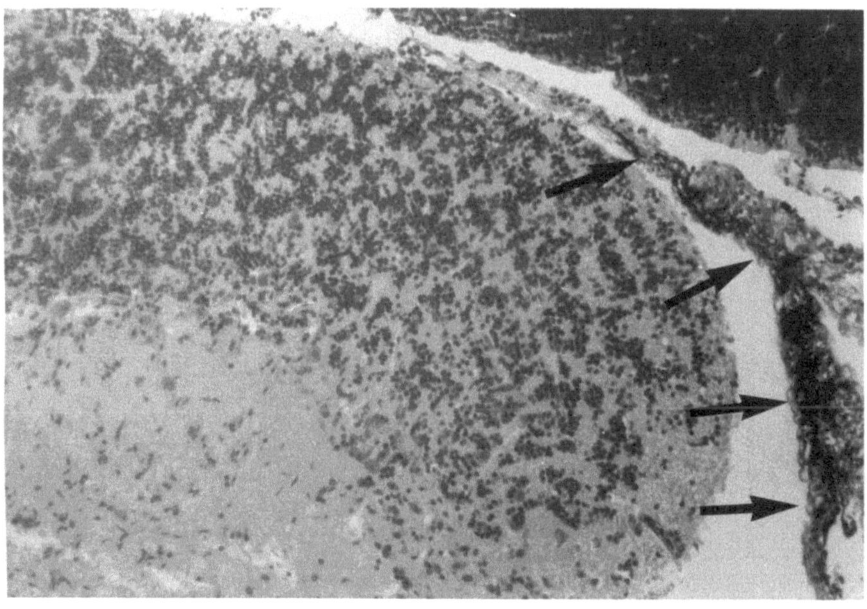

Fig. 7. Central nervous system, X 125. Heavily affected parts of cerebellum.

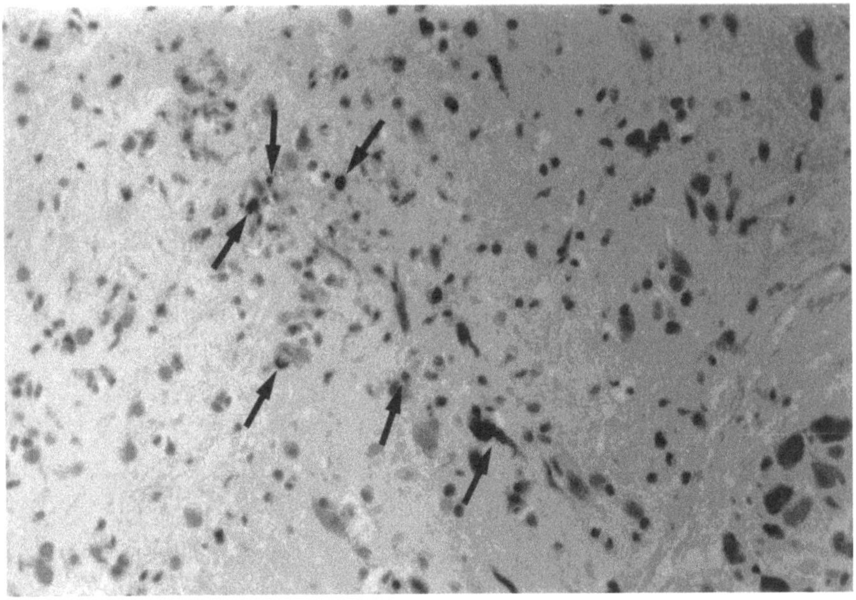

Fig. 8. Central nervous system, diencephalon, X 200. Virus antigen demonstrated in several neuronal cells.

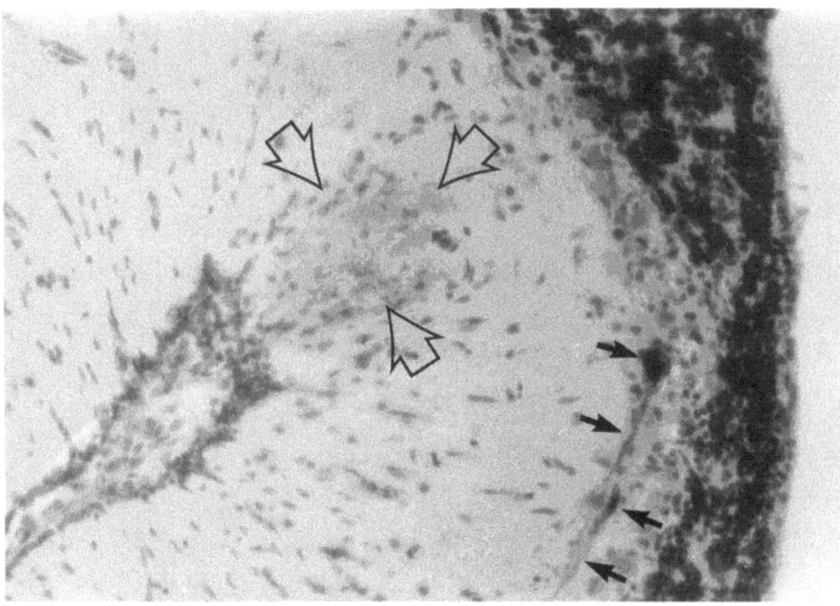

Fig. 9. Central nercous system, cerebellum, X 200. Dark arrows: cell body and axon of affected neurons. Light arrows: region with scattered virus antigen.

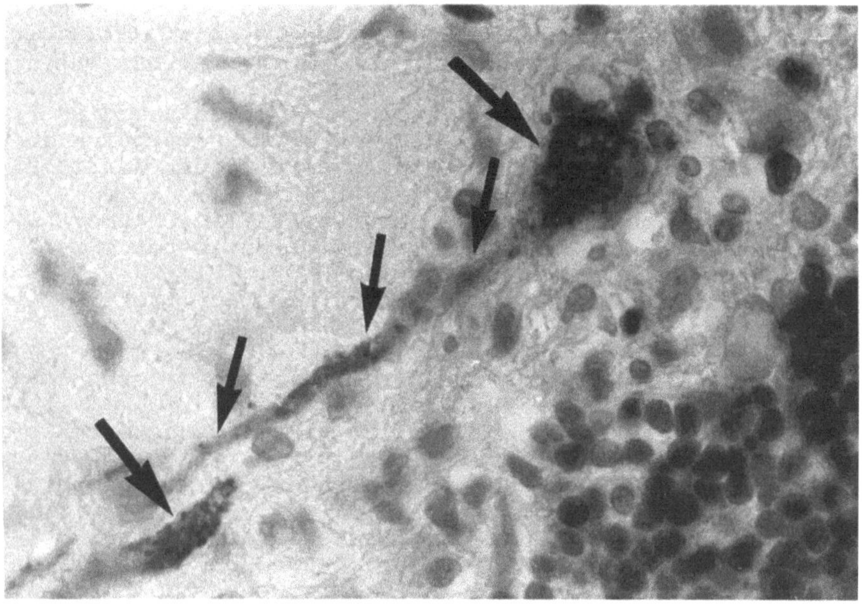

Fig. 10. Central nervous system, cerebellum, X 788. Cell body and axon of neurons affected by virus antigen.

In conclusion, we could demonstrat that VHS virus multiplication takes place in both, the nervous cells of the brain and the tubulus cells of the excretory kidney. Hence we may infer that brain and excretory kidney from trout have to be included in procedures of VHS diagnosis, especially concerning VHS virus carrier fish.

REFERENCES

Bourne JA (1965) Handbook of immunoperoxidase staining methods. Dako Corporation, Santa Barbara

Ghittino P (1965) Viral haemorrhagic septicaemia (VHS) in rainbow trout in Italy. Ann N Y Acad Sci 126: 468-478

Hoffmann R, Wizigmann G (1984) Comparative histological and immunohistological studies in VHS. In: Fish Diseases, Fourth COPRAQ Session, Cadiz, Spain: 29-31

Kölbl O (1978) Infektions- und Immunisierungsversuche mit dem Erreger (Egtved-Virus) der Viralen Hämorrhagischen Septikämie (VHS) der Regenbogenforellen. In: Diagnose und Behandlung von Fischkrankheiten, DVG, München: 61-76

Neukirch M (1984) An experimental study of the entry and multiplication of viral haemorrhagic septicaemia virus in rainbow trout, Salmo gairdneri Richardson, after waterborne infection. J Fish Dis 7: 231-234

Neukirch M, Glass B (1984) Some aspects of virus shedding by rainbow trout (Salmo gairdneri Rich.) after waterborne infection with Viral Haemorrhagic Septicaemia virus. Zentralbl Bakteriol Hyg I Abt Orig A 257: 433-438

Neukirch M (1986) Demonstration of persistent Viral Haemorrhagic Septicaemia (VHS) virus in rainbow trout after experimental waterborne infection. J Vet Med B 33: 471-476

Sternberger LA (1979) Immunocytochemistry, 2nd ed., John Wiley and Sons, New York

Vestergård Jφrgensen PE (1982a) Egtved virus: temperature-dependent immune response of trout to infections with low-virulence virus. J Fish Dis 5: 47-55

Vestergård Jφrgensen PE (1982b) Egtved virus: occurrence of inapparent infections with virulent virus in freeliving rainbow trout, Salmo gairdneri Richardson, at low temperature. J Fish Dis 5: 251-255

Wizigmann G, Hoffmann R (1982) Vergleichende lichtmikroskopische und immunhistologische Untersuchungen bei der viralen hämorrhagischen Septikämie (VHS) der Regenbogenforelle. Zentralbl Vet Med B 29: 782-788

Wolf K, Quimby C (1962) Established eurythermic line of fish cell In vitro. Science 135(3508): 1065-1066

Cloning the Genes of Viral Haemorrhagic Septicaemia of the Trout

J. Bernard and P. de Kinkelin
Institut National de la Recherche Agronomique
Laboratoire d'Ichtyopathologie
Centre de Recherches de Jouy-en-Josas
Domaine de Vilvert
78350 Jouy-en-Josas
France

Introduction

Viral Haemorrhagic Septicaemia virus (VHS-V)is a rhabdovirus responsible for a systemic infection of trout, grayling, white fish and pike, occurring at water temperature lower than 14°C and currently widespread in continental Europe. Its impact is a major matter of concern for economics of trout farming. Thus cloning its genes is interesting from several points of vue such as epidemiological survey by mean of molecular probes, identification of new isolates, comparison between pathogenic and avirulent strains, and production of production of proteins which could be used as vaccines.

Materials and Methods

Virus and cells

The isolate 07/71 belonging to serotype 1 (Le Berre *et al*, 1977) was grown onto *Epithelioma Papulosum Cyprini* (EPC) cells according to conditions previously described (Fijan *et al*, 1983). It was submitted to only two passages in cell cultures before use.

mRNAs

EPC cells grown in a 75 cm^2 flask were infected at a multiplicity of 10, then solubilised 8 hrs later in 2 ml of Guanidium Isothyocyanate buffer (Maniatis, 1982). The viscous suspension was sheared 5 times through a 20 gauge needle then extracted once with phenol-chloroform - isoamylalcool (25 :

24 : 1) and twice with chloroform - isoamylalcool (24 :1). The aqueous phase was adjusted to 2M Li Cl and left at 4°C overnight. The precipitated total RNAs were recovered by centrifugation, dissolved in redistilled water, heated for 5 min to 65°C, then chilled in ice. NaCl was added to a final concentration of 0.5 M and the sample was spotted on hybond-messenger affinity paper (Amersham), then eluted in redistilled water.

λ and pBS library

Total mRNAs were copied as ds cDNa with a cDNA synthesis kit (Amersham) then cloned into the Eco R_I restriction site of λ gt 10 (cDNA cloning kit, Amersham). A probe was prepared by random reverse transcription of genomic VHS-V RNA and hybridised with the λ gt 10 plaques. The positive plaques were amplified and stored as infectious phage and purified DNA for further use.

The DNA from two λgt 10 clones was digested by Eco RI. The inserts were purified by mean of Gene clean (Stratagene), then transfered into the Eco RI restriction site of λ ZAP (Stratagene). The subclones were screened by IPTG-Blu O Gal and again stored as infectious phage and purified DNA for further use.

E. coli XL1-Blue (Bullock et al, 1987) were coinfected with λZAP from one subclone and F_1 phage, which resulted in automatic excision and packaging of the phagemid as described by the manufacturer (Stratagene). The pBS plasmid was rescued by plating the phagemid on XL1-Blue in the presence of tetracycline ampiciline pUC 54.

That clone was a gracious gift from M. Thiry (Société Eurogentec, Campus du Sart Tilman, Bât. B6, 4000 Liège, Belgique).

Other methods

Contaminating RNAses were removed by filtration of buffers on either nylon or nitrocellulose membranes. Double stranded DNAs were sequenced using Boehringer's pUC sequencing kit. T_3 RNA polymerase and rabbit reticulocyte lysate were from Stratagene and T_7 RNA polymerase from Ozyme.

Results

Cloning strategy

The cloning strategy in λ and pBS vectors is described in Materials and Methods and summarised in Fig.1. The study of the different clones and subclones we obtained is still underway.

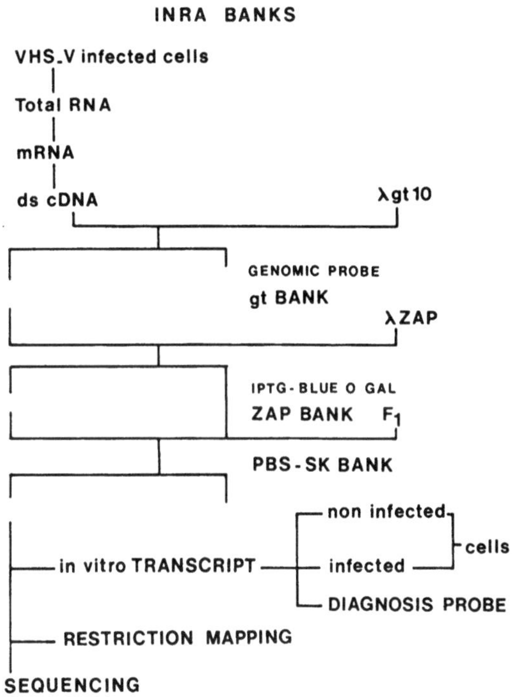

Fig. 1 Cloning strategy in and pBS vectors - As described by the manufacturer, coinfection of XL1 Blue *E. coli* by ZAP and F1 phages resulted in the packaging and secretion of a phagemid DNA which could be rescued as pBluescript (pBS) plasmid.

Rectriction map

One clone, pBS 26, has been more extensively studied because it belongs to a class which is much more frequent than the others in our library. The products of digestion by restriction enzymes is presented in Fig.2. Similar results were obtained with pUC 54 insert (data not shown).

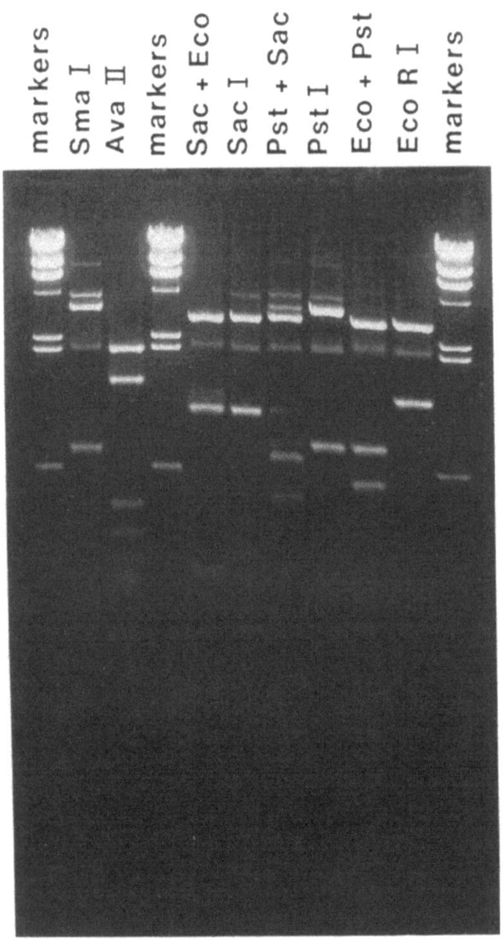

Fig. 2 Digestion of pBS 26 by restriction enzymes - The molecular weight markers (Boehringer) are 23130, 9416, 6557, 4361, 2322, 2027, 564 and 125 BP.

Transcription and translation

Since the inserts inside pBS polylinker (Fernandez *et al*, 1987) are under control of the T_3 and T_7 RNA polymerases promotors, specific RNAs were transcribed *in vitro* with either of the two. we found that the RNAs transcribed by mean of T_3 hybridised with RNAs from VHS-V infected EPC cells but not with the uninfected control. Inversely, RNAs transcribed by mean of T_7 did not hybridised. None of the polypeptides, either IPTG-induced or *in vitro* translated were recognized by rabbit anti VHS-V immune serum (data not shown).

Sequence analysis

Partial sequence analysis of pBS 26 was performed, using M 13 direct, M 13 reverse, SK and KS primers. The results obtained with M 13 direct and KS (Fig.3) on one hand, with M 13 reverse and SK on the other hand, could be merged. Finally the sequence obtained with M 13 direct and KS as primers was homologous to the sequence of the 5' end (messenger sense) from pUC 54.

```
T7 + 1
↓                                       30
GGG CGA ATT GGG TAC CGG GCC CCC CCT CGA GGT CGA CGG TAT CGA

                  ECO RI  insert
                         ↓                                 90
TAA GCT TGA TAT CGA ATT CGT GCA GCG TTT TCA GGC CTG AAT GAT

                                        100
GTT AGG ATT GAC CCC ACC GGT GGA GAG GGA CGG GTA CTT GTA CCT

                    130                                   160
GGT GAA GTG GAG CTC ATC GTG TAT GTT GGT GAA TTT GGT GAG GAA

GAT AGG AAG GTG ATT GTG GAT
```

Fig.3 Partial sequence of pBS 26 - (T7 + 1) : the sequence has been written starting from promotor T7 + 1. The insertion site (Eco RI) has been boxed. (insert) : the sequence starting from that point is identical to that starting from nucleotide 124 on pUC 54.

To avoid duplicate work, only pUC 54 was more extensively sequenced. The insert was approximately 1,400 BP long but only 1,110 nucleotides from the 5'end were identified since downstream of that, numerous stacks impaired the progression of both T_7 DNA polymerase (Sequenase, USB) and klenow enzyme, on both strands of pUC 54.

The complete sequence will be published elsewere. An open reading frame (ORF) was included between nucleotides 109 and 882. The codon usage is presented in Fig.4. The coded protein was rich in Lysine (4), Arginine (15) and Serine (19) but those aminoacids were randomly distributed (data not shown).

Phe	TTT	2	Ser	TCT	2	Tyr	TAT	1	Cys	TGT	1
	TTC	4		TCC	4		TAC	2		TGC	7
Leu	TTA	1		TCA	4	End	TAA	0	End	TGA	2
	TTG	6		TCG	2		TAG	0	Trp	TGG	3
	CTT	1	Pro	CCT	2	His	CAT	1	Arg	CGT	2
	CTC	7		CCC	5		CAC	2		CGC	0
	CTA	2		CCA	2	Gln	CAA	6		CGA	0
	CTG	7		CCG	3		CAG	8		CGG	3
Ile	ATT	5	Thr	ACT	4	Asn	AAT	2	Ser	AGT	3
	ATC	6		ACC	8		AAC	6		AGC	4
	ATA	0		ACA	8	Lys	AAA	0	Arg	AGA	3
Met	ATG	11		ACG	1		AAG	5		AGG	7
Val	GTT	3	Ala	GCT	0	Asp	GAT	6	Gly	GGT	6
	GTC	6		GCC	5		GAC	9		GGC	4
	GTA	4		GCA	8	Glu	GAA	9		GGA	9
	GTG	6		GCG	3		GAG	7		GGG	9

	S	
Acidic (Asp + Glu)	31	(11.9)
Basic (Arg + Lys)	20	(7.7)
Aromatic (Phe + Trp + Tyr)	12	(4.6)
Hydrophobic (Aromatic + Ile + Len + Met + Val)	77	(29.6)
-SH bridges Cys	8	
Met	11	

Fig. 4 Codon usage -

The A from the ATG codon was in position 109, the G from an internal Eco RI site in position 119, and the consensus sequence representing the signal for transcription started in position 54 (data not shown). The comparison of that consensus sequence in VHS-V, Rabies, and Vesicular Stomatitis Virus (VSV) is presented in Fig.5.

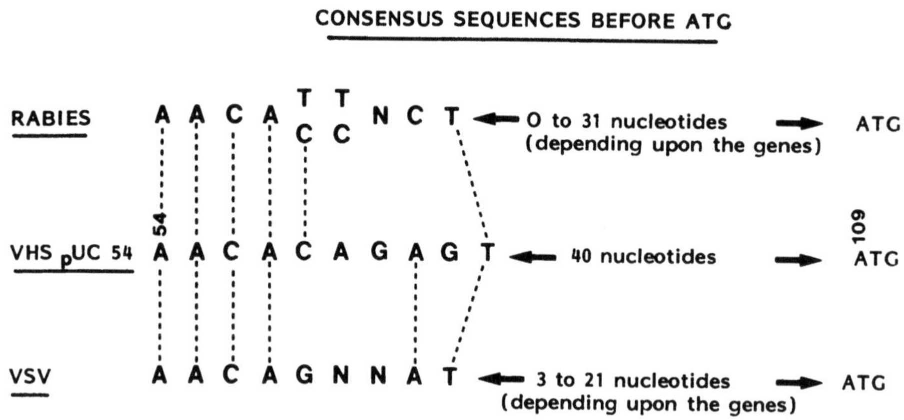

Fig. 5 Comparison of intergenic consensus sequences from VSV and Rabies with VHS -

Discussion

The protein coded by pBS 26 is yet unidentified since : i - the (-) strand was synthesized by T_3 RNA polymerase (it hybridized with messenger RNAs of infected cells, see above). ii - the promotor of the Lac Z gene in pBS-SK vector is in the same orientation as the T_3 promotor. iii -the T_7 RNA polymerase promotor was not in phase with the insert (Fig.3), because part of the λgt 10 insert was missing in pBS 26, due to the presence of an internal Eco RI site.
Nevertheless, it will now be easy to use pBS 26 as a probe to reclone the gene directly into pBS vector and thus avoid such an artifact.
The sequence from pUC 54 did not help identification. It was a priori suspected that the insert represented the N gene, on the basis that this type of clones was the most abundant in our library. Since transcription of the viral genome in infected cells is sequential (Bernard and de Kinkelin, 1985), one could suppose that the first messenger to be transcribed was the most abundant. But the ORF we evidenced was only 774 BP long, equivalent to a coding capacity of 258 aminoacids. Thus the calculated molecular weight of the protein would be 27,864, which is too low for the N protein (minimal published

value : 38,000. Lenoir and de Kinkelin, 1975), even assuming that the value obtained by PAGE is overestimated due to phosphorylation of the native polypeptide. On the other hand, the protein coded by that ORF is more acidic (Fig.4) than should be a protein associated with a nucleic acid or as compared with the N of rabies and VSV. Finally, Lys, Arg and Ser are abundant but not associated as would be expected for phosphorylation sites.

The ATG codon in position 109 of pUC 54 was identified as a signal for the beginning of translation because : i - it was surrounded by the Kozak's box AcA ATG G AA (Kozak, 1987). ii - It was the only one to open a large ORF. The sequence upstream to that ATG is surprisingly long when compared with Rabies and VSV. Comparison with other fish rhabdoviruses is impossible since, to our knowledge, no gene other than G has yet been sequenced. The consensus sequence representing the transcription signal for Rabies and VSV (Tordo *et al*, 1986) was found there (Fig.5) despite the fact that the cloned RNA was messenger and not the genomic.

pUC 54 could not be sequenced downstream from nucleotide 1,110 which was separated from the poly A tail by approximately 290 BP. The signal for the termination of translation is about 518 BP from the poly A tail. Again it should be noted that neither Rabies virus nor VSV have such long non coding sequences on their m RNA.

References

Bernard J. and de Kinkelin P. (1985) Effect of U.V irradiation of viral haemorrhagic septicaemia virus on virus-specific intracellular syntheses. Ann. Virol. (Inst. Pasteur) 136 E : 213-222

Bullock W.O., Fernandez J.M. and Short J.M. (1987) XL1 Blue : a high efficiency plasmid transforming recA *Escherichia coli* strain with Beta-Galactosidase selection

Fernandez J.M., Short J.M., Renshaw M., Huse W. and Sorge J.A. (1987) pBluescript plasmids : multipurpose cloning vectors for sequencing and transcription-Gene

Fijan N. Sulimanovic D., Bearzotti M., Muzinic D., Zwillenberg L.O., Chilmonczyk S., Vautherot J.F. and de Kinkelin P. (1983) Some properties of the Epithelioma Papulosum Cyprini (EPC) cell line from carp Cyprinus Carpio. Ann. Virol. (Inst. Pasteur) 134 E : 207-220

Kozak M. (1987) An analysis of 5'-non coding sequences from 679 vertebrate messenger RNAs. Nucleic Acids Research 15 : 8125-8148

Le Berre M., de Kinkelin P. and Metzger A. (1977) Identification sérologique des Rhabdovirus des Salmonidés. Bull. Off. int. Epiz. 87 : 391-393

Lenoir G. and de Kinkelin P. (1975) Fish rhabdoviruses. Comparative study of protein structure. J. Virol. 16 : 259-262

Maniatis T., Fritsch E.F. and Shambrook J. (1982) Molecular cloning, a laboratory manual. Cold Spring harbour Laboratory, USA

Tordo N., Poch O., Ermine A. and Keith G. (1986) Primary structure of leader RNA and nucleoprotein genes of the rabies genome : segmented homology with VSV. Nucleic Acids Research 14 : 2671-2683.

Rhabdovirus Olivaceus (Hirame Rhabdovirus)

T. Kimura, M. Yoshimizu, N. Oseko, and T. Nishizawa

Laboratory of Microbiology
Faculty of Fisheries
Hokkaido University
Minato 3-1-1
Hakodate, 041
Japan

Hirame rhabdovirus (HRV) is a fish rhabdovirus discovered from a moribund hirame (Japanese flounder, *Paralichthys olivaceus*) cultured in Hyogo Prefecture, Japan, at 1984. Characteristics of HRV have been found sufficiently distinct from other known fish rhabdoviruses; this virus has been named *Rhabdovirus olivaceus*, derived from the scientific name of the host species. This virus is pathogenic for hirame, black sea bream (*Milio macrocephalus*), red sea bream (*Crysophrys major*) and black rock fish (*Sebastes inermis*) among oceanic fishes, and also the salmonid species, especially rainbow trout. HRV has been found widely distributed in Japan.

Isolation

An unknown disease occurred during March 1984 in cultured hirame held in near-shore pens and among ayu (*Plecoglossus altivelis*) fry held in sea-water tanks at Hyogo Prefecture, Japan. Examination of diseased fish failed to show any known bacterial, fungal or parasitic agents. During the course of viral examination, a cytopathic effect (CPE) similar to infectious hematopoietic necrosis virus (IHNV) was observed in RTG-2 cells (Gorie et al. 1985, 1986; Kimura et al. 1986).

Biophysical and Biochemical Properties of HRV

At an optimal incubation of 15° to 20°C, HRV shows distinctive CPE within 2 to 4 days, i.e., rounded cells. The EPC, FHM, BF-2, YHK, STE-317, BB, EK-1, CCO and HF-1 lines were susceptible to HRV infection, while no CPE developed in CHH-1, CHSE-214 and KO-6 cell lines. The EPC and FHM cell lines showed the greatest sensitivity, and the maximum titer of culture-grown virus is about $10^{8.3}$ to $10^{8.8}$ $TCID_{50}$/ml. The loss of infectivity of HRV was slightly less in sea water then in MEM.

The HRV is heat-, ether-, and acid(pH 3)-labile, and does not hemagglutinate human O erythrocyte cells. It is inactivated more than 99% (ID_{99}) by ultraviolet (UV) irradiation of 2.0×10^4 μw·sec/cm^2. The deoxyuridine analogue IUdR was not effective in blocking viral replication.

Under electron microscopy uf ultra-thin sections of HRV infected RTG-2 cells and kidney tissue of HRV infected hirame, virus particles were noted to be bullet-shaped with one round

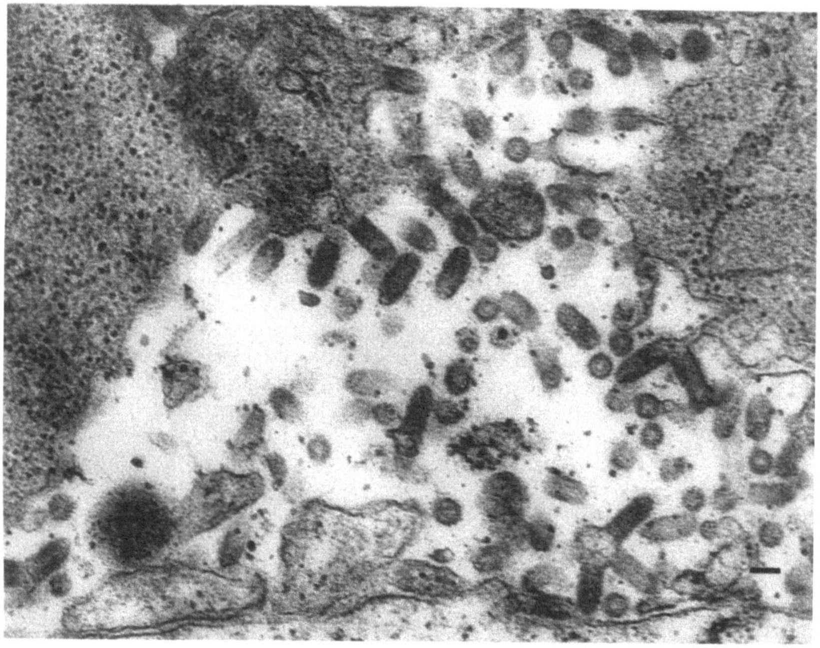

Fig. 1. Electron micrograph of an ultrathin section showing large numbers of bullet shaped virus particles in RTG-2 cells infected with HRV (—— 100 nm)

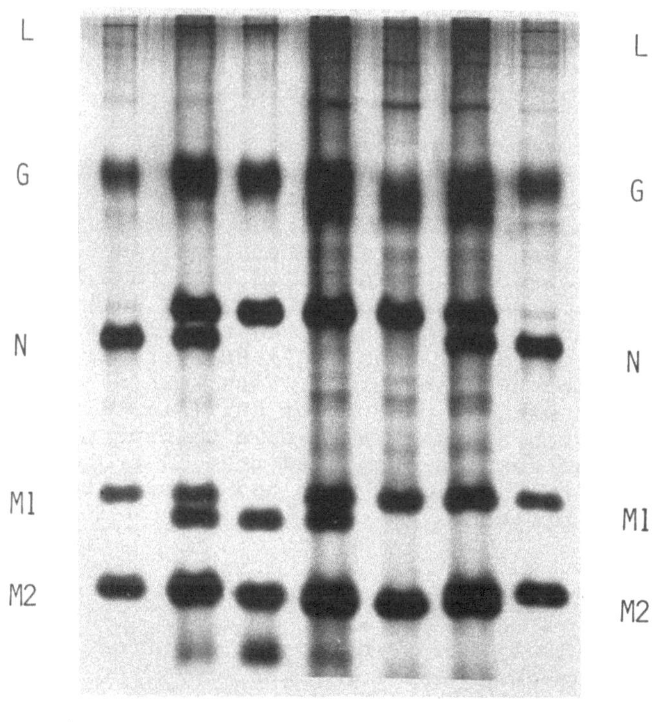

Fig. 2. Comparison of the SDS-PAGE patterns of HRV, VHSV and IHNV structural polypeptides on 10% polyacrylamide gel stained with silver nitrate. A: IHNV, B: IHNV plus VHSV, C: VHSV, D: VHSV plus HRV, E: HRV, F: HRV plus IHNV, G: IHNV

end (Fig. 1). The particles were found to be between 180 nm and 200 nm long and to have a mean diameter of 80 nm.

Antigenic Properties

Eleven isolates of HRV, isolated in 1984 through 1986 from four kinds of fishes at seven different locations in Japan, were neutralized with anti-HRV rabbit serum, and the 50% neutralization dose (ND_{50}) was 1:80 to 160. No neutralization of HRV infectivity occurred with antiserum against IHNV, infectious pancreatic necrosis virus (IPNV), *Oncorhynchus masou* virus (OMV), viral hemorrhagic septicemia virus (VHSV), spring viremia of carp virus (SVCV), pike fry rhabdovirus (PFR), eel virus of America (EVA), or eel virus Europe X (EVEX). Cross neutralization tests indicated that HRV was clearly distinguishable from the six reference rhabdoviruses: IHNV, VHSV, SVCV, PFR, EVA and EVEX.

Properties of Structural Proteins

Purified HRV had five structural polypeptides with molecular weights of 156, 67.8, 46.0, 26.5 and 19.9 KD, respectively. Although VHSV and IHNV also had five structural polypeptides; L, G, N, M1 and M2, PFR, SVCV, EVA and EVEX had only four polypeptides. Therefore, structural proteins of HRV closely resembled those of VHSV or IHNV, but the N protein of VHSV was bigger than that of IHNV. M1 protein of HRV was bigger than that of VHSV, and M2 protein of HRV was smaller than that of VHSV and IHNV, respectively (Fig. 2). Additionally, SDS-PAGE patterns of eleven different HRV strains isolated from four different species of fish at seven places were almost the same (Nishizawa et al., unpublished data).

Antigenically, M1 protein was independent among HRV, IHNV and VHSV, which belonged to the *Lyssavirus* group. But cross-antigenic reactions were observed in G, N and M2 proteins among the viruses. Especially, cross-reaction of G protein was strong compared with N and M2 proteins. Antigenic profiles of structural polypeptides of the *Lyssavirus* group were clearly distinguished from that of PFR, SVCV, EVEC and EVA, which belonged to *Vesiculovirus*. Anti-HRV serum absorbed by IHNV particles did not show the reaction with glycoprotein of HRV, and on the other hand, anti-HRV serum absorbed by Triton X-100 treated HRV particles, which were devoid of M1 protein, did not react with M2 protein of HRV (Fig. 3). However, both of these absorbed antisera were able to neutralize the infectivity of HRV. These results suggested that M1 protein, which was one part of the viral envelope, also had a role with virulence of HRV, such as G protein.

Host Range and Geographic Distribution

February to May in 1985 and 1988, the same disease occurred again among hirame, black sea bream and black rock fish, cultured in sea-water tanks at Hyogo, Kagawa, Hokkaido, Okayama and Mie Prefecture. The cumulative mortality ranged from 3.3% to greater

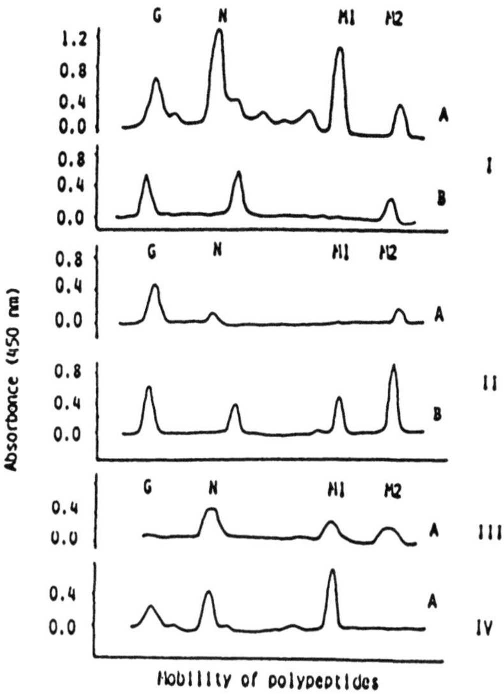

Fig. 3. Densitogram of structural polypeptides of HRV and IHNV stained with antisera. I: antiserum against HRV, II: antiserum against IHNV, · III: antiserum against HRV absorbed by IHNV particles, IV: antiserum against HRV absorbed by MI protein devoid HRV particles. A: HRV, B: IHNV

than 90%. Water temperature and size of infected hirame were 6° to 18°C and 100 g - 300 g in Hyogo, 2° to 12°C and 110 g - 200 g in Hokkaido, 8° to 16°C and 300 g - 700 g in Kagawa, and 15° to 16°C and 0.5 g in Mie Prefecture, respectively.

Pathogenicity to Hirame

Mortality of hirame artificially infected with HRV (IP injection, $10^{3.0}$ and $10^{5.0}$ $TCID_{50}$/fish, body weight 95 g - 265 g) began at 14 days post-injection and 20% had succumbed by day 19 (Gorie and Nakamoto, 1986). The mortality of artificially infected hirame, red sea bream and and black rock fish (body weight 98 - 203 g, 61 - 99 g and 62 - 95 g, HRV was injected intraperitoneally, $10^{6.1}$, $10^{5.7}$ and $10^{5.7}$ $TCID_{50}$/fish) reached 67%, 92% and 97%, respectively (Sorimachi, personal communication).

The mortality of smaller hirame (body weight 0.4 - 0.7 g), red sea bream (body weight 3 - 4 g) and black rock fish (body weight 8 - 12 g), infected with HRV by immersion method (60 min, $10^{3.6}$, $10^{3.3}$ and $10^{5.7}$ $TCID_{50}$/ml, respectively), reached 63%, 52% and 97%, respectively (Sorimachi, personal communication).

Pathogenicity to Salmonid

Mortality among rainbow trout (*Salmo gairdneri*; body weight 8 g) began at 6 days post-injection (IP $10^{3.0}$ $TCID_{50}$/fish) and 60% had succumbed by day 12. The fry of chum (*O. keta*), coho (*O. kisutch*)

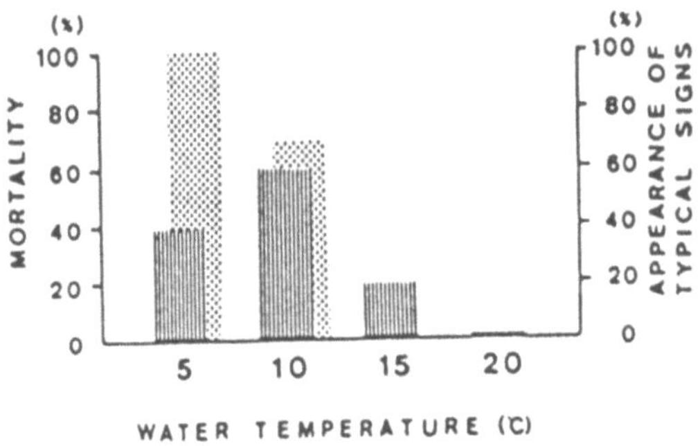

Fig. 4. Effect of the water temperature on cumulative mortality and appearance of the typical signs on the hirame after 15 days artifically infected with HRV; dose, I.P. $10^{3.30}$ $TCID_{50}$/fish.
Dead fish — Fish showing the typical signs

and masu salmon (*O. masou*) and ayu (body weight 0.2 g, 0.5 g, 0.2 g and 0.5 g, respectively), experienced low or no mortalities when exposed to HRV by an immersion method ($10^{3.0}$ $TCID_{50}$/ml, 60 min). The mortality of rainbwo trout artificially infected with HRV (intraperitoneal injection, $10^{3.8}$, $10^{3.8}$ and $10^{4.0}$ $TCID_{50}$/fish, body weight 0.2 g, 0.5 g and 3.0 g, respectively) reached 100%, 55% and 40%, respectively.

Effect of Temperature on HRV Infection

Natural outbreaks of HRV infections disappeared when the water temperature was increased to 15°C. Cumulative mortality of artificially infected hirame (IP $10^{5.3}$ $TCID_{50}$/fish), which were reared at 5°, 10°, 15° and 20°C, were 40%, 60%, 10% and 0%, respectively. The highest virus titer was obtained from the fish of the 5°C test group, followed by the 10°C (Fig. 4; Oseko et al. 1988b).

Disease Signs

The common signs of HRV disease in hirame is congestion of the gonad, focal hemorrhage of skeletal muscle and fins, and accumulation of ascitic fluid (Fig. 5). In red sea bream, black sea bream and rainbow trout, hemorrhage was commonly observed in muscle and fin.

Histopathology

Histopathologically, the kidney indicated necrotic changes by nuclear degeneration of hematopoietic cells and hemorrhage in the interstitial tissue (Fig. 6). The spleen showed necrosis and hemorrhage in the pulp, and skeletal muscle revealed hyperemia and hemorrhage of capillary vessels. Hyperemia and hemorrhage were observed in the intestinal tissue of the seminiferous duct, ovarian lamella and in the connective tissue around the seminal duct and oviduct of the testis and ovary. Mucosa of the alimentary tract showed hyperemia and hemorrhage. No pathological changes were observed in the liver (Oseko et al. 1988a).

Fig. 5. Diseased hirame infected with HRV, showing hemorrhage of skeletal muscle, accumulation of ascitic fluid and congestion of gonad

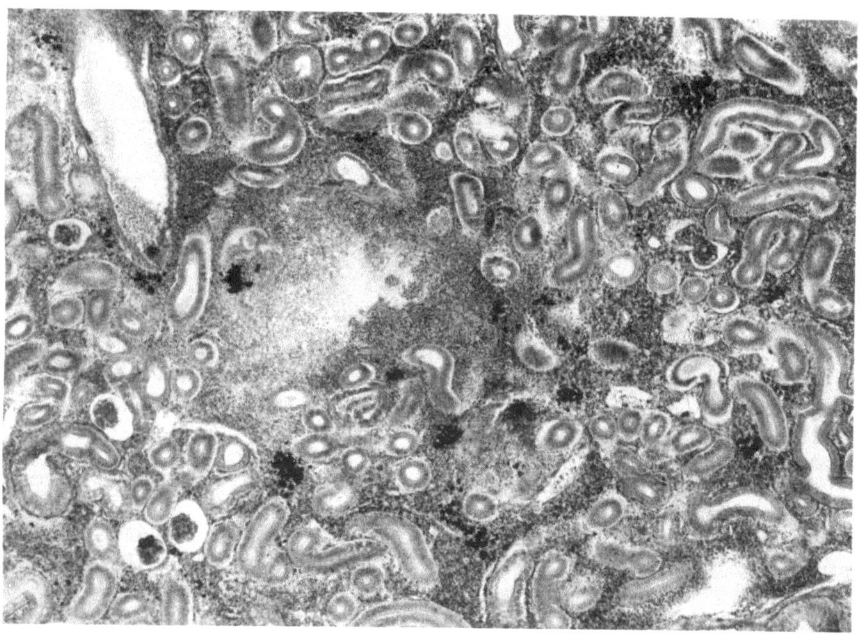

Fig. 6. Section of the kidney of a diseased fish, sampled at Hyogo Prefectural Fisheries Experimental Station, showing necrosis of hematopoietic cells and hemorrhage in the connective tissue. H & E stain

Immune Reactions

Neutralizing activity against HRV in serum samples showed a higher titer than the uninfected fish and this antibody remains about 6 months (Table 1).

In the case of artificially infected fish reared at different temperatures, the neutralizing antibody was generally not very high, but the titer was comparatively higher in the higher water temperature group than that of the lower temperature groups. The increase of immature erythrocytes at 5°C, small lymphocyte-like cells at 20°C were observed in the blood smears (Oseko et al. 1988b).

Table 1. Neutralization titer[a] in sera of hirame cultured at Hokkaido, 4 and 6 months after outbreak of HRV infection

Group	Months after outbreak of disease	Number of fish employed (%)		Neutralization titer
Disease outbreak	4	10	20	1:112
			10	1: 80
			70	1: 66
Uninfected		3	100	1:<10
Disease outbreak	6	5	40	1:112
			20	1: 20
			20	1: 10
			20	1:<10
Uninfected		5	100	1:<10

[a]Completely neutralized the 100 $TCID_{50}$ HRV

Diagnosis

The infectivity of this agent remains unchanged for at least 1 month at 0° to 5°C, and at -20°C or below the infectivity is stable more than 2 years. Virus isolation should be done using the fish transported to the laboratory with ice in a cooler. In the case of small fish, the whole body must be homogenized with 9 volumes of Hank's BSS and then filtered by a 0.45 µm cellulose acetate membrane.

For the purpose of virological survey of mature hirame, eggs are collected by pressing the body; the ovary fluid is added to the same volume of antibiotic solution and reacted at 5°C overnight. Usually RTG-2 or FHM cell lines are employed. Suitable incubation temperature for isolation of HRV is 15°C.

Control and Prevention of HRV Infection

Natural outbreaks of HRV infection disappeared when the water temperature was increased to 15 C. Keeping the water temperature at 15 C is one effective measure of blocking HRV infection. The agents are sensitive to ultraviolet irradiation, ID_{99} is 1.0-2.0 x 10^4 $\mu W \cdot sec/cm^2$ and are also inactivated by iodophore treatment at 50 ppm for 15 min. Eggs contaminated with HRV can be treated and disinfected with iodophore. The eggs must be incubated using the ultraviolet-treated water (Yoshimizu et al. 1987).

References

Gorie S, Nakamoto K, Katashima K (1985) Disease of cultured hirame (Japanese flounder, *Paralichthys olivaceus*)-I. Preliminary report on a disease of marine pen cultured hirame may be viral infection. Bull Hyogo Pref Fish Exp Stn 23, 66-68

Gorie S, Nakamoto K (1986) Pathogenicity of virus isolated from cultured hirame (Japanese flounder). Fish Pathology 21, 177-180

Gorie S, Nakamoto K, Kimura T (1986) *Rhabdovirus olivaceus* (hirame rhabdovirus) infection occurred among tank-cultured hirame at Hyogo Prefecture in January, 1985. Bull Hyogo Pref Fish Exp Stn 24, 43-47

Kimura T, Yoshimizu M, Gorie S (1986) A new rhabdovirus isolated in Japan from cultured hirame (Japanese flounder) *Paralichthys olivaceus* and ayu *Plecoglossus altivelis*. Dis Aquat Org 1, 209-217

Oseko N, Yoshimizu M, Gorie S, Kimura T (1988a) Histopathological study on diseased hirame (Japanese flounder, *Paralichthys olivaceus*) with *Rhabdovirus olivaceus* (hirame rhabdovirus, HRV). Fish Pathology 23, 117-123

Oseko N, Yoshimizu M, Kimura T (1988b) Effect of water temperature on artificial infection of *Rhabdovirus olivaceus* (hirame rhabdovirus, HRV) to hirame (Japanese flounder, *Paralichthys olivaceus*). Fish Pathology 23, 125-132

Yoshimizu M, Oseko N, Nishizawa T, Kimura T (1987) Rhabdovirus infection of hirame (Japanese flounder, *Paralichthys olivaceus*). Fish Pathology 22, 54-55

Rhabdovirus Infection of Ulcerated Fish in South-East Asia

G.N. Frerichs, S.D. Millar, and M. Alexander
Institute of Aquaculture
University of Stirling
Stirling FK9 4LA
Scotland

EPIZOOTIC ULCERATIVE SYNDROME

Epizootiology. The term 'epizootic ulcerative syndrome' has been proposed (FAO, 1986) to describe a severe ulcerative disease condition of uncertain aetiology which has affected a wide variety of freshwater and estuarine fish in the Indo-Pacific region since 1980 (Figure 1).

The origin of the disease is unknown, but it is now generally considered that the syndrome was probably first encountered in southeast Queensland, Australia in 1972 where it was initially referred to as Bundaberg Fish Disease (McKenzie and Hall, 1976). Estuarine fish stocks, particularly sea mullet (*Mugil cephalus*), were affected with large, shallow, circular or oval skin ulcers, invariably associated with an invasive fungal infection. Further outbreaks of a similar condition, now termed 'red spot', occurred in a variety of estuarine fish species in adjacent river systems in succeeding years, progressively extending the geographical range of the disease down the coast to central New South Wales (Rodgers and Burke, 1981). Sea mullet affected with cutaneous lesions consistent with 'red spot' have since been found in southwestern Western Australia (Callinan, 1985), and a diverse range of estuarine and freshwater fish in a number of water systems in the Northern Territory succumbed to the disease in 1986 (Humphrey and Langdon, 1986).

The spread of the ulcerative condition outside Australia has been comprehensively reviewed by Tonguthai (1985) and Roberts et al (1986). Outbreaks first affected freshwater

Figure 1 Occurrence of ulcerative fish disease in the Indo-Pacific Region

fish in the rivers of southern Papua New Guinea in 1975 and freshwater and estuarine species in the Sepik river system in the northern part of the country in 1981. In Indonesia, a severe epizootic of a haemorrhagic condition known locally as 'penyakit merah' (red disease) occurred near Bogor, Java in 1980 and spread rapidly throughout West, Central and East Java during the following two months. However, this disease outbreak primarily affected cyprinid fish and was essentially an acute septicaemic condition without the presence of characteristic skin ulcers, which does not conform with the description of the ulcerative syndrome elsewhere in the region. Ulcerated snakeheads and catfish were subsequently identified in Central Java in 1983 and Kalimantan (Borneo) in 1984. In Malaysia, ulcerative conditions affecting many species of fish associated with rice field cultivation have been recorded since 1979, with a particularly severe epizootic causing high mortalities of fish cultured in paddy fields and submerged cages in Kedah, close to the Thailand border, in 1983. In Thailand itself, the first severe outbreak of the condition occurred in the southern provinces in late 1981 and similar epizootics have recurred throughout the remaining areas of the country every consecutive year up to the present time. A wide range of wild and cultured freshwater fish and some brackish water species have reportedly been affected. By 1984, the disease had become extensively established throughout southeast Asia with outbreaks being reported in all the principal freshwater fishing areas of Kampuchea, the southern and central parts of Lao PDR and southern Burma.

Since the publication of these accounts by Tonguthai (1985) and Roberts et al (1986), severe epizootics of the ulcerative syndrome have occurred in Laguna de Bay, Luzon, Philippines (Natividad, 1986), in the Kelani river system, Western Province, Sri Lanka in 1987 and, most recently, in southern and eastern Bangladesh early in 1988.

Clinical disease. Characteristically, the disease occurs

seasonally during the cooler, drier months from October to March. In all affected species, early lesions are generally seen as single or multiple, irregularly circumscribed, focal areas of acute dermatitis with hyperaemia and oedema rapidly leading to the development of typically large, shallow ulcers with a haemorrhagic or necrotic base (Roberts et al, 1986). The lesions are usually distributed on the dorso-lateral body surface although the dorsal fin, tail or operculum may also be involved. In some species of fish, particularly snakeheads (*Ophicephalus* sp), erosion of the head and jaws may be the principal finding.

Dermal ulceration is followed by progressive necrosis of the underlying muscle and skeletal tissues associated with bacterial infection and, frequently, infiltration with non-septate, highly invasive, fungal hyphae. Ulceration may become so severe that the peduncle is completely lost, the abdominal wall penetrated through to the internal organs, the cranium eroded to expose the brain and the jaws effectively destroyed. The internal organs, however, consistently show only minimal pathological changes except in the very late stages of the disease when bacterial and fungal infections extend from the ulcers to the viscera. Mortality rates in certain species may be very high, particularly when the disease first strikes a previously unaffected area, but recovery with scar formation and anatomical deformity appears to be the more usual outcome in less susceptible species and other locations.

Susceptible species. Numerous species of both freshwater and estuarine fish are reportedly susceptible to the ulcerative disease syndrome. True marine fish appear to be uniformly resistant. Tonguthai (1985) lists no fewer than 80 species which were identified as affected during outbreaks of the disease in Australia, Papua New Guinea, Thailand and Burma between 1972-1985. Further epizootics in these and other countries since 1985 have increased the listing to over 100 species. However, it is recognised that the skin and musculature of fish can respond in only a limited number of ways to different injuries and infections, and the observation of skin ulcers in any particular fish captured during an outbreak of the ulcerative syndrome does not mean *per se* that that species is susceptible to this specific disease condition.

In Australia, 'red spot' is recognised as a disease primarily of estuarine fish although some freshwater species have also been affected. Sea mullet (*Mugil cephalus*) have been principally involved in disease outbreaks in Queensland and Western Australia (Burke and Rodgers, 1981; Callinan, 1985) while in the Northern Territory the condition was first diagnosed in diamond-scale mullet (*Liza diadema*). Flat-tailed mullet (*Liza dussumieri*) and barramundi (*Lates calcarifer*) are among a number of other estuarine species which have been similarly affected. In Papua New Guinea, mullet (*Liza dussumieri*) were severely affected in coastal brackish waters and, at the same time, freshwater eleotrid gudgeons (*Oxyeleotris* sp) were principally involved in the spread of the disease through the inland river systems (Coates, 1984). The same species of mullet in brackish water estuaries and lagoons were also found to be severely affected during the more recent epizootic in Sri Lanka (Costa, pers. comm).

Elsewhere in southeast Asia the disease has predominantly involved freshwater species. It is generally agreed that wild and cultured snakeheads, particularly *Ophicephalus striatus* and to a slightly lesser extent *Ophicephalus punctatus* and *Ophicephalus marulius*, are most susceptible to infection, develop the severest clinical lesions and suffer the highest mortalities. The freshwater eel (*Fluta alba*) and spiny eels (*Macrognathus siamensis*, *Mastamcembelus armatus*) also appear to be readily affected. The susceptibility of other freshwater species is less well established. Catfish (*Clarias* sp, *Macrones* sp, *Heteropneustus* sp), gouramis (*Trichogaster* sp), gobies (*Oxyeleotris* sp,

Glossogobius sp) and a variety of cyprinids (Puntius sp), among many other species, have variously been reported as either susceptible, less susceptible or resistant to the disease.

VIRUS ISOLATION

Materials and Methods. A virological examination was carried out on tissues obtained from diseased fish during outbreaks of the ulcerative syndrome in southeast Asia in 1985, 1987 and 1988. In 1985, material was obtained as part of a fish disease survey carried out between September and December, at a relatively early stage of the epizootic, from a number of affected sites in Thailand, Burma, Malaysia, Indonesia and Lao PDR (Roberts et al, 1986). Further samples were obtained from the Philippines in March 1987, when only very few overtly diseased fish were caught, and from Sri Lanka in February 1988 when the incidence of clinical cases was also rapidly declining.

A total of 94 fish showing varying degrees of severity and stages of ulceration of the head, body and/or tail were selected for virological study. Snakehead fish undoubtedly appeared to be most susceptible to the disease and 64 of the selected specimens were either Ophicephalus striatus (49) or Ophicephalus punctatus (15) species (Table I). An additional 46 fish with no clinical lesions from the same sites were also examined.

During the 1985 disease outbreak, samples of liver and spleen were obtained from almost all of the selected fish, with kidney and lesion material being taken in some instances. In the later investigations liver, spleen and kidney material was taken in all cases. Pooled tissues were placed in small volumes of antibiotic-supplemented viral transport medium (Flow Laboratories Inc.) or Hanks' balanced salt solution (without Ca^{++} or Mg^{++}) + 10% fetal calf serum (FCS) and kept as cool as possible, without freezing, until further processed. Pooled tissue samples were weighed, homogenised with a pestle and mortar and diluted 1:10 (w/v) with Hanks' balanced salt solution (HBSS). Homogenates were clarified by centrifugation at 1200g for 10-15 minutes and the supernatents further diluted 1:5 with HBSS to give a final dilution of 1:50 (w/v) before filtration through a 450 nm membrane filter.

Freshly grown, 60-90% confluent, 25cm^2 monolayer cultures of a cell line (SSN-1) developed from whole snakehead fry (*Ophicephalus striatus*) were prepared in Leibovitz L-15 + 10% FCS growth medium and inoculated with 1ml tissue extract. Initially, BF-2, RTG-2 and EPC established cell line cultures were similarly inoculated but, as the investigation proceeded it appeared that the SSN-1 cells were most suitable for virus isolation and only these cells were used for many of the later examinations. Tissue extracts were allowed to absorb for 30-60 minutes before overlaying the monolayers with 7ml fresh growth medium and incubating at 28°C.

Results. Cytopathic effects (cpe) of varying degrees of severity developed between 1 and 14 days post-inoculation in approximately one-third of cultures. As it was not always possible to distinguish between toxic and virus-specific effects, aliquots of medium from all cultures which developed a cpe, and also a number which appeared normal, were passaged onto fresh SSN-1 monolayers. A repeatedly transmissible cpe was obtained with tissue extracts from three diseased fish from Thailand, two from Burma and one from Sri Lanka (Table II). No such effect was obtained from any clinically normal fish.

Electron microscopy of negatively stained (1% phosphotungstic acid pH 6.8) pellets obtained following ultracentrifugation of infected cell culture fluids revealed morphologically similar rhabdovirus particles in all cases.

TABLE I Ulcerated Fish Examined 1985-1988

Country	Fish Species						
	O striatus	O punctatus	Clarias sp	Puntius sp	Trichogaster sp	Eels	Others
Thailand	35	-	2	8	-	1	1
Burma	5	-	2	1	-	1	-
Indonesia	-	-	1	-	-	-	-
Lao PDR	3	-	1	-	-	2	3
Malaysia	2	-	-	-	1	-	-
Philippines	2	-	-	-	-	-	-
Sri Lanka	2	15	-	-	4	1	1
	49	15	6	9	5	5	5

TABLE II Virus isolations from ulcerated fish

Virus strain	Country	Fish species	Tissues examined
02	Thailand (north)	*Ophicephalus striatus*	L, S, K, Le
04	Thailand (northeast)	*Ophicephalus striatus*	L, K, Le
Ban Pako	Thailand (northeast)	*Ophicephalus striatus*	L, S, K
19	Burma	*Ophicephalus striatus*	L, S, Le
20	Burma	*Fluta alba*	K
SL 11	Sri Lanka	*Ophicephalus punctatus*	L, S, K

L = liver; S = spleen; K = kidney; Le = lesion

ULCERATIVE DISEASE RHABDOVIRUS (UDRV)

The size of the ulcerative disease rhabdovirus (UDRV) particles has been determined as 120 $\pm$ 10 nm x 80 $\pm$ 5 nm (Frerichs et al 1986). Serum neutralization studies have also shown that all six UDRV isolates are strains of the same virus, which is serologically distinct from VHSV, IHNV, EVA, EVX, SVCV, PFRV and perch rhabdovirus (Frerichs et al, 1989).

Although it was found that it could take as long as 14 days for a cpe to first appear in snakehead cell cultures inoculated with infected fish tissue extracts, culture adapted isolates, free from any inhibitory tissue substances, induce a much more rapidly developing cpe closely associated with the release of progeny virus particles from infected cells. As with all other lytic viruses, this varies with the cell substrate used and the temperature of incubation. Studies with two UDRV isolates in SSN-1 cells have shown that at 25°C and 30°C extracellular virus can appear in the culture medium within 12 hours of inoculation. At 20°C the latent period was extended to nearer 24 hours and at 15°C it took as long as 96 hours before virus mulitplication could be identifed (Figure 2). Following this variable latent period, the rate of virus multiplication and concurrent development of cytopathology continued to show the same pattern of temperature dependence. Infected cells became granular, rounded up, detached from the monolayer and lysed. Typically, this effect spread to the whole monolayer, leaving only clumps of degenerated cells and fine, granular debris in the culture medium. Less typically, incomplete destruction of the monolayer followed by regrowth of seemingly normal cells to re-establish confluency has also been observed. At least some of these 'healed' cultures have continued to produce infectious virus, even after subculturing, and may be regarded as persistently infected.

UDRV is notably resistant to heat treatment. Storage in clarified cell culture fluids at up to 25°C for 16 days did

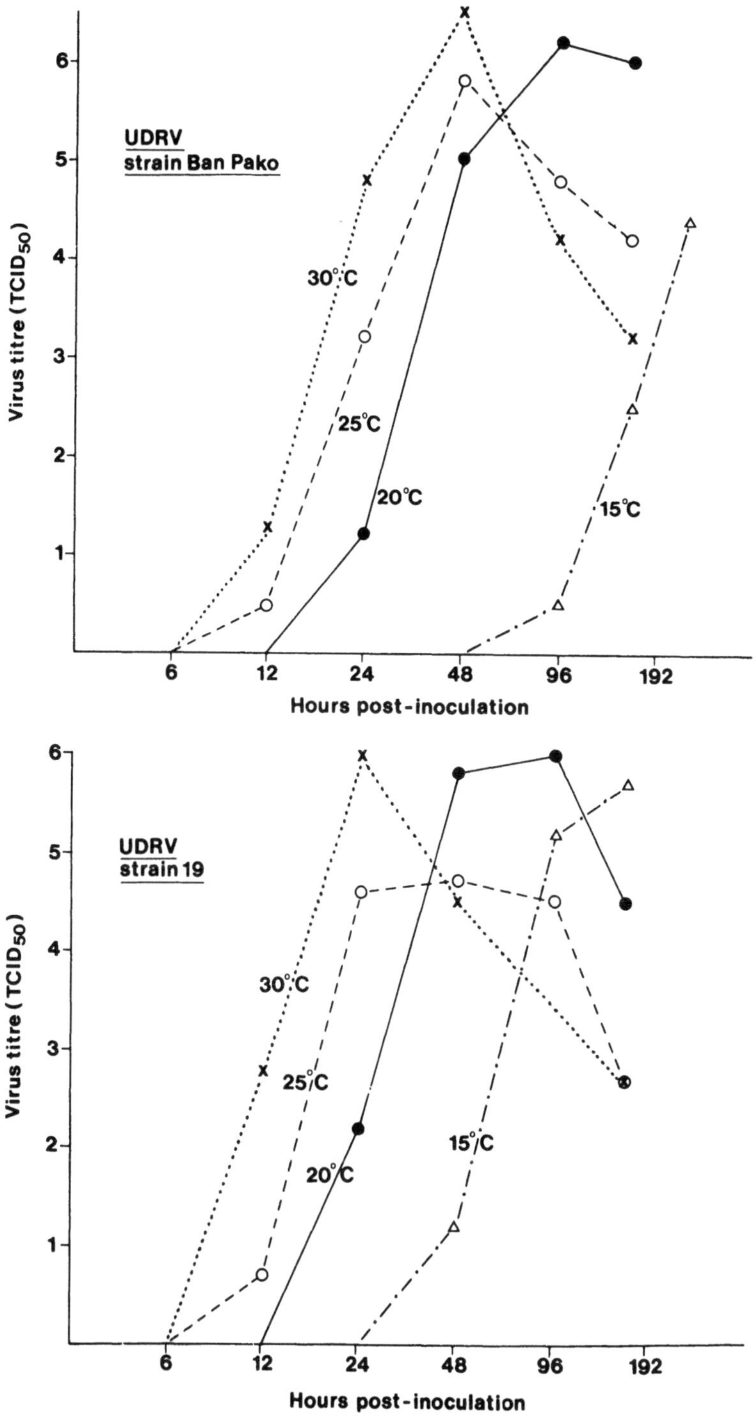

Figure 2 Growth of UDRV at different temperatures

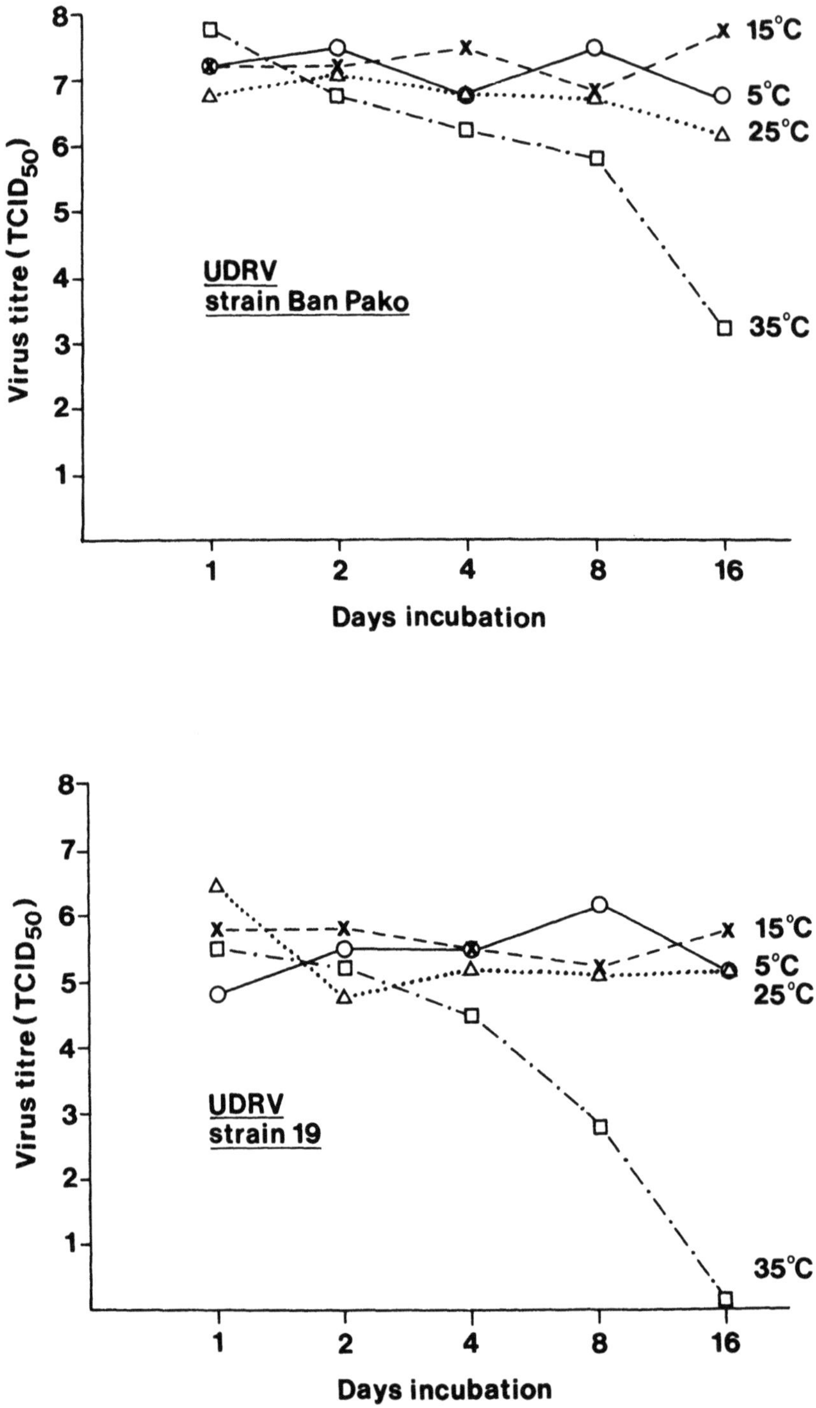

Figure 3 Infectivity of UDRV after storage at different temperatures

not result in any detectable loss in infectivity and at 35°C a 100-fold fall in titre did not occur in less than one week (Figure 3).

DISCUSSION

The pathogenesis of the epizootic ulcerative condition has not yet been established by controlled laboratory studies. Initially, it was considered that pollution of the waterways and ponds with herbicides, particularly paraquat, or organochlorine pesticides used in rice cultivation were responsible for the condition in southeast Asia. Industrial effluents have also been incriminated in some instances. These chemical agents have now been effectively discounted, as outbreaks of disease have occurred in areas where herbicides and pesticides have not been used and where there is no associated industrial development. Furthermore, areas where these compounds have been applied to crops for some years, or where industrial effluents have been discharged into waterways, have not always been affected by the disease.

The possibility of a primary bacterial or fungal infection has also been extensively investigated as *Aeromonas* or *Vibrio* species of bacteria are frequently associated with the lesions in freshwater and estuarine species respectively, and invasive *Saprolegnia* or *Achlya* species of fungi are commonly found in advanced cases of disease (McKenzie and Hall, 1976; Burke and Rodgers, 1981 ; Coates, 1984; Boonyaratpalin, 1987). However, the absence of a septicaemic condition characteristic of epizootic *Aeromonas* and *Vibrio* infections, and freedom from fungal involement in the early stages of ulcer development, indicate that these environmental micro-organisms are essentially opportunistic secondary pathogens. Nevertheless, the pattern of spread of the disease, particularly over the past few years, firmly indicates an infectious agent as the underlying cause of the condition.

A primary viral infection has, from the outset, always been considered a likely possibility and the presence of virus-like particles of varying morphology were observed in the tissues of diseased snakehead fish in Thailand in 1982 (Wattanavijarn et al, 1983). Subsequently, a birnavirus (SGV) was isolated from diseased sand gobies (Oxyeleotris marmoratus) examined in Thailand in 1984 (Hedrick et al, 1986) and an unclassified virus, similar in some respects to IPNV, has been identified in ulcerated snakeheads by Saitanu et al (1986). Although these agents cannot be dismissed as potential primary pathogens, it is considered that the single, distinct species of rhabdovirus which has now been isolated from similarly diseased fish in widely separated geographical areas at different times is most likely to be primary causal agent of the ulcerative syndrome.

The low rate of recovery of UDRV from clinically affected specimens was a disappointing feature of the present investigation. This could have been due to a number of factors including the elimination of the virus from the affected fish at some stage in the disease process, the presence of anti-viral substances or defective interfering (DI) particles in the tissue extracts, the inappropriate selection or storage of tissues for examination or the use of sub-optimal isolation procedures. Substantive proof of the pathogenic role of any of the viruses so far isolated from ulcerated fish must await the outcome of experimental infection studies but, as the disease appears to be mediated by naturally variable environmental conditions which cannot be replicated in the laboratory this is likely to prove to be particularly difficult to fulfil.

ACKNOWLEDGEMENTS

This study was carried out with the support of the United Kingdom Overseas Development Administration.

REFERENCES

Burke J, Rodgers L (1981) Identification of pathogenic bacteria associated with the occurrence of 'red spot' in sea mullet, *Mugil cephalus* L., in south-eastern Queensland. J Fish Dis 4: 153-159

Boonyaratpalin S (1987) Bacterial pathogen involved in the epizootic ulcerative syndrome of fish in Asia. Dept of Fisheries, Ministry of Agriculture and Cooperatives, Bangkok

Callinan RB (1985) Diseases of native Australian fishes. *In* Proc First Australian Workshop on Diseases of Fish and Shellfish. Humphrey JD, Langdon JS (eds) p 102-117. Australian Fish Health Reference Laboratory, Benalla, Victoria

Coates D (1984) An ulcer-disease outbreak amongst the fresh water fish population of the Sepik River system, with notes on some freshwater fish parasites. Report no. 84-02, Dept of Primary Industry, Fisheries Research and Surveys Branch, Port Moresby, Papua New Guinea

FAO (1986) Report of the expert consultation on fish diseases in the Asia-Pacific region. Food and Agriculture Organisation, Bangkok

Frerichs GN, Hill BJ, Way K (1989) Ulcerative disease rhabdovirus; cell-line susceptibility and serological comparison with other fish rhabdoviruses. J Fish Dis (in press)

Frerichs GN, Millar SD, Roberts RJ (1986) Ulcerative rhabdovirus in fish in South-East Asia. Nature 322(6076):216

Hedrick RP, Eaton WD, Fryer JL, Groberg WG, Boonyaratpalin S (1986) Characteristics of a birnavirus isolated from cultured sand goby *Oxyeleotris marmoratus*. Dis aquat Org 1: 219-225

Humphrey JD, Langdon JS (1986) Ulcerative disease in Northern Territory Fish. Internal Report Australian Fish Health Reference Laboratory

McKenzie RA, Hall WTK (1976) Dermal ulceration of mullet (*Mugil cephalus*). Aust vet J 52: 230-231

Natividad JM (1986) Fish Health Project (Philippines) Annual Report. Bureau of Fisheries and Aquatic Resources, Quezon City, Philippines

Roberts RJ, Macintosh DJ, Tonguthai K, Boonyaratpalin S, Tayaputch N, Phillips MJ, Millar SD (1986) Field and laboratory investigations into ulcerative fish diseases in the Asia-Pacific region. Food and Agriculture Organisation, Bangkok

Rodgers LJ, Burke JB (1981) Seasonal variation in the prevalence of 'red spot' disease in estuarine fish with particular reference to the sea mullet, *Mugil cephalus* L. J Fish Dis 4: 297-307

Saitanu K, Wongsawang S, Sunyasootcharee B, Sahaphong S (1986) Snakehead fish virus isolation and pathogenicity studies. *In* The First Asian Fisheries Forum Maclean JL, Dizon LB, Hosillos (eds) p 327-330. Asian Fisheries Society, Manila

Tonguthai K (1985) A preliminary account of ulcerative fish diseases in the Indo-Pacific region. Dept of Fisheries,

Ministry of Agriculture and Cooperatives, Bangkok
Wattanavijarn W, Rattanaphani R, Tesprateep T, Tangtrongpiroj J, Thirapatsakun T, Ousavaplangchai L, Sukolapong V, Eongpakornkeaw A, Vetchangarun S (1983) Virus-like particles in the skeletal muscle, capillaries, and spleen of sick snakehead fish (Ophicephalus striatus) during a disease epidemic. Thai J Vet Med 13: 122-130

Comparison of Infectious Hematopoietic Necrosis in Natural and Experimental Infections of Spawning Salmonids by Infectivity and Immunohistochemistry

T. Yamamoto*, C.K. Arakawa, W.N. Batts, and J.R. Winton
U.S. Fish and Wildlife Service
National Fishery Research Center
Seattle, W.A., USA 98115

Introduction

Infectious hematopoietic necrosis (IHN) continues to be a serious virus disease of salmonids with epizootics recorded in both wild and hatchery populations (Williams and Amend 1976; Carlisle et al 1979; Groberg and Fryer 1983; Saft and Pratt 1986; Traxler 1987; Follett et al 1987; Meyers et al 1988). While originally enzootic in western North America, the virus appears to be spreading further (Sano et al 1977; de Kinkelin et al 1987; Bovo et al 1987). In hatchery outbreaks occurring in regions where the virus is not enzootic, it is often possible to trace the virus to the importation of infected fingerlings or contaminated eggs. In regions where the virus is widespread among stocks of fish, the source of virus infection is more difficult to establish particularly in watersheds where there are anadromous salmonids. Although salmonid fish surviving infection as fry and returning from the ocean to spawn are considered to be parental carriers of IHNV, there is very little data to support this hypothesis. Amend (1975) in the study of rainbow trout reported that in a population surviving infection and assayed a few years later found that a few trout were carrying virus. This is the study often cited as evidence for the carrier status of returning salmonids. LaPatra et al (1987) presented data that indicated IHNV has been transmitted horizontally through water from adult chinook salmon (*Oncorhynchus tshawytscha*) to adult coho salmon (*O. kisutch*) at a hatchery in northern California. They suggested that horizontal transmission may be an important means for perpetuating IHN. However, the actual mechanisms for persistence and transmission of IHN among fish in a watershed is likely to be complex and involve multiple species and age groups as well as intermediate vectors and/or reservoirs.

This study reexamines IHNV infections among spawning wild sockeye

*Department of Microbiology
University of Alberta
Edmonton Alberta
Canada T6G 2E9

salmon in the Cedar River near Seattle, Washington. The infection rate on these fish approaches 100% of the population (Mulcahy et al 1982, 1983a, and 1984) and the virus is horizontally transmitted to progeny in the water (Mulcahy et al 1983b). We assayed the IHNV titer of selected organs and used mouse monoclonal antibodies for the histologic examination of IHNV antigen localization of infected tissues by immunofluorescence and immunoperoxidase. Since adult sockeye known to be free of IHNV are not readily available and are difficult to handle because of their size, we used adult wild spawning kokanee (land-locked form of *O. nerka*) obtained from a region known to be free of IHN to study the response of sexually mature sockeye salmon to IHN infection.

Materials and Methods

Spawning Fish

Adult sockeye salmon collected from the spawning grounds at Cedar River, Washington were obtained with a beach seine net during October and November of 1987. Ten female fish were sampled at intervals toward the latter part of the spawning period and organs taken in the field for virus infectivity and immunohistochemical assays.

Spawning kokanee measuring approximately 20 cm were obtained at Lake Whatcom Hatchery, located in northwestern Washington State as they arrived to spawn in an adjacent stream. The kokanee were transported to the National Fishery Research Center in Seattle, and the infection experiment conducted without delay because of the natural mortality of the spawning fish.

Infection of Kokanee

Spawning kokanee (20 in each lot) were infected by immersion for 30 min in 5 L of pathogen-free sand filtered, UV-treated Lake Washington water. Two different strains of IHNV were used at two different levels. One strain (CRV) had been isolated from Cedar River sockeye fry. The other strain (DWKV), was isolated from steelhead trout at Dworshak National Fish Hatchery, Idaho. Both isolations were made in 1986, passed twice in EPC cells and stored at -70°C. The two duplicate CRV challenge water titers were assayed to contain an average of 1.4×10^2 and 1.5×10^4 PFU/mL and the DWKV water titers were 7.0×10^1 and 9.6×10^3 PFU/mL. The virus was removed by dilution and each lot of fish was maintained in 3 ft diameter circular tanks at 12°C and observed daily for mortalities. Uninfected control fish were also kept in duplicate tanks following immersion in minimum essential medium (MEM).

The gill and spleen of each kokanee dying on the 3rd to the 8th day were pooled for virus assay.

Tissue Samples

Each sockeye was numbered and approximately 1 g of organ tissues (kidney, spleen, and gill, and brain) were removed aseptically, placed in Stomacher (Tekmar) bags and kept on ice for virus assay. Ovarian fluid and blood were collected in sterile tubes.

Sockeye tissues for histology were placed in plastic scintillation vials containing 10 mL of either 4% buffered formaldehyde, paraformaldehyde-lysine-periodate (PLP) (McLean and Nakane 1974), 0.1% glutaraldehyde-4% paraformaldehyde (GP) (Gorbsky and Borisy 1985) or 3% glutaraldehyde to determine the effect of different fixatives on IHNV antigen preservation. The fixed organs were embedded in paraffin and sectioned at 4-5 μ thickness.

Kokanee spleen and gill tissues from each fish were pooled and processed on the day of death for virus infectivity as was done for the sockeye tissues.

Virus Assay

Organs in blender bags were homogenized in MEM containing Tris buffer, 100 IU/mL penicillin and 100 μg/mL streptomycin with sufficient volume to provide an initial 10^{-1} dilution. Samples were clarified by centrifugation at 1000 x g for 10 min. The supernatant was assayed for virus using a modification of the plaque procedure described by Burke and Mulcahy (1980). To EPC cell monolayers in Lux 8 well multiplates (Miles Scientific) were added replicate 100 μL of log dilutions of of each sample. Virus was adsorbed for 30 min before the cells were overlayed with 0.75% methylcellulose in MEM containing 100 IU penicillin,100 μg streptomycin, 64 μg gentamycin sulfate, and 2.5 μg amphotericin B/mL. The plates were incubated for 7 days at 15°C when the cells were fixed and stained with an 18% formaldehyde-0.5% crystal violet mixture, washed and dried before counting the plaques.

Immunohistochemistry of Sockeye Organs

Tissue sections of gill, spleen, kidney, and brain as well as ovarian fluid smears were examined for IHNV antigen by immunofluorescence and immunoperoxidase staining. The tissue sections or smears were initially reacted for 1.5-3 h at room temperature with a neutralizing mouse monoclonal antibody designated 193-110/B4, specific for viral surface glycoprotein of most isolates of IHNV (Arakawa et al 1986; Winton et al 1988).

For immunofluorescence the sections were rinsed with 0.01M phosphate buffered saline (PBS) and the anti-mouse fluorescein (Hyclone) diluted 1/200 was added for 30 min at room temperature. After rinsing with PBS, the stained sections were lightly counterstained with 0.5 % evans blue, rinsed with PBS, and coverslipped with glycerol containing the anti-oxidant 1,4 diazo-bicyclo-(2,2,2)-octane (DABCO; Sigma), pH adjusted to 8.6-9.0. The stained sections were examined with a Zeiss microscope equipped with a halogen epi-illuminator lamp.

An amplified anti-mouse streptavidin-horseradish peroxidase conjugate (BioGenex) and 6-amino-9-ethylcarbazole was used to visualize IHNV antigen in histologic sections by light microscopy. The tissue sections and smears were reacted with the same mouse monoclonal antibody described above, rinsed with PBS, and followed by the horseradish conjugate for 30 min. After rinsing with PBS the enzyme was detected by the addition of 2 drops of a mixture of: 1 drop of a 3% hydrogen peroxide solution and 1 drop of a 2.2% solution of AEC in 2.5 mL of 0.1 M acetate buffer, pH 5.2. The AEC chromogen was converted by the enzyme within 5-8 min to a red, water insoluble precipitate. The sections, after rinsing with distilled water, were lightly counterstained with Meyer's hematoxylin to visualize the background tissue cells. After rinsing, the sections were covered with 1-2 drops of Crystal Mount (Biomedia) and allowed to dry over-night. A drop of Permount was added, covered with a coverslip to optimize their optical clarity and permanence, and examined by light microscopy.

Results

Virus Infectivity in Organs and Body Fluids of Adult Sockeye Salmon

Virtually all of the 30 spawning female sockeye assayed had infectious virus in one or more organs or body fluids examined. The materials tested included the gill, spleen, kidney, brain, blood and ovarian fluid. Levels of IHNV are shown in Fig. 1. It can be noted that there was considerable variation from one individual to another. The ovarian fluids had the highest levels but also exhibited the widest variation from one individual to another. The spleens had the next highest levels with 10 of the samples exceeding 10^6 PFU/ml. The gills and the kidneys also contained high levels of virus. The other materials tested - namely the brains and the blood, contained the lowest levels of virus.

A comparison of the average virus titers in the organs and body fluids on a

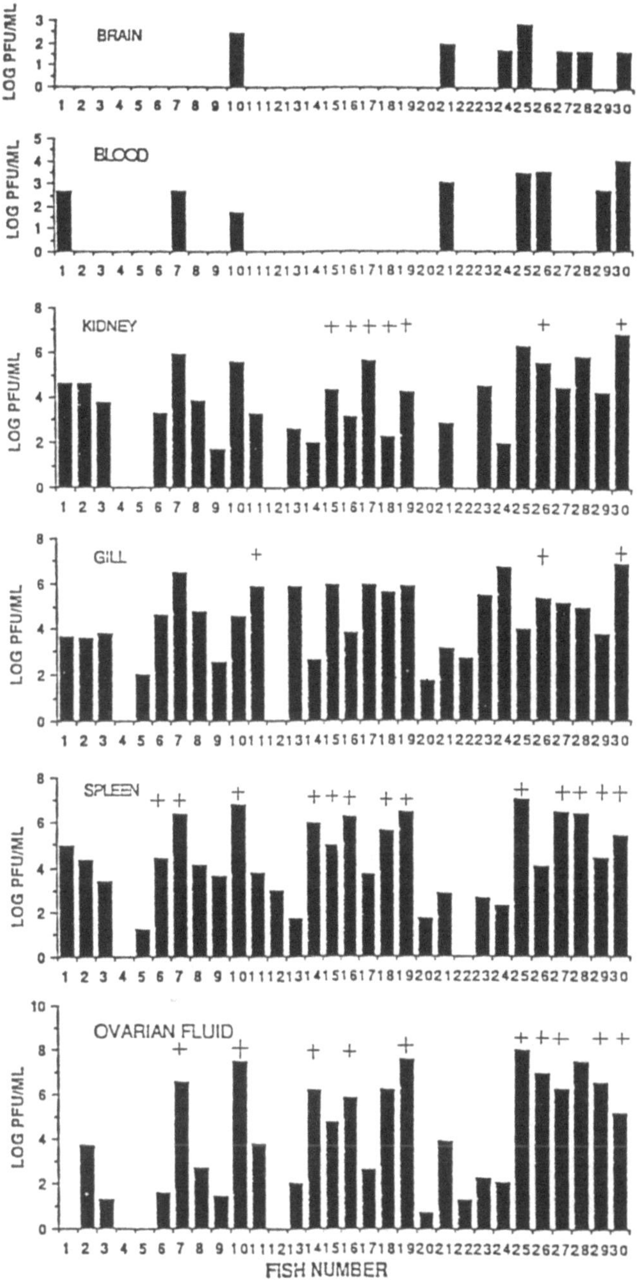

Fig 1 IHNV titers in tissues of spawning sockeye salmon. Each organ or body fluid for each fish is shown separately. A + sign indicates that virus antigen was detected in tissue sections by either immunofluorescence or immunoperoxidase

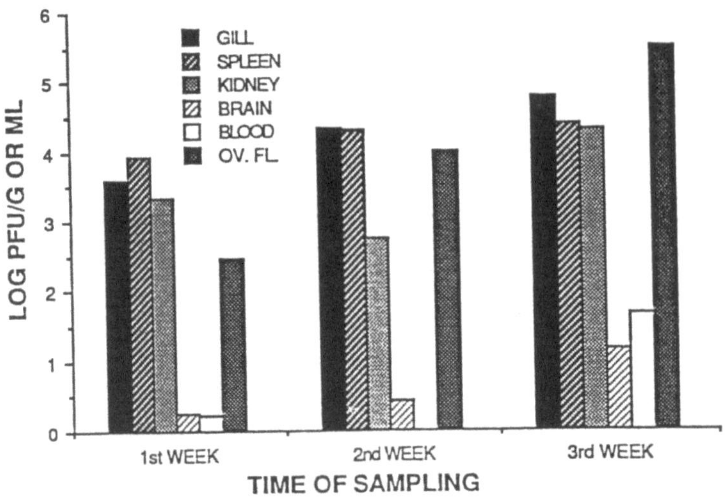

Fig 2. Weekly averaged IHNV titers of sockeye salmon organs. Each of the tissues sampled for the 3 weeks was averaged to show the change in virus titers of the spawning population.

weekly basis shows (Fig. 2) that there was a significant increase from the first to the third week of the study. The importance of this continued increase is that it would result in larger amounts of virus being shed into the water as the spawning period progresses.

Preservation and Visualization of IHNV in Infected Cell Monolayers and Tissues

Initial experiments compared selected fixatives for the preservation of IHNV antigens in infected coverslips and histologic sections by light microscopy. The fixatives tested included unbuffered 10% formaldehyde, neutral buffered formaldehyde, PLP, GP, and glutaraldehyde. Formalin at a concentration of 4% and buffered at pH 7.0 with 0.01M PBS was found to maximally preserve the antigens particularly if the fixed tissues were stored at 4°C. PLP was satisfactory in the preservation of antigen but tissue disintegration was observed after approximately 1 month of storage. GP was also effective in antigen preservation but introduced an undesirable amount of background staining. This nonspecific staining could be eliminated by removal of the endogenous peroxidase activity by pretreatment of the sections with 3% hydrogen peroxide. Glutaraldehyde fixation was unsuitable since it inactivated IHNV antigen as well as increasing nonspecific background fluorescence and background AEC staining.

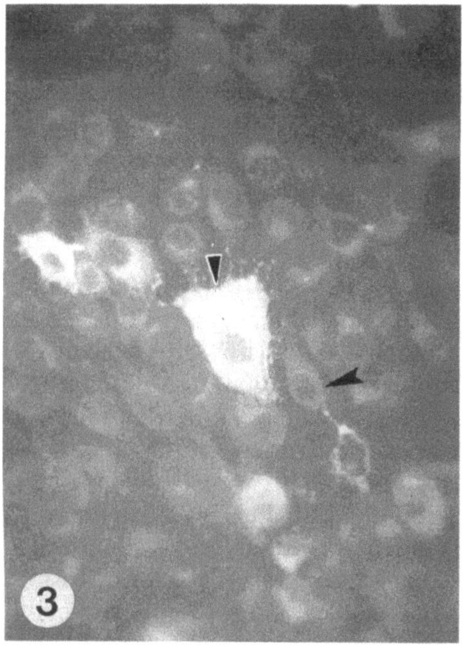

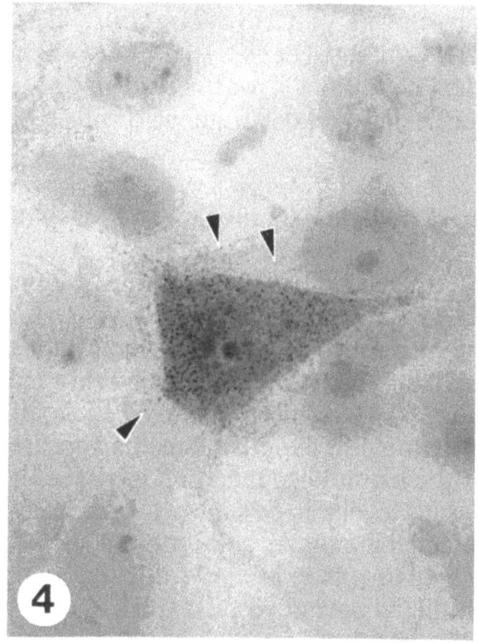

Fig. 3. CHSE-214 cell monolayer infected at low multiplicity with 193-110 IHNV for 22 h, fixed in buffered formalin, reacted with mouse monoclonal antibody and stained for immunofluorescence. White highlighted arrow shows the fluorescein yellow-green fluorescence of the primary infected cell. The dark arrow points to the evans blue reddish fluorescence of the uninfected cells (x 40)

Fig. 4. Cells infected, fixed, and reacted with monoclonal antibody as in Fig. 3. The cells were then reacted with anti-mouse peroxidase and visualized with AEC as the chromogen. Arrows point to red colored dots that are interpreted to be virions attached to cells adjacent to the primary infected cells (x 100)

The peroxidase-AEC method was considered to be slightly superior to immunofluorescence for examination of infected tissue sections. Its greater advantage was the capability to scan the sections at lower magnification with ordinary transmitted illumination and ready visualization of the sectioned cells. The red color of AEC localized virus antigen was clearly contrasted with the light blue coloration of the surrounding tissue. The immunofluorescence method requires observation at higher magnification and observation in a darkened room (Fig. 3). The specificity of the mouse monoclonal antibody (193-110/B4) for different isolates of IHNV was initially examined using monolayers of CHSE-214 cells. The antibody was found to react with all IHNV-infected cells tested. The sensitivity and specificity was such that under ideal conditions it was considered possible to detect individual particles on the surface of monolayer infected cells (Fig. 4).

Immunologic Examination of Sockeye Organs and Body Fluids

Sockeye tissues examined by immunologic methods observed to contain IHNV antigens are shown by + signs in Fig. 1. Virus lesions on gill filaments, although difficult to detect when stained by ordinary histologic stains such as by hematoxylin-eosin (Fig. 5), could readily be detected by immunofluorescence (Fig. 6) or by peroxidase-AEC labelling (Figs. 7 and 8). When stained by hematoxylin, the gill filaments appeared somewhat abnormal, which is typical of the natural degeneration of tissue in the spawning fish. Virus lesions in gill tissues tended to be highly localized. Some individual filaments were massively infected but others adjacent to it showed no evidence of infection. A number of gill filaments exhibited infection along cell layers immediately interior to the epithelial pavement cells (Figs. 6, 7 and 8). In the spleens, IHNV infected cells were mostly in the peripheral epidermal layers (Fig. 9), although some affected generalized regions were also seen. Infections of the kidney (Fig.10) were localized in the interstitial tissues between the tubules. Ovarian fluid smears revealed the presence of large numbers of infected cells. Virus in the brain tissue of spawning sockeye were not detectable by immunohistologic methods although low titers were found in some fish by infectious assay (Fig. 1).

Comparison of the sensitivities of the fluorescein and the peroxidase-AEC methods revealed that the latter was able to detect slightly more infected tissues, but the sensitivities were not markedly different. Usually, positive tissues examined by one method were observed to be positive by the other in adjacent tissue sections. The sensitivity of the immunohistologic methods for IHNV compared to virus assay was in the order of 10^5 to 10^6 PFU per g of tissue or per mL of body fluid. This comparison can be in noted in Fig. 1.

IHNV in Leeches and Sockeye Salmon

Leeches associated with spawning sockeye were collected on each of the three sampling dates and examined for IHNV. In the first week, 6 leeches were present on 9 fish examined, 10 leeches on 14 fish on the second week, and 12 leeches on 10 fish on the third sampling. The percent weekly average of leeches

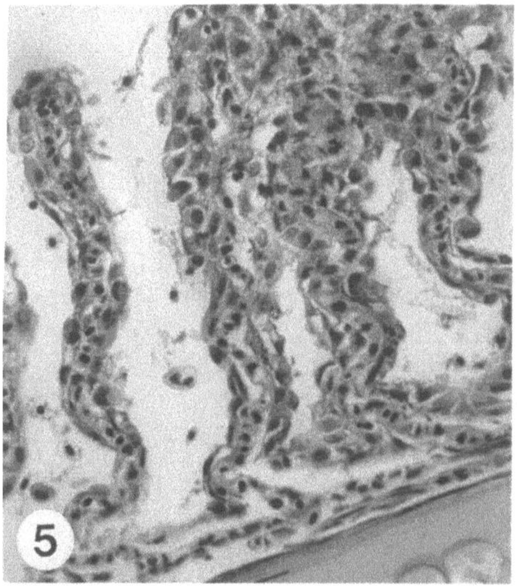

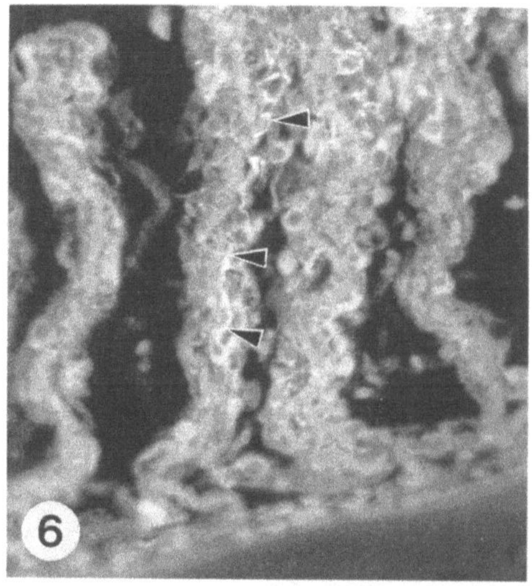

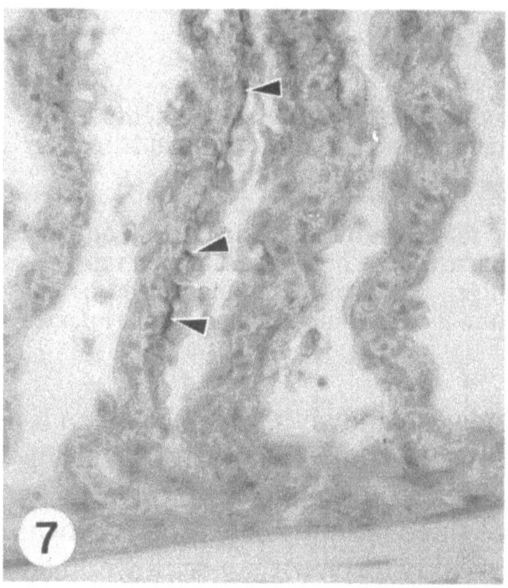

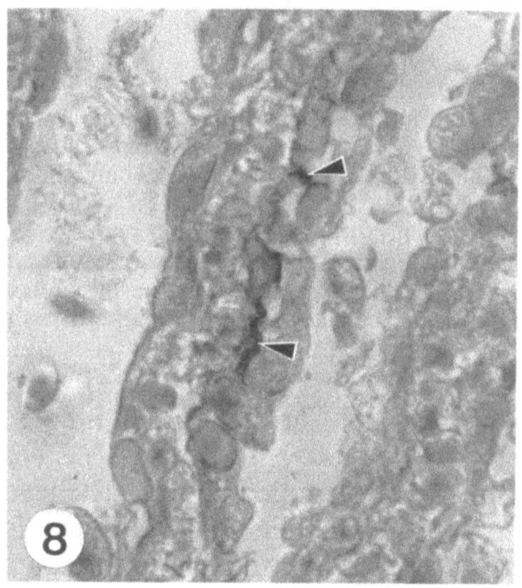

Fig. 5. Wild spawning sockeye gill tissue fixed with neutral buffered formaldehyde, sectioned, and stained with hematoxylin-eosin. The gill filaments are in a state of deterioration and the epithelial cells are hypertropied (x 300)

Fig. 6. Section from the same gill tissue as in Fig. 5 stained for immunofluorescence as in Fig. 3. The fluorecein stain is shown interior to the epithelial layer of cells. The tissue is counterstained with evans blue (x 300)

Fig. 7. Adjacent section to those in Figs. 5 and 6. The virus antigen is detected with anti-mouse horseradish peroxidase and AEC. The red color of the AEC is indicated by highlighted arrows. The background tissue is lightly stained blue with hematoxylin (x 300)

Fig. 8. Higher magnification of gill tissue showing the location of IHNV antigens in the secondary lamella. The site of multiplication appears to be in the cell layer internal to the primary epithelial layer (x 700)

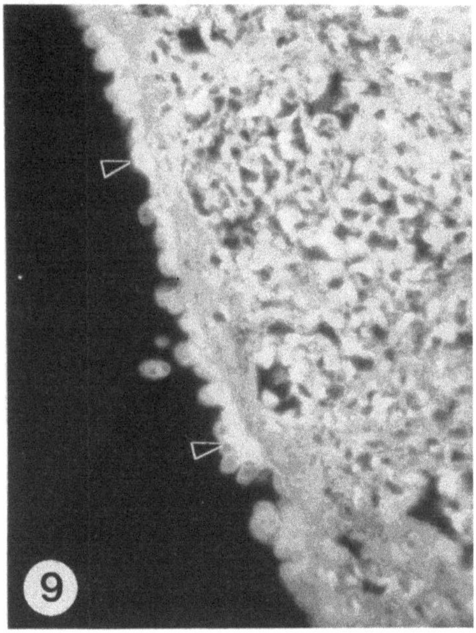

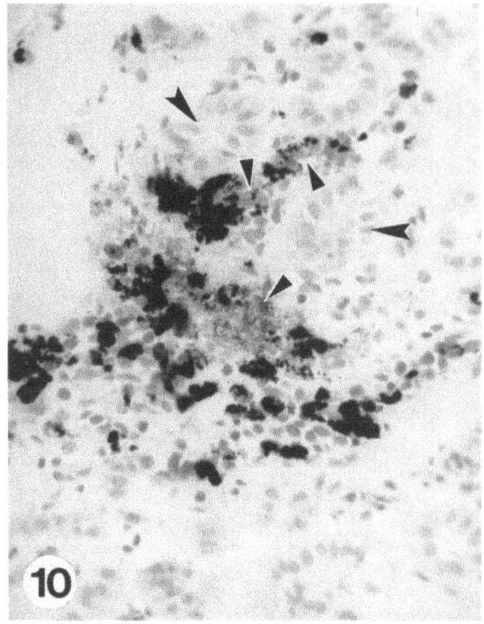

Fig. 9. Immunofluoresecence of wild spawning sockeye spleen. The fluorescense of the epithelial layer of the spleen is shown by highlighted arrows. The background is slightly overstained with evans blue and lacks contrast in the black and white print (x 300)

Fig. 10. Section of wild spawning sockeye kidney reacted with monoclonal antibody and antigen visualized with anti-mouse horseradish peroxidase. The highlighted arrows show the location of the red color indicating the site of viral antigen and is readily distinguished from the dark pigmented renal bodies. The dark arrows point to the renal tubules that are not infected (x300)

containing IHNV for each of the sampling dates are shown in Fig. 11. It can be noted that the number of positive leeches increased from 67% to 100% during the three week sampling period. The levels of virus present in these leeches was compared to virus titers in the gill titers for the three samping dates and is shown in Fig. 12. The graph shows that the average weekly leech titers were lower than those of the host gill titers but increased during the spawning period.

A comparison of the virus levels of individual leech titers to those of the host showed little, if any, relationship. The respective sockeye blood and leech titers are shown in Fig. 13. Only 5 of these blood samples yielded virus and, of those, the virus titers were lower than the respective titers detected in the leeches. However, when a comparison of the gill titers was made with that of levels in the respective leeches (Fig. 14) there also appeared to be little correlation. For example, fish number 12 showed no detectable virus in the gill but its associated leech had a low, but significant, level. Conversely, fish number 24 had a high titer in the gill tissue but a comparatively low level in the

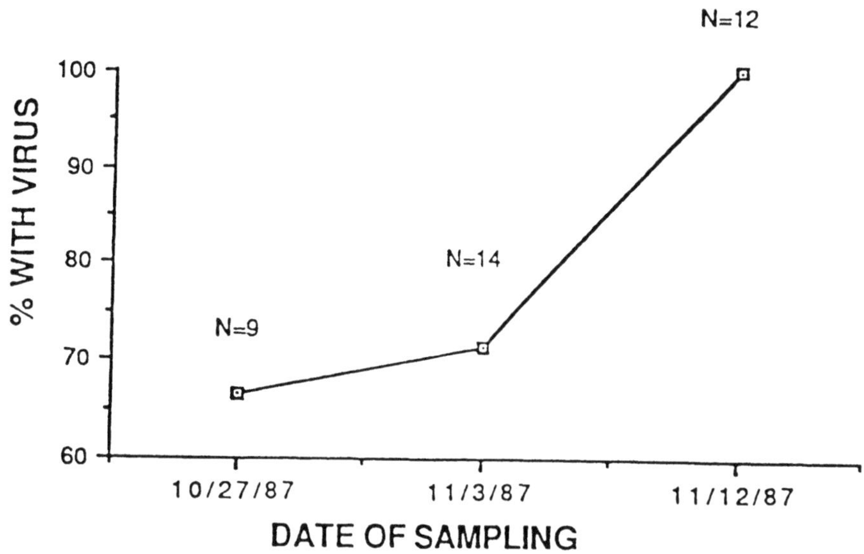

Fig 11. Percent of leeches associated with spawning sockeye salmon and found to contain IHNV for each of the three sampling days. The total numbers of leeches for each week are indicated

leech. A comparison of virus titers of leeches and other organs of the respective host also exhibited little relationship in virus levels (not shown). When the virus levels of the three sampling dates were averaged, it was evident that the leech virus levels had increased from low to high levels during this period (Fig. 12). These data indicate that during the spawning period there is an increase in sockeye virus levels as well as in the attached leeches.

Experimental Infection of Spawning Kokanee

The cumulative mortality for IHNV infected and control spawning kokanee salmon are shown in Fig. 15. The values are an aggregate of data from fish in both 10^2 and 10^4 infected and control duplicate tanks. The death rates were the same in the infected and uninfected populations (100% in both groups) by the 8th day and indicated that the kokanee were not dying due to the virus infection but due to the natural mortality following spawning.

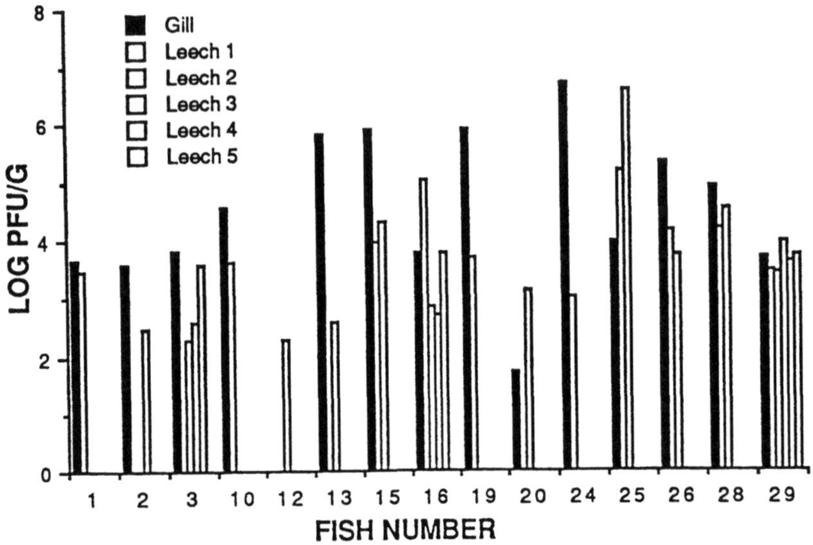

Fig 12. Relationship of averaged IHNV titers of sockeye salmon gill tissues compared to leech IHNV titers during the three sampling dates

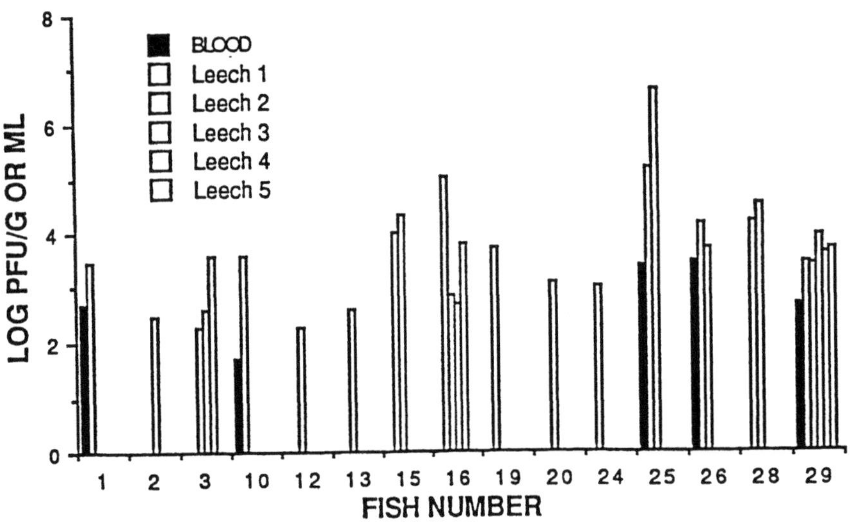

Fig. 13. Comparison of IHNV titers in sockeye salmon blood and in associated leeches

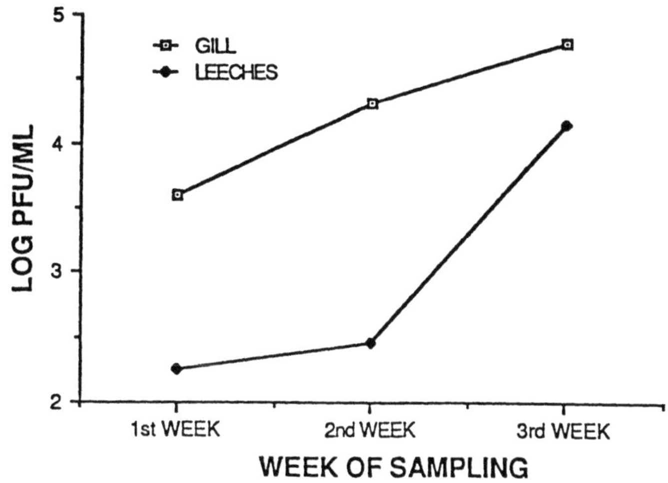

Fig 14. Comparison of IHNV titers in sockeye salmon gill and in associated leeches

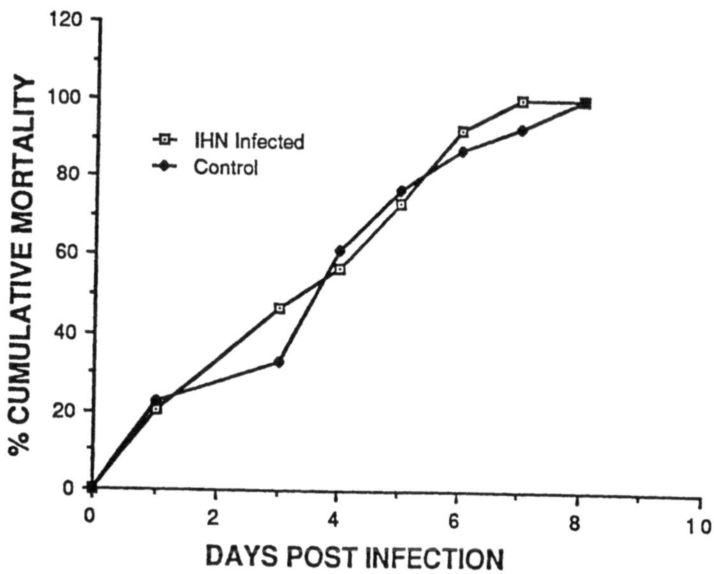

Fig. 15. Mortalities of experimentally infected and control uninfected spawning kokanee salmon

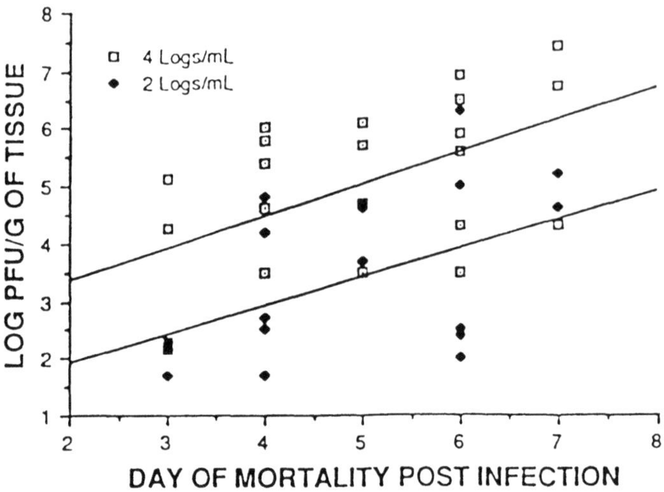

Fig 16. IHNV titers of averaged gill and spleen organs of experimentally infected spawning kokanee salmon at time of death

The titer of the virus in the gill and spleen tissue pools varied considerably from one individual to another as shown in Fig. 16. Control fish had no detectable virus (not shown). The computer generated lines indicate the difference in the average virus content of the organs for the two levels of infection. The data indicate that adult spawning kokanee salmon, immersed for a relatively short period of time in water containing IHNV can be readily infected with virus becoming detectable in the target organs and increasing to high levels before natural mortality occurs.

A similar experiment was conducted at the same time with another strain of IHNV (DWKV) isolated from steelhead trout. Multiplication of this strain, compared to CRV strain, was less rapid in kokanee salmon. The infection rate was more variable and the titers were just beginning to increase on days 7 and 8 when no more fish remained for analysis (data not shown).

Discussion

Our study of IHNV infections in spawning sockeye salmon in Cedar River confirms the earlier studies of Mulcahy et al (1982,1983a, and 1984) who found that virus infections persist at high frequencies, approaching 100% of the spawning population. As in past years, we also noted an increase in the IHNV titers as the spawning season progressed with the ovarian fluids attaining the highest titers (Fig. 1). The reason for this increase in titers during the spawning season is not clear. One explanation for this is that the fish spawning later in the run have been holding longer in the river, allowing more time for the virus to replicate. A second explanation for the increased levels could be due to the fact that later-spawning fish are challenged with a higher dose of virus in the water shed by earlier infected fish. The latter premise is supported by the results of the experimental infection of spawning kokanee that were readily infected by immersion (Fig. 16) and which showed a dose-response relationship so that higher levels of virus in the water gave higher virus titers in the organs.

Virus titers in the blood of sockeye salmon in the Cedar River remained surprisingly low throughout the spawning period in view of the high titers present in the various organs. This low level of detectable virus in the blood could be explained by the presence of complement dependent neutralizing antibodies recently reported by Hattenberger-Baoudouy and de Kinkelin (1988) in freeze-stored or complement inactivated sera of IHNV infected rainbow trout.

The application of immunohistochemical methods for the detection of IHNV antigen was useful in determining the location and nature of the infection. The availability of mouse monoclonal antibodies to IHNV has markedly improved the histologic study of virus infections since the cross-reacting tissue activities of polyclonal sera are eliminated. Faisal and Ahne (1980) used the immunoperoxidase method in frozen sections for the detection of another fish rhabdovirus, spring viremia of carp. Our studies demonstrate that the immunoperoxidase method, particularly with AEC as the chromogen, is suitable for the study of these virus infections in formalin fixed tissues. A major advantage in the use of the immunologic approach is that infections that do not exhibit obvious histopathology can be readily detected. Because of the lack of virus induced cytopathology of spawning sockeye, the detection of infected tissues would have otherwise been impossible to distinguish from the natural degenerative processes. The use of the immunofluorescent method for the detection of virus infections in the organs was limited because of the presence of high levels of auto-fluorescing materials in tissues of these fish.

The spleen, a major target organ in acutely infected young salmonid populations (Amend et al 1969; Yasutake and Amend 1972), was also a site of infection in the spawning sockeye. However, there was not the massive focal viral replication usually observed in infections of the young. Infection of the spawning sockeye spleen was less generalized, being observed in only a few fish, but appeared as small focal infections near the periphery or, more often, at the outer cell layer and the connective tissue surrounding the spleen.

Because of the high levels of virus in the ovarian fluid, it appeared that of all the organs, the ovary, or at least the body fluid released with the eggs from the ovary, was an important source of IHNV contributing to the high virus loads in the water. Examination of the cells in the ovarian fluid by immunoperoxidase-AEC staining showed the presence of large numbers of cells in varying stages of degeneration, almost all containing viral antigen. In these fish, it appeared that the infection spread from the ovary to the cell layers in the abdominal cavity, in this instance the spleen, after the disruption of the membrane wall of the ovary.

The lack of destruction of kidney tissues in infected adult sockeye is in contrast to that of younger fish where infection often leads to massive degeneration of the interstitial hematopoietic tissues as seen in hematoxylin-eosin stained sections (Amend and Yasutake 1969; Yasutake and Amend 1972) and, more recently, by immunoperoxidase techniques (Yamamoto et al 1988).

Gill tissues of spawning sockeye contained relatively high virus titers in many fish but histologic examination revealed that the tissues had highly localized lesions rather than generalized infections. Some infected filaments appeared to have the infection limited to the non-differentiated serosal cell layer described by Laurent and Dunel (1980) which lies internal to the epithelial pavement cell layer.

The first report of IHNV in leeches associated with sockeye salmon was a research bulletin by Mulcahy (1986) who reported that 60-100% of the leeches collected at Cedar River contained virus. He observed some leeches with high titers on fish with minimal titers in the blood, and suggested that IHNV must either replicate in the leech or that the leech must concentrate and store the virus from the fish blood. The role of leeches in the replication and transmission of IHNV requires further study. It is likely that leeches, or other invertebrates and other salmonids both may interact to maintain IHNV in the watershed. Ahne (1985) reported that both the fish louse *Argulus foliaceus* and the leech *Piscicola geometra* appear to act as mechanical vectors of another fish rhabdovirus, spring viremia of carp virus. He found that virus levels in the vectors were lower than that of the host blood. In contrast to his study, the leeches examined from the sockeye salmon in the Cedar River often had high virus titers when the blood of the host had a low titer (shown in Fig. 13). This observation, in itself, would suggest that the virus may be replicating in the salmon leech. However the high gill titers (shown in Fig. 14) would indicate that perhaps the leeches may be obtaining virus from the gill tissue.

The ready infectability of spawning kokanee provides evidence that some species of sexually mature salmonids are highly susceptible to IHNV infection with the virus increasing rapidly following a single short contact with water borne virus. Thus, it appears that large scale IHNV infections of spawning sockeye may require only an initial low level input of virus to begin the infection. Replication and release of the virus creates a sufficiently high waterload to spread to the entire population.

Whether the initial input into the spawning run is from a latent carrier returning from the marine environment or from another species or age group that has maintained a focus of infection in fresh water is unknown. Since IHN epizootics can occur in presmolts (Williams and Amend 1976) and in smolts (Burke and Grischkowski 1984), it is possible that infection could be transmitted from smolts to returning adult fish. Among sockeye which normally have a lake as a nursery environment, (e.g. Lake Washington for the Cedar River) this process could be relatively efficient. Leeches could also play a role in this transmission either as mechanical or active vectors.

Acknowledgement

We wish to thank Dr. J. L. Fryer, Department of Microbiology, Oregon State University, OR for monoclonal antibodies used in the immunohistochemical studies.

References

Ahne W (1985) *Argulus foliaceus* L and *Piscicola geometra* L. as mechanical vectors of spring viraemia of carp virus (SVCV). J Fish Dis 8:241-242

Amend DF (1975) Detection and transmission of infectious hematopoietic necrosis virus in rainbow trout. J Wildl Dis 110:472-487

Amend DF, Yasutake WT, Mead RW (1969) A hematopoietic virus disease of rainbow trout and sockeye salmon. Trans Amer Fish Soc 98:796-804

Arakawa CK, Lannan CN, Winton JR (1986) Monoclonal antibodies recognize strains of IHNV. Amer Fish Soc Fish Health Section Newsletter 14:1

Bovo G, Giorgetti G, Jorgensen PEV, Olesen NJ (1987) Infectious haematopoietic necrosis: first detection in Italy. Bull Eur Ass Fish Pathol 7(5):124

Burke JA, Grischkowsky RS (1984) An epizootic caused by infectious haematopoietic necrosis virus in an enhanced population of sockeye salmon, *Oncorhynchus nerka* (Walbaum), smolts at Hidden Creek, Alaska. J Fish Dis 7:421-429

Burke JA, Mulcahy D (1980) Plaquing procedure for infectious hematopoietic necrosis virus. Appl Environ Microbiol 39:872-876

Carlisle JC, Schat KA, Elston R (1979) Infectious haematopoietic necrosis in rainbow trout, *Salmo gairdneri* Richardson, in a semi-closed system. J Fish Dis 2:511-517

Faisal M, Ahne W (1980) Use of immunoperoxidase technique for detection of fish virus antigens. In: Ahne W (ed) Fish Diseases Third COPRAQ-Session Springer-Verlag, pp 182-192

Follett JE, Thomas JB, Hauck AK (1987) Infectious haematopoietic necrosis virus in moribund and juvenile chum, *Oncorhynchus keta* (Walbaum), and chinook, *O. tshawytscha* (Walbaum) salmon and spawning adult salmon at an Alaskan hatchery. J Fish Dis 10:309-313

Gorbsky G, Borisy GG, (1985) Microtubule distribution in cultured cells and intact tissues: improved immunolabeling resolution through the use of reversible embedment cytochemistry. Proc Nat Acad Sci 82:6889-6893

Groberg WJ Jr, Fryer JL (1983) Increased occurrences of infectious hematopoietic necrosis virus in fish at Columbia River basin hatcheries: 1980-1982. Oregon State University, Sea Grant College Program. Corvallis, Oregon, pp 16

Hattenberger-Baudouy A-M, Kinkelin P de (1988) Serological evidence for infectious hematopoietic necrosis (IHN) in rainbow trout from an outbreak in France. Abs AFS Fish Health Section International Fish Conference. Vancouver, Canada. July 19-21, p 6

Kinkelin P de, Hattenberger N-M, Torchy C, Liefrig F (1987) Infectious haematopoietic necrosis (IHN): first report in Europe. Abs Third International EAFP Conference. Bergen, p 57

LaPatra SE, Fryer JL, Wingfield W, Hedrick RP (1987) Transmission of infectious hematopoietic necrosis virus (IHNV) between adult species of salmon: management strategies for anadromous broodstock. Amer Fish Soc Fish Health Newsletter 15(2):7

Laurent P, Dunel S (1980) Morphology of gill epithelia in fish. Amer J Physiol 238:R147-R157

McLean IW, Nakane PK (1974) Periodate-lysine-paraformaldehyde fixative. A new fixative for immunoelectron microsocopy. J Histochem Cytochem 22:1077-1083

Meyers TR, Thomas J, Follett J, Saft R (1988) IHN; Trends in prevalence and the "farming around" approach in Alaskan salmon culture. AFS Fish Health Section International Fish Conference. Vancouver, Canada. July 19-21, p 4

Mulcahy, D (1986) Isolation of infectious hematopoietic necrosis virus from salmon leeches (*Piscicola salmositica*). Research Information Bulletin U.S. Fish and Wildlife Service, Washington, D.C.

Mulcahy D, Burke J, Pascho R, Jenes CK (1982) Pathogenesis of infectious hematopoietic necrosis virus in adult sockeye salmon (*Oncorhynchus nerka*) Can J Fish Aquat Sci 39:1144-1149

Mulcahy D, Pascho RJ, Jenes CK (1983a) Titre distribution patterns of infectious haematopoietic necrosis virus in ovarian fluids of hatchery and feral populations. J Fish Dis 6:183-188

Mulcahy DM, Pascho RJ, Jenes CK (1983b) Detection of infectious haematopoietic necrosis virus in river water and demonstration of waterborne transmission. J Fish Dis 6:321-330

Mulcahy D, Jenes CK, Pascho R (1984) Appearance and quantification of infectious hematopoietic necrosis virus in feral sockeye salmon (*Oncorhynchus nerka*) during their spawning migration. Arch Virol 80:171-181

Saft RR, Pratt KM (1986) Effect of infectious hematopoietic necrosis on sockeye salmon culture in Alaska. Publ. No 66. Alaska Dept Fish Game. Juneau, Alaska, pp 1-27

Sano T, Nishimura T, Okamoto N, Yamazaki T, Hanada H (1977) Studies on viral diseases of Japanese fishes.VI: infectious hematopoietic necrosis (IHN) of salmonids in the mainland of Japan. J Tokyo Univ Fish 63:81-85

Traxler G (1986) An epizootic of infectious haematopoietic necrosis in 2 year-old kokanee, *Oncorhynchus nerka* (Walbaum) at Lake Cowichan, British Columbia. J Fish Dis 9:545-549

Winton JR, Arakawa CK, Lannan CN, Fryer JL (1988) Neutralizing monoclonal antibodies recognize antigenic variants among isolates of infectious hematopoietic necrosis virus. Dis Aquat Org In press

Williams IV, Amend DF (1976) A natural epizootic of infectious hematopoietic necrosis in fry of sockeye salmon (*Oncorhynchus nerka*) at Chilko Lake, British Columbia. J Fish Res Board Can 33:1564-1567

Yamamoto T, Clermont T, Batts WN, Arakawa CK (1988) The progression of infectious hematopoietic necrosis virus multiplication in selected salmonids following immersion infection. Abs AFS Fish Health Section International Fish Conference Vancouver, Canada. July 19-21, p 3

Yasutake WT, Amend DF (1972) Some aspects of pathogenesis of infectious hematopoietic necrosis (IHN). J Fish Biol 4:261-264

Yoshimizu M, Sami M, Kimura T (1988) Survivability of infectious hematopoietic necrosis virus (IHNV) in fertilized eggs of masou (*Oncorhynchus masou*) and chum salmon (*O. Keta*) Abs AFS Fish Health Section International Fish Conference Vancouver, Canada. July 19-21, p 15

The Delayed Appearance of Infectious Hematopoietic Necrosis Virus (IHNV) in Steelhead Trout (*Salmo gairdneri*)

S.E. LaPatra, W.J. Groberg, J.S. Rohovec, and J.L. Fryer
Department of Microbiology
Nash Hall 220
Oregon State University
Corvallis Oregon 97331-3804 USA

INTRODUCTION

Infectious hematopoietic necrosis (IHN) is a nontreatable viral disease of certain species of salmonid fish (Pilcher and Fryer, 1980). The disease, once confined to limited geographical areas, has in the past decade become widely disseminated. The virus has been responsible for important economic losses of cultured and wild fish in the Pacific Northwest of the USA (Groberg and Fryer, 1983). Disease caused by IHNV has increased significantly and control has become important to the successful culture of salmonids. Because there are no vaccines or drugs available, control depends on preventing horizontal and vertical transmission of the virus. This is done by egg incubation and rearing in virus-free water and propagation of eggs from virus-free parents (Groberg and Fryer, 1983). The efficacy of these methods of control is equivocal and often difficult to assess.

Horizontal transmission is generally accepted and waterborne infections can be accomplished in the laboratory. Evidence for vertical transmission in natural outbreaks is circumstantial and only one report documents such an event under laboratory conditions (Mulcahy and

---

Present Address:

*Oregon Department of Fish and Wildlife
Department of Microbiology
Oregon State University
Corvallis Oregon 97331-3804 USA

Pascho, 1985). We have made numerous attempts, as have others, to clearly demonstrate the importance of this route, but without success. Evidence supporting vertical transmission includes the ability of virus to adhere to salmonid sperm cells (Mulcahy and Pascho, 1984) providing a mechanism for entry into the egg, and that developing embryos support replication after direct inoculation (R. A. Watanabe, Oregon State University, personal communication).

In an effort to clearly define the role of vertical and horizontal transmission in the epizootiology of IHN, we have conducted studies at Round Butte Hatchery (Deschutes County, Oregon, USA) operated by the Oregon Department of Fish and Wildlife. The virus is enzootic in the watershed and annual outbreaks of IHN in cultured steelhead trout are common. These conditions provided an opportunity for transmission studies in a hatchery environment. One such study involved storage of ova for 8 d so that the viral status (IHNV-negative or positive) of parents was known prior to fertilization.

MATERIALS AND METHODS

Cell Culture: The *Epithelioma papulosum cyprini* (EPC) cell line from common carp (*Cyprinus carpio*) (Fijan et al., 1983) was continuously cultured in Eagle's minimum essential medium (MEM) with Earle's salts (Gibco Laboratories, Grand Island, NY) supplemented with 10% fetal calf serum (Hyclone Laboratories, Logan, UT), antibiotics and glutamine (penicillin, 100 IU/ml; streptomycin, 100 μg/ml; and glutamine, 2 mM; all from Gibco). The growth medium was buffered with 1M tris HCl (Sigma Chemical Co., St. Louis, MO) to a pH of 7.5 (MEM-10-Tris). Cells were grown at 22°C.

Virus Detection and Quantitation: Plaque assay procedures were similar to those of Burke and Mulcahy (1980). Briefly, assays were performed using confluent EPC monolayers grown in plastic multi-well tissue culture plates. Fish tissues were diluted in Hank's balanced salt solution (HBSS; Gibco). Replicate inocula undiluted and as serial log dilutions to 10^{-6} were added to monolayers in individual wells and allowed to adsorb for 60 min. The inoculum was removed and 1% methyl cellulose (Sigma) dissolved in double strength MEM plus 5% fetal calf serum (MEM-5) was added. Following 7 to 10 d of incubation at 16°C, cells were fixed with formalin and stained with 1% crystal violet solution. Plaques were counted and pfu/ml or g were determined in replicate wells containing 10-300 plaques.

Virus Detection in Spawning and Post Spawning Adult Fish: Ovarian fluids were collected from individual steelhead trout at spawning and inoculated directly onto EPC cell cultures. Each adult fish was identified by tagging so they could be individually reexamined 9 or 15 d later. Thus, virus concentrations in OF were compared for individual fish at spawning and post-spawning.

Virus Detection in Ovarian Fluid from Stored Ova: Ova were stored after eggs and OF from individual adult female steelhead trout were spawned into plastic bags which were inflated with oxygen and placed at 4°C. Ovarian fluid was collected at spawning and at 48 h intervals for 14 d during storage and virus concentrations determined. Two experiments were performed using the eggs and OF from 16 and 10 fish, respectively.

Ovarian fluid from 60 adult female steelhead trout were examined similarly. At spawning (0 d) OF was obtained from each fish, the ova stored as described, and OF reexamined 8 d post-spawning. Concentrations of virus were determined in OF taken at the time of spawning and in OF taken from eggs post-spawning.

Virus Detection in Stored Ovarian Fluid: Ovarian fluid was collected from 100 adult steelhead trout at spawning and the ova were stored at 4°C. Cell-free OF was obtained after passage through a 0.2 µm membrane filter and filtered and non-filtered OF were assayed for virus concentrations. Filtered and non-filtered OF were then mixed 1:1 with an antibiotic solution and stored at 4°C. Eight days post-spawning OF was collected from all stored ova and virus concentrations determined. Filtered and non-filtered OF taken at the time of spawning (0 d) were reassayed for virus. Virus concentrations in OF taken at the time of spawning were then compared to concentrations detected post-spawning.

Culture of Cells from Ovarian Fluid and Virus Detection: The cells from OF were cultured by decanting 5 ml of OF from each of 24 spawning adult steelhead trout into tissue culture flasks (Corning Glass Works, Corning, NY). The fluid was incubated at 4°C for 24 h to allow cells to settle and attach. The supernatant was removed and an equal volume of MEM-10 containing antibiotics and glutamine (penicillin, 800 IU/ml; streptomycin, 800 µg/ml; gentamycin, 500 IU/ml; fungizone, 400 IU/ml; and glutamine 2 mM; all from Gibco) and buffered with 7.5% $NaHCO_3$ to pH 7.5 (MEM-10-antibiotic-bicarb) was added. The flasks were incubated at 4°C for 24 h, medium removed, and fresh medium added to each flask. This procedure was repeated 5x to remove extraneous virus, red blood cells, and microbial contaminants. To ascertain if virus was replicated in cultured OF cells, virus concentrations in medium were determined after rinsing and incubating at 4°C for selected intervals. As a control, cultures of cells from OF were established from 24 adult steelhead trout from the North Santiam River in Oregon. This fish stock has never been determined to be IHNV carriers in 6 yr of monitoring.

Methods to Enhance Detection of the Delayed Appearance of IHNV: Ovarian fluid from 24 spawning adult steelhead trout were treated by different methods. A portion of each was added to wells on tissue culture plates (Becton Dickinson Labware, Lincoln Park, NJ) containing different media preparations: 1) mixed 1:1 with MEM-10-antibiotic-bicarb; 2) mixed 1:1 with an antibiotic solution (Amos, 1985); or 3) incubated for 24 h, the supernatant removed, and MEM-5 overlay added. Incubation was 4°C. Virus concentrations in each were determined after 8 and 16 d and compared to concentrations detected at the time of spawning.

RESULTS

Virus Detection in Adults Post-Spawning: The delayed appearance of IHNV was demonstrated *in vivo* on two occasions. In the first observation, no virus was detected in OF from 30 adult steelhead trout females at spawning. Fifteen days later, 5 of 25 remaining fish were positive for IHNV (five fish died, escaped or yielded insufficient fluid for assay between examination dates). In the second observation, 3 of 11 virus-negative fish were positive for IHNV 9 d post-spawning.

Virus Detection in Ovarian Fluid from Stored Ova: Ovarian fluid obtained at spawning and at 48 h intervals from stored ova showed 2 of 16 egg lots changed from IHNV-negative to positive. The two lots of eggs which had undetectable virus at spawning were IHNV positive day 4 and 16 post-spawning. The test was repeated and the same shift from virus-negative to positive was detected in 2 of 10 egg lots. Additionally, two lots determined to have low concentrations of virus at the time of spawning exhibited increases in virus concentrations when examined on days 8 and 14 post-spawning (Table 1). In another test, OF from 9 of 60 egg lots converted from virus-negative to positive. Virus concentrations detected ranged from 10^1 to 10^4 pfu/ml. Two other egg lots showed a 10^1 to 10^2 increase in virus concentration. These determinations were made using OF collected at spawning (0 d) and 8 d post-spawning from eggs stored at 4°C.

Virus Detection in Stored Ovarian Fluid: Ovarian fluids collected at the time of spawning (0 d) were reexamined after 8 d storage at 4°C. Results showed 17 of 100 OF converted from IHNV-negative to positive or increased in virus concentration after the storage period. Ovarian fluid obtained from stored ova 8 d post-spawning showed a similar

Table 1. IHNV concentrations detected in OF obtained from stored ova at spawning (0 d) and at 2 d intervals over the 14 d storage period.

Assay Interval (days)	IHNV Concentrations[1] in OF from Stored Ova — Egg Lot Number			
	2	3	5	6
0	1.8	NEV[2]	NEV	2.8
2	1.5	NEV	NEV	2.8
4	1.5	NEV	NEV	2.6
6	1.0	NEV	NEV	2.9
8	3.5	NEV	NEV	2.8
10	2.8	1.0	1.3	2.3
12	2.8	NEV	C[3]	2.3
14	4.2	1.3		3.1

[1]IHNV concentrations represented as $\log_{10}$ pfu/ml.
[2]NEV = no evidence of virus in ovarian fluid (OF).
[3]C = bacterial contamination.

conversion. However, cell-free preparations of these same OF taken at the time of spawning which were then reexamined post-spawning did not show this delayed appearance of virus (Table 2).

Table 2. IHNV concentrations detected in filtered and non-filtered OF examined at spawning (0 d) and after an 8 d storage period at 4°C. Concentrations of virus were also compared for OF obtained from stored ova 8 d post-spawning.

	IHNV Concentrations[1] at Spawning (0 d)		IHNV Concentrations 8 d Post-spawning		
Fish No.	OF	Filtered OF	OF	Filtered OF	OF from Stored Ova
1	NEV[2]	NEV	4.0	1.0	2.7
2	0.7	0.7	2.6	NEV	NEV
3	1.7	NEV	2.4	NEV	1.5
8	NEV	NEV	2.2	NEV	1.0
12	NEV	NEV	1.3	NEV	0.7
16	1.7	1.5	2.0	NEV	1.3
19	NEV	NEV	0.7	NEV	NEV
21	NEV	NEV	1.7	NEV	1.2
22	NEV	NEV	1.4	0.7	1.6
24	1.7	1.8	2.5	1.2	2.2
36	NEV	NEV	1.7	NEV	.17
37	NEV	NEV	2.3	NEV	1.0
38	NEV	NEV	1.8	NEV	NEV
43	NEV	0.7	1.2	0.7	0.7
47	NEV	NEV	1.4	NEV	NEV
49	NEV	NEV	1.4	NEV	1.2
60	1.8	NEV	3.0	NEV	2.7

[1] IHNV concentrations represented as $\log_{10}$ pfu/ml.
[2] NEV = no evidence of virus in ovarian fluid (OF).

Culture of Ovarian Fluid Cells and Virus Detection: Culture of cellular components from OF in tissue culture flasks and assaying culture medium for IHNV concentrations during incubation intervals showed 11 of 24 cultures had a constant production of virus. Five OF cell cultures showing this were seeded with OF having virus concentrations of 10^7 pfu/ml at the time of spawning and three were seeded with OF containing concentrations of 10^5 to 10^6 pfu/ml. By rinsing the cultures daily for 5 d, virus concentrations were reduced to 10^1 to $10^{2.5}$ pfu/ml. Assaying culture medium after selected intervals of incubation showed virus concentrations increasing up to $10^{4.4}$ pfu/ml. Three other cultures were seeded with OF containing 10^1 to 10^3 pfu/ml and had no detectable virus after the rinses, but with incubation, virus concentrations increased to $10^{3.6}$ pfu/ml (Figure 1). Control cell cultures taken from OF showed no evidence of virus throughout the assay interval.

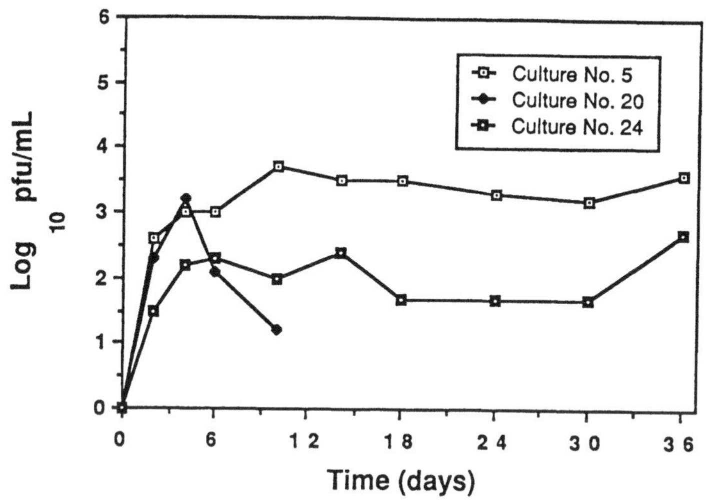

Figure 1. IHNV concentrations detected in culture medium from cells cultured from ovarian fluid (OF) after rinsing and incubating for selected intervals. Cells were obtained from OF containing 10^1 to 10^3 pfu/ml at spawning.

<u>Methods to Enhance Detection of the Delayed Appearance of IHNV</u>: The most consistent method for culturing cells from OF and detecting the delayed appearance of virus used a 24-well plate inoculated with 0.5 ml OF at the time of spawning. This was incubated for 24 h, the sample removed, and 0.5 ml of a 1% methyl cellulose MEM-5 overlay added back to each well. IHNV concentrations were then determined by assaying medium taken from this plate 8 and 16 d post-spawning (Table 3).

Table 3. Method for the enhanced detection of the delayed appearance of IHNV. IHNV concentrations detected in OF taken at the time of spawning (0 d) and in medium from cells cultured from OF assayed 8 and 16 d post-spawning.

IHNV Concentrations[1]		
OF at Spawning (0 d)	Medium from Cells Cultured from OF	
	8 d	16 d
NEV[2]	0.7	1.6
NEV	2.9	3.1
1.7	3.8	3.5
NEV	NEV	0.7
0.7	NEV	2.1
0.7	NEV	3.0

[1]IHNV concentrations represented as $\log_{10}$ pfu/ml.
[2]NEV = no evidence of virus in ovarian fluid (OF).

DISCUSSION

Initial observations of the delayed appearance of IHNV were made in post-spawning steelhead trout. Fish determined to be virus free at the time of spawning were found to be virus carriers when OF was reexamined 9 or 15 d post-spawning. Similar observations were made by Mulcahy and Pascho (1986) in prespawning sockeye salmon which had been individually tagged, penned, and collected daily over a 6 d period. They suggested that post-spawning fish be examined to increase the sensitivity for identifying IHNV carriers.

We reported the delayed appearance of IHNV in OF collected from stored ova of steelhead trout held for delayed fertilization (LaPatra and Groberg, 1985). Ovarian fluid collected at the time of spawning and determined to be virus-negative became virus-positive when reexamined after the 8 d storage period. Results reported here suggest cells present in the OF may be responsible for the delayed appearance of

IHNV. Supporting evidence are those instances when higher concentrations of virus were detected in OF after storage than were detected in filtered (cell-free) OF stored under identical conditions. Further, cells removed from the OF and cultured in vitro replicated the virus. Mulcahy and Batts (1987) used low-speed centrifugation to remove cells from OF prior to addition of culture medium at 15°C. They subsequently assayed the medium usually after only 7 d post-spawning. Incubation of OF at 4°C with repeated washing with media, we were able to establish cultures of cells from OF which maintained a constant production of virus for up to three months. Morphologically most of the cells appeared to be either epithelial or leukocytes. Results suggest that a persistent infection had been established in some of these cells and may have been responsible for the delayed appearance of IHNV.

Cell lines reported to be persistently infected with IHNV are epithelial (Engelking and Leong, 1981; Watanabe, 1983). Viral antigens have been detected by immunofluorescence on trout lymphocytes infected with viral hemorrhagic septicemia virus, another rhabdovirus which causes disease in trout (Enzmann, 1981) and trout lymphocytes infected with infectious pancreatic necrosis virus (Yu and Macdonald, 1982). We have also detected fluorescing cells from OF of IHNV carrier adults and from peripheral blood of clinically infected juvenile fish (LaPatra and Rohovec, 1987; LaPatra et al., submitted) but we have not been able to correlate a specific cell with the delayed appearance of IHNV. It would seem likely that leukocytes such as macrophages and lymphocytes could be involved because these cell types have been implicated in latent and persistent viral infections in other hosts (Kenyon and Piper, 1977; Paul et al., 1979; Narayan et al., 1983). Peripheral blood leukocytes of trout can also support replication of IHNV (J. R. Winton, U. S. Fish and Wildlife Service, personal communication).

Methods to enhance detection of the delayed appearance of IHNV include: 1) at the time of spawning seed multi-well tissue culture plates with OF from individual female fish. Incubate at 4°C for 24 h. Remove the fluid portion of OF and add 1% methyl cellulose MEM-5 overlay to each well and incubate at 4°C. Assay medium from cell cultures obtained from OF to detect the delayed appearance of IHNV; 2) when collecting OF use a device which allows cells to enter but not eggs. Immediately express OF into a tube containing the antibiotic mixture used to decontaminate tissue samples used for detection of fish virus (Amos, 1985). Store at 4°C and assay post-spawning. When monitoring for virus in different stocks of steelhead trout, the delayed appearance of IHNV could not be detected unless some portion of the population was already virus-positive. The percentage of fish which could then be demonstrated to convert from virus-negative to virus-positive was always low. While delayed appearance may provide information about the biology of IHNV it does not appear to be a major concern for diagnosticians unless IHNV-free eggs are required.

ACKNOWLEDGEMENTS

This research was sponsored by the Bonneville Power Administration (Agreement No. DE-A179083BP11987) with G. R. Bouck as the Contracting Officer's Technical Representative and a grant from the Oregon Department of Fish and Wildlife, Agreement PL89-304, the Anadromous Fish Act. Oregon Agricultural Experiment Station Technical Paper No. 8607.

REFERENCES

Amos KH (ed) (1985) Procedures for the detection and identification of certain fish pathogens. Fish Health Section American Fisheries Society Corvallis Oregon

Burke JA, Mulcahy D (1980) Plaquing procedure for infectious hematopoietic necrosis virus. Appl Environ Microbiol 39:872-876

Engelking HM, Leong JC (1981) IHNV persistently infects chinook salmon embryo cells. Virology 109:47-58

Enzmann PJ (1981) Rapid identification of VHS-virus from trout by immunofluorescence. Develop Biol Standard 49:57-62

Fijan N, Sulimanovic D, Bearzotti M, Muzinie D, Zwillenberg LD, Chilmonczyk S, Vantherot JF, de Kindelin P (1983) Some properties of the Epithelioma papulosum cyprini (EPC) cell line from carp Cyprinus carpio. Ann Virol (inst Pasteur) 134E:207-220

Groberg WJ, Fryer JL (1983) Increased occurrences of infectious hematopoietic necrosis virus in fishes at Columbia River Basin hatcheries. Oregon State Univ Sea Grant College Prog ORESU-T-83-002

Kenyon SJ, Piper CE (1977) Cellular basis of persistent lymphocytosis in cattle infected with bovine leukemia virus. Infect Immun 16:891-897

LaPatra SE, Groberg WJ (1985) Delayed appearance of IHNV in stored summer steelhead eggs. Am Fish Soc Fish Health Sect Newsl 13:3

LaPatra SE, Rohovec JS (1987) Indirect fluorescent antibody test for IHNV. Am Fish Soc Fish Health Sect Newsl 15:2

LaPatra SE, Roberti KA, Rohovec JS, Fryer JL (1988) Fluorescent antibody test for the rapid detection of infectious hematopoietic necrosis. J Aquat Animal Health (submitted)

Mulcahy DM, Pascho RJ (1984) Adsorption to fish sperm of vertically transmitted viruses. Science (Wash DC) 225:333-335

Mulcahy DM, Pascho RJ (1985) Vertical transmission of infectious hematopoietic necrosis virus in sockeye salmon (Oncorhynchus nerka): isolation of virus from dead eggs and fry. J Fish Dis 8:393-396

Mulcahy DM, Pascho RJ (1986) Sequential tests for infectious hematopoietic necrosis virus in individuals and populations of sockeye salmon (Oncorhynchus nerka). Can J Fish Aquat Sci 43:2515-1519

Mulcahy D, Batts WN (1987) Infectious hematopoietic necrosis virus detected by separation and incubation of cells from salmonid cavity fluid. Can J Fish Aquat Sci 44:1071-1075

Narayna O, Kennedy-Stoskopf S, Sheffer D, Griffin DE, Clements JE (1983) Activation of carpine arthritis-encephalitis virus expression during maturation of monocytes to macrophages. Infect Immun 41:67-73

Paul PS, Mengeling WL, Brown TT (1979) Replication of porcine parvovirus in peripheral blood lymphocytes, monocytes and peritoneal macrophages. Infect Immun 25:1003-1007

Pilcher KS, Fryer JL (1980) The viral diseases of fish: a review through 1978. CRC Critical Reviews in Microbiology 7:287-364

Watanabe T (1983) Persistent infection in YNK cells with infectious hematopoietic necrosis virus (IHNV). Bull Japan Soc Sci Fish 49:157-163

Yu KK-Y, Macdonald RD (1982) Replication of infectious pancreatic necrosis virus in trout leucocytes and detection of the carrier state. J Fish Dis 5:401-410

Preliminary Characterization of a Virus Causing Infectious Anemia Among Stocks of Salmonid Fish in the Western United States

C.K. Arakawa*, D.A. Hursh, C.N. Lannan, J.S. Rohovec, and J.R. Winton*
Department of Microbiology
Oregon State University
Corvallis, Oregon 97331-3804 USA

INTRODUCTION

Since 1982, anemias occurring in stocks of yearling coho (Oncorhynchus kisutch) and chinook salmon (Oncorhynchus tshawytscha) have been associated with serious losses at hatcheries in the Pacific Northwest, USA. The anemia is often accompanied by infections with external fungus (e.g. Saprolegnia) or the bacterial pathogens Cytophaga psychrophila or Renibacterium salmoninarum (Holt and Rohovec 1984, Leek 1987). The losses associated with the anemia are thought to be caused by these secondary infections.

Blood smears that were made from anemic fish and stained with Giemsa or pinacyanol chloride showed erythrocytic inclusions ranging in size from 1 to 8 um and varying in number per cell. Thin sections of infected red blood cells (RBC) examined by electron microscopy revealed virus particles approximately 70 nm in diameter. The virions were scattered in the cytoplasm of the RBC or contained within membrane bound organelles. These virus particles were morphologically distinct from the iridovirus, erythrocytic necrosis virus (ENV), which is also associated with anemia (Holt and Rohovec 1984, Leek 1987). Evidence suggests that the etiological agent of this new anemic disease, termed erythrocytic inclusion body syndrome (EIBS) by Leek (1987), is a previously undescribed virus infecting salmon. The purpose of this study was to experimentally transmit the disease to healthy fish, to determine the blood parameters associated with infection, and to investigate the nature of the virus associated with EIBS.

---

*U.S. Fish and Wildlife Service
National Fisheries Research Center
Building 204, Naval Station
Seattle, Washington 98115 USA

MATERIALS AND METHODS

Examination of blood

Blood was collected in heparinized capillary tubes by severing the caudal peduncle of anesthetized fish. Smears made from a drop of blood were air dried, fixed with absolute methanol for 5 min, then stained with Giemsa or pinacyanol chloride (Yasutake 1986). The blood smears were examined by light microscopy for the presence of viral inclusions in the erythrocytes, bacterial infection, and general appearance of the blood cells.

Blood cells were collected from infected fish for examination by electron microscopy. The cells were fixed for 1 hr in Balanced Salt Solution (BSS) containing 3% glutaraldehyde, pelleted, resuspended in BSS, and embedded. Thin sections of infected RBC were cut, stained and examined.

Hematological tests were performed using standard methods as outlined by Wedemeyer and Yasutake (1977). These included red and white blood cell count (RBC and WBC), hematocrit (HCT), hemoglobin (HGB), mean corpuscular volume (HCV), mean corpuscular hemoglobin (MCH), mean corpuscular hemoglobin concentration (MCHC), osmotic fragility, and WBC differential.

Virology

Kidney and spleen tissues from infected fish were aseptically removed and homogenized in BSS. The homogenate was clarified by centrifugation (4000 x g) and diluted 1:10 in an antibiotic mixture (1000 U penicillin, 1000 ug/ml Streptomycin, 500 IU/ml Nystatin, 250 IU/ml Gentamicin in minimal essential medium). After overnight incubation in antibiotics, samples were inoculated on monolayer cultures of fish cell lines. These included: chinook salmon embryo (CHSE-214), chum salmon heart (CHH-1), coho salmon embryo (CSE-119), sockeye salmon embryo (SSE-5), rainbow trout hepatoma (RTH-149), fathead minnow (FHM), brown bullhead (BB), and epithelioma papulosum cyprini (EPC). The cultures were incubated at 18 C and observed daily for the development of cytopathic effect (CPE). After 14 days a "blind passage" was performed by removing samples from each culture and inoculating a second set of the same cell lines. These subcultures were incubated at 18 C for an additional 3 weeks.

Bacteriology

Kidney tissue was inoculated onto brain heart infusion (BHI) agar, coho blood agar (5% coho blood in BHI), chocolate agar (5% coho blood in BHI heated at 80 C for 15 min), Ogawa egg medium, Mueller-Hinton agar and thioglycollate broth. Three sets of media were prepared and incubated aerobically, anaerobically, and with 5% CO_2 for 2 weeks at 18 C.

Experimental transmission

Yearling chinook salmon suffering from a natural outbreak of EIBS were collected from the Little White Salmon National Fish Hatchery, Cook, WA, USA. Homogenates of kidney tissue and blood from these fish were injected intraperitoneally (i.p.) into healthy chinook salmon. Injected fish were held for 6 weeks in 12 C, pathogen-free fresh water at the Hatfield Marine Science Center. At 2 week intervals, 20 fish were sampled for determination of blood parameters and the presence of inclusions.

Homogenates of kidney tissue and blood were also passed through 0.45 and 0.22 um pore-size membrane filters before i.p. injection into coho salmon. Thermal stability of the virus was tested by heating homogenates for 1 hr at 60 C. Sensitivity to chloroform was determined by the method of Rovozzo and Burke (1973) using tissue homogenates instead of culture supernatant. Ten fish were injected with each preparation, held 4 weeks in 12 C pathogen-free fresh water, and sampled for the presence of inclusions.

RESULTS

Examination of blood

Blood values of yearling chinook salmon suffering an outbreak of EIBS at the Little White Salmon National Fish Hatchery showed a moderate anemia and elevated WBC counts (Table 1). While the hematocrit was not as low as that seen during other outbreaks of the disease, the general blood picture was that of an erythrocytic anemia.

Blood smears from affected fish stained with pinacyanol chloride or Giemsa revealed cytoplasmic inclusions in the erythrocytes ranging in size from 1 to 8 um and varying in number per cell (Fig 1A). When examined by electron microscopy, the infected blood cells contained spherical virus

Table 1. Blood values of yearling chinook salmon (*Oncorhynchus tshawyshcha*) collected during a natural outbreak of infectious anemia (EIBS) at the Little White Salmon National Fish Hatchery.

No. sampled	20
RBC (x $10^6/mm^3$)	1.2
WBC (x $10^4/mm^3$)	3.6
WBC:RBC	1:28
Hematocrit (%)	37
Hemoglobin (g/100ml)	7.5
Mean Corpuscular Volume (g/dl)	331
Mean Corpuscular Hemoglobin (pg)	66
Mean Corp. Hem. Conc. (%)	20
Osmotic Fragility (%NaCl)	0.45
Lymphocytes (/100 WBC)	94
Neutrophils (/100 WBC)	1
Thrombocytes (/100)	5
Inclusions in erythrocytes	++

particles approximately 70 nm in diameter with a 38 nm electron-dense center. The virions were observed scattered in the cytoplasm of the cell (Fig 1B), associated with a viroplasm-like material (Fig 1C), or contained within membrane-bound organelles (Fig 1D).

Culture of infectious agent

Although others have recovered *C. psychrophila* and *R. salmoninarum* from fish during outbreaks of EIBS, no bacteria were isolated consistently or observed in stained blood or kidney smears during this study.

None of the eight fish cell lines inoculated with homogenates of tissue or blood from infected fish showed cytopathic effect during the initial 14 day incubation. No cytopathology was seen in any of the subcultures (blind passage) during an additional 3 weeks incubation.

Transmission

The disease could be transmitted to coho and chinook salmon following i.p. injection of blood or kidney homogenates from infected fish. Typical inclusions were observed in erythrocytes of experimentally infected fish 14 days post-injection. Blood values were determined for control and infected chinook at week 2, 4, and 6 post-injection (Table 2).

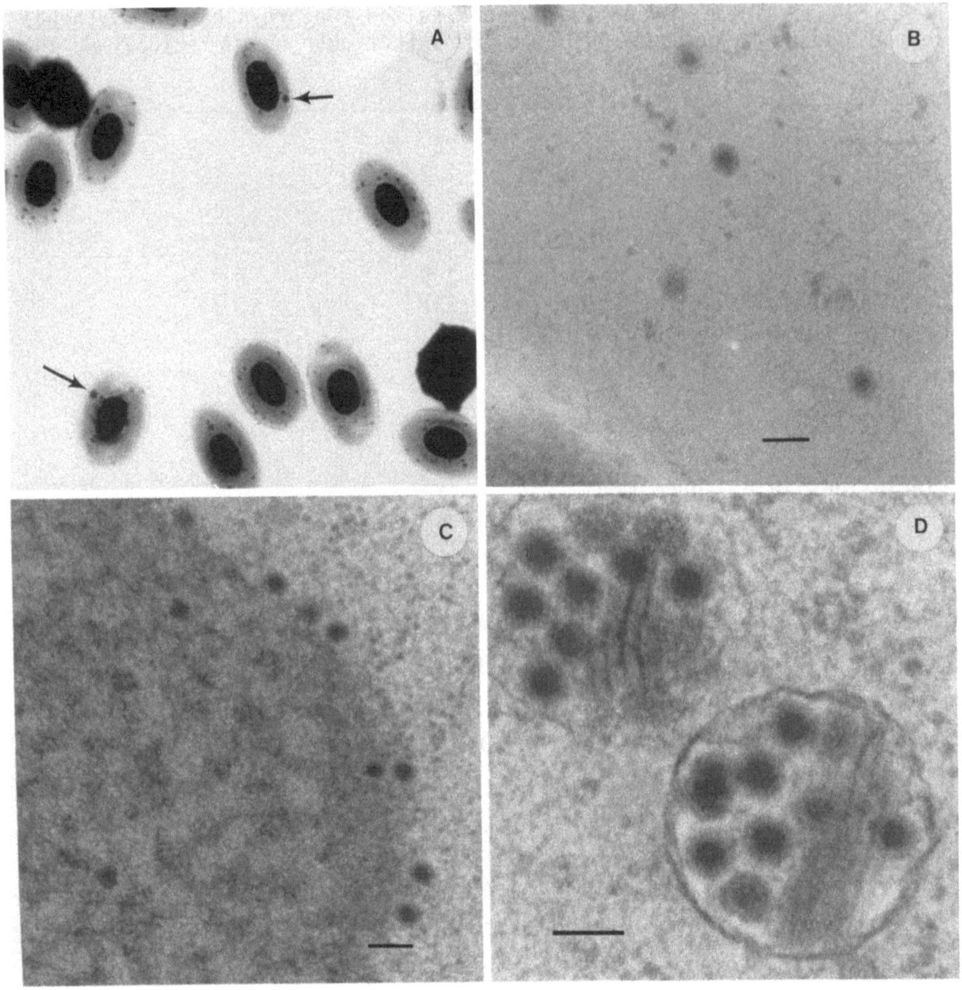

Figure 1. Salmon erythrocytes infected with the EIBS virus: (A) pinacyanol chloride stain of RBC showing typical inclusions (arrows); (B) virions in the cytoplasm of an erythrocyte; (C) virions associated with an electron-dense material that may represent the inclusion body; (D) virions within membrane-bound organelles. Bars represent 100 nm.

Table 2. Blood values for yearling chinook salmon infected by intraperitoneal injection of the virus causing infectious anemia (EIBS).

	CONTROL			INJECTED		
	WEEK 2	WEEK 4	WEEK 6	WEEK 2	WEEK 4	WEEK 6
Number sampled	20	20	20	20	20	20
Fork length (cm)	10.5	12.0	12.0	11.0	11.5	12.5
Weight (g)	13.2	18.0	20.0	13.1	16.7	18.7
RBC (x $10^6/mm^3$)	1.2	1.2	1.3	1.2	1.1	1.1
WBC (x $10^4/mm^3$)	1.2	1.7	1.5	1.7	1.8	1.6
WBC:RBC	1:53	1:43	1:52	1:41	1:38	1:40
Hematocrit (%)	44	46	46	40	42	46
Hemoglob (g/100ml)	7.7	8.1	8.6	7.0	7.4	7.9
Mean Corp Vol (g/dl)	363	392	370	359	410	434
Mean Corp Hemoglob (pg)	63	69	69	62	72	74
Mean Corp Heme Conc (%)	17	17	19	18	18	17
Osmotic Frag (% NaCl)	0.35	0.35	0.35	0.30	0.35	0.35
Lymphocytes (/100 WBC)	56	43	ND	71	44	ND
Neutrophils (/100 WBC)	4	3	ND	8	5	ND
Thrombocytes (/100 WBC)	40	54	ND	21	51	ND
Inclusions	-	-	-	+	+	+

ND = Not done

Typical EIBS inclusions were seen in the RBC of coho salmon injected with homogenates of infected kidney tissue and blood passed through 0.45 and 0.22 um pore-size membrane filters. This helps confirm a viral etiology for the disease. No inclusions were seen in blood cells of control fish or in fish receiving chloroform treated homogenates. This suggests the virion contains essential lipids and probably has an envelope. Infectivity was reduced, but not abolished, following heating at 60 C for 1 hr (Table 3).

Table 3. Transmission of EIBS by intraperitoneal injection of filtered, heated, or chloroform-extracted homogenates of kidney tissue or blood from infected fish into healthy coho salmon.

INJECTED MATERIAL	INCLUSIONS
Untreated homogenate	++
0.45 um filtrate	++
0.22 um filtrate	++
Heat treated	+
Chloroform treated	-
Control	-

\- = no inclusions observed
\+ = less than one inclusion per field
++ = greater than one inclusion per field

DISCUSSION

Although anemias have been observed for several years among stocks of salmonid fish reared at hatcheries in the western portion of North America, the "erythrocytic inclusion body syndrome" was recognized as a distinct disease only recently. In the initial reports of EIBS, Holt and Rohovec (1984) and Leek (1987) were careful to differentiate the virus particles seen by electron microscopy in fish suffering EIBS from the virus causing viral erythrocytic necrosis (VEN). While VEN occurs widely among marine fish including salmonids (Evelyn and Traxler 1978, Rohovec and Amandi 1981) and produces a similar erythrocytic anemia, the virus causing the disease is thought to be a member of the Iridoviridae, based on morphology and presumed DNA genome (Reno et al 1978). Our results confirmed that the virus associated with EIBS was morphologically distinct from that causing VEN.

Because we were able to transmit EIBS to healthy coho salmon by injection of bacteria-free 0.22 um filtrates and were able to find similar virus particles in the blood of artificially infected fish, we believe the virus observed by electron microscopy is the causative agent of EIBS. Considering the size and shape of the virion, the presence of an electron dense center, the possession of a lipid envelope, and the association with membranes of internal organelles, the EIBS virus seems most like a member of the family Togaviridae.

Togaviruses are positive sense, single-stranded RNA viruses that replicate in the cytoplasm of infected cells. The virions are spherical particles, 40-70 nm in diameter, and have a lipid containing envelope (Fields et al 1985). There are four genera within the family including serious pathogens of homeothermic vertebrates (yellow fever, rubella). Many members have insect vectors or reservoirs (Fenner et al 1987). No togavirus infections of fish were included in a review by Wolf (1984).

Although our attempt to replicate the EIBS virus in several fish cell lines was not successful, transmission of the disease to healthy coho or chinook salmon was accomplished by intraperitoneal injection of either blood or kidney tissue homogenates from anemic fish. In our study, typical erythrocytic inclusions were observed in red blood cells of artificially infected fish and the virus was seen by electron microscopy. The disease progressed slowly in both coho and chinook with both anemia and inclusion bodies becoming maximal 4 weeks after injection, but never reaching the level of severity seen in natural outbreaks. After 6 weeks, the fish appeared to be recovering from the disease as hematocrits rose and the RBC morphology returned to normal.

Piacentini et al (1988) have artificially infected coho salmon with the EIBS virus and determined a specific and consistent time course for the resulting disease. Fish recovering from EIBS were less susceptible to reinfection and sera from these fish was used to passively immunize other salmon. In their study, the virus could also be transmitted by waterborne exposure. Rainbow (*Salmo gairdneri*) and cutthroat (*Salmo clarki*) were shown to be susceptible.

The pathogenesis of the virus causing EIBS is poorly understood and the impact of the disease on natural populations of coho and chinook salmon is unknown. Landolt et al (1977) described a virus in the erythrocytes of moribund adult rainbow trout suffering hemorrhage, hypoxia, and death. The disease was accompanied by subcapsular hematomas in the liver and free blood in the body cavity. Histologically, a diffuse necrotic hepatitis, eosinophilic intracytoplasmic inclusion bodies in hepatocytes and renal tubular epithelium, disseminated intravascular coagulation, and skeletal muscular dystrophy were noted. By electron microscopy, virions were observed that were morphologically similar to the EIBS virus.

The anemia caused by the EIBS virus occurs primarily in yearling coho and chinook salmon, often with secondary infections of external fungus, cold water disease and bacterial kidney disease. Our results and those of Piacentini et al (1988) suggest that EIBS alone does not cause direct mortality in coho or chinook salmon. Rather, it appears that the anemia weakens the ability of fish to control existing infections with these pathogens or predisposes the fish to new, secondary infections.

ACKNOWLEDGMENTS

The authors thank Dr. Tosh Yasutake, National Fisheries Research Center for providing Figure 1A. This paper is a result of research sponsored by Oregon State University Agricultural Experiment Station, Animal Health and Disease Funds No. 95-113 Section 1433, Project 51. The U.S. Government is authorized to produce and distribute reprints for governmental purposes notwitnstanding any copyright notation that may appear hereon. Oregon State University Agricultural Experiment Station Technical Paper Number 8690.

REFERENCES

Evelyn TPT, Traxler GS (1978) Viral erythrocytic necrosis: natural occurrence in Pacific salmon and experimental transmission. J Fish Res Board Can 35:903-907

Fenner F, Bachmann PA, Gibbs EPJ, Murphy FA, Studdert MJ, White DO (1987) Veterinary Virology. Academic Press

Fields BN, Knipe DM, Channock RM, Melnick JL, Roizman B, Shope RE (eds) (1985) Virology. Raven, New York

Holt R, Rohovec J (1984) Anemia of coho salmon in Oregon. AFS Fish Health Sec Newsl 12:4

Landolt ML, MacMillan JR, Patterson M (1977) Detection of an intra-erythrocytic virus in rainbow trout (Salmo gairdneri). Fish Health News 6:4-6

Leek SL (1987) Viral erythrocytic inclusion body syndrome (EIBS) occurring in juvenile spring chinook salmon (Onchorynchus tshawytscha) reared in freshwater. Can J Fish Aquat Sci 44:685-688

Piacentini SC, Fryer JL, Rohovec JS (1988) Epizootiology of erythrocytic inclusion body syndrome. In: Abstracts, International Conference on Fish Health, Vancouver, Canada p 154

Reno PW, Phillipon-Fried M, Nicholson BL, Sherburne SW (1978) Ultra-structural studies of piscine erythrocytic necrosis (PEN) in Atlantic herring (Clupea harengus harengus). J Fish Res Board Can 35:148-154

Rohovec JS, Amandi A (1981) Incidence of viral erythrocytic necrosis among hatchery reared salmonids in Oregon. Fish Pathol 15:135-141

Rovozzo GC, Burke CN (1973) A manual of basic virological techniques. Prentice-Hall, New Jersey

Wedemeyer GA, Yasutake WT (1977) Clinical methods for the assessment of the effects of environmental stress on fish health. USFWS Tech Paper #89, Washington, D.C.

Wolf K (1984) Fish viruses: Their biology, classification, hosts, pathology and control. In: Kurstak E, Marusyk RG (eds) Control of Virus Diseases, Marcel Dekker, New York

Yasutake WT (1986) Standardization of stain used for diagnosing erythrocytic inclusion body syndrome (EIBS). AFS Fish Health Sec Newsl 15:7

Section IV Immunology of Fish

Detection and Quantification of Anti-Viral Immunoglobulin of Rainbow Trout Serum

T. Watanabe
Laboratory of Fish Pathology
Department of Veterinary Medicine
Nihon University
Kameino 1866, Fujisawa 252
Japan

Viral diseases have been the most significant disease for salmonid fish culture for more than 20 years, since infectious pancreatic necrosis (IPN) was reported by Wolf et al. in 1960. IPN and infectious hematopoietic necrosis (IHN) (Amend and Chambers, 1970) has also been a great menace to salmonid fish culturists in Japan. Immunological methods have been used to detect and identify viral disease successfully. The antibody of immunized and infected fish sera has also been investigated for the purpose of vaccine development. Recently, enzyme-linked immunosorbent assay (ELISA) has been used frequently for the detection and measurement of fish antibodies against pathogens (Bortz et al. 1984; Kodama et al. 1985; Hamilton et al. 1986). We developed the method for the purification of fish immunoglobulin (Ig) using affinity chromatography (Watanabe and Suzuki, 1986). In this presentation, we describe ELISA for the quantification of rainbow trout *Salmo gairdneri* Ig and fluorescein antibody method for the detection of anti-virus antibody of salmonid fish sera. And we also discuss the possibility of quantification of anti-virus Ig in fish sera.

Materials and Methods

Sera: Blood was obtained from 700-2,000 g rainbow trout cultured in a commercial fish farm. Blood was clotted at 4°C for 24 h and centrifuged at 3,000 rpm for 15 min. Sera were stored at -80°C until use.

Preparation of anti-rainbow trout Ig rabbit IgG: Rainbow trout Ig was purified from serum by using an affinity chromatocolumn filled with anti-rainbow trout (RT) Ig rabbit IgG conjugated Sepharose 4B as described previously (Watanabe and Suzuki, 1986). Rabbits were immunized with 1-2 mg of RT-Ig by repeated injections at intervals of 10 days (4 times). Sera were collected on the 7th day of the last injection. These anti-RT-Ig sera were salted out with 18% (weight/volume) sodium sulfate and centrifuged at 3,000 rpm for 30 min. The precipitate was dissolved in distilled water and dialyzed to 0.07 M phosphate buffer (pH 6.3). Then, anti-RT-Ig rabbit IgG was purified by applying it to a DEAE-Sephacel chromatocolumn at 4°C.

Conjugation of peroxidase with Fab': Anti-RT-Ig rabbit IgG (18 mg) was digested with 1 mg of pepsin at 37°C for 16 h. After stopping the reaction with 1N NaOH, the reactant was passed through Sephacryl S-200 with 0.1 M borate buffer (pH 8.0). F(ab')2 fractions were pooled, concentrated, dialyzed to 0.1 M acetate buffer (pH 5.0), and reacted with 10 mM 2-mercaptoethylamine at 37°C for 90 min. Fab'-SH (5 mg) was mixed with 6 mg maleimide ester of horseradish peroxidase (HRP) and incubated at room temperature for 60 min as described by Yoshitake et al. (1982). After centrifugation at 10,000 rpm for 10 min, the supernatant of the reaction mixture was applied to Sephacryl S-200 chromatocolumn with 0.1 M phosphate buffer (pH 6.5). HRP labeled Fab' (HRP-Fab') was obtained in the fractions with the largest molecular weight of elutions.

Conjugation of FITC to rabbit IgG: Anti-RT-Ig rabbit IgG (10-20 mg/ml) in 0.25 M carbonate buffer (pH 9.0) was incubated with 1 mg of fluorescein isothiocyanate (FITC) at 4°C for 24 h. FITC conjugated rabbit IgG (FITC-IgG) was purified by passing through a Sephadex G-25 chromatocolumn with phosphate buffered saline (PBS).

ELISA: Wells of microtiter plates were coated with 0.20 μg/ml of anti-RT-Ig rabbit IgG in PBS at room temperature for 2 h. After incubation with 1% ovalbumin and washed with 0.1% Tween 20 supplemented PBS (PBS-T), 50 μl of serially diluted rainbow trout serum or purified RT-Ig in PBS-T was added to wells, incubated at room temperature for 1 h, and washed with PBS-T. Then, 50 μl of HRP-Fab' (1 μg) were added to each well. Plates were incubated at room temperature for 2 h and washed with PBS-T for 3 times, and then, 50 μl of o-phenylenediamine solution (0.4 mg/ml in 0.1 M citrate - 0.2 M phosphate buffer [pH 5.0] containing 0.006% hydrogen peroxide) was added. After 30 min at 25°C, the reaction was terminated with 2 M sulfuric acid. The absorbance at 492 nm was monitored photometrically.

Fluorescein antibody method: Normal, IHNV infected, or IPNV infected YNK cells (Watanabe et al. 1978) were grown on glass coverslips. Two days later, the cells were fixed in cold acetone, and stored at -20°C until use. Serially diluted rainbow trout serum was applied to the cells for 30 min at room temperature. After washed with PBS, the cells were overlaid for another 30 min with FITC-IgG solution. After washed with PBS, the cells were examined with a fluorescence microscope. The values of anti-virus titer were expressed as the reciprocal of highest dilution of sera showing a specific fluorescein in virus infected cells.

Fluorescein-linked immunosorbent assay (FLISA): FLISA for the measurement of anti-virus rainbow trout Ig was performed under the method of modified ELISA. Wells were coated with IHNV solution, which was purified with 15-60% sucrose density gradient centrifugation, and post-coated with ovalbumin. Wells were incubated with diluted rainbow trout serum as described above. FITC-IgG was added to the wells preceding incubation in a dark box for 1 h. After washing with PBS, FITC-IgG was combined with RT-Ig resulting in an antigen-antibody reaction that was excited at 490 nm and emitted light was measured photometrically.

Results and Discussion

The quantitative measurement of Ig concentration was performed using the sera obtained from rainbow trout with healthy appearance. There is neither external nor internal abnormality in used fish. Seasonal change of rainbow trout serum Ig was examined on 4 groups of sera obtained from 3-year old rainbow trout in April, June, September, and December. As shown in Figure 1, the sera of female fish just after spawning (December) showed significantly higher Ig concentration ($p<0.05$ and 8.65 ± 2.33 mg/ml) than other 3 groups showing the Ig concentration of 2.50 ± 1.20 mg/ml (April), 4.58 ± 2.34 mg/ml (June), and 2.84 ± 1.17 mg/ml (September), respectively. Artificial and natural stress may increase in spawning season, especially in commercial hatcheries. Then, the relationship between serum Ig concentration and sex maturation and/or spawning requires further investigation.

When comparing between 2-year old fish and 3-year old fish, the serum Ig concentrations of 2-year old fish are significantly higher ($p<0.05$) than those of 3-year old fish as shown in Figure 2. The mean of serum Ig concentration of 3 groups (60 fish), which had significantly lower Ig concentration, is 4.36 ± 3.34 mg/ml. This value is higher than that of yellowtail serum Ig reported by Matsubara et al. (1985) and lower than that of human serum Ig (Calvanico, 1984).

Fluorescein antibody method is one of the most sensitive methods for the detection of antibody against pathogenic viruses. We have been detecting and investigating the anti-IHNV and anti-IPNV antibody in rainbow trout sera for several years using this method. As shown in Figure 3A, the cytoplasm of IHNV-infected YNK cells, near a nucleus, was stained intensively with rainbow trout serum. Figure 3B shows the cytoplasm of IPNV-infected cells also stained with rainbow trout sera (anti-IPNV antibody positive).

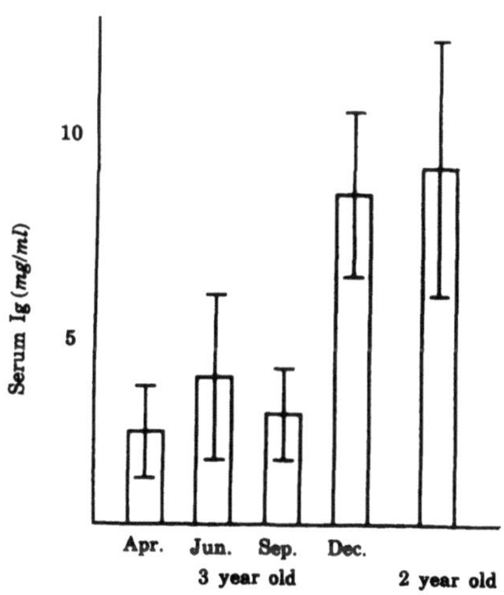

Fig. 1. Seasonal change of rainbow trout serum Ig concentration

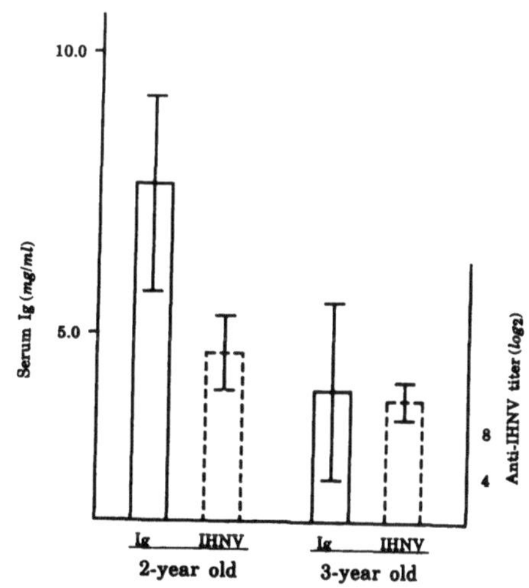

Fig. 2. Serum Ig concentration (——) and anti-IHNV titer (---) of rainbow trout

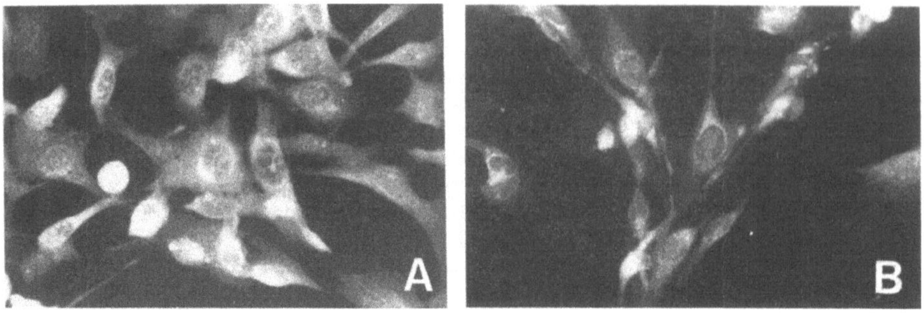

Fig. 3. Immunofluorescent photomicrograph of IHNV-infected (A) and IPNV-infected (B) YNK cells stained with a 2-year old rainbow trout serum

The sera of 2-year-old fish, which had two or three times higher Ig concentration than those of 3-year-old fish, had also 4-8 times higher anti-IHNV titer as shown in Figure 2. The anti-IPNV titer were very low and most of them were 1:4 or less in both groups.

We tried to measure the serum concentration of anti-IHNV Ig of 2-year-old rainbow trout, the sera of which had a high titer of anti-IHNV antibody. FLISA was performed using IHNV virions purified with sucrose density gradient centrifugation as antigen. The preliminary study showed that the 2-year-old rainbow trout serum, with the highest anti-IHNV antibody titer (1:256), had 66 μg/ml anti-IHNV Ig. The result indicates that anti-IHNV Ig of the serum enhanced the serum Ig level, and that this fact alone could not explain the high Ig concentration of 2-year-old fish sera. Further investigation is required on antibody against bacterial and fungal diseases.

In this work, we reported (1) the measurement of serum Ig concentration of rainbow trout using ELISA, (2) the detection of anti-virus antibody using fluorescein antibody method, and (3) the possibility of measurement of anti-IHNV Ig in serum using FLISA. ELISA and FLISA are very useful methods for detecting and quantitating antibody or Ig against a pathogen. These methods require larger amounts of antigen and enzyme (or fluorescein) conjugated antibody than the fluorescein antibody method. The purification of viral antigen also takes much time. The recovery rate of IHNV from tissue culture supernatant is not so high that large amounts of cultures are necessary to perform ELISA or FLISA. Further investigations and efforts are required to resolve these difficulties.

References

Amend DF, Chambers VC (1970) Morphology of certain viruses of salmonid fishes. I: In vitro studies of some viruses causing hematopoietic necrosis. J fish Res Board Can 27, 1285-1293

Bortz BM, Kenny GE, Pauley GB, Garcia-Ortigoza E, Anderson DP (1984) The immune response in immunized and naturally infected rainbow trout (*Salmo gairdneri*) to diplostomum as detected by enzyme-linked immunosorbent assay (ELISA). Dev Comp Immunol 8, 813-822

Calvanico NJ (1984) Structure and function of immunoglobulins in "Molecular Immunology (eds. Atassi MZ, van Oss CJ, Absolom DR)". Marcel Dekker Inc, New York & Basel, pp 141-174

Hamilton AJ, Fallon MJM, Alexander J, Canning EU (1986) A modified enzyme-linked immunosorbent assay (ELISA) for monitoring antibody production during experimental *Aeromonas salmonicida*-infection in rainbow trout (*Salmo gairdneri*). Dev Comp Immunol 10, 443-448

Kodama H, Honda A, Moustafa M, Mikami T, Izawa H (1985) Detection of antibody in rainbow trout against *Aeromonas salmonicida* by enzyme-linked immunosorbent assay. Fish Pathol 20, 237-242

Matsubara A, Mihara S, Kusuda R (1985) Quantitation of yellowtail immunoglobulin by enzyme-linked immunosorbent assay (ELISA). Bull Japan Soc Sci Fish 51, 921-925

Watanabe T, Suzuki Y (1986) Purification of fish immunoglobulin using affinity chromatography. Bull Japan Soc Sci Fish 52, 1283

Watanabe T, Kobayashi N, Sato Y, Ishizaki Y (1978) Continuous cell line derived from the kidney of Yamame, *Oncorhynchus masou*. Bull Japan Soc Sci Fish 44, 415-418

Wolf K, Sniesko SF, Dunber CE, Pyle E (1960) Nature of infectious pancreatic necrosis virus in trout. Proc Soc Exp Biol Med 104, 105-108

Yoshitake S, Imagawa M, Ishihara E (1982) Efficient preparation of rabbit Fab'-horseradish peroxidase conjugates using maleimide compounds and its use for enzyme immuno-assay. Analytical Letters 15, 147-160

Change of Immunological Indices in Carps (*Cyprinus carpio L.*) Following Rhabdovirus Carpio Injection

Y.D. Temnikhanov, I.A. Balakhnin, E.V. Kompanets, V.V. Litvinenko, and G.A. Strilko
Ministry of the USSR Fishing
Ukrainian Research Institute of Fishing Industry
Ukrainian SSR Academy of Sciences
Insitute of Zoology
Kiev
USSR

At present, it is obvious that spring viremia of carp (SVC) is a viral disease inflicting considerable damage to fish farming, especially to carp breeding (Lobuntsov 1986; Lobuntsov and Rudikov 1979). Despite controversial viewpoints regarding the etiology of this pathology, a number of authors agree that it is triggered by a virus and then developed due to bacteria acting as a secondary infection (Osadchaya 1984; Osadchaya and Prosyanaya 1984; Osadchaya et al. 1977). By lysing the cells, the virus causes pathological changes in organs and tissues of fish which create favourable conditions for secondary microflora development (Rudikov 1985). It is established that 1-year-old carps infected by live virus acquire immunity, but unlike adult species, their defence reaction is not often dependent on neutralizing activity of serum antibodies (Fijan et al. 1977; Fijan et al. 1977). In the natural environment, viral infection often leads to the extinction of part of the stock, because the temperature optimum of viral replication is lower than the temperature range favourable for the production of antibodies (Ahne 1980; Ahne 1984).

As to the antibacterial defence, the carps react to parenteral administration of antigen from *Aeromonas hydrophila* by producing the immunity based on the presence in serum of high-titre agglutinins which protect the carp from infection by live *Aeromonas* culture (Lobuntsov 1986).

Observations prove that after contracting the virus, the fish become more susceptible to bacterial infections. Here, the mechanism of the disease onset is largely monitored by stress factors (Wedemeyer et al. 1976).

Today, the concept of anti-infectious immunity in fish has not been adequately studied. In this connection it is necessary to investigate a number of immunological indices of the humoral immunity in healthy and in virus infected carps prior to and after adding rhabdovirus in order to further develop the necessary immunopreventive measures against viral, bacterial and virobacterial infections.

Materials and Methods

The study involved 2-year-old specimens of *Cyprinus carpio* weighing about 250 g obtained in autumn, 1987, from the breeding farm of the 5th Kiev Heat and Power Plant (122 pcs). These were divided into 2 groups, i.e. those fish after having had the disease and having ulcers on the body surface (100 pcs) (1) and clinically sound species (22 pcs) (2). Before the experiment, the fish were adapted for 2 weeks in plastic baths (1 m^3, 15-16°C) with a continuous supply of aerated water without feeding. Both groups prior to and after rhabdovirus administration were assayed for antibodies against rhabdovirus (indirect hemagglutinin reaction using a specially developed erythrocytic diagnosticum) (Litvinenko and Osadchaya 1986), natural antibodies against erythrocytes of the O(1) human-donor group (Balakhnin and Davydov 1988; Vikhman 1979), and bacterial activity of blood serum (BABS) using a nephelometer (Smirnova and Kuzmina 1966). Physiological conditions were estimated in fish serum by refractometric determination of the protein level; the glucose level was established by the orthotoluidine method (Kushmanova and Ivchenko 1983) and survival rate of species during the experiment (120 days) measured. The fish were infected by rhabdovirus (*Rhabdovirus carpio*-strain KT-120 b 30 passage) obtained from SVC carps in 1979. The virus with a titre of log 10 6.5 TCD_{50}/ml was transmitted intraperitoneally by 0.5 ml/specimens. Initially, blood samples in fish were taken after 15 days from delivery to the laboratory. Next, blood sampling was repeated 15 days after rhabdovirus administration. Statistical analysis of the results was made according to standard methodology (Rokitsky 1973).

Results

The comparison of blood serum indices from 2-year-old healthy and SVCD carps has shown that before rhabdovirus administration, both groups differed significantly in the frequency of antiviral antibody occurrence (see Table 1). E.g., the share of species having the above antibodies among SVCD carps was one order higher than in healthy fish (51.5% and 4.5%, respectively); moreover, the antibody titres among SVCD carps varied in the range of $\log_2 =$ 0-7. On the one hand, antibody titres among healthy carps varied insignificantly ($\log_2 = 0$-1), and on the other hand, incidence of natural heterohemagglutinins in healthy carp blood serum was 2.3 times higher than in those species which had the disease. According to other parameters (BABS, protein and glucose levels), there were no significant differences between the compared groups prior to rhabdovirus administration.

After transmission of rhabdovirus to healthy carps, the incidence of specimens having antiviral antibodies in the given group increased more than tenfold. At the same time maximum titres among some of the healthy fish increased from $\log_2 = 1$ to $\log_2 = 4$, whereas the incidence of natural antibodies and BABS among those groups of fish did not change significantly. The titre of natural antibodies among fish of both groups varied in the range $\log_2 = 0$-5. The only parameter revealed was the decrease of protein and glucose levels in serum of the fish, which agrees with the data from literature (Lukianenko 1971).

Another situation was observed in SVCD carps after administration of the virus in the same dose. The share of the specimens showing antiviral antibodies in this group dropped from 51.5% to 27.6%; accordingly, there is a noticeable decrease of the maximum titres among these fish from $\log_2 = 7$ to $\log_2 = 3$. The analysis has

Table 1. Immunological characters of carp prior to and after rhabdovirus administration

Fish groups	Indices (M ± m)					
	Share of natural antibodies (%)	BABS (%)	Protein (%)	Share of antibodies against virus (%)	Glucose level mg (%)	Survival rate, share (%)
Healthy species	$77.27 \pm 5.35^{+++}$ (22)	30.78 ± 2.30 (21)	$1.75 \pm 0.14^{+++}$ (21)	$4.54^{+} \pm 4.43^{+++}$ (22)	$104.29 \pm 5.88^{+++}$ (21)	100
	77.27 ± 5.35 (22)	27.00 ± 14.07 (10)	$0.78 \pm 0.10^{+++}$ (21)	$52.63^{+} \pm 11.47$ (19)	$47.50 \pm 5.98^{+++}$ (20)	100
SVCD	$33.68 \pm 4.85^{+++}$ (95)	$42.73 \pm 2.05^{+++}$ (84)	$2.11 \pm 0.16^{+++}$ (22)	$51.48 \pm 4.97^{+++}$ (100)	90.69 ± 4.49 (29)	100^{+++}
	33.00 ± 8.37 (30)	0^{+++} (29)	$0.73 \pm 0.07^{+++}$ (29)	27.58 ± 8.30 (29)	-	$32.67 \pm 8.17^{+++}$

Note: + $P < 0.05$; +++ $P < 0.001$; number of examined fish in parentheses

shown that 18 initially positive fish yielded only 5 (27.7%) specimens which retained antibodies. Here, only 3 (27.3%) of 11 refractory carps acquired antibodies.

BABS analysis gave particularly interesting results. Thus, unlike the healthy group, the effect of the virus on SVCD carps led to a complete loss of the initial BABS levels by sera of these fish. At the same time, the group of SVCD carps manifested a significant threefold decrease of the survival rate, while healthy carps infected by the virus remained alive during the entire experiment. Most specimens showed a significant reduction of the protein and glucose level by the end of the experiment.

Discussion

The growing incidence of species with acquired antirival antibodies among healthy fish is a good alternative in comparison to SVCD carps which, on the contrary, manifest a decrease in the share of specimens with antibodies against rhabdovirus. The latter circumstance is explained by the neutralizing ability of the virus with respect to antiviral antibodies. The immunosuppressive effect of the viral antigen observed in fish having the corresponding antibodies indicates the need for preliminary investigation of fish populations aimed at detecting antibodies in their sera against the virus prior to undertaking therapeutic and preventive measures.

Of particular interest is the fact that the bacterial properties are lost by blood sera of SVCD carps after infection with the virus and a reduction of the survival rate of these specimens occurs. According to the data from the literature, the BABS level and survival rate of carps in hibernation are connected by a positive relation, and a high initial level of BABS may ensure a high survival rate, if this index remains relatively constant (Mikryakov 1980; Mikryakov 1984). Our findings prove that in viral infection the BABS level often decreases to zero and, eventually, this causes the death of most specimens.

The fact that the surviving carps previously infected with the virus finally showed zero BABS indices may only prove that the survival rate of carps, apart from BABS, may also be explained by a number of other resistance factors.

Conclusions

In most healthy carps, intraperitoneal administration of rhabdovirus results in the occurrence of antiviral antibodies. In diseased carps, the same procedure causes extinction of the specimens having the above antibodies.

In most cases the effect of rhabdovirus on carps, not yet recovered from the disease, is followed by a complete loss of the bacterial activity of blood serum, eventually leading to the extinction of most of the specimens.

In the future, when developing and using biopreparations (such as live viral vaccines), it is important to consider not only the physiological state, but also to examine carps for the availability of antiviral antibodies in blood.

References

Ahne W (1984) Important viral diseases in European fish cultures. "Fish, pathogens and environment in European polyculture", Szarvas Hungary, 3-15

Ahne W (1980) Spring viremia of carp (SVC). Studies on immunization of carp. Fish diseases. III COPRAQ Session, Springer-Verlag Berlin, 26-27

Balakhnin IA, Davydov ON (1988) Protective action of natural antibodies in 1-year-old carps during invasions. Gidrobiol Zh 24, 39-43 (In Russian)

Fijan N, Petrinec Z, Stancl Z, Teskeredzic E (1977) Vaccination of carp against spring viraemia: Comparison of intraperitoneal and peroral application of live virus to fish kept in ponds. Bull Off Int Epiz 87, 441

Fijan N, Petrinec Z, Stancl Z, Dorson M, Leberre M (1977) Hyperimmunization of carp with *Rhabdovirus carpio*. Bull Off Int Epiz 87, 439

Kushmanova OD, Ivchenco GM (1983) Guideline for laboratory works on biochemistry. M Meditsina, 272 (In Russian)

Litvinenko VV, Osadchaya YF (1986) Development of erythrocytic diagnosticum and indirect hemagglutination for estimation of antibody accumulation level in rabbit and fish serum against *Rhabdovirus carpio*. Rybnoye khozyaistvo. Current report, series: Economic utilization of internal water reservoirs. Iss M, TsNIITERKh 5, 6-13 (In Russian)

Lobuntsov KA (1986) Results and prospectives of fish bacterial disease analysis. Profilaktika, lechenie i diagnostika infektsionnykh boleznei rib. M, Gosagroprom SSSR, 56-61

Lobuntsov KA, Rudikov NI (1979) Etiologic structure of infectious dropsy in carp. Trudy VIEV 49, 146-152 (In Russian)

Lukianenko VI (1971) Fish immunobiology. M, Pishchevaya promyshlennost, 1-364 (In Russian)

Mikryakov VR (1984) Regularities of immune system functions in fresh-water fish. Author's abstract of Doc Sci (Biol) Thesis, M, AN SSSR, 37 (In Russian)

Mikryakov VR, Grechanov IG (1980) A method for estimation of survival rate in fish during herbination. Author's Certif No 709042, Bull Goskomizobreteny 2, 9 (In Russian)

Osadchaya YF (1984) Fish diseases induced by rhabdoviruses in the Ukraine. "Fish, pathogens and environment in European polyculture", Szarvas Hungary, 36-47 (In Russian)

Osadchaya YF, Prosyanaya VV (1984) Results of bacteriologic and virologic studies of pathologic material from carps with infectious dropsy. Rybnoye khozyaistvo Kiev 19, 104-107 (In Russian)

Osadchaya YF, Prosynaya VV, Rudenco AP (1977) Infectious dropsy of carp as a complex virobacterial pathology. Rybnoye khozaistvo M, MRKh 6, 31-34 (In Russian)

Rokitsky PF (1973) Biological statistics. Minsk, Vyssaya shkola, 1-315 (In Russian)

Rudikov NI (1985) Viruses and viral pathologies in fish. Achievements of science and technology. Ikhtiologia, M, VINITI 1, 6-92 (In Russian)

Smirnova OV, Kuzmina TA (1966) Bacterial activity estimation of blood serum using photonehlometry. ZhMEI, M, Meditsina 4, 8-11 (In Russian)

Vikhman AA (1979) On quantitative estimation of normal hemagglutinins in organs and tissues of carp. Fish diseases and their control. M, MRKh, 69-81 (In Russian)

Wedemeyer GA, Meyer FP, Smith L (1976) Environmental stress and fish diseases. By TFH Publications, Inc, Itd, 1-192

Immunotechnology and Viral Diseases of Fish

D.P. Anderson
U.S. Fish and Wildlife Service
National Fish Health Research Laboratory
Route 3, Box 700
Kearneysville, WV 25430
USA

INTRODUCTION

The techniques of immunology and serology have become more sophisticated and are greatly influencing how we diagnose and control viral diseases in fishes. In the last few years we have seen more sensitive and quicker ways to perform assays, making possible diagnostic surveys for detecting the presence of small amounts of antigen or antibody in large populations. In addition, new developments in vaccine preparation and administering techniques promise to bring immunization against viral agents inco regular use in fish farms and aquaculture facilities.

This review gives a brief summary of the state-of-the-art of fish viral diagnosis and immunization.

DIAGNOSIS OF VIRAL PRESENCE

Diagnosis of viral fish disease agents at present is mostly done by characterization of plaque formation on different cell line monolayers. Serological procedures, while often used in research, are limited in routine use by the accessibility to specific antisera (Ahne 1981). Cell lines needed for optimal culture of viral agents are also being developed from different fish species (Nicholson 1987). The recent development of the labeled antibody assays, beginning with the fluorescent antibody techniques and enzyme labeled immunosorbent antibody (ELISA), are especially promising (Kurstak 1986), and have been used in fish research and diagnosis (Bortz et al. 1984). Optimizing cell cultures to grow virus particles quickly enables suspect samples to be infected on the cell lines and growth can be seen within a few days (Swanson, 1981). Questions always arise when infections of viruses in fish are very low. Antibody amplification .by anti-antibodies and enzyme building such as biotin-avidin techniques are becoming standardized on a commercial basis, what is needed from the commercial groups are specific antibody to the viral agents. Goat antibody specific for the Renibacterium salmoninarum and to trout immunoglobulin are now available commercially through Kirkegaard and Perry Research Laboratories, Inc., Rockville, MD. Also BioMed, Inc. Bellevue, WA offers antisera to selected major bacterial fish pathogens.

Hybridoma cell lines producing monoclonal antibody have been made against most major fish disease pathogens such as IPNV (Wolski et al. 1986), IHNV (Winton et al. 1988) and VHSV (McAllister and Owens, 1988) and are being used for research purposes. The advantages of using the monoclonals as tools for detecting disease agent antigens include the greater specificity of the antibody against defined epitopes and the ability to hold the cell lines for retrieval to again work with the same antibody. The complete specificity is a disadvantage sometimes, however, as polyclonal antibodies also have the ability to cover more idiosyncrasies of antigenic variation. Hybridomas and antibody-producing cells from fish or other animals can be frozen for later use in verification of antigenic similarities (Dixon et al. 1987).

IMMUNIZATION

Immunization of fish against viral diseases is a possibility for commercial application in the near future. Presently formalin-killed vaccines give promising results with infectious pancreatic necrosis virus (Sano et al, 1981). The attentuated vaccines are effective also with viral hemorrhagic septicemia (de Kinkelin et al. 1981) and spring viremia of carp. The release of living viral agents into the field draws criticism from some fish pathologists who fear return to virulence and/or recombination in field environments with other viral strains.

Immunization of fish against viral diseases will follow research from medical and veterinary research and is primarily dependent upon economical feasibility. While killed and attenuated vaccines have proven successful, barriers remain. Recombinate vaccines using genetic engineering techniques may become available in the future enabling bulk antigen production and greater purity. Already infectious hematopoietic virus genome segments have been isolated.

IMMUNOSTIMULATION

Presently we are investigating the stimulation of the immune response by adjuvants, immunostimulants and biological response modifiers. FK-565, a product from Japan (Kitao et al, 1986) has shown promise in elevating the immune response, particularly in the opsinization of the cellular activity. Grayson et al. (1987) have worked with the immunopotentiating agents Bacille Calmette Guerin (BCG) and saponin Quil-A to stimulate the immune response in immersion vaccination of rainbow trout to Yersinia ruckeri bacterins. Presently we are using rainbow trout in experiments testing levamisole, an immunostimulant used in treatment of ruminants for helminth infections, and known to stimulate the nonspecific defense mechanisms in carp (Siwicki 1987).

In some cases, stimulation of the nonspecific defense mechanisms may be sufficient by such drugs as levamisole, so that inclusion of immunogens against the specific agent may not be necessary, especially if that agent is not exactly defined. For instance, Vibrio sp. are notorious for strain differences, and there is still some uncertainty about cross protection in fish immunization to single strains when exposed to variants. Immunostimulants may be effectively used under these conditions.

CONCLUSIONS

Problems remain in diagnosis of viral agents when the infectivity rate and antigenic load is low. In these cases more sensitive assays become used, and as a result, there is an increased risk of confusion of cross reactions with similar antibodies or antigens. In addition to missing antigens because of low concentrations, antigenic variation may occur as mutations arise in the environments and the specific antibody probes may not recognize new variants.

The large AIDS vaccine research programs promise to add information on new approaches to immunization against viruses. All avenues of immunization are being investigated, including the use of live attenuated or recombinant viruses, whole inactivated preparations, synthetic peptides, natural and recombinant DNA products, anti-idiotype antibodies, and passive immunization (Koff and Hoth 1988).

At present, no drugs are cleared in the United States for use in treating viral infections in food fish. Certainly, some of the medical and veterinary drugs would be effective against fish viral agents, but procedures for approval are necessarily thorough and expensive.

Problems in the future will arise as fish are intensively cultured and eggs, sperm and other products are shipped throughout the world. The spread of viral fish agents is controlled by governmental regulations on shipping of suspected viral materials. As fish farming and aquaculture become more popular, shipping routes are more difficult to control, and enforcement on a practical basis is tenuous. Also, unrecognized microorganisms causing no disease problems in indigenous environments may be transported by fish or fish products and result in serious disease outbreaks in virgin fish populations.

Responsible vigilance concerning the spread of viral agents must continue. Immunotechnology will give use better diagnostic methods for detecting fish viral disease agents and developing immunogens for inducing disease protection.

The initial research and methodology may be complex, requiring trained personnel and committed investments, however the benefits will be great.

FISH VIRAL IMMUNOTECHNOLOGY

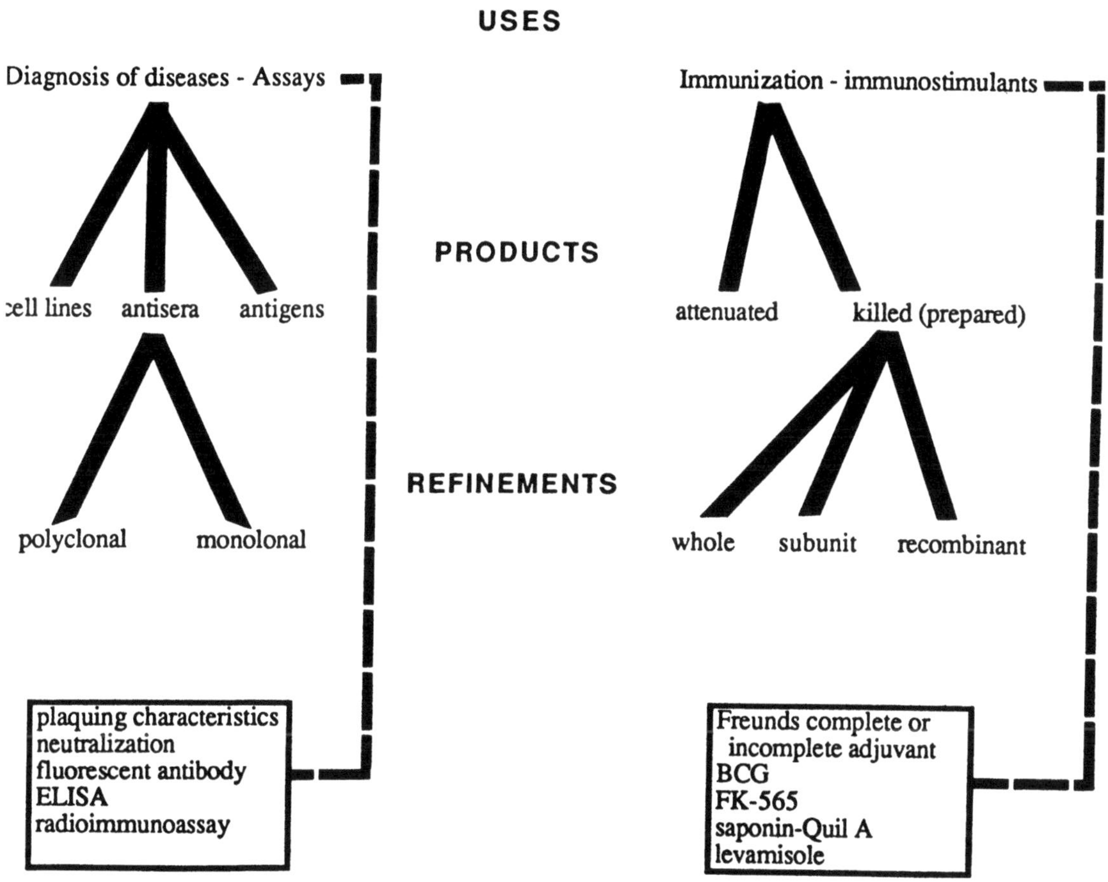

References

Ahne W (1981) Serological techniques currently used in fish virology. Devel Biol Std 49:3-27

Bortz BM, Kenny GE, Pauley GB (1984) The immune response in immunized and naturally infected rainbow trout (Salmo gairderni) to Diplostomum spathaceum as detected by enzyme-linked immunosorbent assay (ELISA). Devel Comp Immun 8:813-822

de Kinkelin P, Bearzotti M (1981) Immunization of rainbow trout against viral hemorrhagic septicemia (VHS) with a thermoresistant variant of the virus. Devel Biol Std. 49:431-439

Dixon OW, Anderson DP, Kimenai R, Lizzio EF (1987) Longevity of rainbow trout antibody-producing cells after in vitro immunization culture In Kuroda Y, Kurstak E, Maramorosch (eds) Invertebrate and fish tissue culture Japan Scien Soc Press Tokyo 228-231

Grayson TH, Williams RJ, Wrathmell AB, Munn CB, Harris JE (1987) Effects of immunopotentiating agents on the immune response of rainbow trout, Salmo gairdneri Richardson, to ERM vaccine. J Fish Biol. Supp A:195-202

Koff WC, Hoth DF (1988) Development and testing of AIDS vaccines. Sci 241:426-432

Kurstak E (1986) Enzyme immunodiagnosis. Academic Pr NY 235p

McAllister PE, Owens WJ (1988) Monoclonal antibody produced against viral hemorrhagic septicemia virus. Res Info Bull US Fish & Wildlife 88-46

Nicholson BL (1987) Fish cell cultures: an overview. In Kuroda Y, Kurstak E, Maramorosch K (eds) Invertebrate and fish tissue culture. Japan Scien Soc Press Tokyo 191-194

Sano T, Tanaka K, Fukuzaki S (1981) Immune response in adult trout against formalin killed concentrated IPNV. Devel Biol. Std. 49:68-70

Siwicki A (1987) Immunomodulating activity of levamisole in carp spawners Cyprinus carpio L. J Fish Biol Supp A:245-246

Swanson RN (1981) Use of the indirect fluorescent antibody test to study the pathogenesis of infectious pancreatic necrosis virus infection in trout. Devel Biol. Std. 49:71-77

Winton JR, Arakawa CK, Lannan CN, Fryer JL (1988) Neutralizing monoclonal antibodies recognize antigenic variants among isolates of infectious hematopoietic necrosis virus. Dis Aquatic Organisms 4:199-204

Wolski SC, Roberson BS, Hetrick FM (1986) Monoclonal antibodies to the Sp strain of infectious panreatic necrosis virus. Vet Immun Immunopath 12:373-381

Immunostimulation by Levamisole in Rainbow Trout (*Salmo gairdneri*) in Vivo

D.P. Anderson, A.K. Siwicki[1], O.W. Dixon, and E.F. Lizzio[2]
U.S. Fish and Wildlife Service
National Fish Health Research Laboratory
Route 3, Box 700
Kearneysille, WV 25430 USA

INTRODUCTION

The use of immunostimulants for boosting the defense mechanisms and protection against diseases in fish is of increasing interest to fish culturists. Several promising adjuvants, drugs and biological response modifiers have been tested in experiments with fish. Evelyn (1974) injected Freunds complete adjuvant with formalin-killed Renibacterium salmoninarum and found an elevated antibody titer compared to control fish. Kitao et al (1986) showed that the phagocytic activity was increased by the drug, FK-565, and later (Kitao et al. 1987) showed that it also caused a slight elevation of numbers of antibody-producing cells and serum antibody titers when given with specific bacterins. BCG (Bacille Calmette Guerin) extract and saponin Quil-A also have been shown to stimulate the nonspecific immune response in rainbow trout (Grayson et al. 1987).

Levamisole, the levo-isomer of tetramisole, has long been recommended in veterinary practice as an anthelmintic for ruminants. It has been shown to stimulate the precursor T-lymphocyte compartment of the immune system, and is also known

---

[1]Inland Fisheries Institute in Olsztyn
Department in Zabieniec near Warsaw
05-500 Piaseczno, Poland

[2]Immunohematology Branch
Division of Blood and Blood Products
Office of Biologics
Food and Drug Administration
Bethesda, MD 20205 USA

to increase macrophage engulfment activity as well as raise neutrophil enzyme levels (Desiderio et al 1986). Siwicki (1987) presented data showing that when carp (Cyprinis carpio) are fed levamisole, the nonspecific defense mechanisms are stimulated and the fish are less susceptible to disease outbreaks in field challenges.

In these studies, we continue the experiments with levamisole on rainbow trout placed on selected time regimens, giving levamisole before, with or after injections of Y. ruckeri O-antigen or DNP-Ficoll. The nonspecific defense parameters, including the phagocytic indexes, neutrophil enzyme activities, numbers of adherent cells and percent of adherent cells are assayed. Changes in the specific immune response are monitored by the passive hemolytic plaque assay enumerating numbers of antibody-producing cells. Consideration is given to the possible use of administering levamisole to trout during stressful periods or combining levamisole with specific immune regimens to heighten the nonspecific defenses and specific immune protection.

MATERIALS AND METHODS

Animals

Young rainbow trout, stock hatchery fish, weighing 20 to 30 g were transported to the National Fish Health Research Laboratory-Leetown, Kearneysville, West Virginia, from White Sulphur Springs (WV) National Fish Hatchery and held in 340 l fiberglass tanks fed with spring water (11-13°C). The chemistry of the water has been described by Warren (1963). The fish were fed a commercial pelletized dry feed (#PR 11-30, Ziegler Brothers, Inc. Gardners, PA[/1]). The trout were 200-250 g when these experiments were initiated.

[/1]Reference to trade name or supplier does not imply Government endorsement of a commercial product.

Antigens

The Yersinia ruckeri O-antigen was prepared in our laboratory as given in Anderson et al (1979). The DNP-(dinitrophenyl)-Ficoll was obtained from Biosearch, Inc. and assayed to average 38 DNP haptens per molecule on the long chain carbohydrate. Both antigens have been used extensively in fish immunization experiments (Anderson et al 1979, 1982).

Immunostimulant

Levamisole was obtained from Pittman-Moore, Inc. The product was diluted in phosphate buffered saline for injection at 5ug/fish.

Experimental design

In the first experiment, 5 groups of 5 fish each were placed on separate time regimens of administration of the immunostimulant in relation to the single injection of 10ug of the Y. ruckeri O-antigen. Levamisole doses were 5ug at each injection time. Group I was injected with the antigen, then given injections of levamisole 2, 4 and 6 days later; Group II was given one injection of levamisole simultaneously with the antigen; Group III was injected 6, 4 and 2 days before the antigen; Group IV was injected with antigen only; Group V was a negative control (no immunostimulant, no antigen).

In the second set of experiments, 100ug of DNP-Ficoll was used for each antigen injection and 5 ug for each levamisole injection. The immunization regimens for groups I, II, III and IV were as above; Group V was injected with levamisole 6, 4 and 2 days before phosphate buffered saline injection, Group VI was a negative control.

Assays of nonspecific defenses and specific immune response

Fish were sampled on day 10 after antigen exposure, and the spleen cell suspensions were made by scraping the organ through a wire grid. After filtering out debris through a glass wool matrix, the cells were washed once by centrifugation at 1000 rpm and aliquoted for the different assays. The nitro blue tetrazolium (NBT) assay was done according to Sigma (1978); briefly, 0.1 ml cell suspension was mixed with the NBT reagent, held for 20 min in separate microtiter wells, then transfered with a plastic-tipped pipette to a glass microscope slide where the suspension was allowed to dry. Slides were examined within 24 hrs for dark granules in neutrophils, indicative of formate-positive cells. Aliquots of 0.1 ml were also mixed with 0.1 ml of standardized suspension of *Staphlococcus aureus* for the phagocytic indexes and adherent cell assays. Slides were as above air dried, stained with safranin, and read within 24 hrs (van Oss 1973).

The passive hemolytic plaque assay, counting the numbers of antibody-producing cells, was done according to Anderson et al. (1979). Again, 0.1 ml of the cell suspension was mixed with the melted agar and 0.1 ml of DNP-labeled or *Y. ruckeri* O-antigen labeled sheep red blood cells. Numbers of leukocytes were counted in the cell suspension within six hrs after fish had been sampled.

RESULTS

The results from the assay of the nonspecific defense parameters in relationship to the timing of the injection of the immunostimulant and the antigen are given in Tables 1 and 2. In summary, Groups I and II, which were given levamisole either after or with the antigen injections, responded the highest in all nonspecific defense parameters and had the highest specific immune responses. If levamisole was given before the antigen (Group III), the nonspecific defense parameters were moderately elevated, whereas the numbers of specific antibody-producing cells were greatly reduced. Both of the antigens injected without the addition of the immunostimulant increased the nonspecific parameters, but the specific immune response was lower than the first two groups. Levamisole given alone stimulated some background immune activity against the DNP-Ficoll antigen as shown in Figure 2, Group V. Standard deviation is given for the nonspecific assays; standard error for the specific assays.

Nitroblue Tetrazolium Assays

Neutrophils staining positive for dark granules of cytoplasmic formate were tabulated by percent of positive cells per microscopic field. Groups I and II had comparable results, showing the potency of the levamisole when given before or with the antigen. The NBT assay results from administering the immunostimulant before the antigen (Group III) were comparable to those of injecting the immunostimulant or antigen alone. Noteworthy are the results of antigen or levamisole administered alone in comparison to the low NBT numbers from negative control fish.

Phagocytic Indexes

Spleen phagocytic cells were incubated with *Staphylococcus aureus* for information concerning nonspecific engulfment activities. Safranin-stained cells and bacteria were examined by microscope. The results from each group follow the patterns set by the NBT assay. Groups I and II were elevated, and Groups given the levamisole before the antigen were comparable to those given the levamisole or antigens alone.

Adherence Indexes and Percent Adhering Cells

S. aureus cells adhering to splenic leukocytes were viewed by microscope, counted and the percent calculated. The results confirmed the patterns above, demonstrating the importance of post or simultaneous injection of immunostimulant and antigen for a heightened total response.

Specific Immune Response

The passive hemolytic plaque assay was used to demonstrate numbers of splenic antibody-producing cells (PFC) from each fish sampled. The results confirmed the potency of the DNP-Ficoll in comparison to the *Y. ruckeri* O-antigen for rainbow trout. The specific immune assays also showed the abrogration of the numbers of PFC if the antigen injection was preceded by 3 injections of levamisole (Group III in both experiments).

Table 1. Nonspecific defense parameters from rainbow trout immunized with Yersinia ruckeri O-antigen and/or injected with levamisole (LV) on different immunization regimens. Group I, 3 injections of LV after antigen injection; Group II, LV and antigen injected at same time; Group III, 3 injections of LV before antigen injection; Group IV, injected with antigen only; Group V, negative controls.

Fish group	Percent NBT positive cells	Phagocytic index	Adherence index	Percent adherent cells
I	43.8 ± 3.7	17.4 ± 0.7	16.0 ± 1.1	61.2 ± 3.4
II	48.4 ± 2.7	17.9 ± 0.5	16.4 ± 0.9	61.6 ± 3.0
III	37.2 ± 2.9	13.8 ± 0.9	13.6 ± 0.6	54.4 ± 7.3
IV	36.2 ± 1.9	13.3 ± 0.8	12.1 ± 0.6	53.6 ± 4.8
V	31.3 ± 3.6	11.0 ± 0.2	10.3 ± 0.7	43.0 ± 2.0

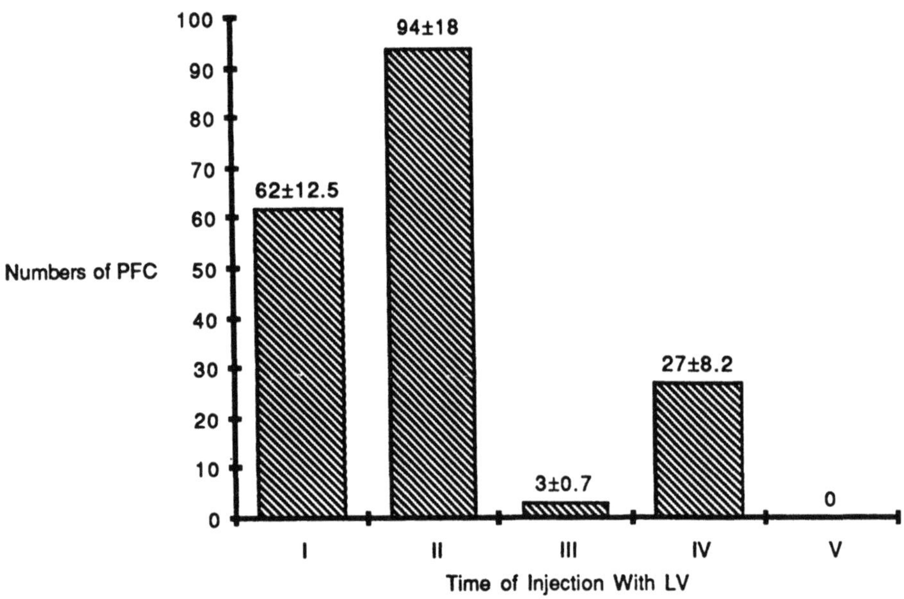

Figure 1. Numbers of PFC from groups of trout injected with Yersinia ruckeri O-antigen and/or levamisole (LV). Group I, 3 injection of LV after antigen injection; Group II, antigen and LV injected simultaneously; Group III, 3 injections of LV before antigen injection; Group IV, injection of antigen only; Group V, negative control.

Table 2. Nonspecific defense parameters from rainbow trout immunized with DNP-Ficoll and/or injected with levamisole (LV) on different immunization regimens. Group I, 3 injections of LV after antigen injection; Group II, Lv and antigen injected at same time; Group III, 3 injections of LV before antigen injection; Group IV, injected with antigen only, Group V, injected with LV only; Group VI, negative controls.

Fish group	Percent NBT positive cells	Phagocytic index	Adherence index	Percent adherent cells
I	67.4 ± 4.4	18.5 ± 0.6	17.2 ± 0.3	79.2 ± 2.4
II	63.2 ± 2.8	16.6 ± 0.3	15.7 ± 0.4	71.0 ± 3.2
III	54.0 ± 4.0	14.4 ± 0.4	13.8 ± 0.5	65.0 ± 4.6
IV	49.6 ± 1.5	13.0 ± 0.4	13.0 ± 0.6	60.4 ± 2.7
V	52.0 ± 3.7	13.3 ± 0.4	13.4 ± 0.5	62.8 ± 5.7
VI	32.4 ± 5.7	9.8 ± 0.5	9.1 ± 0.6	46.0 ± 2.7

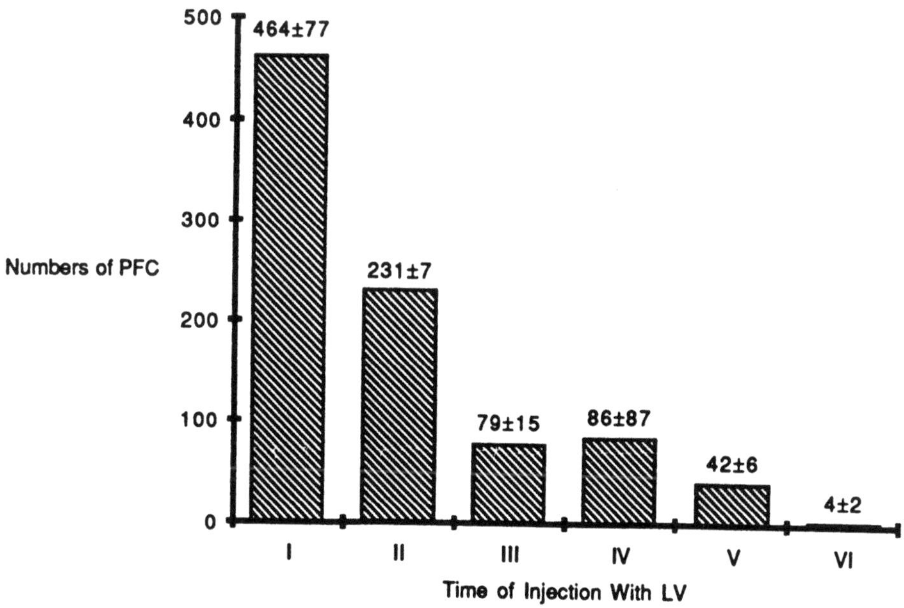

Figure 2. Numbers of PFC from groups of trout immunized with DNP-Ficoll and/or levamisole (LV). Group I, 3 injections of LV after antigen injection; Group II, LV and antigen injected simultaneously; Group III, 3 injections of LV before antigen injection; Group IV, injection of antigen only; Group V, 3 injections of LV before PBS; Group VI, negative control.

In contrast, there was an increase in the specific immune response if the levamisole was given after or combined with the antigen injection.

DISCUSSION

Levamisole acts on the nonspecific defense mechanisms of the rainbow trout as shown by the increased phagocytic activity, elevated neutrophil enzyme production, and observations of more nonspecific particle adherence to the leukocytes when the drug is administered after or with the antigen. These parameters were also moderately elevated if the immunostimulant was given alone. The specific immune system was elevated with the addition of levamisole with or after antigen injection. However, if injections of levamisole at doses used here were given 6, 4 or 2 days prior to the immunogens, the immune response was greatly reduced as assayed by counting the numbers of PFC. These results were confirmed by using either DNP-Ficoll or *Y. ruckeri* 0-antigens. The numbers of specific PFC were always higher in these experiments with the DNP-Ficoll antigen, and indeed, this is also reflected in the nonspecific defense assays. Both of these antigens were active stimulators of the nonspecific assays used here with *S. aureus*, even when given without levamisole. This information is valuable for the eventual practical application of the immunostimulants in inducing protection of fish against diseases in fish farms and aquaculture facilities.

On some Polish farms, levamisole is presently experimentally used to protect carp (*Cyprinus carpio*) broodfish during stressful periods (Siwicki, 1987). If a diagnosis of a state of immunosuppression is made by utilizing the nonspecific assays as were described here, the immunostimulant is injected or fed to the fish. Similar procedures may be recommended for salmonid culture in the future.

Levamisole is presently approved in the United States for veterinary use as a treatment of ruminants for helminth infections. We may consider the use of levamisole alone or have the immunostimulant integrated with immunization regimens for stimulation of the nonspecific defense mechanisms and specific immune protection.

Acknowledgements

Part of this work was financed by the Food and Agriculture Organization of the United Nations, Project No. POL. 86/044/A/01/12. We also thank Ms. Jenni Keith, University of Washington, School of Fisheries, for graphics and proofreading.

References

Anderson DP, Roberson RS, Dixon OW (1979) Cellular response in rainbow trout Salmo gairdneri Richardson to Yersinia ruckeri O-antigen monitored by passive haemolytic plaque assay test. J Fish Dis 2:169-178

Anderson DP, Roberson RS, Dixon OW (1982) Immunosuppression induced by a corticosteroid or an alkylating agent in rainbow trout (Salmo gairdneri) administered a Yersinia ruckeri bacterin. Devel Comp Immun Supp 2:197-204

Desiderio JV, Rankin BA (1986) Immunomodulators. In Kirk KW (ed) Current Veterinary Therapy IX. W.B. Saunders Phila 1093-1097

Evelyn TPT (1971) The agglutinin response in sockeye salmon vaccinated intraperitoneally with a heat-killed preparation of the bacterium responsible for salmonid kidney disease. J Wildlife Dis 7:328-335

Grayson TH, Williams RJ, Wrathmell AB, Munn CB, Harris JE (1987) Effects of immunopotentiating agents on the immune response of rainbow trout, Salmo gairdneri Richardson, to ERM vaccine. J. Fish Biol. Supp A:245-246

Kitao T, Yoshida T (1986) Effect of an immunopotentiator on Aeromonas salmonicida infection in rainbow trout (Salmo gairdneri). Vet Immun Immunopath 12:287-296

Kitao T, Yoshida T, Anderson DP, Dixon OW, Blanch A (1987) Immunostimulation of antibody-producing cells and humoral antibody to fish bacterins by a biological response modifier. J Fish Biol Supp A:87-91

Sigma Chemical Company (1978) Nitroblue tetrazolium (NBT) reduction in neutrophils. Sigma Technical Bulletin No.840, St. Louis, MO. 13 pp.

Siwicki AJ (1987) Immunomodulating activity of levamisole in carp spawners *Cyprinus carpio* L. J Fish Biol Supp A:245-246

van Oss CJ, Fuji H, Wicher K, Rabin B, Kite J (1973) Phagocytosis. *In* Rose NR, Bigazi PE (eds) Methods in Immunodiagnosis. John Wiley & Sons NY 157-159

Warren JW (1963) Kidney disease of salmonid fishes and the analysis of hatchery water. The Prog Fish Cult 25:121-131

Nonspecific Defense Mechanisms and Specific Immune Protection of Trout Against Viral Agents

D.P. Anderson and A.K. Siwicki*

U.S. Fish and Wildlife Service
National Fish Health Research Laboratory
Route 3, Box 700
Kearneysville, WV 25430 USA

Introduction

Fish protect themselves against invasive disease agents with nonspecific defense barriers and mechanisms and a specific immune response. The nonspecific defense barriers already in place include physical epithelial shield of the scales, skin, and the mucus. If an infectious agent is entrapped by the mucopolysaccharide complexes of the mucus, it may be sluffed from the fish or held to be digested by the mucous lytic enzymes. If the agent does breach these initial barriers, the serum lysozymes, lectins and complement components may coat the particles and opsinize it for further destruction (Fletcher, 1981).

Other factors of the nonspecific defense system may be activated by invading stimuli (Ellis 1981). Phagocytic cells, including the tissue macrophages, circulatory monocytes and neutrophils, adhere to and engulf invasive particles. In most cases, the microorganisms are destroyed by digestive and lytic enzymes at this point; however, inflammation may result at a tissue-damaged site, resulting in the migration of leukocytes to wound areas and the elevation of serum components, including C-reactive protein and ceruloplasmin. Viral agents may induce the production of interferon in fish, a product giving a warning response from invaded cells to surrounding cells for the initiation of mechanisms that will

---

*Inland Fisheries Institute
Zabieniec near Warsaw
05-500 Piaseczno, Poland

make further invasion and proliferation of the virus difficult (Dorson and de Kinkelin, 1974; de Kinkelin and Dorson, 1982). Interferon, produced by viral-infected cells, alarms other cells to activate intracellular endonucleases that will destroy viral DNA and RNA and will shut down physiological pathways that viruses use for replication.

The specific immune response, including the production and release of antibody into the circulation and the opsinization and activation of phagocytic cells, T-lymphocytes, and natural killer cells, requires a longer period of time to reach full effectiveness after fish have been exposed to invasive materials or actively immunized. Different species of fish are widely divergent in which organs are involved (Anderson et al. 1987). In this paper we are detailing some results from experiments with fish injected with Infectious Pancreatic Necrosis Virus (IPNV) and following the nonspecific parameters of neutrophil enzyme levels, phagocytic indexes, adherence indexes, and percent adherent cells. The specific immune response to viruses and bacteria is also discussed. By following the kinetics of these parameters, we may be able to anticipate and interrupt the course of the viral diseases in fish.

Table 1. Nonspecific defense barriers and mechanisms in trout.

In place	Respond to stimulus
skin	phagocytic cells
scales	neutrophil enzyme release
mucus	C-reactive protein
mucopolysaccharides	interferon
lytic enzymes	natural killer cells
serum	
complement	
lectins	
lytic enzymes	

Nonspecific response studies

Our studies followed the kinetics of the nonspecific response parameters after the injection of 1×10^8 units of IPNV into 250g rainbow trout (Salmo gairdneri). We measured the response of neutrolphil enzyme levels by the nitro blue tetrazolium test, the phagocytic response by calculating phagocytic indexes, and adherence of leukocytes to bacterial cells by adherence indexes and percent adherent cells. Fish were sampled before and on days given in Figure 1 by first excising the spleen and making a cell suspension. Since whole spleens were used, and the amount of cells/spleen relatively consistent, it was easiest to draw 0.1 ml samples from stock 0.5 ml suspension and counting numbers of leukocytes was done within 24 hours. For example 0.1 ml of the spleen suspension was mixed with 1×10^8 Staphylococcus aureus and allowed to incubate for 20 min at room temperature in a plastic microtiter well. Likewise 0.1 ml could be taken from this spleen stock and mixed with 0.1 ml NBT stock solution. Samples were then taken from the appropriate wells and placed on slides for observation by microscope within 24 hours. The results show that the phagocytic activity was not remarkably changed in marked contrast to the NBT tests and the adherent cell indexes and percent adherent cells (Figures 1 & 2). This is in marked contrast to that shown with fish injected with bacteria which show an increase in phagocytic activity (Ellis, 1981).

Specific immune response

A specific immune response of rainbow trout is illustrated in Figure 3; showing the numbers of antibody-producing cells (APC) after injection with with Yersinia ruckeri bacterin, patterns from other studies. Several notable studies have followed the kinetics of antibody-production and protection after injection of viral vaccines (Sano et al. 1981, de Kinkelin and Dorson 1982), and the same patterns emerge. The procedure for demonstration of APC by the passive hemolytic plaque assay, essentially follows the first steps given for

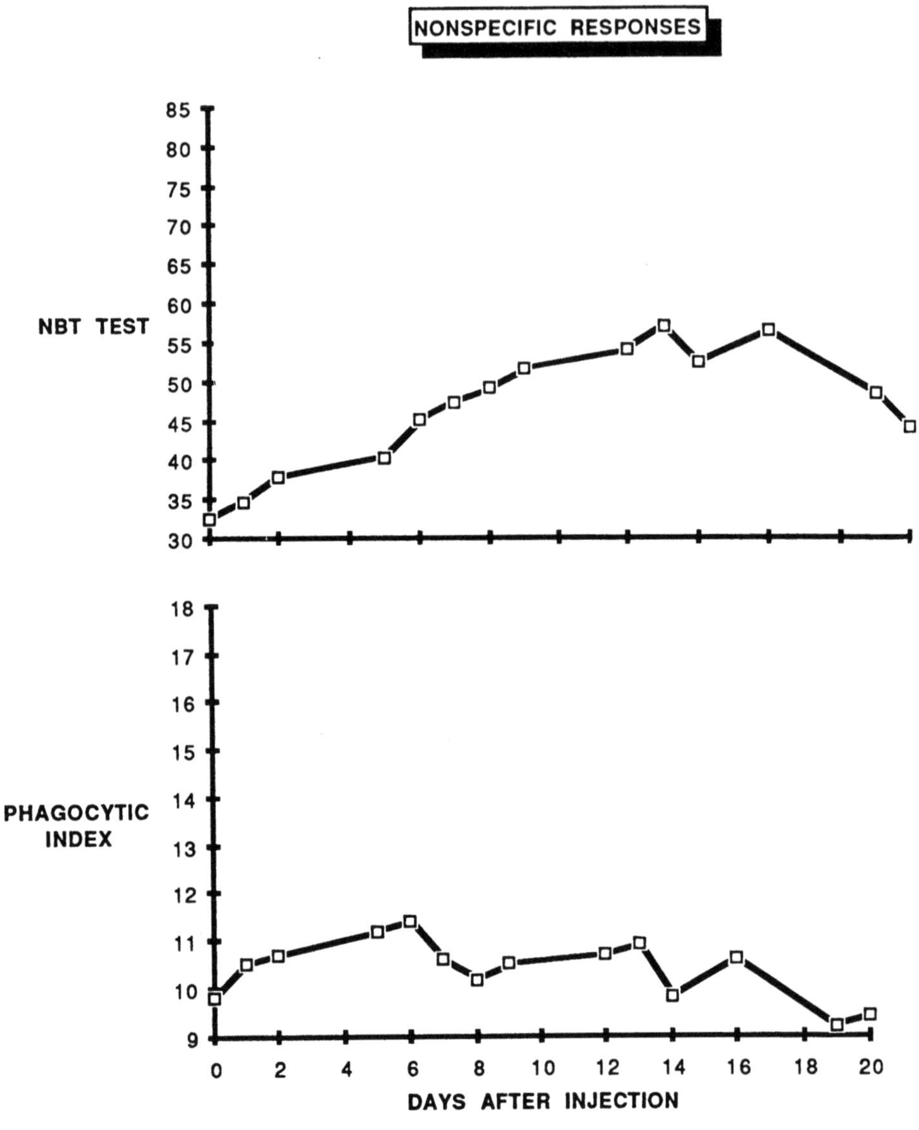

Figure 1. Rainbow trout injected with Infectious Pancreatic Necrosis Virus show the nonspecific responses as measured by the nitroblue tetrazolium assay (top) and phagocytic index (bottom).

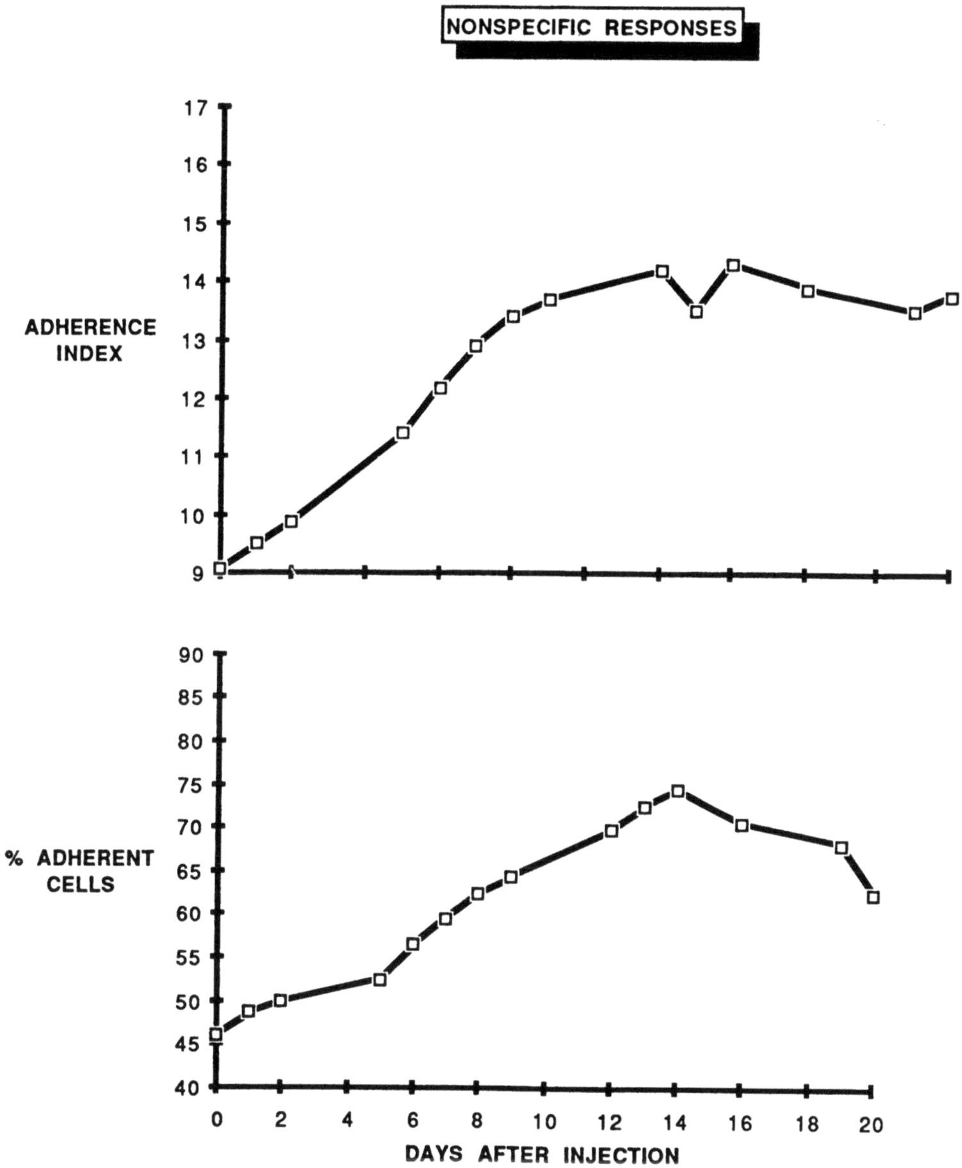

Figure 2. Rainbow trout injected the Infectious Pancreatic Necrosis Virus show the nonspecific defense responses as measured by the numbers of leukocytes showing bacterial (*Streptococcus aureus*) adherence (top) and the percent of bacterial adherent cells (bottom).

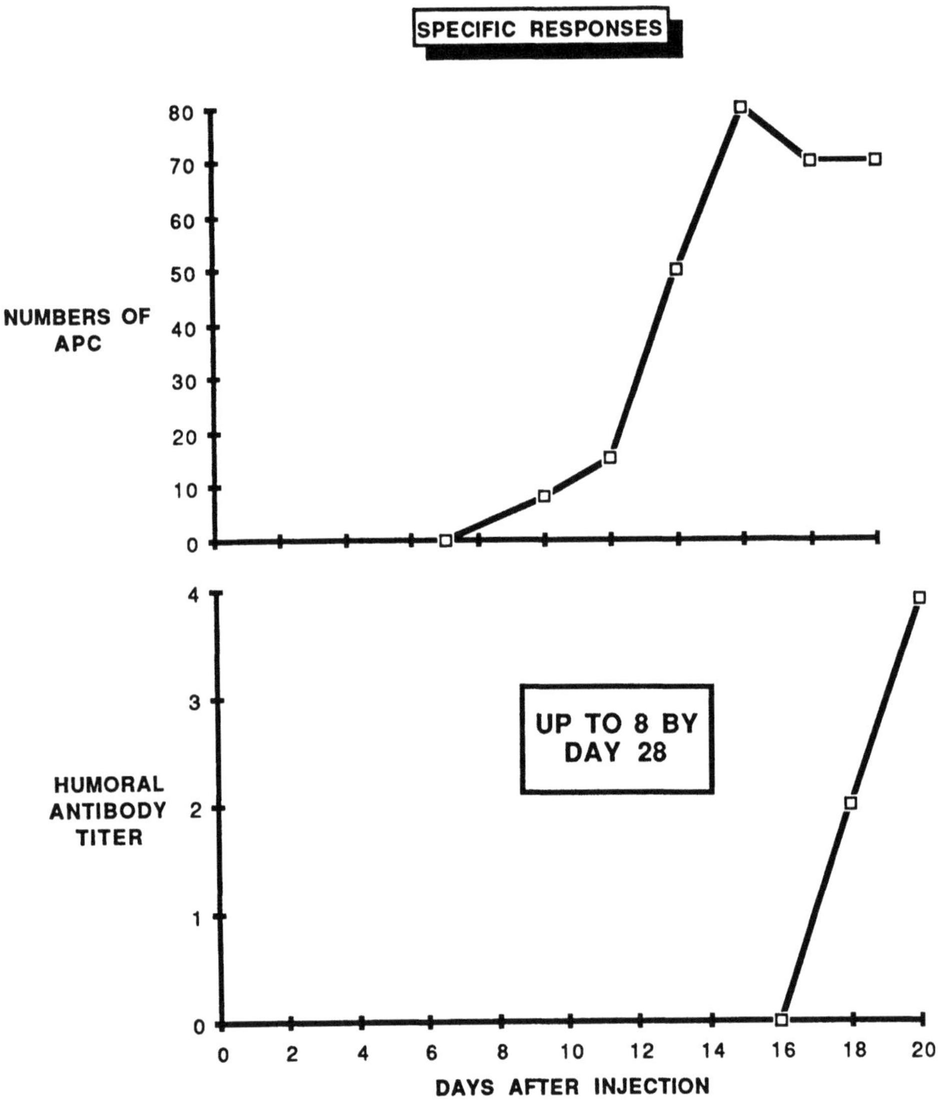

Figure 3. Rainbow trout injected with *Yersinia ruckeri* bacterin gives the specific immune response as measured by the rise in the numbers of antibody-producing cells (top) and the appearance of humoral antibody titers (bottom).

the nonspecific assays. An aliquot of 0.1 ml of a spleen cell suspension is mixed with 0.1 ml of antigen-labeled sheep red blood cells (SRBC) and 0.2 ml of warm, liquid agarose in culture media, then set to harden for a complement overlay to show the lysis of the labeled SRBC surrounding the specific APC. Kinetics of the antibody released into the serum is shown in the Figure 3.

Conclusions

Monitoring basic health parameters in fish farms and aquaculture facilities is now done by using physical growth patterns and possibly taking blood samples for hematocrits and differential smears. Following the nonspecific defense mechanisms of neutrophil enzyme levels and phagocytic and adherence indexes may be desirable. Divergence from recognized healthy patterns by these parameters could warn of viral infections to the managers so that alternative strategies may be necessary to improve the health of the fish. The NBT assay, phagocytic and adherence indexes as outlined here are simple and quick to set up; however, some experience in reading the microscope slides is necessary.

The lateness of the specific immune response gives reason for the fish to have strong nonspecific defense systems. However, monitoring the numbers of of specific APC and titers of humoral antibody gives credence to immunization programs. While simple agglutination assays are easy for the biologist to do when bacterins are given, doing cell culture and neutralization assays are more complex for following antibody to viral immunizations. The antibody titers are found 2-3 weeks after immunization. The presence of APC by the passive hemolytic plaque assay, gives earlier information. However, labeling SRBC with the IPN virus needed in these assays has not been successfully done.

The nonspecific immune mechanisms can also be monitored to show the effectiveness of immunization in fish. The assays done without the necessarily using specific immunogen; later confirmation can be done by specific antigen-antibody assays.

Indeed by confirming the elevations of the nonspecific defense mechanisms with the later specific immune response, fish managers can be more knowledgeable about the day-to-day health and immune status of their fish.

Acknowledgements

Part of this work was financed by the Food and Agriculture Organization of the United Nations, Project No. POL. 86/044/A/01/12. We thank Ms. Jenni Keith, University of Washington, School of Fisheries, Seattle, WA for graphics and Dr. Philip McAllister, National Fish Health Research Laboratory, Kearneysville, WV for IPNV isolate.

References

Anderson DP, Kitao T, Yoshida T, Dixon OW (1987) Investigations of antibody-producing cells from spleens and anterior kidneys of rainbow trout, striped bass and goldfish. In Kuroda Y, Kurstak E, Maramorosch K (eds) Invertebrate and fish tissue culture Japan Scien Soc Press Tokyo 232-235

de Kinkelin P, Dorson M (1982) Interferon synthesis in trout and carp after viral infection. Devel Comp Immun Supp 2:167-174

Dorson M, de Kinkelin P (1974) Mortality and interferon production in rainbow trout (Salmo gairdneri). In Mawdesley-Thomas (ed) Diseases of fish. Academic Press New York 129-140

Ellis AE (1981) Non-specific defense mechanisms in fish and their role in disease processes. Devel Biol Std 49:337-352

Fletcher TC (1981) The identification of non-specific humoral factors in the plaice (Pleuronectes platessa L.) Devel Biol Std 49:321-327

Sano T, Tanaka K, Fukuzaki S (1981) Immune response in adult trout against formalin killed concentrated IPNV. Devel Biol Std 49:63-70

Modulation of Cell-Mediated Immune Responses by Lymphocystis Disease in the Gilthead Sea Bream *Sparus Aurata*

M. Faisal
Dept. of Avian and Aquatic Animal Medicine
Faculty of Veterinary Medicine
University of Alexandria
Egypt

Introduction

Lymphocystis disease – caused by the iridovirus Fish Lymphocystis Disease Virus, FLDV – is a recurrent slowly progressing dermal infection of teleosts (Flügel, 1985; Wolf, 1984). Despite the viral etiology of this disease had been proposed 6 decades ago by Weissenberg (1920) and later confirmed by Walker (1962), little is known on the virus-host interactions in general and the host immune responses in particular.

Studies on the virus-host relationship in mammals showed a modulatory role of viral infections on both juvenile and antigen-induced immune response (Drummond et al. 1985; Tsutsumi et al. 1986; Yasumoto et al. 1987). It is not known, however, whether the previously mentioned observation also applies to fish. The aim of the present investigation was therefore to study the effect of Lymphocystis disease on 2 cell-mediated immune responses, namely nonspecific cytotoxicity (NCC) and proliferative responses to the mitogens phytohemagglutinin, concanavalin A and *Escherichia coli* lipopolysaccharides.

Materials and Methods

Fish: Normal and infected gilthead seabream *Sparus aurata* L. (Pisces: Spariidae) were caught from El-Anfoushy and El-Max coastal areas of west Alexandria, Egypt. The adult fish (17.4 cm ± 4.3, 170 g ± 20.4) were anaesthetized (200 mg/l MS222, Sandoz) immediately after capture. The heparinized blood was collected by puncture of the caudal vein. Spleens and anterior kidney were obtained under aseptic conditions.

Light and electron microscopy: Nodules of the hypertrophied cells were removed, fixed in 2.5% glutaraldehyde in cacodylate buffer (0.05 M, pH 7.2) and embedded in epon. The semithin sections were stained with toluidine blue; he ultrathin sections with lead citrate and uranyl acetate.

*Present address: Dept. of Microbiology and Immunology (Factor 12-538) School of Medicine, University of California at Los Angeles, Los Angeles, CA 90024, USA

Medium: All media and salt solutions were adjusted to marine teleost isotonicity by the addition of NaCl to 0.2 M as described for marine fish by Hinuma et al. (1980) as well as McKinney and Schmale (1988). Cell cultures were performed in a complete medium (CM) that consisted of RPMI-1640 (Gibco) supplemented with 10% heat inactivated fetal calf serum (FCS) and antibiotics (100 units penicillin, 100 µg streptomycin/ml). For cell washing, Hank's balanced salt solution was used.

Cell preparation: Peripheral blood was diluted 3 times with HBSS, overlaid on Histopaque 1077 (Sigma), centrifuged for 20 min at 450 g (room temperature) and washed 3 times with HBSS. The anterior kidney and spleenic tissues were gently pressed through 80-gauge mesh stainless steel sieves. The resulting cell suspension was further overlaid on Histopaque 1077, centrifuged and cells at the interface were washed 3 times with HBSS and finally suspended in CM. Cell counts and cytospin preparations were performed by standard technique.

Assessment of Nonspecific Cytotoxitity: Cytotoxicity was assessed by the ^{51}Chromium-release assay using the murine cell line YAC-1 (Kiessling et al. 1975) as the target. Chromation of the target cells was achieved by incubating 5×10^6 cells with 100 µCi of $^{51}Cr\text{-}Na_2CrO_4$ for 60°min at 37 C. Following three washes, the target cells were diluted to 10^5 cells/ml in medium, and delivered (100 µl) to each of the wells of a 96-well, round bottom Linbro-plate (Flow Laboratories, McLean, VA). Effector cells were added in a volume of 100 µl at a ratio of 100:1, 50:1, 25:1 and 10:1. The plates were centrifuged (200 x g, 5 min, at room temperature), and incubated for 240 min at 28°C. The incubation was terminated by centrifuging as before. One hundred µl of the supernatant were removed from each well for counting in a Beckman gamma counter. Percent cytotoxicity was computed as:

$$100 \times \frac{(\text{experimental CPM} - \text{spontaneous CPM})}{(\text{maximum release} - \text{spontaneous CPM})}$$

and the data transformed into lytic units (LU_{20}) per 10^6 cells with the aid of the computerized non-linear equation program described by Pross et al. (1981).

Conjugate formation: Effector cells (1×10^4) were incubated with 1×10^5 YAC-1 target cells in 1 ml of CM in 12 x 75 mm glass tubes, sedimented by centrifugation at 200 g for 5 min and incubated for 1 h at 4°C. Pellets were resuspended, and cytocentrifuged smears were stained with Giemsa. Percentage of conjugate forming cells (CFCs) was determined by counting 200 leucocytes (bound and free) in triplicate samples.

Assessment of lymphocyte proliferation: Phytohemagglutinin (PHA, Difco Laboratories, Detroit, MI) (2 mg/ml), concanavalin A (Con A, Pharmacia, Sweden) (1 mg/ml), and lipopolysaccharide from *E. coli* (LPS, Difco) (4 mg/ml) were reconstituted as stock solutions in Hank's balanced salt solution (HBSS, Gibco). Isolated leucocytes (10^5/200 µl/well) were incubated in triplicate in 96-well plates

(Flow) in the presence of mitogen (1 µg/ml PHA; 2.5 µg/ml Con A; 100 µg/ml LPS) for 4 days at 28°C. On day 3, the cultures were supplemented with 1 µCi/200 µl of ^{3}H-thymidine (5 Ci/mM) for 24 h. The cells were harvested onto glass fiber filters, and the radioactivity assessed in a Beckman liquid scintillation counter, using Beckman Ready-Safe (Beckman, Fullerton, CA) as scintillant.

Statistical analysis: Mean and standard deviation were calculated. The statistical analysis was performed using the student-t-test.

Results

Lymphocystis Disease in Sparus aurata Population

Lymphocystis infected fish comprised 9.2% (46 out of 500) of *Sparus aurata* population caught from West Alexandria in December, 1985, and November, 1987. The fish were divided according to the severity of the infection into 3 groups. The first group (21 fish) showed slight infection with small foci of hypertrophic cells on soft fins of caudal or pectoral fins. The second group (17 fish) showed moderate infection with the lymphocystis outgrowths covering between 10-50% of the body surface. In some fishes, dense aggregation of the hypertrophied dermal cells was observed (Fig. 1). The third group consisted of 8 fish that showed clustering of hypertrophied cells that covered a large portion of the entire surface of fins.

Morphological Characteristics

In semithin sections, lymphocystis cells appeared enlarged (200-400 µm in diameter), had a 11-13 µm thick hyaline structureless capsule, a large vesicular nucleus and granulated basophilic inclusion body (Fig. 2). In ultrathin sections, virus particles

Fig. 1. Gilthead seabream *Sparus aurata* infected with Fish Lymphocystis Disease Virus

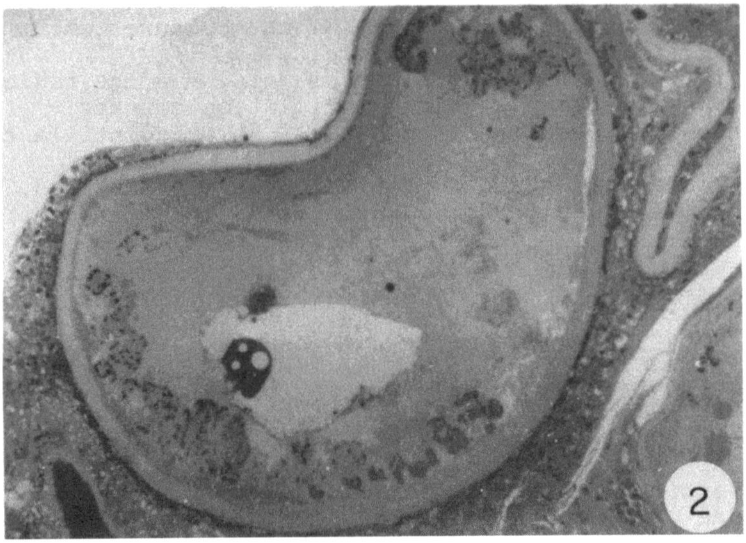

Fig. 2. Semithin section in a lymphocystis cell (x 1,000)

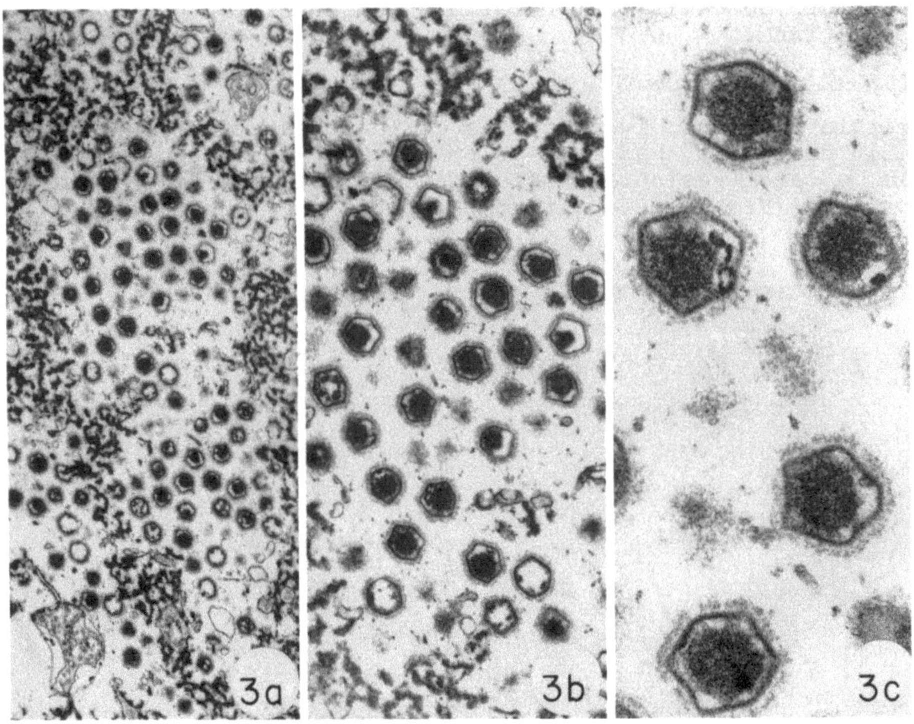

Fig. 3a-c. Virus particles in the cytoplasm of lymphocystic cells. a) x 12,000; b) x 27,000; c) x 58,500

were found scattered in the cytoplasm but not in the nucleus or the inclusion bodies. These particles usually showed a hexagonal outline and, occasionally, a pentagonal shape (Fig. 3a). Diameters measured from 50 regular size virions gave 210 nm between opposite sides or 230 nm between opposite vertices. Empty small (110-150 nm) particles were seldom observed (Fig. 3b). A central electron dense nucleoid (5-9 nm in diameter) was observed in most virus particles. Uniform halos (17-49 nm in thickness) surrounded the virions (Fig. 3c). The virus etiology was confirmed in all of the infected fish.

Nonspecific Cytotoxic Cell (NCC) Activity

Spontaneous chromium release did not exceed 8-15% of the total which indicates the resistance of YAC-1 cells to the hyperosmotic medium. In non-infected *Sparus aurata*, anterior kidney leucocytes (AKLs) showed the most potent cytotoxic activity (38.7 ± 9.3 lytic units at 40% lysis, LU_{40}) followed by spleenic leucocytes (SL) which possessed 19.8 ± 6.4 LU_{40} while peripheral blood leucocytes (PBL) showed minimal figures of 8.9 ± 3.4 LU_{40}. These values, however, changed by infection. In groups 1 and 2, PBLs showed augmented ($p < 0.001$) NCC-activity, AKLs cytotoxic activity was significantly ($p < 0.001$) suppressed, while that of spleen was unchanged (Fig. 4). An overall suppression was observed in heavily infected fish (group 3).

As shown in Figure 5, the % of conjugate forming cells decreased in the anterior kidney by infection ($p < 0.001$) and increased in the peripheral blood ($p < 0.001$). No significant changes were, however, observed with spleenic cells.

Proliferative Responses to Mitogens

Leucocytes of non-infected *Sparus aurata* responded to in vitro stimulation by mitogens. The values of 27 individual fish are expressed as means ± standard deviation displayed in Table 1.

Table 1. Response of *Sparus aurata* leucocytes to mitogens as measured by ^{3}H-thymidine uptake (counts per minute x 10^3)

Tissue	PHA	Mitogen Con A	LPS
Anterior kidney	80 ± 14.8[a]	119.8 ± 23.6	115.7 ± 21.9
Spleen	67.9 ± 10.9	210.4 ± 20.5	94.3 ± 18.9
Peripheral blood	52.4 ± 7.3	90.9 ± 14.3	74.9 ± 11.6

[a]Mean ± standard deviation

As shown in Figure 6, fish in group 3 showed lower ($p < 0.001$) mitogenic responses to PHA and Con A as compared to control uninfected fish. Their response to LPS was, however, unaffected. On the other hand, in groups 1 and 2, responses to PHA and Con A were suppressed ($p < 0.02$) but were surprisingly elevated to LPS ($p < 0.001$).

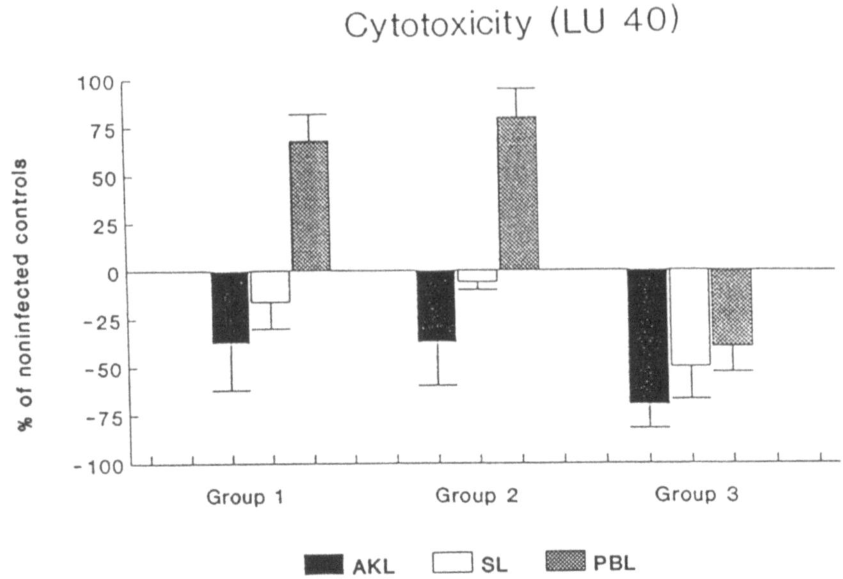

Fig. 4. Modulation of cytotoxic cell activity in different tissues by Lymphocystis Disease. Data (means ± standard deviation) are expressed as percent of the values given by non-infected fish leucocytes

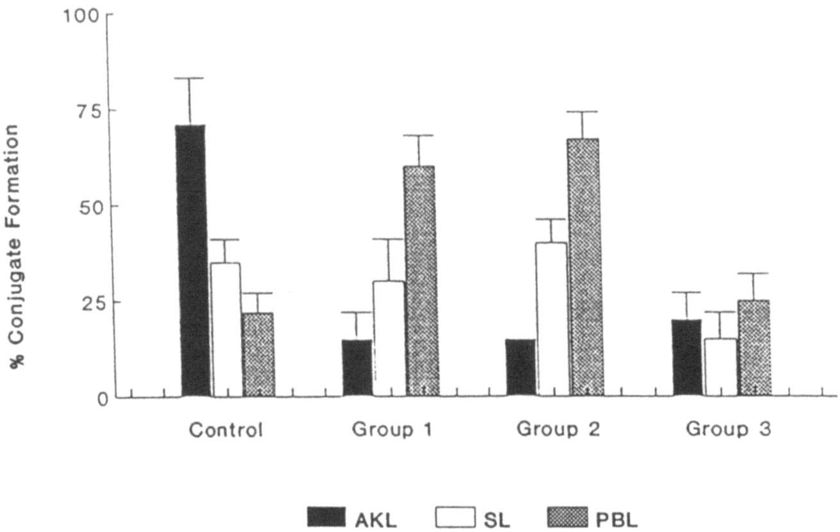

Fig. 5. Conjugate formation in infected and control uninfected *Sparus aurata*

Fig. 6. Modulation of proliferative responses to mitogens by Lymphocystis Disease. Data (means ± standard deviation) are expressed as percent of the values given by non-infected fish leucocytes

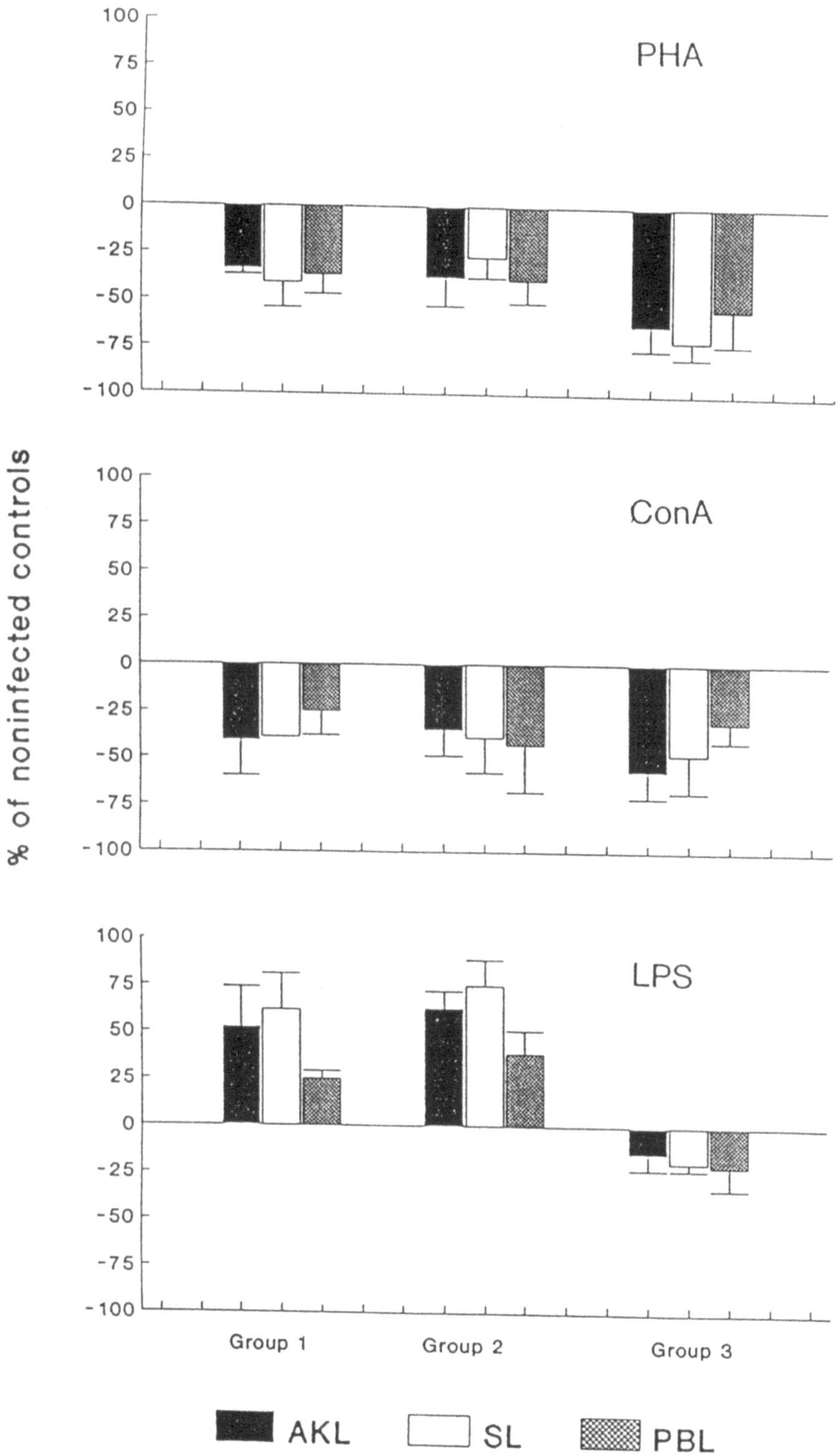

PHA
ConA
LPS
% of noninfected controls
100
75
50
25
0
-25
-50
-75
-100
Group 1
Group 2
Group 3
AKL
SL
PBL

Discussion

The morphological appearance of the hypertrophied dermal cells as well as the virus particles show a similarity to Fish Lymphocystis Disease Virus as described by Berthiame et al. (1984), Midlidge and Malsberger (1968) as well as Zwillenberg and Wolf (1968). Although Paperna et al. (1982) reported lymphocystis disease in *Sparus aurata* cultured in the Red Sea at Eilat, this study is considered the first report on lymphocystis disease affecting *S. aurata* in Egypt.

Our data indicate a significant modulation of two innate cell-mediated immune responses by FLDV infection. Regarding the NCC, the pattern of its tissue distribution resembles that described in other freshwater species, e.g. channel catfish *Ictalurus punctatus* (Graves et al. 1984) or tilapia (Faisal et al., in press). In groups 1 and 2, the significant augmentation of the peripheral blood leucocytic cytotoxicity was accompanied by a suppression in that of the anterior kidney. The parallel change in the number of conjugate forming cells points to a possible migration of NCCs from the anterior kidney into the peripheral blood and possibly to sites of infection in the integument. One of the known functions of nonspecific cytotoxic cells in mammals is the attack of host cells expressing viral antigens (Welsch, 1986). Another possibility for the NCC-mobilization is the stress caused by continuous irritation of the skin. As Ghoneum et al. (1987) reported, some stress paradigms induce a redistribution of tissue cytotoxic cells. Whether the observed NCC-modulation in our study is attributed specifically to the virus or nonspecifically to the dermal irritation needs further investigation.

The consistent suppression of the mitogenic responses to the migogens PHA and Con A in infected fish indicates an aberration of the functional activity of immunocompetent cells. In teleosts, PHA and Con A are known to stimulate a subpopulation of lymphocytes that functionally react as the mammalian T cells (Caspi et al. 1984; Etlinger et al. 1976; Sizemore et al. 1984). As T cells are extremely important in antigen recognition and eliciting the immune responses, their ablation indicates a poor immune status of the host. Whether this resulted from the viral infections or that the infection could settle only in originally immunocompromised fish is not clear at present. The augmented mitogenic response to LPS reflects the activation of macrophages and B lymphocyte equivalent cells in fish (Kaatari and Irwin, 1985; Kaatari and Yui, 1987). Activation of B lymphocytes is mostly a sequence of T cell dysfunction (Roitt et al. 1985). The overall suppression in immune responses in group 3 is most probably due to the severe exhaustion of fish and malnutrition.

Studies of virus-host interaction in mammals indicated that immunological changes accompanying viral infections play a key role in determining their pathogenicity, disease course and recurrency (Oldstone, 1979). Little information is available on the effect of viral infection on the teleostean immune cells. The obtained results of the present study encourage further research in the topic.

Acknowledgement. The authors express their thanks to Prof. W.J. Hargris for the critical review of the manuscript.

References

Berthiame L, Alain R, Robin J (1984) Morphology and ultrastructure of lymphocystis disease virus, a fish iridovirus, grown in tissue culture. Virol 135, 10-19

Caspi RR, Shahrabani R, Kehati-Dan T, Avtalion RR (1984) Heterogeneity of mitogen-responsive lymphocytes in carp (*Cyprinus carpio*). Dev Comp Immunol 8, 61-70

Drummond JE, Shah KV, Donnenberg AD (1985) Cell-mediated immune responses to BK virus in normal individuals. J Med Virol 17, 237-247

Etlinger HM, Hodgins HO, Chiller JM (1976) Evolution of the lymphoid system. I. Evidence for lymphocyte heterogeneity in rainbow trout revealed by the organ distribution of mitogenic responses. J Immunol 116, 1547-1553

Faisal M, Cooper EL, Ahmed II, Peters G (in press) Nonspecific cytotoxicity of tilapia leucocytes. J Appl Ichthyol

Flügel RM (1985) Lymphocystis disease virus. Curr Topics Microbiol Immunol 116, 133-150

Ghoneum M, Gill G, Assanah P, Stevens E (1987) Susceptibility of natural killer activity of old rats to stress. Immunol 60, 461-470

Graves SS, Evans DL, Cobb D, Dawe DL (1984) Nonspecific cytotoxic cells in fish (*Ictalurus punctatus*). I. Optimum requirements for target cell lysis. Dev Comp Immunol 8, 293-302

Hinuma S, Abo T, Kumagai K, Hata M (1980) The potent activity of freshwater fish kidney cells in cell killing. I. Characterization and species-distribution of cytotoxicity. Dev Comp Immunol 4, 653-666

Kaatari SL, Irwin MJ (1985) Salmonid spleen and anterior kidney harbor populations of lymphocytes with different B cell repertoires. Dev Comp Immunol 9, 433-444

Kaatari SL, Yui MA (1987) Polyclonal activation of salmonid B lymphocytes. Dev Comp Immunol 11, 155-165

Kiessling R, Klein E, Wigzell H (1975) Cytotoxic cells with specificity for mouse Moloney leukemia cells. Specificity and distribution according to genotype. Eur J Immunol 5, 112-118

McKinney EC, Schmale MC (1988) Proliferative and cytotoxic responses of healthy bicolor damselfish. Dev Comp Immunol 12, 207-212

Midlige FH, Malsberger RG (1968) In vitro morphology and maturation of lymphocystis virus. J Virol 2, 830-835

Oldstone BA (1979) Immune responses, immune tolerance and viruses. Comprehensive Virol 15, 1-36

Paperna I, Sabanai I, Coloroni A (1982) An outbreak of lymphocystis in *Sparus aurata* L. in the Gulf of Aqaba, Red Sea. J Fish Dis 5, 433-437

Pross HF, Baines MG, Rubin P, Schragge P, Patterson MS (1981) Spontaneous human lymphocyte-mediated cytotoxicity against tumor target cells. IX. The quantitation of natural killer cell activity. J Clin Immunol 1, 51-63

Roitt I, Brostoff J, Male D (1985) Immunology. Gower Medical Publishing, London

Sizemore RC, Miller NW, Cuchens MA, Lobb CJ, Clem LW (1984) Phylogeny of lymphocyte heterogeneity: The cellular requirements for in vitro mitogenic responses of channel catfish leukocytes. J Immunol 133, 2920-2924

Tsutsumi H, Bernstein JM, Riepenhoff-Talty M, Cohen E, Orsini F, Ogra PL (1986) Immune responses to herpes simplex virus in patients with recurrent herpes labialis. I. Development of cell-mediated cytotoxic responses. Clin Exp Immunol 66, 507-515

Walker R (1962) Fine structure of lymphocystis virus of fish. Virology 18, 503-505

Weissenberg R (1920) Lymphocystisstudien (Infektiöse Hypertrophie von Stützgewebszellen bei Fischen). I. Die reifen Geschwülste bei Kaulbarsch und Flunder. Lymphocystisgenese beim Kaulbarsch. Arch Mikrosk Anat Entwick Mech 94, 55-134

Welsh RM (1986) Regulation of virus infections by natural killer cells: A review. Natural Immunity and Cell Growth Regulation 5, 169-199

Wolf K (1984) Disease caused by microorganisms. Agents: Virales. In: Diseases of Marine Animals, vol IV (1) Introduction, Pisces. O Kinne (ed) Biologische Anstalt Helgoland, Hamburg

Yasumoto S, Hayashi Y, Aurelian L (1987) Immunity to herpes simplex virus type 2. Suppression of virus-induced immune responses in ultraviolet B-irradiated mice. J Immunol 139, 2788-2793

Zwillenberg LO, Wolf K (1968) Ultrastructure of lymphocystis virus. J Virol 2, 393-399

Aquatic Birds: An Evolutionary Repository of Ortho- and Paramyxoviruses?

K.F. Shortridge
Department of Microbiology
University of Hong Kong
Pathology Building
Queen Mary Hospital Compound
Hong Kong

INTRODUCTION

Much of the information on the ecology of avian orthomyxoviruses (influenza viruses) and paramyxoviruses (e.g. Newcastle disease virus) is recent. The recognition that the pandemic influenza virus that emerged through Hong Kong in 1968 (Chang, 1969) was antigenically related to a virus that had been isolated some years previously from a domestic duck (Laver and Webster, 1973) led to virus surveillance studies of avian and other species that commenced around 1974. These were initially fostered by the World Health Organization to see if there is a finite number of influenza viruses in nature, to gain insight into the ecology of these viruses and to see if it is possible to isolate a pandemic virus in advance of its appearance in man.

Until these studies, the only viruses from these two families about which a modest amount of information was known were those that caused fowl plague, equine and swine influenza and Newcastle disease (fowl pest). Surveillance unearthed a vast range of influenza viruses in avian species notably the domestic duck and extended the number of known paramyxoviruses.

The impetus to study lower vertebrates in the same way just does not exist. Most work has been on those of economic importance such as fish which are intensively raised under artificial conditions. Attempts to isolate viruses from reptiles have arisen mainly from sporadic attempts in the 60's and 70's to explain the ecology of arboviruses. More recently, perhaps stemming from an increased awareness of environmental issues and the need to know more about diseases in the animal world, reptiles have been the subject of some virological studies (see Jacobson, 1986).

This report takes account of the orthomyxoviruses and paramyxoviruses of birds, their successful association with aquatic birds and attempts to evaluate in an evolutionary sense the role that reptiles may have played in handing down these viruses.

THE VIRUSES

Orthomyxoviridae

There are three types of influenza virus A, B and C based on the antigenicity of the internal ribonucleoprotein. Viruses that infect birds are all type A. The surface of the influenza virus is covered with two kinds of spiked glycoprotein projections responsible for subtype specificity. One is the haemagglutinin (H) so called because it agglutinates erythrocytes, the other is the neuraminidase (N) an enzyme that frees the virus after it has adsorbed onto a cell. A virus is subtyped on the basis of these H and N antigens. The RNA of influenza A viruses is composed of eight discrete segments, two of which code for H and N, so that in dual infection of a susceptible cell, the 16 segments may reassort to give rise to a possible 254 combinations of RNA.
The surveillance studies undertaken since 1974 have extended the known number of H subtypes to 13 which are now designated H1 to H13, and N subtypes to 9 now designated N1 to N9. Thus, influenza A viruses are subtyped H1N1, H1N2, ... H1N9; H2N1, ... H2N9, and so on.

Paramyxoviridae

There are three genera of Paramyxoviridae. The viruses that infect birds belong to the genus Paramyxovirus, Newcastle disease virus (NDV) being the prototype virus. In contrast to influenza A viruses, its RNA is not segmented and it therefore cannot reassort. NDV also has two kinds of spiked projections one of which jointly carries both H and N activities. The second projection is responsible for haemolytic activity. Surveillance has extended the number of avian paramyxoviruses to 9, these now being designated avian-PMV1 to avian-PMV9 with NDV as avian PMV-1.

AVIAN HOSTS

Influenza

Influenza Isolations have been made from most of the major families of wild birds - Corvidae, Fringillidae, Ploceidae, Motacillidae, Scolopacidae, Columbidae, Sturnidae, Laridae, Phasianidae, Hirundinidae, Charadriidae, Ardeidae, Procellariidae, Rallidae, Alcidae, Accipitridae, Phalacrocoracidae (Lvov, 1978; Lvov and Zhdanov, 1983; Hinshaw and Webster, 1982). However, by far the majority of isolations have been from waterfowl of the family Anatidae (Table 1).

In the case of domestic and game poultry, the duck has provided the bulk of the isolates. The most extensive study on this has been in longitudinal surveillance over the period 1975-80 on domestic poultry of Hong Kong and southern China; that study amplified the importance of the duck as an influenza host (Table 2) (Shortridge 1982a and b). All the then known H subtypes, H1 to H12, apart from H8, were recorded amongst the isolates . Serological studies indicated the presence of the H8 subtype (Shortridge, unpublished data) and Shen (1983) subsequently confirmed its presence by isolation. The most common H subtypes were H4, H3 and H6 with isolation rates of 29%, 25% and 22%, respectively. All isolations were made from apparently healthy poultry. In the case of ducks, isolations were made all year round but were generally more frequent in the warmer, summer months.

Table 1. A GUIDE TO THE AVIAN HOST RANGE OF INFLUENZA A VIRUSES

<u>Wild birds</u>

Corvidae, Fringillidae, Ploceidae, Motacillidae, Scolopacidae, Columbidae, Sturnidae, Laridae, Phasianidae, Hirundinidae, Charadriidae, Ardeidae, Procellariidae, Rallidae, Alcidae, Accipitridae, Phalacrocoracidae and Anatidae (waterfowl especially migratory birds)

<u>Domestic birds</u>

Anatidae (domestic duck); other major poultry: chicken, turkey, goose Minor poultry: pheasant, quail, partridge, Muscovy duck, malland, Guinea fowl

Table 2. SITE OF ISOLATION OF INFLUENZA A VIRUSES FROM DOMESTIC POULTRY AT A HONG KONG DRESSING PLANT, NOVEMBER 1975 - OCTOBER 1980 (Shortridge, 1983)

Type of poultry	No. of swabs	Virus isolations Number	Percent	Ratio
Duck	8737[a]	564	6.5	16
Goose	1353	15	1.1	3
Chicken	1708	7	0.4	1

[a] combined totals of cloacal and tracheal swabs

The overall isolation rate from ducks was 6.5% (Table 2) being about twice as high from the cloaca over the trachea (data not shown). The importance of water is exemplified by the isolation rates from ducks, geese and chickens which were 16:3:1, respectively. This suggested that infection is mainly intestinal and waterborne (Markwell and Shortridge, 1982). Intestinal multiplication has been demonstrated experimentally (Kida et al., 1980; Webster et al., 1980; Higgins et al., 1987). The domestic duck is intensively raised on flooded rice fields and ponds in southern China providing an ideal opportunity for continuity of influenza viruses (Shortridge, 1983) through faecally contaminated water (Markwell and Shortridge, 1982).

Migrattory water birds are the other major host for influenza viruses but isolations are mainly made when these birds are assembled on lakes and bays prior to and during migrations (Hannoun and Devaux, 1980; Hinshaw et al., 1980). Very high isolation rates may be made from the birds and lake water at this time. This is not surprising as the birds are living on or near what is, in effect, a virus containing, faecally contaminated soup (Hinshaw et al., 1979; Markwell and Shortridge, 1982). It is of interest that migratory birds probably help to sustain the influenza gene pool in southern China as they pass through the region between the two hemispheres (Shortridge, 1983).

Outbreaks of disease in domestic poultry attributable to or associated with influenza viruses occur mainly in turkeys and chikkens (Alexander, 1982a; APHIS, 1984; 1983; Bahl et al., 1979; Lang, 1982; Pomeroy, 1982). The ability of influenza viruses to cause disease was shown by Lang (1982) who recorded 66 outbreaks in turkeys in Ontario, Canada over the period 1960 to 1980. Similar observations have been made in California and Minnestoa, USA. It is generally felt that these outbreaks are the result of introduction of virus by wild birds (Lipkind, 1982). The fact that many outbreaks in turkeys occur on farms that lie on migratory bird routes (Bahl et al., 1979; Halvorson, et al., 1985; Pomeroy, 1982) suggest introductions of virus from an outside source.

A similar situation prevails with chickens but infections have been fewer in recent years since much of the large scale raising nowadays takes place indoors. The strict separation of chickens (and turkeys) from wild and feral birds seems essential (Lang, 1982).

The main point to emerge is that the domestic duck of southern China is the major host or reservoir of influenza viruses; studies elsewhere highlight the importance of influenza viruses in causing disease in non-aquatic domestic and other birds.

Paramyxoviruses

The most important avian paramyxovirus is Newcastle disease virus (NDV), now designated as avian PMV-1. This virus comprises an antigenically related group of pathotypes which can produce a complete spectrum of virulence for most species of birds ranging from inapparent infection to 100% mortality. Symptoms include varying degrees of respiratory and neurological signs and enteritis. Other avian PMVs cause mild symptoms or none at all (Alexander, 1980).

NDV is important economically as a cause of morbidity and mortality in chickens and is the number one disease of poultry; birds must be vaccinated in those areas where the disease is endemic.

In studies on avian PMVs conducted in parallel with influenza surveillance on domestic poultry of southern China (Table 2,3), the domestic duck was shown to be the major reservoir of these PMVs (Table 3). The slightly higher the isolation rate from chikkens reflects their susceptibility to the virus and the presence of vaccine strains (data not shown). The isolations from ducks were from apparently healthy birds, the isolation rate from the cloaca being about four times that from the trachea (data not shown). Isolations were made all year round but in contrast to influenza, they were generally more frequent in the cooler, winter months. The view has been expressed that the duck which is

Table 3. ISOLATION OF AVIAN PARAMYXOVIRUSES FROM DOMESTIC POULTRY SAMPLED AT A HONG KONG DRESSING PLANT, NOVEMBER 1975 - OCTOBER 1980 (Foo, 1982)

Type of poultry	No. of samples[a]	Isolation frequencies (%) Avian-PMV1 (Newcastle disease Virus)	Other avian PMVs
Duck	8737	2.6	0.5
Goose	1353	1.6	0.3
Chicken	1708	3.1	0.1

[a] combined totals of cloacal and tracheal swabs

TABLE 4. ISOLATION OF PARAMYXOVIRUSES FROM REPTILE HOSTS

Host	Symptoms	Reference
Fer-de-lance (*Bothrops atrox*)	Lethargy, terminal respiratory disease	Foelsch and Leloup (1976) Clark et al. (1979)
Rock rattle snake (*Crotalus lepidus*)	Progressive central nervous disease, pneumonia	Jacobson et al. (1980)
Family Viperidae members of genera *Crotalus*, *Vipera*, *Bothrops*, *Trimeresurus*, *Bitis*	Gaping mouth; convulsions; mucoid exudate throughout respiratory tract	Jacobson et al.(1981)
Viperids and elaphids (cobras)	Not recorded but indicative of paramyxovirus infection	Jacobson et al.(1986)
Ottoman pit vipers (*Vipera xanthina xanthina*)	Pneumonia, mucous in mouth	Potgieter et al. (1987)

raised in close proximity to domestic poultry in the region is the major source of infection for chickens (Riddell-Swan and Shortridge, 1985).

While other birds may carry avian PMV-1, (Alexander et al., 1984), their role in the spread of disease to domestic chickens is not always that clear because of spread through the agency of man (Lancaster and Alexander, 1975). Details on spread and transmission are considered elsewhere (Alexander, 1982b; Lancaster and Alexander, 1975; Shortridge, 1982c).

As with influenza, the main point to emerge is that the domestic duck of southern China is the major host or reservoir of avian PMVs, especially avian PMV-1.

REPTILIAN HOSTS

Paramyxoviruses have been isolated from viperid snakes with respiratory symptoms mainly from dieoffs in collections (Table 4). There is a noticeable absence of isolations from other snakes held in the vicinity. Jacobson (1986) did, however, report that several elaphids in a major dieoff in a collection of several hundred viperids also died with clinical signs typical of paramyxovirus infection.

Preliminary studies indicate that the virus isolated from Ottoman vipers is antigenically related to parainfluenza 2 virus (Potgieter et al., 1987). These data, albeit somethat limited, presently indicate that reptile paramyxoviruses may have a restricted host range. On the other hand, serological studies using the haemagglutination inhibition (HI) test, suggest that they may infect a wide variety of species (Jacobson, 1986). However, the interpretation of HI serology in such animals must always be treated with caution because of false positives arising from the possible incomplete removal or inactivation of nonspecific inhibitors of haemagglutination (NSI). Avian sera contain little NSI activity, if any, for avian ortho- and paramyxoviruses (Shortridge, unpublished data) and it remains to be seen if the same applies to reptile sera.

REPTILE-BIRD ANCESTRAL LINKS?

The recognition that waterfowl, notably the domestic duck of southern China, are host to a vast reservoir of influenza and paramyxoviruses begs the question of these virus's ancestral hosts. It is perhaps appropriate to take account of the link between reptiles and birds notwithstanding the limitation of the geological record and its interpretation.

There has been little doubt that birds originated from reptiles. The major area of concern has been which group of archosaurs, namely the crocodiles, thecodonts or theropods, is most closely related to the avian lineage. Renewed interest in this field points strongly to the theropods, i.e. the predaceous dinosaurs, being the most likely ancestor to the class Aves. Osteological chambers that were once regarded as uniquely avian, e.g. furcula, hypotarsus, have now been identified in various theropod dinosaurs and it has been argued by Benton (1985), Thulborn (1984) and Thulborn and Hamley (1982) that some of these animals were actually more bird-like than Archaeopteryx. An outline phylogeny of diapsids is shown in Figure 1.

The early radiations from the thecodonts are far from clear. This is reflected in Olsen's grouping of birds into "basal" and "higher" land bird assemblages, and "water" bird assemblages (Olson, 1985).

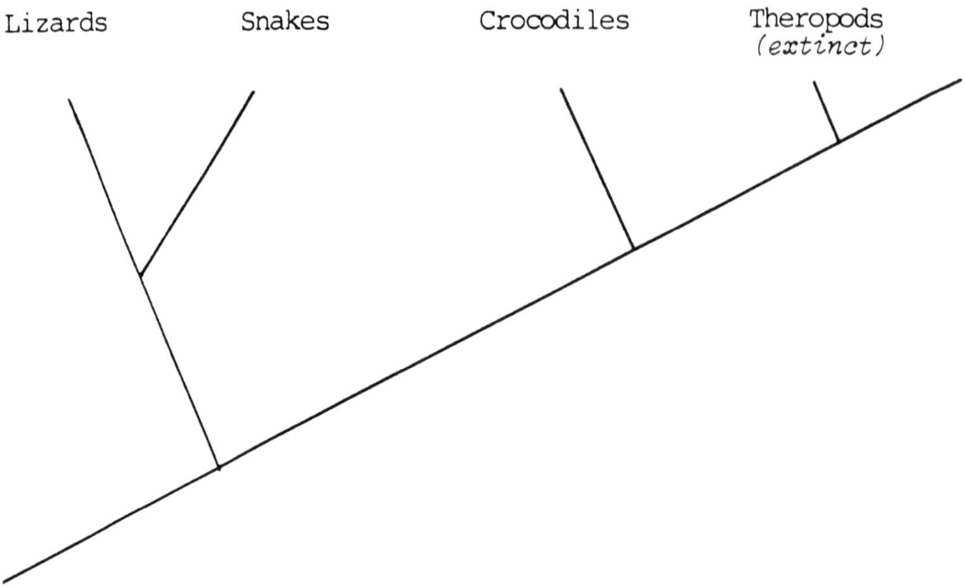

Figure 1. An outline phylogeny of diapsids (adapted from Benton (1986), Thoulborn (1984) and Hamley, 1982)

Figure 2 gives a somewhat simplified phylogeny adapted from this based on splitting points rather than geological time. The earliest basal land birds are considered to have arisen before the other birds and to have continued. Nevertheless, there is evidence that in early avian history there do appear to be aquatic, water-loving birds but whether these represent a primitive stage is open to question (Feduccia, 1980) since the possibility that they are an artefact of the fossil record cannot be excluded (Thulborn, personal communication). They do, however, represent a possible link for the water birds that followed. Thus, it is relevant to note that the most conclusive fossil evidence of an early separation comes from recent work which traces the origin of the Anseriformes (waterfowl) back to a group of highly colonial Charadriformes (shorebirds) that adapted to shallow saline lakes and evolved a unique filter feeding mechanism (Olson and Feduccia, 1980). Put simply, the waterfowl comprise a very successful group of birds with a long evolutionany history that includes the most vagile birds and the sedentary Muscovy duck and its domestic varieties.

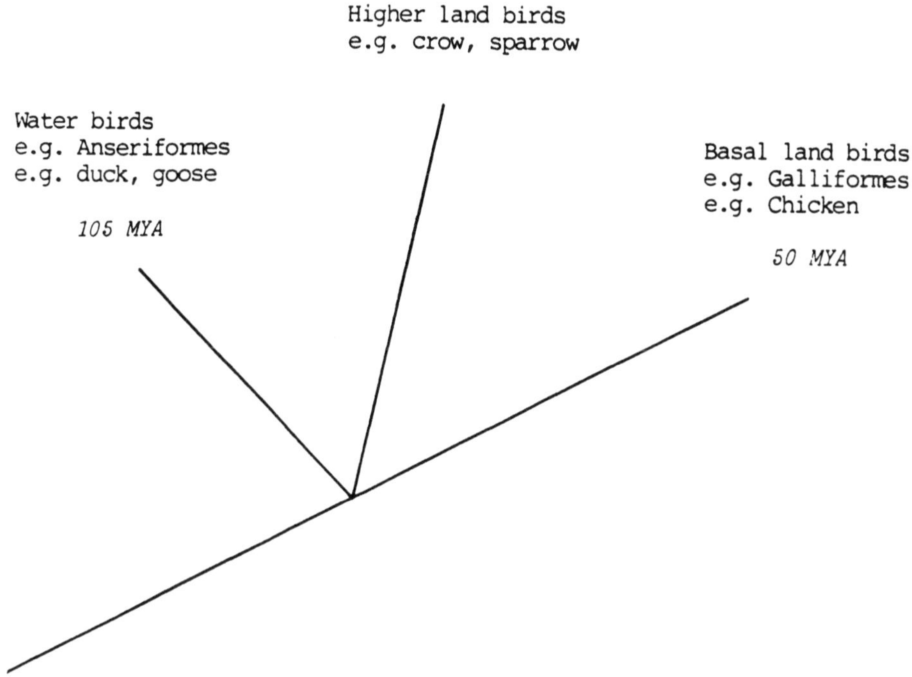

Figure 2. An outline phylogeny of recent birds adapted from Olson (1985) and Molnar and Archer (1984) including representative age estimates by DNA-DNA hybridization (Sibley and Ahlquist, 1986)

(MYA = million years ago)

When the fossil record is viewed against emerging data based on DNA-DNA hybridization (of modern birds), the aquatic domestic duck appears to be much older than, say, land based domestic poultry such as chickens and turkeys. Making interpretations from the data of Sibley and Ahlquist (1986), this difference is at least 55 million years. These data are based on evidence which suggests that the average rate of DNA evolution (base pairs) is the same in all lineages of birds. Sibley and Ahlquist (1986) assert that it takes about 4.5 million years for the genomes of two species to become different at 1% of their base pairs after they have diverged from a common ancestor.

DISCUSSION

Although evidence suggests that there is a reptile-theropod - early land bird (?) - water bird lineage, it does not necessarily follow that each step in the lineage gave rise to the next's pathogens, in this case influenza and paramyxoviruses. The theropods, now extinct, are the missing link in tracing viral ancestors. Any attempt to trace the ancestral origin of these viruses must be speculative; this is compounded by the paucity of isolations from modern reptiles.

While influenza and paramyxoviruses are remarkably successful in their relationship with present day aquatic birds, there is no way of telling whether a similar successful relationship existed with the theropods. Influenza and paramyxoviruses cause short-lived infections and can only be sustained in animals that live in high population density with high turnover such as the domestic duck of southern China (Shortridge, 1982a and 1983).
Whether they could have sustained in theropods in the same way is speculative. However, it is noteworthy that dinosaur remains have very recently been discovered in China which suggest that these animals lived in large herds that migrated across an ancient land mass that joinet North America and Asia (Anon, 1988) Of course, these viruses may have undergone evolution but available biological and molecular data does not afford any insight into this.
It is of interest that two taxonomically related human viruses, namely measles and mumps, can cause serious infections. Measles has the ability to spread rapidly especially in non-immune populations causing extensive morbidity and mortality while mumps can give rise to sterility in a small proportion of adolescent males. The possibility that a virus infection could have led to the demise of the dinosaurs is worthy of further throught!

The phylogeny as given in Figure 1 places snakes much further away from birds than crocodiles. This might account in an evolutionary sense for the very limited number of influenza and paramyxovirus isolations from them (Table 4). The fact that these isolations have been from die-offs in collections (see Jacobson, 1986) and that free living snakes generally do not live in communities suggest that the prospect for many isolations is not high. Even in the case of togaviruses and flaviviruses which are transmitted by insects and might therefore be expected to have a higher order of occurrence in reptiles, the isolation rate is low. This is influenced by a number of factors including season and temperature (see Shortridge and Oya, 1984) which might explain the low isolation rate. Attempted isolation from the garter snake Thamnopsis sp. which undergoes a yearly migration in vast numbers in Canada might prove useful. In so far as the crocodiles being a closer living relative is concerned, this would be technically formidable. There are large high-density crocodile farms in some tropical countries and it would be worthwhile finding out if clinical manifestations similar to those described in Table 4 occur in the crocodiles there as well as attempting virus isolation from them.

The study of viruses of reptiles is relatively new; the very low isolation rate may also reflect an inappropriate site of sampling and isolation system. As it is now known that both avian influenza and paramyxoviruses can be isolated with relative ease from the cloaca and faecal matter of ducks, the intestinal lining and contents of snakes should be investigated for virus. The liklihood of virus replicating in the intestine would depend in part on its inherent stability and time for it to reach the lower intestine unless, of course, it might also reach them by venereal route. Avian influenza viruses are more stable than mammalian ones at low pH; while passage into the lower intestine of ducks can occur in a matter of hours (Webster et al., 1978). The successful use of snake embryos (Clark et al., 1979) and reptile cell cultures (Jacobson et al., 1981) at lower temperatures for virus isolation illustrate the need for this aspect to be explored further.

The isolation of influenza-like viruses from ticks (Haig et al., 1965) with segmented negative sense single stranded RNA (Clerx et al., 1983) that can reassort in both vertebrate (Jones et al., 1987) and invertebrate (Davies et al., 1987) hosts adds another dimension to the ecological perspective of influenza (and parainfluenza) viruses. One of these influenza-like viruses, Thogoto virus, causes abortion in sheep but is able to spread rapidly in large flocks suggesting that it was unlikely to have been entirely tick transmitted (Davies et al., 1984).
This raises the question whether invertebrate vectors which may feed on reptiles can transmit influenza and paramyxoviruses in the same way as they do togaviruses and flaviviruses (Shortridge and Oya, 1984). This suggests that the net needs to be cast wider for attempted virus isolation studies. Experimental infection of reptiles and ducks with a range of influenza and paramyxoviruses may provide informative data on ancestral host links.

All in all, there can be little doubt that influenza and paramyxoviruses are remarkably successful parasites of the aquatic duck which is probably an evolutionary repository for them. If their prime natural habitat in the duck is the intestine, then infection of other birds and animals via the respiratory route may give rise to recognizable infection best exemplified by disease outbreaks in domestic turkeys and chickens. It is conceivable that a similar situation also exist in the lower vertebrates. Increased virus surveillance of reptiles via intestinal and respiratory routes will be informative and may help shed light on the ancestral origin of these successful avian viruses.

ACKNOWLEDGEMENTS

Thanks are due to Drs G. Ingram, W. Longmore and R. for discussions on the evolution of birds.

REFERENCES

ALEXANDER, D.J. (1980) Avian paramyxoviruses. Vet. Bull., 50, 737-752.

ALEXANDER, D.J. (1982a) Current situation of avian influenza in poultry in Great Britain. In: Proc. 1st Int. Symp. Avian Inf., Beltsville, Md, 1981, R.A. Bankowski ed.,pp. 35-45. Carter Composition Corp., Richmond, Va, USA

ALEXANDER, D.J. (1982b) Avian influenza - recent developments. Vet. Bull., 52, 341-359

ALEXANDER, D.J., G. PARSONS and R. MARSHALL (1984) Infection of fowls with Newcastle disease virus contaminated with pigeon faeces. Vet. Rec., 115, 601-602

ANIMAL AND PLANT HEALTH INSPECTION SERVICE (APHIS FACTS) (1984) Avian influenza eradication 1983-84. US Dept. Agricult. 1984 Hyattsville, Md, USA.

ANONYMOUS (1988) Bones betray the herd mentality of dinosaurs. New Scientist, No. 1626, 18 August, 1988

BAHL, A.K., A. LANGSTON, R.A. Van DEUSEN, B.S. POMEROY, J. NEWMAN, D.; KARUNAKARAN and D. HALVORSON (1979) Prevention and control avian influenzy in turkeys. Proc. ann. Meet. US Anim. Hlth Assn, 83,355-363

BENTON, M.J. (1985) Classification and phylogeny of the diapsid reptiles. Zool. J. Linn. Soc., 84, 96-164.

CHANG, W.K. (1969) National influenza experience in Hong Kong. Bull. Wld. Hlth. Org., 41, 349-351.

CLARK, H.F., F.S. LIEF, P.D. LUNGER, D. WATERS, P. LELOUP, D.W. FOELOSCH and R.W. WYLER (1979) Fer-de-lance virus: a probable paramyxovirus isolated from a reptile. J. gen. Virol., 44, 405-418

CLERX, J.P.M., F. FULLER and D.H.C. BISHOP (1983) Tick-borne viruses structurally similar to orthomyxoviruses. Virology, 127, 205-219.

DAVIES, F.G., R.K. SOI and R.N. WARIRU (1984) Abortion in sheep caused by Thogoto virus. Vet. Rec., 115, 654.

DAVIES, C.R., L.D. JONES, B.M. GREEN and P.A. NUTTALL (1987) In vivo reassortment of Thogoto virus (a tick-borne influenzalike virus following oral infection of Rhipicephalus appendiculatus ticks. J. gen. Virol., 68, 2331-2338.

FEDUCCIA, A. (1980) The Age of Birds. Cambridge, Mass., Harvard University Press

FOELSCH, D.W. and P. LELOUP (1976) Fatale endemische infektion in einem Serpentarium. Tieraeztl. Praxis, 4, 527-536.

FOO, S.Y. (1982) Virus surveillance studies in Hong Kong. In Newcastle Disease and its Control in Southeast Asia K.F. Shortridge ed., pp.37-39. Hong Kong University Press, Hong Kong.

HAIG, D.A., J.P. WOODALL and D. DANSKIN (1965) Thogoto virus a hitherto undescribed agent isolated from ticks in Kenya. J. gen. Microbiol., 38, 389-394

HALVORSON, D.A., C.J. KELLEHER and D.A. SENNE (1985) Epizootiology of avian influenza: Effect of season on incidence in sentinel ducks and domestic turkeys in Minnesota. Appl. environ. Microbiol., 49, 914-919.

HANNOUN, C. and J.M. DEVAUX (1980) Circulation enzootique permanente de virus grippaux dans la baie de la Somme. Comp. Immun. Microbiol. infect. Dis., 3, 177-183.

HIGGINS, D.A., K.F. SHORTRIDGE and P.L.K. NG (1987) Bile immunoglobulin of the duck (Anas platyrhynchos) II. Antibody response in influenza A virus infections. Immunology, 62, 499-504.

HINSHAW, V.S., and R.G. WEBSTER (1982) The natural history of influenza A viruses. In Basic and Applied Influenza Research (A.S. Beare, ed) pp. 79-104. CRC Press, Boca Raton, Fl., USA.

HINSHAW, V.S., R.G. WEBSTER and B. TURNER (1979) Water-borne transmission of influenza A viruses? Intervirology, 11, 66-68.

HINSHAW, V.S., R.G. WEBSTER and B. TURNER (1980) The perpetuation of orthomyxoviruses and paramyxoviruses in Canadian waterfowl. Canad. J. Microbiol., 26, 622-629.

JACOBSON, E.R. (1986) Viruses and viral associated diseases of reptiles. In Maintenance and Reproduction of Reptiles in Captivity. Vol II. (V.L. Bels and A.P. Van den Sande, eds). Acta Zoologica et Pathologica Antverpiensia, No. 79, 73-90.

JACOBSON, E., J.M. GASKIN, C.F. SIMPSON and T.G. TERRELL (1980) Paramyxo-like virus infection of a rock rattlesnake. J. Amer. vet. med. Assoc., 177, 796-799.

JACOBSON, E., J.M. GASKIN, D. PAGE, W.O. IVERSON and J.W. JOHNSON (1981) Illness associated with paramyxo-like virus infection in a zoologic collection of snakes. J. Amer. vet. med. Assoc., 179, 1227-1230.

JONES, L.D., C.R. DAVIES, B.M. GREEN and P.A. NUTTALL (1987) Reassortment of Thogoto virus (a tick-borne influenza-like virus) in a vertebrate host. J. gen. Virol., 68, 1299-1306. KIDA, H., R.

YANAGAWA and Y. MATSUOKA (1980) Duck influenza lacking evidence of disease signs and immune response. Infect. Immun., 30, 547-553.

LANCASTER, J.E. and D.J. ALEXANDER (1975) Newcastle disease: Virus and Spread. Monograph No. 11, Ottawa, Ontario. Canada Department of Agriculture. pp. 79.

LANG, G. (1982) A review of influenza in Canadian domestic and wild birds. In Proc. 1st int. Symp. Avian Inf., Beltsville, Md, 1981 (R.A. Bankowski, ed.) pp. 21-27. Carter Composition Corp., Richmond, Va, USA.

LAVER, W.G. and R.G. WEBSTER (1973) Studies on the origin of pandemic influenza III. Evidence implicating duck and equine influenza viruses as possible progenitors of the Hong Kong strain of human influenza. Virology, 51, 383-391.

LIPKIND, M., Y. WEISMAN, Y. SHIHMANTER and D. SHOHAM. (1982) Review of the three year studies on the ecology of avian influenza viruses in Israel. In Proc. 1st int. Symp. Avian Inf., Beltsville, Md, 1981 (R.A. Bankowski, ed.) pp. 69-78. Carter Composition Corp., Richmond, Va, USA.

LVOV, D.K. (1978) Circulation of influenza viruses in natural biocoenosis. In Viruses and Environment, (E. Kurstak and K. Maramorosch, eds) pp. 351-380. Academic Press, New York.

LVOV, D.K. and V.M. ZHDANOV (1983) Persistence of genes of epidemical influenza viruses in natural populations in the USSR. Med. Biol., 61, 83-91.

MARKWELL, D.D. and K.F. SHORTRIDGE (1982) Possible waterborne transmission and maintenance of influenza viruses in domestic ducks. Appl. environ. Microbiol., 43, 110-116.

MOLNAR, R. and M. ARCHER (1984) Feeble and not so feeble flapping fliers: A consideration of early birds and bird-like reptiles. In Vertebrate Zoogeography and Evolution in Australia (M. Archer and G. Clayton, eds) pp. 407-419. Hesperian Press, Australia.

OLSON, S.L. (1985) The fossil record of birds. In Avian Biology. Vol. VIII (D.S. Farner, J.R. King and K.C. Parkes, eds) pp. 79-239. Academic Press, London.

OLSON, S.L. and A. FEDUCCIA (1980) Presbyornis and the origin of the Anseriformes. Smithsonian Contrib. Zool., No. 323, 1-24.

POMEROY, B. (1982) Avian influenza in the United States (1964-1970). In Proc. 1st int. Symp. Avian Inf. Beltsville, Md, 1981 (R.A. Bankowski, ed.) pp. 13-17. Carter Composition Corp., Richmond, Va, USA.

POTGIETER, L.N.D., R.E. SIGLER and R.G. RUSSELL (1987) Pneumonia in Ottoman vipers (Vipera xanthina xanthina) associated with a parainfluenza 2-like virus. J. wildl. Dis., 23, 355-360.

RIDDELL-SWAN, J.M. and K.F. SHORTRIDGE (1985) Newcastle disease in Hong Kong. In Veterinary Viral Diseases. Their Significance in Southeast Asia and the Western Pacific, Geelong, 1984 (A.J. Della-Porta, ed.) pp. 326-329. Academic Press, Sydney.

SHEN, G.Z., C.M. WU, Y.F. CHEN, J.B. LIU, W.S. ZHANG, J.B. FENG, R.T. GUO, Z.Y. LIN, and Z.Q. CHEN, (1983) First report of influenza virus infection in Hua-Nan ducks. Chinese J. Epidemiol. 4, 219-222.

SHORTRIDGE, K.F. (1982a) Avian influenza A viruses of southern China and Hong Kong: Ecological aspects and implications for man. Bull. Wld Hlth Org., 60, 129-135.

SHORTRIDGE, K.F. (1982b) Epidemiology of avian influenza and sources of infection in domestic species. In Proc. 1st int. Symp. Avian Inf. Beltsville, Md, 1981 (R.A. Bankowski, ed.) pp. 54-68. Carter Composition Corp., Richmond, Va, USA.

SHORTRIDGE, K.F. (1982c) Newcastle Disease and its Control in Southeast Asia (K.F. Shortridge ed.) Hong Kong University Press, Hong Kong. pp. 107.

SHORTRIDGE, K.F. (1983) Pandemic influenza: Application of epidemiology and ecology in the region of southern China to prospective studies. In Origin of Pandemic Influenza Virus, Beijing, 1982 (W.G. Laver, ed.) pp. 191-200. Elsevier Science Publishing Co., Inc., New York.

SHORTRIDGE, K.F. and A. OYA (1984) Arboviruses. In Diseases of Amphibians and Reptiles, (G.L. Hoff, F.L. Frye and E.R. Jacobson, eds) pp. 107-148. Plenum Press, New York and London.

SIBLEY, C.G. and J.E. AHLQUIST (1986) Proc. 19th Congress Ornithology, Ontario, In press.

THULBORN, R.A. (1984) The avian relationships of Archaeopteryx. Zool. J. Linn. Soc., 82, 119-158.

THULBORN, R.A. and T.L. HAMLEY (1982) The reptilian relationships of Archaeopteryx. Aust. J. Zool., 30, 611-634.

WEBSTER,R.G., M. YAKHNO, V.S. HINSHAW, W.J. BEAN and K.G. MURTI (1978) Intestinal influenza: Replication and characterization of influenza viruses in ducks. Virology, 84, 268-278.

MIX
Papier aus verantwortungsvollen Quellen
Paper from responsible sources
FSC® C105338

If you have any concerns about our products,
you can contact us on
ProductSafety@springernature.com

In case Publisher is established outside the EU,
the EU authorized representative is:
Springer Nature Customer Service Center GmbH
Europaplatz 3, 69115 Heidelberg, Germany

Printed by Libri Plureos GmbH
in Hamburg, Germany